Recent Titles in This Series

(*Continued in the back of this publication*)

Theory of Entire and Meromorphic Functions
Deficient and Asymptotic Values and Singular Directions

Translations of

MATHEMATICAL
MONOGRAPHS

Volume 122

Theory of Entire and Meromorphic Functions
Deficient and Asymptotic Values and Singular Directions

Zhang Guan-Hou

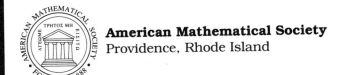

American Mathematical Society
Providence, Rhode Island

整函数和亚纯函数理论

——亏值、渐近值和奇异方向

张 广 厚 著

Translated from the Chinese by Chung-Chun Yang

1991 *Mathematics Subject Classification.* Primary 30D30, 30D35.

Library of Congress Cataloging-in-Publication Data

Zhang, Guan-Hou, 1937–1987.
    [Cheng han shu ho yeh ch'un han shu li lun. English]
    Theory of entire and meromorphic functions: deficient and asymptotic values and singular directions/Zhang Kuan-hua.
        p. cm.—(Translations of mathematical monographs; v. 122)
    ISBN 0-8218-4589-6
    1. Functions, Entire. 2. Functions, Meromorphic. I. Title. II. Series.
QA353.E5Z4313   1993   93-43
515′.98—dc20   CIP

Contents

Preface

This is a monograph about the theory of entire and meromorphic functions. It sums up basically the development of this theory ever since the 1950s, with the discussion focus centered on the relationship among three main concepts, namely the deficient value, the asymptotic value and the singular direction.

In 1929, by examining some examples, R. Nevanlinna recognized that there is an intrinsic relationship between the problem of exceptional values (deficient values are exceptional value under a certain kind of implication) and the asymptotic value theory. Moreover, he anticipated that the study of their relationship might help to clarify some of the profound problems of the theory of entire and meromorphic functions [32a]. Concretely speaking, he conjectured that a deficient value is simultaneously an asymptotic value [32a]. However, this conjecture was negated later [37a, 3a]. In 1978, the author restudied this problem and found that there is a close relationship among the number of deficient values, the number of asymptotic values and the number of singular directions for a function of finite lower order. He obtained some general results [43c]. In recent years, new developments have been made for this study [43g, h, j]; for some other important kinds of functions, their corresponding related formulas are also obtained. The main content of this monograph is to introduce these studies. It is written in the sense that both systematic property and the reference of some results of recent studies are taken into consideration.

This book is divided into six chapters. The first chapter introduces Nevanlinna's fundamental theory, including mainly the famous First Fundamental Theorem, the Second Fundamental Theorem, as well as the related formulas of deficiencies. All these are the foundation of the whole book.

Chapter 2 introduces the Theory of Singular Directions, including mainly the proof of the existence of the Julia direction and the Borel direction, as well as some significant properties. In addition, this chapter includes the discussion and proof of the existence of the Nevanlinna direction under a certain meaning. This discussion is, in fact, a recent study made by Li Yu-Nian and the author himself.

Chapter 3 introduces the Deficient Theory. Deficient values are the main issue of the recent study of the Value Distribution Theory. The results obtained for this issue are particularly fruitful and wonderful. However, due to limited space, we can include in this chapter a relatively comprehensive and systematic discussion regarding merely the study of the number of deficient values. Particularly, here we include also the famous results due to Edrei-Fuchs and Weitsman, etc., with the proof of most of these results newly presented.

At the end of each of the above three chapters, we write an annotated note introducing some important results and corresponding books(^{1}) that have not been covered in the main text as supplements.

Chapter 4 introduces the fundamental theory of asymptotic values and some new results of the study. Included are mainly the classical Iversen Theorem and the famous result regarding the proof of Denjoy's conjecture made by Ahlfors, as well as the estimation of the growth property of a function along the asymptotic path and the estimation of the length of the asymptotic path, all attributed to the author's effort.

The first four chapters lay the foundation for, simultaneously, Chapters 5 and 6, with these last two chapters being the focus of the whole book. Chapter 5 discusses the relationship among the number of deficient values, the number of asymptotic values and the number of Julia directions of an entire function whose lower order is finite, and also the relationship among the number of deficient values, the number of asymptotic values and the number of Julia directions of an entire function which consists of finite Julia directions or with its zeros accumulating around the neighborhood of finite half straight lines. Chapter 6 takes into consideration the case of a meromorphic function and discusses the corresponding problems of a meromorphic function whose lower order is finite or having a maximal deficiency sum. However, here we have to replace the number of asymptotic values with the number of directly transcendental singularities of an inverse function.

Readers intending to peruse this book need only have the knowledge of the complex function theory and of the real function theory provided in the mathematics department of a university. This book, however, is especially suitable for those postgraduate students who study the theory of entire and meromorphic functions under their advisors' guidance.

Finally, here I shall like to express my wholehearted gratitude to Professor Chuang Chi-tai. It is under his encouragement that I became determined to write this book, and through several discussions with him, the outline of this book was drawn. I shall also like to thank Associate Professor He Yu-zan for his meticulous checking and proofreading of the manuscripts. He put

---

(^{1}) If readers want to have a relatively comprehensive understanding of the problem regarding the theory of entire and meromorphic functions in recent study and its progress, they can refer to [21d, f], [2a] and [8a].

forth lots of valuable opinions for amendments. Furthermore, I shall like to thank comrade Wu Peng-cheng who transcribed, proofread and examined the manuscripts several times. Without his assistance, surely it would have taken a longer time to finish writing this book. Finally, I shall also like to thank Associate Professor Li Yu-nian for his help. Before the publication of this book, he spared his busy time to seriously examine and proofread the entire book once.

Zhang Guan-Hou

CHAPTER 1

The Nevanlinna Theory

This chapter is essentially a lead for the rest of the book, and will briefly introduce parts of the Nevanlinna Theory. Included are the well-known First and Second Fundamental Theorems, which constitute the basis of the Nevanlinna Theory. These two theorems were first established in 1925 by R. Nevanlinna who initiated the contemporary study concerning the value distribution theory of meromorphic functions.

§1.1. The Poisson-Jensen formula

1.1.1. The Poisson-Jensen formula.

THEOREM 1.1. *Suppose that* $f(z)$ *is meromorphic on the disk* $|z| \leq \rho$ $(0 < \rho < +\infty)$, *and that* a_i $(i = 1, 2, \ldots, n)$ *and* b_j $(j = 1, 2, \ldots, m)$ *are the zeros and poles of* $f(z)$ *in the disk* $|z| < \rho$, *respectively, where each zero and pole appears as often as its multiplicity indicates. Then we have for the arbitrary value* $z = re^{i\theta}$ $(0 \leq r < \rho, \ 0 \leq \theta < 2\pi)$:

$$
\log|f(z)| = \frac{1}{2\pi} \int_0^{2\pi} \log|f(\rho e^{i\varphi})| \frac{\rho^2 - r^2}{\rho^2 - 2\rho r \cos(\theta - \varphi) + r^2} \, d\varphi
$$
$$
+ \sum_{i=1}^n \log\left|\frac{\rho(z - a_i)}{\rho^2 - \bar{a}_i z}\right| - \sum_{j=1}^m \log\left|\frac{\rho(z - b_j)}{\rho^2 - \bar{b}_j z}\right|. \tag{1.1}
$$

R. Nevanlinna called formula (1.1) the Poisson-Jensen formula.

PROOF. (1) First we assume that $f(z)$ in the disk $|z| \leq \rho$ has no zeros and poles. Hence, for any selected point $z_0 = r_0 e^{i\theta_0}$ $(0 \leq r_0 < \rho, \ 0 \leq \theta_0 < 2\pi)$, function

$$
\log f(z) \cdot \frac{\rho^2 - |z_0|^2}{\rho^2 - \bar{z}_0 z}
$$

We refer to [32a] and [21c] for the basic references of this chapter.

is regular,* in the disk $|z| \leq \rho$. We obtain, according to the Cauchy Formula,

$$\log f(z_0) = \frac{1}{2\pi i} \int_{|\zeta|=\rho} \log f(\zeta) \cdot \frac{(\rho^2 - |z_0|^2)}{(\rho^2 - \overline{z}_0 \zeta)(\zeta - z_0)} d\zeta. \tag{1.2}$$

On the circumference $|\zeta| = \rho$, if we denote $\zeta = \rho e^{i\varphi}$, it follows that $d\zeta = i\rho e^{i\varphi} d\varphi$ and

$$(\rho^2 - \overline{z}_0 \zeta)(\zeta - z_0) = \rho e^{i\varphi} \{\rho^2 - 2\rho r_0 \cos(\theta_0 - \varphi) + r_0^2\}.$$

Consequently formula (1.2) leads to:

$$\log f(z_0) = \frac{1}{2\pi} \int_0^{2\pi} \log f(\rho e^{i\varphi}) \cdot \frac{(\rho^2 - r_0^2) d\varphi}{\rho^2 - 2\rho r_0 \cos(\theta_0 - \varphi) + r_0^2}.$$

Further equalizing the real part of both sides, we have

$$\log |f(z_0)| = \frac{1}{2\pi} \int_0^{2\pi} \log |f(\rho e^{i\varphi})| \frac{(\rho^2 - r_0^2) d\varphi}{\rho^2 - 2\rho r_0 \cos(\theta_0 - \varphi) + r_0^2}.$$

Taking the arbitrary characteristic of point z_0 into account, we prove that formula (1.1) is valid.

(2) Suppose merely that $f(z)$ in the disk $|z| < \rho$ has no zeros and poles, but there is the possibility that some finite zeros and poles are on the circumference $|z| = \rho$. Under this assumption, we can still prove the validity of formula (1.1). Indeed, since the integrand in formula (1.2) has only the logarithmic singularity, the integration still makes sense. We need only to change slightly the integral line $|\zeta| = \rho$ around each zero and pole, and then by means of the limiting process, we can prove that formula (1.2) holds and, in turn, the validity of formula (1.1) is assured.

(3) Now we consider the general case and let

$$\psi(z) = f(z) \cdot \prod_{j=1}^m \left\{ \frac{\rho(z - b_j)}{\rho^2 - \overline{b}_j z} \right\} \Big/ \prod_{i=1}^n \left\{ \frac{\rho(z - a_i)}{\rho^2 - \overline{a}_i z} \right\}, \tag{1.3}$$

where function $\psi(z)$ is regular in the disk $|z| < \rho$ and has no zeros. We then apply formula (1.1) to $\psi(z)$, and notice that when $\zeta = \rho e^{i\varphi}$, $|a| < 1$, we have

$$\left| \frac{\rho(\zeta - a)}{\rho^2 - \overline{a}\zeta} \right| = \left| \frac{\rho e^{i\varphi} - a}{\rho - \overline{a} e^{i\varphi}} \right| = \left| \frac{\rho - a e^{-i\varphi}}{\rho - \overline{a} e^{i\varphi}} \right| = 1,$$

and hence,

$$\log |\psi(z)| = \frac{1}{2\pi} \int_0^{2\pi} \log |f(\rho e^{i\varphi})| \frac{(\rho^2 - r^2) d\varphi}{\rho^2 - 2\rho r \cos(\theta - \varphi) + r^2}.$$

Using formula (1.3) we obtain formula (1.1). Consequently Theorem 1.1 is proved completely.

Translator's note: The terms "regular" and "holomorphic" both are synonymous with "analytic".

When $f(0) \neq 0, \infty$ if we let $z = 0$ in formula (1.1), we get

$$\log|f(0)| = \frac{1}{2\pi} \int_0^{2\pi} \log|f(\rho e^{i\varphi})| d\varphi - \sum_{i=1}^n \log \frac{\rho}{|a_i|} + \sum_{j=1}^m \log \frac{\rho}{|b_j|}. \qquad (1.4)$$

Formula (1.4) is called the Jensen formula.

1.1.2. The Jensen-Nevanlinna formula. We denote

$$n(r, a) = n(r, f = a) = \begin{cases} n(r, f), & a = \infty, \\ n(r, \frac{1}{f-a}), & a \neq \infty \end{cases}$$

as the number of roots of the equation $f(z) = a$ in the disk $|z| < r$ $(0 < r \leq \rho)$ (multiple roots being counted with their multiplicities), and

$$n(0, a) = n(0, f = a) = \begin{cases} n(0, f), & a = \infty, \\ n(0, \frac{1}{f-a}), & a \neq \infty \end{cases}$$

as the number of roots of the equation $f(z) = a$ around the origin $z = 0$ (multiple roots being counted multiply), and also let

$$N(r, a) = N(r, f = a)$$
$$= \begin{cases} N(r, f) = \int_0^r \frac{n(t,a)-n(0,a)}{t} dt + n(0, a) \log r, & a = \infty, \\ N(r, \frac{1}{f-a}) = \int_0^r \frac{n(t,a)-n(0,a)}{t} dt + n(0, a) \log r, & a \neq \infty. \end{cases}$$

One then finds

$$\sum_{i=1}^n \log \frac{\rho}{|a_i|} = \int_0^\rho \left(\log \frac{\rho}{t} \right) dn(t, 0)$$

$$= \left(\log \frac{\rho}{t} \right) n(t, 0) \Big|_0^\rho + \int_0^\rho \frac{n(t, 0)}{t} dt \qquad (1.5)$$

$$= \int_0^\rho \frac{n(t, 0)}{t} dt = N(\rho, 0),$$

and

$$\sum_{j=1}^m \log \frac{\rho}{|b_j|} = \int_0^\rho \frac{n(t, \infty)}{t} dt = N(\rho, \infty). \qquad (1.6)$$

On the other hand, we define

$$\log^+ x = \begin{cases} \log x, & x \geq 1, \\ 0, & 0 \leq x < 1. \end{cases}$$

Therefore, when $x \geq 0$, we have

$$\log x = \log^+ x - \log^+ \frac{1}{x}.$$

Let

$$m(r, a) = m(r, f = a)$$
$$= \begin{cases} m(r, f) = \frac{1}{2\pi} \int_0^{2\pi} \log^+ |f(re^{i\theta})| d\theta, & a = \infty, \\ m(r, \frac{1}{f-a}) = \frac{1}{2\pi} \int_0^{2\pi} \log^+ \frac{1}{|f(re^{i\theta})-a|} d\theta, & a \neq \infty. \end{cases}$$

It follows that

$$\frac{1}{2\pi} \int_0^{2\pi} \log |f(\rho e^{i\varphi})| \, d\varphi$$

$$= \frac{1}{2\pi} \int_0^{2\pi} \log^+ |f(\rho e^{i\varphi})| \, d\varphi - \frac{1}{2\pi} \int_0^{2\pi} \log^+ \frac{1}{|f(\rho e^{i\varphi})|} \, d\varphi \qquad (1.7)$$

$$= m(\rho, \infty) - m(\rho, 0).$$

According to formulas (1.4), (1.5), (1.6) and (1.7), one derives

$$\frac{1}{2\pi} \int_0^{2\pi} \log^+ |f(\rho e^{i\varphi})| \, d\varphi + \int_0^\rho \frac{n(t, \infty)}{t} \, dt$$

$$= \frac{1}{2\pi} \int_0^{2\pi} \log^+ \frac{1}{|f(\rho e^{i\varphi})|} \, d\varphi \qquad (1.8)$$

$$+ \int_0^\rho \frac{n(t, 0)}{t} \, dt + \log |f(0)|,$$

and

$$m(\rho, \infty) + N(\rho, \infty) = m(\rho, 0) + N(\rho, 0) + \log |f(0)|. \qquad (1.9)$$

When $f(0) = 0$ or ∞, we may assume that at a neighborhood of $z = 0$ $f(z)$ has the Laurent expansion as follows:

$$f(z) = c_s z^s + c_{s+1} z^{s+1} + \cdots, \qquad c_s \neq 0.$$

Let $f_1(z) = z^{-s} f(z)$, then $f_1(z)$ is meromorphic in the disk $|z| \leq \rho$ with $f_1(0) = c_s$ and

$$\frac{1}{2\pi} \int_0^{2\pi} \log |f_1(\rho e^{i\varphi})| \, d\varphi = \frac{1}{2\pi} \int_0^{2\pi} \log |f(\rho e^{i\varphi})| \, d\varphi - s \log \rho,$$

$$s = n(0, 0) - n(0, \infty).$$

Applying formula (1.8) to $f_1(z)$ we get

$$\frac{1}{2\pi} \int_0^{2\pi} \log^+ |f(\rho e^{i\varphi})| d\varphi + \int_0^\rho \frac{n(t, \infty) - n(0, \infty)}{t} \, dt + n(0, \infty) \log \rho$$

$$= \frac{1}{2\pi} \int_0^{2\pi} \log^+ \frac{1}{|f(\rho e^{i\varphi})|} \, d\varphi + \int_0^\rho \frac{n(t, 0) - n(0, 0)}{t} \, dt \qquad (1.10)$$

$$+ n(0, 0) \log \rho + \log |c_s|,$$

$$m(\rho, \infty) + N(\rho, \infty) = m(\rho, 0) + N(\rho, 0) + \log |c_s|.$$

Formula (1.10) is called the Jensen-Nevanlinna formula.

§1.2. The characteristic function

1.2.1. Definition of the characteristic function. Suppose that $f(z)$ is a meromorphic function nonidentically vanishing in the disk $|z| < R$ $(0 < R \leq +\infty)$. Replace ρ by r $(0 < r < R)$ in formula (1.9), and assume that

$$T(r, f) = m(r, f) + N(r, f). \tag{1.11}$$

Then formulas (1.9) and (1.10) can be rewritten respectively as

$$T(r, f) = T\left(r, \frac{1}{f}\right) + \log|f(0)|, \tag{1.12}$$

and

$$T(r, f) = T\left(r, \frac{1}{f}\right) + \log|c_s|. \tag{1.13}$$

R. Nevanlinna called $T(r, f)$ the characteristic function of $f(z)$ [32a].

1.2.2. Cartan's identical relation. We find a new formula to express $T(r, f)$ and prove two important properties of $T(r, f)$. According to formula (1.4), for an arbitrary complex number a ([1]) we have

$$\frac{1}{2\pi} \int_0^{2\pi} \log|a - e^{i\theta}|\, d\theta = \begin{cases} \log|a|, & |a| \geq 1, \\ \log|a| - \log|a| = 0, & |a| < 1, \end{cases}$$

that is

$$\frac{1}{2\pi} \int_0^{2\pi} \log|a - e^{i\theta}|\, d\theta = \log^+|a|. \tag{1.14}$$

First we assume that $f(0) \neq \infty$. When $f(0) \neq e^{i\theta}$, applying formulas (1.7) and (1.9) to $f(z) - e^{i\theta} (0 \leq \theta < 2\pi)$ we obtain

$$\frac{1}{2\pi} \int_0^{2\pi} \log|f(re^{i\varphi}) - e^{i\theta}|\, d\varphi$$

$$= N\left(r, \frac{1}{f - e^{i\theta}}\right) - N(r, f) + \log|f(0) - e^{i\theta}|,$$

where $0 < r < R$. Fixed value r and integrated with respect to θ, results in

$$\frac{1}{2\pi} \int_0^{2\pi} d\theta \frac{1}{2\pi} \int_0^{2\pi} \log|f(re^{i\varphi}) - e^{i\theta}|\, d\varphi$$

$$= \frac{1}{2\pi} \int_0^{2\pi} N\left(r, \frac{1}{f - e^{i\theta}}\right) d\theta - N(r, f)$$

$$+ \frac{1}{2\pi} \int_0^{2\pi} \log|f(0) - e^{i\theta}|\, d\theta.$$

From formulas (1.12) and (1.14), one then deduces

$$T(r, f) = \frac{1}{2\pi} \int_0^{2\pi} N\left(r, \frac{1}{f - e^{i\theta}}\right) d\theta + \log^+|f(0)|. \tag{1.15}$$

([1]) In this book, complex numbers will include ∞.

This is the so-called Cartan's identical relation [9b].

Next we assume that $f(0) = \infty$, and that c_s is the first nonzero coefficient of the expansion of $f(z)$ in a neighborhood around $z = 0$. An application of formulas (1.7) and (1.10) to $f(z) - e^{i\theta}$ yields

$$\frac{1}{2\pi} \int_0^{2\pi} \log|f(re^{i\varphi}) - e^{i\theta}|\, d\varphi = N\left(r, \frac{1}{f - e^{i\theta}}\right) - N(r, f) + \log|c_s|,$$

$$T(r, f) = \frac{1}{2\pi} \int_0^{2\pi} N\left(r, \frac{1}{f - e^{i\theta}}\right) d\theta + \log|c_s|. \tag{1.16}$$

When θ is fixed, $N(r, e^{i\theta})$ is a nondecreasing function of r. Hence we deduce from formulas (1.15) and (1.16) that $T(r, f)$ is a nondecreasing function of r, and also

$$\frac{dT(r, f)}{d\log r} = \frac{1}{2\pi} \int_0^{2\pi} n(r, e^{i\theta})\, d\theta.$$

Consequently we conclude that $T(r, f)$ is a convex function of $\log r$.

1.2.3. Some inequalities of the characteristic function. Suppose that a_1, $a_2, \ldots, a_p$ are arbitrarily p complex numbers; then

$$\log^+\left|\prod_{\nu=1}^p a_\nu\right| \leq \sum_{\nu=1}^p \log^+|a_\nu|, \tag{1.17}$$

and

$$\log^+\left|\sum_{\nu=1}^p a_\nu\right| \leq \sum_{\nu=1}^p \log^+|a_\nu| + \log p. \tag{1.18}$$

Correspondingly, for any p arbitrary meromorphic functions $f_1(z), f_2(z), \ldots, f_p(z)$ in $|z| < R$ $(0 < R \leq +\infty)$ we have

$$m\left(r, \sum_{\nu=1}^p f_\nu\right) \leq \sum_{\nu=1}^p m(r, f_\nu) + \log p,$$

$$m\left(r, \prod_{\nu=1}^p f_\nu\right) \leq \sum_{\nu=1}^p m(r, f_\nu),$$

where $0 < r < R$. On the other hand, when $f_\nu(0) \neq \infty$ $(\nu = 1, 2, \ldots, p)$ we have

$$N\left(r, \sum_{\nu=1}^p f_\nu\right) \leq \sum_{\nu=1}^p N(r, f_\nu),$$

$$N\left(r, \prod_{\nu=1}^p f_\nu\right) \leq \sum_{\nu=1}^p N(r, f_\nu).$$

Hence, according to formula (1.11) we obtain

$$T\left(r, \sum_{\nu=1}^{p} f_{\nu}\right) \leq \sum_{\nu=1}^{p} T(r, f_{\nu}) + \log p, \qquad (1.19)$$

and

$$T\left(r, \prod_{\nu=1}^{p} f_{\nu}\right) \leq \sum_{\nu=1}^{p} T(r, f_{\nu}). \qquad (1.20)$$

When $R > 1$, if only $1 \leq r < R$ is considered, we can ignore the condition of $f_{\nu}(0) \neq \infty$ $(\nu = 1, 2, \ldots, p)$.

1.2.4. The relationships between the maximum modulus and the characteristic function of a regular function. Suppose that $f(z)$ is a function regular in the disk $|z| < R$ $(0 < R \leq +\infty)$. Let

$$M(r, f) = \max_{|z|=r} |f(z)|,$$

then the following inequality holds:

$$T(r, f) \leq \log^{+} M(r, f) \leq \frac{\rho + r}{\rho - r} T(\rho, f) \qquad (0 < r < \rho < R). \qquad (1.21)$$

In fact, taking into consideration the regularity of $f(z)$, we have

$$T(r, f) = m(r, f) = \frac{1}{2\pi} \int_{0}^{2\pi} \log^{+} |f(re^{i\theta})| \, d\theta \leq \log^{+} M(r, f),$$

meaning that the inequality on the left-hand side of formula (1.21) holds. In order to prove the inequality on the right-hand side, first we notice that when $M(r, f) \leq 1$, it is obvious that the inequality on the right-hand side of formula (1.21) is valid. We can, therefore, assume that $M(r, f) > 1$. Selecting the point $z_0 = re^{i\theta}$, such that $|f(z_0)| = M(r, f)$. Then based on the Poisson-Jensen formula we derive

$$\log^{+} M(r, f) = \log^{+} |f(z_0)|$$

$$\leq \frac{1}{2\pi} \int_{0}^{2\pi} \log |f(\rho e^{i\varphi})| \frac{(\rho^2 - r^2) \, d\varphi}{\rho^2 - 2\rho r \cos(\theta - \varphi) + r^2}$$

$$\leq \frac{\rho + r}{\rho - r} \cdot \frac{1}{2\pi} \int_{0}^{2\pi} \log^{+} |f(\rho e^{i\varphi})| \, d\varphi$$

$$= \frac{\rho + r}{\rho - r} m(\rho, f) = \frac{\rho + r}{\rho - r} T(\rho, f).$$

Thus the inequality on the right-hand side of formula (1.21) is valid.

§1.3. The Ahlfors-Shimizu characteristic

1.3.1. The Ahlfors-Shimizu characteristic. We denote $k(a, b)$ or $|a, b|$ as the chordal distance between points a and b on the sphere. Concretely speaking, when both a and b are finite points we have

$$k(a, b) = |a, b| = \frac{|a - b|}{\sqrt{1 + |a|^2} \sqrt{1 + |b|^2}};$$

when either a or b (e.g., b) is ∞, then

$$k(a, \infty) = |a, \infty| = \frac{1}{\sqrt{1 + |a|^2}}.$$

Let $f(z)$ be a nonconstant meromorphic function in the disk $|z| < R$ $(0 < R \leq +\infty)$. When $0 < r < R$, we set

$$m_0(r, a) = \frac{1}{2\pi} \int_0^{2\pi} \log \frac{1}{k\{f(re^{i\theta}), a\}} d\theta,$$

$$s(r) = \frac{1}{\pi} A(r) = \frac{1}{\pi} \iint_{|z|<r} \frac{|f'(z)|^2}{(1 + |f(z)|^2)^2} r \, dr \, d\theta, \qquad z = re^{i\theta},$$

$$T_0(r, f) = \int_0^r \frac{s(t)}{t} dt,$$

where $A(r)$ denotes the area of image when $f(z)$ maps the disk $|z| \leq r$ to the Riemann sphere. $T_0(r, f)$ is called the Ahlfors-Shimizu characteristic [1a, 36a].

THEOREM 1.2.[*] *For every complex value a and arbitrary value r, $0 < r < R$, we have*

$$T_0(r, f) = N(r, a) + m_0(r, a) - m_0(a), \tag{1.22}$$

where $m_0(a)$ is a constant unrelated to r, with its precise definition as follows:

$$m_0(a) = \begin{cases} m_0(0, a), & f(0) \neq a, \\ \log|c_{-m}|, & f(0) = a = \infty, \\ \log \frac{1+|a|^2}{|c_s|}, & f(0) = a \neq \infty. \end{cases} \tag{1.23}$$

Here c_{-m} denotes the first nonzero coefficient of the expansion of $f(z)$ in a neighborhood around point $z = 0$, which is the pole of order m of $f(z)$; and c_s is the first nonzero coefficient of the expansion of $f(z)$ in a neighborhood around point $z = 0$, which is the zero of order s of $f(z) - a$.

PROOF. Let a and b be arbitrarily two distinct finite complex numbers. By the argument principle we find

$$\frac{dm_0(r, a)}{dr} - \frac{dm_0(r, b)}{dr} = \frac{1}{2\pi} \int_0^{2\pi} \frac{d}{dr} \log \left| \frac{f - b}{f - a} \right| d\theta$$

$$= \frac{1}{2\pi r} \int_{|z|=r} d \arg \left(\frac{f - b}{f - a} \right)$$

$$= \frac{n(r, b) - n(r, a)}{r}.$$

---

[*] *Translator's note*: Formula (1.22) is the so-called Ahlfors-Shimizu form of the First Fundamental Theorem.

We derive further from $\frac{dN(r,a)}{dr} = \frac{n(r,a)}{r}$ that

$$\frac{dm_0(r,a)}{dr} + \frac{dN(r,a)}{dr} = \frac{dm_0(r,b)}{dr} + \frac{dN(r,b)}{dr}.$$

By means of integration on both sides and taking formula (1.23) into consideration we see that

$$m_0(r,a) + N(r,a) - m_0(a) = m_0(r,b) + N(r,b) - m_0(b).$$

When a or b is ∞, apparently this formula is valid. Therefore $m_0(r,a) + N(r,a) - m_0(a) + N(r,a) - m_0(a)$ is a quantity depending on r but unrelated to a. Hence, if we denote $d\omega(a)$ the area element, $\iint_K d\omega(a) = \pi$ on the Riemann sphere K, then it follows that

$$N(r,a) + m_0(r,a) - m_0(a)$$
$$= \frac{1}{\pi} \iint_K \{N(r,a) + m_0(r,a) - m_0(a)\} \, d\omega(a)$$
$$= \frac{1}{\pi} \iint_K N(r,a) \, d\omega(a) + \frac{1}{\pi} \iint_K m_0(r,a) \, d\omega(a)$$
$$- \frac{1}{\pi} \iint_K m_0(a) \, d\omega(a).$$

Notice that

$$\iint_K \log \frac{1}{k\{f(z),a\}} \, d\omega(a)$$

is a constant unrelated to point z and $f(z)$; therefore

$$\frac{1}{\pi} \iint_K m_0(r,a) \, d\omega(a) - \frac{1}{\pi} \iint_K m_0(a) \, d\omega(a) = 0.$$

On the other hand we have

$$\frac{1}{\pi} \iint_K N(r,a) \, d\omega(a) = \frac{1}{\pi} \int_0^r \frac{dt}{t} \iint_K n(t,a) \, d\omega(a)$$
$$= \frac{1}{\pi} \int_0^r \frac{A(t)}{t} \, dt = \int_0^r \frac{s(t)}{t} \, dt = T_0(r,f).$$

Hence formula (1.22) holds. Theorem 1.2 is thus proved.

1.3.2. The relationship between $T(r,f)$ and $T_0(r,f)$. We let $a = \infty$ in (1.22); then

$$T_0(r,f) = N(r,\infty) + m_0(r,\infty) - m_0(\infty). \tag{1.24}$$

Notice that

$$\log^+ |f| \leq \log \sqrt{1 + |f|^2} \leq \log^+ |f| + \frac{1}{2} \log 2,$$

$$m(r,f) \leq m_0(r,\infty) \leq m(r,f) + \frac{1}{2} \log 2,$$

and taking formula (1.24) into account we obtain

$$T(r,f) \leq T_0(r,f) + m_0(\infty) \leq T(r,f) + \frac{1}{2} \log 2. \tag{1.25}$$

It follows further from formula (1.23) that when $f(0) \neq \infty$,

$$|T(r, f) - T_0(r, f) - \log^+ |f(0)|| \leq \frac{1}{2} \log 2;$$

and when $f(0) = \infty$,

$$|T(r, f) - T_0(r, f) - \log |c_{-m}|| \leq \frac{1}{2} \log 2.$$

§1.4. The First Fundamental Theorem

1.4.1. The First Fundamental Theorem. Let $f(z)$ be a nonconstant meromorphic function in the disk $|z| < R$ $(0 < R \leq +\infty)$ and a be a finite complex number; then $f(z) - a$ at a neighborhood around point $z = 0$ has the following expansion:

$$f(z) - a = c_s z^s + c_{s+1} z^{s+1} + \cdots, \qquad c_s \neq 0.$$

By virtue of the above assumption, for any arbitrary value r, $0 < r < R$, we have according to formula (1.13):

$$T(r, f - a) = T\left(r, \frac{1}{f-a}\right) + \log |c_s|.$$

On the other hand, since

$$m(r, f - a) \leq m(r, f) + \log^+ |a| + \log 2,$$
$$m(r, f) \leq m(r, f - a) + \log^+ |a| + \log 2,$$

and $N(r, f - a) = N(r, f)$, then we obtain

$$|T(r, f - a) - T(r, f)| \leq \log^+ |a| + \log 2. \tag{1.26}$$

Accordingly we conclude that

$$\left| T(r, f) - T\left(r, \frac{1}{f-a}\right) - \log |c_s| \right| \leq \log^+ |a| + \log 2. \tag{1.27}$$

Theorem 1.3 is thus proved.

THEOREM 1.3. *For any arbitrary finite complex number* a *we have*

$$T\left(r, \frac{1}{f-a}\right) = T(r, f) - \log |c_s| + \varepsilon(r, a),$$

where $|\varepsilon(r, a)| \leq \log^+ |a| + \log 2.$

R. Nevanlinna called Theorem 1.3 the First Fundamental Theorem [32a]. Suppose that a $(a \neq 0)$ and b are two constants. Let

$$f_1(z) = f(z) - a, \quad f_2(z) = af(z), \quad f_3(z) = \frac{1}{f(z)};$$

then according to formulas (1.13), (1.19) and (1.20), the differences between any two characteristic functions among $T(r, f_1)$, $T(r, f_2)$ and $T(r, f_3)$

as well as $T(r, f)$ are all bounded functions. Moreover, we consider the function

$$F(z) = \frac{af(z) + b}{cf(z) + d},$$

where a, b, c and d are constants, and that $ad - bc \neq 0$. Since F can be resolved into three combinations mentioned above, the difference between characteristic functions $T(r, F)$ of $F(z)$ and $T(r, f)$ is also a bounded function.

1.4.2. Growth of a transcendental meromorphic function. Let $f(z)$ be a transcendental meromorphic function on the open plane $|z| < +\infty$, meaning that it will not be degenerated into a rational function; then

$$\lim_{r \to +\infty} \frac{T(r, f)}{\log r} = +\infty. \tag{1.28}$$

In the following, we shall prove two different cases:

(1) Suppose that $f(z)$ has no poles. Then $f(z)$ has the expansion

$$f(z) = \sum_{n=0}^{\infty} a_n z^n, \qquad |z| < +\infty,$$

and has infinitely many nonzero coefficients. According to Cauchy's inequality we have

$$|a_n| r^n \leq M(r, f) \qquad (|z| = r > 0, \; n = 1, 2, \ldots).$$

In turn, for any arbitrary positive integer p, it follows that

$$\lim_{r \to +\infty} \frac{M(r, f)}{r^p} = +\infty.$$

Accordingly

$$\lim_{r \to +\infty} \frac{\log M(r, f)}{\log r} = +\infty.$$

On the other hand, we let $\rho = 2r$ in formula (1.21) and derive $\log^+ M(r, f) \leq 3T(2r, f)$. Hence, the validity of formula (1.28) is proved.

(2) Suppose that $f(z)$ has poles. When $f(z)$ has infinitely many poles, formula (1.28) holds, since

$$T(r, f) \geq N(r, f) \geq N(r, f) - N(\sqrt{r}, f)$$

$$\geq \frac{1}{2} n(\sqrt{r}, f) \log r (r > 1).$$

Now assume that $f(z)$ has only finite poles b_j $(j = 1, 2, \ldots, k)$, with their orders being m_j $(j = 1, 2, \ldots, k)$, respectively. Let

$$p(z) = \prod_{j=1}^{k} (z - b_j)^{m_j}, \qquad g(z) = p(z) f(z),$$

then $g(z)$ is a transcendental entire function (i.e. without any poles). Hence we obtain

$$\lim_{r \to +\infty} \frac{T(r, g)}{\log r} = +\infty.$$

On the other hand, it follows from formulas (1.19) and (1.20) that

$$T(r, g) \leq T(r, p) + T(r, f)$$

$$\leq \log^+ r \sum_{j=1}^{k} m_j + \sum_{j=1}^{k} m_j \log^+ |b_j|$$

$$+ \log 2 \sum_{j=1}^{k} m_j + T(r, f).$$

Consequently, formula (1.28) is valid.

1.4.3. Examples. (1) Let

$$f(z) = c \cdot \frac{z^p + \cdots + a_p}{z^q + \cdots + b_q}, \qquad (c \neq 0)$$

be a rational function. If $p > q$, $f(z) \to \infty$ when $z \to \infty$. Hence for any arbitrary finite complex number a, when r is sufficiently large, $m(r, a) = 0$. On the other hand, since equation $f(z) = a$ has p roots, when t is sufficiently large, $n(t, a) = p$. Therefore,

$$N(r, a) = \int_0^r \frac{n(t, a) - n(0, a)}{t} dt + n(0, a) \log r$$

$$= p \log r + O(1),$$

$$T(r, f) = m(r, a) + N(r, a) + O(1) = p \log r + O(1).$$

Analogously, when $p < q$, for any arbitrary complex number a $(a \neq 0)$ we have

$$T(r, f) = q \log r + O(1),$$

$$N(r, f) = q \log r + O(1), \qquad m(r, a) = O(1),$$

and when $p = q$, for any arbitrary complex number a $(a \neq c)$ we get

$$T(r, f) = q \log r + O(1),$$

$$N(r, a) = q \log r + O(1), \qquad m(r, a) = O(1).$$

Hence, provided that $a \neq f(\infty)$, it follows that

$$T(r, f) = d \log r + O(1),$$

$$N(r, f) = d \log r + O(1), \qquad m(r, a) = O(1),$$

where $d = \max(p, q)$. When $a = f(\infty)$, if infinitely far away around a point, equation $f(z) = a$ has a multiple root of order α $(0 < \alpha < d)$, then

$$T(r, f) = d \log r + O(1),$$

$$N(r, f) = (d - \alpha) \log r + O(1), \qquad m(r, a) = \alpha \log r + O(1).$$

To sum up, we find that when $f(z)$ is a rational function, then it must be

$$T(r, f) = O(\log r).\tag{1.29}$$

(2) Let $f(z) = e^z$, $z = re^{i\theta}$. A simple calculation gives

$$m(r, 0) = m(r, \infty) = \frac{r}{\pi}, \qquad N(r, 0) = N(r, \infty) = 0,$$
$$T(r, f) = \frac{r}{\pi}.\tag{1.30}$$

When $a \neq 0, \infty$, if point z_0 is a root of the equation $e^z = a$, then $z_0 + 2k\pi i$ $(k = 0, \pm 1, \pm 2, \ldots)$ are all the roots of this equation. We then obtain

$$n(t, a) = \frac{t}{\pi} + O(1), \qquad N(r, a) = \frac{r}{\pi} + O(\log r).$$

Further, according to formula (1.30) we conclude that $m(r, a) = O(\log r)$.

(3) Let $f(z) = \int_0^z e^{-t^q} dt$, $q \geq 2$, and $a_k = e^{2\pi k i/q} \int_0^\infty e^{-t^q} dt$, $k = 1, 2, \ldots, q$; then when z tends to ∞ in the angular region $|\arg z - (2\pi k)/q| < \pi/(2q) - \varepsilon$ $(\varepsilon > 0)$, we have

$$f(z) - a_k = -\int_z^\infty e^{-t^q} dt = -\frac{e^{-z^q}}{qz^{q-1}} + \frac{q-1}{q} \int_z^\infty \frac{e^{-t^q}}{t^q} dt$$
$$= -\frac{e^{-z^q}}{qz^{q-1}} \{1 + o(1)\}.$$

And hence

$$m(r, a_k) = \frac{1}{2\pi} \{1 + o(1)\} r^q \int_{-\pi/(2q)}^{\pi/(2q)} \cos q\theta \, d\theta$$
$$= \{1 + o(1)\} \frac{r^q}{q\pi}, \qquad k = 1, 2, \ldots, q.$$

Analogously, when z tends to ∞ in the angular region

$$|\arg z - ((2k-1)\pi/q)| < \pi/(2q) - \varepsilon,$$

it follows that

$$f(z) = \int_0^z e^{-t^q} dt = \frac{e^{-z^q} \{1 + o(1)\}}{qz^{q-1}} + O(1).$$

And therefore $m(r, f) = \{1 + o(1)\} r^q/\pi$. As for $a \neq a_k$ $(k = 1, 2, \ldots, q)$ and ∞, based on the Second Fundamental Theorem discussed later, we conclude that $m(r, a) = o(r^q)$. In view of the above discussion and an application of the First Fundamental Theorem, we consequently obtain

$$m(r, a) = \{1 + o(1)\}\frac{r^q}{\pi}, \qquad N(r, a) = 0, \quad a = \infty;$$

$$m(r, a) = \{1 + o(1)\}\frac{r^q}{q\pi},$$

$$N(r, a) = \left(1 - \frac{1}{q}\right)\frac{r^q}{\pi}, \quad a = a_1, \ldots, a_q$$

$$m(r, a) = o(r^q), \quad N(r, a) = \{1 + o(1)\}\frac{r^q}{\pi}, \quad a \neq a_1, \ldots, a_q.$$

1.4.4. Orders. The growth of the characteristic function $T(r, f)$ discussed above displays certain intrinsic properties of the meromorphic function $f(z)$. Recall that E. Borel introduced the concept of orders for entire functions in terms of the maximum modulus of functions [6a]. Here we exhibit R. Nevanlinna's definition of the order for the meromorphic function [32a]: Suppose that $f(z)$ is a function meromorphic on the open plane $|z| < +\infty$. If λ and μ denote the order and lower order of $f(z)$, respectively, then

$$\varlimsup_{r \to +\infty} \frac{\log^+ T(r, f)}{\log r} = \begin{cases} \lambda, \\ \mu. \end{cases}$$

Clearly, $0 \leq \mu \leq \lambda \leq +\infty$. It follows that when $\lambda = 0$, $0 < \lambda < +\infty$ and $\lambda = +\infty$, $f(z)$ is a meromorphic function of zero order, finite positive order and infinite order, respectively. We can later see that these three kinds of meromorphic functions are characteristically different from one another. Meanwhile, we have to differentiate more precisely in some aspect on the growth of $T(r, f)$, such as the concepts of its type and class. However, since these concepts are not needed in this book, we will not introduce them here. We may also differentiate further correspondingly the lower order μ. Indeed, this book concentrates mainly on functions having a finite lower order μ.

To define λ and μ, based on formulas (1.21) and (1.25), we find that the values of λ and μ remains unchanged if $T_0(r, f)$ is used or if $T(r, f)$ is replaced by $\log M(r, f)$ when $f(z)$ is an entire function.

Finally, to define the order λ and lower order μ of the meromorphic function $f(z)$ in the disk $|z| < 1$, concretely speaking we have

$$\varlimsup_{r \to 1} \frac{\log^+ T(r, f)}{\log \frac{1}{1-r}} = \begin{cases} \lambda, \\ \mu. \end{cases}$$

§1.5. Lemma on the logarithmic derivative

1.5.1. Two lemmas.

LEMMA 1.1. *For any arbitrary finite complex number z and value r, $0 < r < +\infty$, if we denote E_k the set of all values θ $(0 \leq |\theta| < \pi)$ that satisfies* *the inequality* $|z - re^{i\theta}| < kr$ $(0 < k < 1)$, *then we have*

$$\int_{E_k} \log \frac{r}{|z - re^{i\theta}|} d\theta < \pi k \left\{ \log \frac{1}{k} + 1 \right\}. \tag{1.31}$$

PROOF. Without any loss of generality, we may assume that z is a positive number. Hence, when $\theta \in E_k$, it follows that $|z - re^{i\theta}| \geq r \sin \theta$. Let θ_0 be the smallest positive root of equation $\sin \theta = k$; then E_k lies in the interval $[-\theta_0, \theta_0]$. Notice when $\frac{\pi}{2} \leq |\theta| < \pi$, $|z - re^{i\theta}| > r$. Therefore,

$$\int_{E_k} \log \frac{r}{|z - re^{i\theta}|} d\theta \leq 2 \int_0^{\theta_0} \log \left(\frac{1}{\sin \theta} \right) d\theta$$

$$\leq 2 \int_0^{\theta_0} \log \frac{\pi}{2\theta} d\theta = 2\theta_0 \left\{ \log \left(\frac{\pi}{2\theta_0} \right) + 1 \right\}.$$

Also notice that $\theta_0 \{ \log \frac{\pi}{2\theta_0} + 1 \}$ in the interval $(0, \frac{\pi}{2})$ is an increasing function of θ_0, and that $\sin \theta_0 = k > \frac{2}{\pi}\theta_0$, $\theta_0 < \frac{\pi}{2}k < \frac{\pi}{2}$. Replacing θ_0 by $\frac{\pi}{2}k$, one obtains immediately formula (1.31). Hence, Lemma 1.1 is proved.

LEMMA 1.2. *Let* $z_1, z_2, \ldots, z_n$ *be* n $(n \geq 1)$ *points on the open plane* $|z| < +\infty$, *and that* $\delta(z)$ *be the shortest distance between point* z *and these* n *points. Then*

$$\frac{1}{2\pi} \int_0^{2\pi} \log^+ \frac{r}{\delta(re^{i\theta})} d\theta \leq 2 \log n + \frac{1}{2}.$$

PROOF. Let E_ν be the set of all values θ $(0 \leq |\theta| < \pi)$ satisfying the inequality $|re^{i\theta} - z_\nu| < \frac{r}{n}$ and $E = \bigcup_{\nu=1}^n E_\nu$. When $x \geq n$, let $\log_0 x = \log x$; and when $0 < x < n$, let $\log_0 x = 0$. Hence when $\theta \in E$, we have $\delta(re^{i\theta}) < \frac{r}{n}$. It follows accordingly that

$$\log^+ \frac{r}{\delta(re^{i\theta})} = \log_0 \frac{r}{\delta(re^{i\theta})} \leq \sum_{\nu=1}^n \log_0 \frac{r}{|re^{i\theta} - z_\nu|}.$$

Applying Lemma 1.1, where we let $k = \frac{1}{n}$, we obtain

$$\int_E \log^+ \frac{r}{\delta(re^{i\theta})} d\theta \leq \sum_{\nu=1}^n \int_0^{2\pi} \log_0 \left| \frac{r}{re^{i\theta} - z_\nu} \right| d\theta \leq \pi(\log n + 1). \tag{1.32}$$

On the other hand, if we denote CE the complementary set of E, then when $\theta \in CE$, we have $\delta(re^{i\theta}) \geq \frac{r}{n}$. Therefore,

$$\int_{CE} \log^+ \frac{r}{\delta(re^{i\theta})} d\theta \leq \int_{CE} \log n \, d\theta \leq 2\pi \log n.$$

Combining this with formula (1.32) yields

$$\int_0^{2\pi} \log^+ \frac{r}{\delta(re^{i\theta})} d\theta \leq 2\pi \log n + \pi(\log n + 1) = \pi(3 \log n + 1),$$

$$\frac{1}{2\pi} \int_0^{2\pi} \log^+ \frac{r}{\delta(re^{i\theta})} d\theta \leq 2 \log n + \frac{1}{2}.$$

Thus Lemma 1.2 is proved.

The above two lemmas are attributed to W. Hayman's efforts [21c].

1.5.2. Lemma on the logarithmic derivative. The so-called "lemma on the logarithmic derivative", that is Lemma 1.3 to be mentioned below, plays a crucial role in the Nevanlinna Theory.

LEMMA 1.3. *Suppose* $f(z)$ *is a function meromorphic in the disk* $|z| < R$ $(0 < R \leq +\infty)$, *and that* $f(0) \neq 0, \infty$; *then when* $0 < r < \rho < R$, *we have*

$$
m\left(r, \frac{f'}{f}\right) \leq 4\log^+ T(\rho, f) + 4\log^+ \log^+ \frac{1}{|f(0)|} + 5\log^+ \rho
$$
$$
+ 6\log^+ \frac{1}{\rho - r} + \log^+ \frac{1}{r} + 14 .
$$
(1.33)

PROOF. Let $\rho' = (\rho + r)/2$ and select arbitrarily a point z_0 in the disk $|z| < \rho'$, such that $f(z_0) \neq 0, \infty$. We find that $\log f(z)$ is regular at a neighborhood around z_0. With reference to the Poisson-Jensen formula and the inequality

$$
\frac{\rho'^2 - r^2}{\rho' - 2\rho' r \cos(\theta - \varphi) + r^2} = \text{Re} \left\{ \frac{\rho' e^{i\varphi} + z}{\rho' e^{i\varphi} - z} \right\}, \qquad z = r e^{i\varphi},
$$

we have at a neighborhood around z_0 the following expression:

$$
\log f(z) = \frac{1}{2\pi} \int_0^{2\pi} \log|f(\rho' e^{i\varphi})| \frac{\rho' e^{i\varphi} + z}{\rho' e^{i\varphi} - z} d\varphi
$$
$$
+ \sum_{i=1}^n \log \frac{\rho'(z - a_i)}{\rho'^2 - \overline{a}_i z} - \sum_{j=1}^m \log \frac{\rho'(z - b_j)}{\rho'^2 - \overline{b}_j z} + ic,
$$

where c is a real constant. Taking derivatives on both sides, we get

$$
\frac{f'(z)}{f(z)} = \frac{1}{2\pi} \int_0^{2\pi} \log|f(\rho' e^{i\varphi})| \frac{2\rho' e^{i\varphi}}{(\rho' e^{i\varphi} - z)^2} d\varphi
$$
$$
+ \sum_{i=1}^n \left(\frac{\overline{a}_i}{\rho'^2 - \overline{a}_i z} - \frac{1}{a_i - z} \right)
$$
(1.34)
$$
+ \sum_{j=1}^m \left(\frac{1}{b_j - z} - \frac{\overline{b}_j}{\rho'^2 - \overline{b}_j z} \right) .
$$

Notice that formula (1.34) is valid at a neighborhood around z_0 and, in particular, at point z_0. Since z_0 is arbitrary, we conclude that formula (1.34) is valid everywhere except at the zeros and poles of $f(z)$. Furthermore, at the zeros and poles of $f(z)$, formula (1.34) assumes ∞ on both sides of (1.34). Hence formula (1.34) is valid everywhere in the disk $|z| < \rho'$. Let $n(\rho') = n(\rho', 0) + n(\rho', \infty)$ and $\delta(z)$ be the shortest distance at which point

z reaches the zeros and poles of $f(z)$ in the disk $|z| < \rho'$; then we have

$$\frac{1}{|a_i - z|} \leq \frac{1}{\delta(z)}, \qquad \frac{1}{|b_j - z|} \leq \frac{1}{\delta(z)}.$$

Besides, we also obtain

$$|\rho'^2 - \overline{a}_i z| \geq \rho'^2 - \rho' r = \rho'(\rho' - r),$$

$$\left| \frac{\overline{a}_i}{\rho'^2 - \overline{a}_i z} \right| \leq \frac{1}{\rho' - r}, \qquad \left| \frac{\overline{b}_j}{\rho'^2 - \overline{b}_j z} \right| \leq \frac{1}{\rho' - r}$$

and

$$\left| \frac{1}{2\pi} \int_0^{2\pi} \log |f(\rho' e^{i\varphi})| \frac{2\rho' e^{i\varphi}}{(\rho' e^{i\varphi} - z)^2} \, d\varphi \right|$$

$$\leq \frac{2\rho'}{(\rho' - r)^2} \cdot \frac{1}{2\pi} \int_0^{2\pi} |\log |f(\rho' e^{i\varphi})|| \, d\varphi$$

$$= \frac{2\rho'}{(\rho' - r)^2} \left\{ m(\rho', f) + m\left(\rho', \frac{1}{f}\right) \right\}.$$

Hence we derive from formula (1.34) that

$$\left| \frac{f'(z)}{f(z)} \right| \leq \frac{2\rho'}{(\rho' - r)^2} \left\{ m(\rho', f) + m\left(\rho', \frac{1}{f}\right) \right\}$$

$$+ n(\rho') \left\{ \frac{1}{\delta(z)} + \frac{1}{\rho' - r} \right\}.$$

Further according to formulas (1.11) and (1.12) we deduce

$$\left| \frac{f'}{f} \right| \leq \frac{4\rho'}{(\rho' - r)^2} \left\{ T(\rho', f) + \log^+ \frac{1}{|f(0)|} \right\}$$

$$+ \frac{n(\rho')}{r} \left\{ \frac{r}{\delta(z)} + \frac{r}{\rho' - r} \right\}.$$

Taking $\log^+$ on both sides and considering also formulas (1.17) and (1.18) we get

$$\log^+ \left| \frac{f'(z)}{f(z)} \right| \leq \log^+ \rho' + 2 \log^+ \frac{1}{\rho' - r}$$

$$+ 2 \log 2 + \log^+ T(\rho', f)$$

$$+ \log^+ \log^+ \frac{1}{|f(0)|} + \log 2 \qquad (1.35)$$

$$+ \log^+ \frac{n(\rho')}{r} + \log^+ \frac{r}{\delta(z)}$$

$$+ \log^+ \frac{r}{\rho' - r} + 2 \log 2.$$

In the following we estimate $n(\rho')$. Notice that

$$N(\rho, f) \geq \int_{\rho'}^{\rho} \frac{n(t, f)}{t} \, dt \geq n(\rho', f) \cdot \frac{\rho - \rho'}{\rho},$$

$$N\left(\rho, \frac{1}{f}\right) \geq n\left(\rho', \frac{1}{f}\right) \cdot \frac{\rho - \rho'}{\rho}.$$

Then

$$n(\rho') \leq \frac{\rho}{\rho - \rho'} \left\{ N(\rho, f) + N\left(\rho, \frac{1}{f}\right) \right\}$$

$$\leq \frac{2\rho}{\rho - \rho'} \left\{ T(\rho, f) + \log^+ \frac{1}{|f(0)|} \right\},$$

$$\log^+ n(\rho') \leq \log^+ \rho + \log^+ \frac{1}{\rho - \rho'} + \log^+ T(\rho, f)$$

$$+ \log^+ \log^+ \frac{1}{|f(0)|} + 2 \log 2.$$

Substituting this formula into formula (1.35) and considering also $\rho' = (\rho + r)/2$, then we conclude from Lemma 1.2 that

$$m\left(r, \frac{f'}{f}\right) \leq 4 \log^+ T(\rho, f) + 4 \log^+ \log \frac{1}{|f(0)|} + 5 \log^+ \rho$$

$$+ 6 \log^+ \frac{1}{\rho - r} + \log^+ \frac{1}{r} + 14,$$

meaning that Lemma 1.3 is valid.

Lemma 1.3 is basically attributed to R. Nevanlinna's efforts [32a]. Indeed, among the formulas worked out by him, the original item should contain item $\log^+(1/(|f(0)|))$. Here item $\log^+(1/|f(0)|)$, is replaced merely by $\log^+ \log^+(1/|f(0)|)$. This relatively precise formula was proved by G. Valiron [39d] and its importance can be seen in the next chapter which is about the theory of singular direction.([2])

Finally, when $f(0) = 0$ or ∞, we make a supplementary illustration as follows: Assume $f(z)$, at a neighborhood around $z = 0$, has the following expansion:

$$f(z) = c_s z^s + c_{s+1} z^{s+1} + \cdots, \qquad c_s \neq 0.$$

Let $f_1(z) = z^{-s} f(z)$, then $f_1(z)$ conforms to the condition of Lemma 1.3, meaning that when $0 < r < \rho < R$, we have

$$m\left(r, \frac{f_1'}{f_1}\right) \leq 4 \log^+ T(r, f_1) + 4 \log^+ \log^+ \frac{1}{|f_1(0)|} + 5 \log^+ \rho$$

$$+ 6 \log^+ \frac{1}{\rho - r} + \log^+ \frac{1}{r} + 14.$$

([2]) See the proof of Theorem 2.4.

Notice that

$$\frac{f_1'(z)}{f_1(z)} = -\frac{s}{z} + \frac{f'(z)}{f(z)},$$

$$m\left(r, \frac{f'}{f}\right) \le m\left(r, \frac{f_1'}{f_1}\right) + \log^+ |s| + \log^+ \frac{1}{r} + \log 2,$$

and $T(\rho, f_1) \le T(\rho, f) + |s| \log^+ (1/r)$, we conclude that

$$m\left(r, \frac{f'}{f}\right) \le 4\log^+ T(\rho, f) + 4\log^+ \log^+ \frac{1}{|c_s|}$$

$$+ 5\log^+ \rho + 6\log^+ \frac{1}{\rho - r} \tag{1.36}$$

$$+ 6\log^+ \frac{1}{r} + 5\log^+ |s| + 19.$$

1.5.3. Borel Lemma.

LEMMA 1.4. (1) *Let* $T(r) \ge 1$ *be a continuous, nondecreasing function on the interval* $[r_0, +\infty)$ $(r_0 \ge 0)$; *then*

$$T\left(r + \frac{1}{T(r)}\right) < 2T(r), \tag{1.37}$$

but probably excluding on the interval $[r_0, +\infty)$ *an exceptional set* E_0 *of value* r, *which has linear measure* meas $E_0 \le 2$.

(2) *Let* $T(r) \ge 1$ *be a continuous, nondecreasing function on the interval* $[r_0, R]$ $(r_0 \ge 0, R < +\infty)$; *then*

$$T\left(r + \frac{R - r}{eT(r)}\right) < 2T(r), \tag{1.38}$$

but probably excluding on the interval $[r_0, R)$ *an exceptional set* E_0 *of value* r, *and that* $\int_{E_0} dr/(R - r) \le 2$. *Particularly, if* ρ *and* ρ' *conform to the conditions* $r_0 < \rho < R$ *and* $R - \rho' < (R - \rho)/e^2$, *then there exists in the interval* (ρ, ρ') *certain values* r, *such that formula is valid.*

PROOF. First we try to prove (1). Suppose r_1 is the smallest value on $(r_0, +\infty)$ that does not satisfy formula (1.37). If such an r_1 does not exist, it is obvious that (1) is valid. In the following we use the induction method: Let r_n be defined and $r_n' = r_n + (1/T(r_n))$. Then we also define r_{n+1} as the smallest value not satisfying formula (1.37) on the interval $[r_n', +\infty)$. We obtain according to this method a sequence $\{r_n\}$. Taking the continuity of $T(r)$ into account, when $r = r_n$ $(n = 1, 2, \ldots)$, formula (1.37) is not valid. Therefore, $r_n \in E_0$. Further, based on the definition of r_{n+1}, we conclude that there exist no values of E_0 in the interval (r_n', r_{n+1}). Hence, E_0 is included in the set of intervals $\{[r_n, r_n'] | n = 1, 2, \ldots\}$. Now we prove that r_n cannot tend to a finite value r if the sequence $\{r_n\}$ contains infinitely many elements. In fact, otherwise, let $r_n \to r$. Then according to $r_n < r'_n \leq r_{n+1}$, we have

$$r'_n - r_n = \frac{1}{T(r_n)} \geq \frac{1}{T(r)} > 0,$$

which leads to a contradiction. Finally we prove $\sum(r'_n - r_n) \leq 2$. Indeed, since $r_n \in E_0$, it follows that

$$T(r'_n) = T\left(r_n + \frac{1}{T(r_n)}\right) \geq 2T(r_n).$$

Furthermore $T(r_{n+1}) \geq T(r'_n) \geq 2T(r_n) \geq \cdots \geq 2^n T(r_1) \geq 2^n$. Consequently we obtain

$$\sum_{n=1}^{\infty}(r'_n - r_n) = \sum_{n=1}^{\infty} \frac{1}{T(r_n)} \leq \sum_{n=1}^{+\infty} 2^{1-n} = 2$$

meaning that (1) is proved.

Next we try to prove (2). Let

$$\rho = \log \frac{1}{R-r}, \qquad r = R - e^{-\rho}, \qquad \rho_0 = \log \frac{1}{R-r_0}.$$

Then $T_1(\rho) = T(R - e^{-\rho})$ on $[\rho_0, +\infty)$ satisfies (1). Suppose that E is a set of values ρ satisfying the inequality

$$T_1\left(\rho + \frac{1}{T_1(\rho)}\right) \geq 2T_1(\rho)$$

on the interval $[\rho_0, +\infty)$, and that E_0 is a set of values r corresponding to E. We have, according to (1),

$$\int_{E_0} \frac{dr}{R-r} = \int_E d\rho \leq 2.$$

Now suppose that value r does not belong to the set E_0; we have therefore $T(r') < 2T(r)$. Here r' is defined by the equality

$$\log \frac{1}{R-r'} = \log \frac{1}{R-r} + \frac{1}{T(r)}.$$

Notice when $0 \leq x \leq 1$, we have $1 - e^{-x} = xe^{-\theta} \geq x/e$, and we may conclude that

$$r' = r + (R-r)\{1 - e^{-\frac{1}{T(r)}}\} \geq r + \frac{R-r}{eT(r)}.$$

It follows that formula (1.36) is valid. If $r_0 < \rho < \rho' \leq R$ and $R - \rho' < (R-\rho)/e^2$, then

$$\int_\rho^{\rho'} \frac{dr}{R-r} = \log \frac{R-\rho}{R-\rho'} > 2.$$

Hence there exists definitely on the interval (ρ, ρ') some values r that do not belong to E_0, such that

$$T\left(r + \frac{R-r}{eT(r)}\right) < 2T(r).$$

Thus Lemma 1.4 is proved completely.

Lemma 1.4 is basically attributed to E. Borel's efforts [6a].

§1.6. The Second Fundamental Theorem

1.6.1. The Second Fundamental Theorem.

THEOREM 1.4. *Let* $f(z)$ *be a nonconstant meromorphic function in the disk* $|z| < R$ $(0 < R \leq +\infty)$, $f(0) \neq 0, \infty$, $f'(0) \neq 0$, *and* $a_1, a_2, \ldots, a_q$ $(q \geq 2)$ *be* q *distinct finite complex numbers, and that* $|a_i - a_j| \geq \delta$ $(\delta > 0, 1 \leq i \neq j \leq q)$. *Then for any arbitrary value* r, $0 < r < R$, *we have*

$$m(r, \infty) + \sum_{i=1}^{q} m(r, a_i) \leq 2T(r, f) - N_1(r) + S(r, f), \tag{1.39}$$

where

$$N_1(r) = N\left(r, \frac{1}{f'}\right) + 2N(r, f) - N(r, f'),$$

$$S(r, f) = m\left(r, \frac{f'}{f}\right) + m\left(r, \sum_{i=1}^{q} \frac{f'}{f - a_i}\right) + q \log^+ \frac{3q}{\delta} \tag{1.40}$$

$$+ 2\log 2 + \log \frac{1}{|f'(0)|},$$

and that $S(r, f)$ *satisfies the following condition:*

(1) *When* $R = +\infty$, *if* $f(z)$ *is of finite order, then*

$$S(r, f) = O(\log r). \tag{1.41}$$

If $f(z)$ *is of infinite order, then*

$$S(r, f) = O\{\log[rT(r, f)]\}, \tag{1.42}$$

but probably excluding an exceptional set E_0 *of* r *which has linear measure* meas $E_0 \leq 2$, *and that* E_0 *depends only on* $T(r, f)$ *and is unrelated to* a_i *and* q *in particular.*

(2) *When* $0 < R < +\infty$, *we obtain*

$$S(r, f) = O\left\{\log^+ T(r, f) + \log \frac{1}{R - r}\right\}, \tag{1.43}$$

but probably excluding an exceptional set E_0 *of* r, *and also* $\int_{E_0} dr/(R-r) < +\infty$.

R. Nevanlinna called Theorem 1.4 the Second Fundamental Theorem [32a].

PROOF. We consider function

$$F(z) = \sum_{i=1}^{q} \frac{1}{f(z) - a_i} .$$

If there exists a value i $(1 \le i \le q)$, such that $|f(z) - a_i| < \delta/(3q)$ holds at a point z, then when $1 \le j \ne i \le q$, we have

$$|f(z) - a_j| \ge |a_i - a_j| - |f(z) - a_i| \ge \delta - \frac{\delta}{3q} \ge \frac{2}{3}\delta,$$

meaning that we get

$$\frac{1}{|f(z) - a_j|} \le \frac{3}{2\delta} \le \frac{1}{2q} \cdot \frac{1}{|f(z) - a_i|} .$$

Consequently,

$$|F(z)| \ge \frac{1}{|f(z) - a_i|} - \sum_{j \ne i} \frac{1}{|f(z) - a_j|}$$

$$\ge \frac{1}{|f(z) - a_i|} \left\{ 1 - \frac{q-1}{2q} \right\} \ge \frac{1}{2} \cdot \frac{1}{|f(z) - a_i|} ,$$

$$\log^+ |F(z)| \ge \log^+ \frac{1}{|f(z) - a_i|} - \log 2 .$$

Notice that when $1 \le j \ne i \le q$, it follows that

$$\log^+ \frac{1}{|f(z) - a_j|} \le \log^+ \frac{2}{\delta} .$$

Hence we find that

$$\log^+ |F(z)| \ge \sum_{j=1}^{q} \log^+ \frac{1}{|f(z) - a_j|} - q \log^+ \frac{2}{\delta} - \log 2$$

$$\ge \sum_{j=1}^{q} \log^+ \frac{1}{|f(z) - a_j|} - q \log^+ \frac{3q}{\delta} - \log 2 . \tag{1.44}$$

On the other hand, if for each value i $(1 \le i \le q)$, $|f(z) - a_i| \ge \delta/(3q)$ holds, then the estimation

$$\log^+ |F(z)| \ge \sum_{i=1}^{q} \log^+ \frac{1}{|f(z) - a_i|} - q \log^+ \frac{3q}{\delta} - \log 2$$

is obviously valid. Combining this with formula (1.44) we assure that this estimation holds in whatever conditions. Hence we have

$$m(r, F) \ge \sum_{i=1}^{q} m(r, a_i) - q \log^+ \frac{3q}{\delta} - \log 2, \tag{1.45}$$

which is an estimation of the lower bound of $m(r, F)$. To find the estimation of the upper bound of $m(r, F)$, first we get

$$
m(r, F) = m\left(r, \frac{1}{f} \cdot \frac{f}{f'} \cdot f' \cdot F\right)
$$

$$
\leq m\left(r, \frac{1}{f}\right) + m\left(r, \frac{f}{f'}\right) + m(r, f' \cdot F).
$$

Based on formula (1.12) we further obtain

$$
m\left(r, \frac{f}{f'}\right) = m\left(r, \frac{f'}{f}\right) + N\left(r, \frac{f'}{f}\right)
$$

$$
- N\left(r, \frac{f}{f'}\right) + \log\left|\frac{f(0)}{f'(0)}\right|,
$$

$$
m\left(r, \frac{1}{f}\right) = T(r, f) - N\left(r, \frac{1}{f}\right) + \log\frac{1}{|f(0)|}.
$$

Hence, we derive

$$
m(r, F) \leq T(r, f) - N\left(r, \frac{1}{f}\right) + \log\frac{1}{|f(0)|} + m\left(r, \frac{f'}{f}\right)
$$

$$
+ N\left(r, \frac{f'}{f}\right) - N\left(r, \frac{f}{f'}\right)
$$

$$
+ m(r, f'F) + \log\left|\frac{f(0)}{f'(0)}\right|.
$$

Combining this with formula (1.45) yields

$$
\sum_{i=1}^{q} m(r, a_i) + m(r, \infty)
$$

$$
\leq m(r, F) + m(r, f) + q\log^+\frac{3q}{\delta} + \log 2
$$

$$
\leq T(r, f) - N\left(r, \frac{1}{f}\right) + N\left(r, \frac{f'}{f}\right) - N\left(r, \frac{f}{f'}\right) \qquad (1.46)
$$

$$
+ m\left(r, \frac{f'}{f}\right) + m(r, f'F) + \log\frac{1}{|f'(0)|}
$$

$$
+ T(r, f) - N(r, f) + q\log^+\frac{3q}{\delta} + \log 2.
$$

According to Jensen formula we have

$$N\left(r, \frac{f'}{f}\right) - N\left(r, \frac{f}{f'}\right)$$

$$= \frac{1}{2\pi} \int_0^{2\pi} \log\left|\frac{f(re^{i\theta})}{f'(re^{i\theta})}\right| d\theta - \log\left|\frac{f(0)}{f'(0)}\right|$$

$$= \frac{1}{2\pi} \int_0^{2\pi} \log|f(re^{i\theta})| d\theta - \log|f(0)|$$

$$- \frac{1}{2\pi} \int_0^{2\pi} \log|f'(re^{i\theta})| d\theta + \log|f'(0)|$$

$$= N\left(r, \frac{1}{f}\right) - N(r, f) - N\left(r, \frac{1}{f'}\right) + N(r, f').$$

Consequently formula (1.46) gives

$$\sum_{i=1}^{q} m(r, a_i) + m(r, \infty)$$

$$\leq 2T(r, f) - \left\{2N(r, f) - N(r, f')N\left(r, \frac{1}{f'}\right)\right\} + S(r, f),$$

where

$$S(r, f) = m\left(r, \frac{f'}{f}\right) + m\left(r, \sum_{i=1}^{q} \frac{f'}{f - a_i}\right)$$

$$+ q \log^+ \frac{3q}{\delta} + \log 2 + \log \frac{1}{|f'(0)|}.$$

In the following we make an estimation on $S(r, f)$. Let a be a finite complex number. When $r \to R$, according to Lemma 1.3 we have

$$m\left(r, \frac{f'}{f - a}\right) \leq 4 \log^+ T(\rho, f - a) + 5 \log^+ \rho$$

$$+ 6 \log^+ \frac{1}{\rho - r} + O(1) \qquad (1.47)$$

$$\leq 4 \log^+ T(r, f) + 5 \log^+ \rho$$

$$+ 6 \log^+ \frac{1}{\rho - r} + O(1), \qquad (0 < r < \rho < R).$$

Further discussion involves the differentiation of two cases: (1) $R = +\infty$. Suppose $f(z)$ is of finite order, i.e. it consists of positive real number k, such that $T(r, f) = O(r^k)$ $(r \to +\infty)$. Letting $\rho = 2r$ in formula (1.47), it follows that

$$m\left(r, \frac{f'}{f - a}\right) \leq O(\log r).$$

And hence

$$S(r, f) = m\left(r, \frac{f'}{f}\right) + m\left(r, \sum_{i=1}^{q} \frac{f'}{f - a_i}\right) + O(1)$$

$$\leq \sum_{i=1}^{q} m\left(r, \frac{f'}{f - a_i}\right) + O(\log r) = O(\log r).$$

Suppose $f(z)$ is of infinite order. Let $\rho = r + (1/T(r, r))$ in formula (1.47). According to Lemma 1.4, if only value r does not belong to the exceptional set E_0, meas $E_0 \leq 2$, we may obtain

$$m\left(r, \frac{f'}{f - a}\right) \leq 4\log^+ 2T(r, f) + 5\log^+ r + 6\log^+ T(r, f) + O(1)$$

$$= O\{\log[rT(r, f)]\}.$$

And therefore

$$S(r, f) = O\{\log[rT(r, f)]\}.$$

(2) $R < +\infty$. Let

$$\rho = r + \frac{R - r}{eT(r, f)}$$

in formula (1.47). According to Lemma 1.4, if only value r does not belong to the exceptional set E_0, $\int_{E_0} dr/(R - r) < +\infty$, we may find

$$m\left(r, \frac{f'}{f - a}\right) \leq 4\log^+ 2T(r, f) + 6\log\frac{eT(r, f)}{R - r} + O(1)$$

$$= O\left\{\log\frac{T(r, f)}{R - r}\right\}.$$

And hence

$$S(r, f) = O\left\{\log\frac{T(r, f)}{R - r}\right\}.$$

Theorem 1.4 is thus proved completely.

When $f(0) = 0$ or ∞ or $f'(0) = 0$, we make a supplementary illustration as follows: Suppose $f(z)$ at a neighborhood around $z = 0$ has the expansion $f(z) = c_s z^s + c_{s+1} z^{s+1} + \cdots$, $c_s \neq 0$. Hence $f'(z) = sc_s z^{s-1} + (s+1)c_{s+1} z^s + \cdots$. Examining the proof of Theorem 1.4, we see that under this case, the only change of $S(r, f)$ is to replace $\log(1/|f'(0)|)$ by $\log(1/|sc_s|)$, which then yields

$$S(r, f) = m\left(r, \frac{f'}{f}\right) + m\left(r, \sum_{i=1}^{q} \frac{f'}{f - a_i}\right) + q\log^+\frac{3q}{\delta}$$

$$+ 2\log 2 + \log\frac{1}{|sc_s|}.$$

(1.48)

Therefore, formulas (1.41), (1.42) and (1.43) are still valid.

1.6.2. Applications. (1) We prove the following well-known Picard Theorem.

THEOREM 1.5. *Suppose $f(z)$ is a transcendental meromorphic function on the open plane $|z| < +\infty$. Then for any complex value a (including ∞) equation $f(z) = a$ has infinitely many roots, but probably with the exception of at most two values a. We call these exceptional values the Picard exceptional values of f.*

PROOF. Assume that Theorem 1.5 does not hold, meaning that there exist three distinct complex numbers a_ν ($\nu = 1, 2, 3$), such that equation $f(z) = a_\nu$ has only finitely many roots. We obtain, therefore, $N(r, a_\nu) = O(\log r)$, $\nu = 1, 2, 3$. Applying the First and Second Fundamental Theorems we further derive $T(r, f) \le O\{\log[r \cdot T(r, f)]\}$, but probably excluding an exceptional set E_0 of value r which has linear measure $\text{meas}\, E_0 \le 2$. Hence we conclude that

$$\lim_{r \to +\infty} \frac{T(r, f)}{\log r} < +\infty,$$

meaning that $f(z)$ degenerates into a rational function, and thus gives a contradiction.

(2) Next we prove the following well-known Borel Theorem.

THEOREM 1.6. *Suppose $f(z)$ is a meromorphic function on the open plane $|z| < +\infty$ with its order being λ, $0 < \lambda < +\infty$; then for any arbitrary complex value a we have*

$$\varlimsup_{r \to +\infty} \frac{\log^+ n(r, a)}{\log r} = \lambda,$$

but there may have at most two exceptional values a_1 and a_2, such that

$$\varlimsup_{r \to +\infty} \frac{\log^+ n(r, a)}{\log r} < \lambda, \qquad i = 1, 2.$$

We call these exceptional values the Borel exceptional values. Clearly Picard exceptional values must be the Borel exceptional values.

PROOF. First for any arbitrary complex number a, we have

$$n(r, a) \le \frac{1}{\log 2} \int_r^{2r} \frac{n(t, a)}{t} \, dt \le \frac{1}{\log 2} N(2r, a)$$

$$\le \frac{1}{\log 2} \{T(2r, f) + O(1)\},$$

and hence

$$\varlimsup_{r \to +\infty} \frac{\log^+ n(r, a)}{\log r} \le \varlimsup_{r \to +\infty} \frac{\log T(2r, f) + O(1)}{\log 2r} \le \lambda.$$

Suppose Theorem 1.6 is not valid, meaning that there exist three distinct complex numbers a_ν ($\nu = 1, 2, 3$), such that

$$\varlimsup_{r \to +\infty} \frac{\log^+ n(r, a_\nu)}{\log r} < \lambda, \qquad \nu = 1, 2, 3.$$

When r is sufficiently large we have

$$N(r, a_\nu) = \int_0^r \frac{n(t, a_\nu) - n(0, a_\nu)}{t} dt + n(0, a_\nu) \log r$$
$$\leq n(r, a_\nu) \log r + O(1).$$

An application of the First and Second Fundamental Theorems derives

$$T(r, f) \leq \sum_{\nu=1}^{3} N(r, a_\nu) + S(r, f)$$

$$\leq \log r \cdot \sum_{\nu=1}^{3} n(r, a_\nu) + O(\log r).$$

Consequently we conclude that

$$\varlimsup_{r \to +\infty} \frac{\log T(r, f)}{\log r} < \lambda,$$

meaning that the order of $f(z) < \lambda$ and, in turn, leads to a contradiction.

(3) Now we introduce the most significant application of the Second Fundamental Theorem, i.e. to derive the deficiency relation. Suppose $f(z)$ is a function meromorphic in the disk $|z| < R$ $(0 < R \leq +\infty)$. We denote

$$\bar{n}(r, a) = \bar{n}(r, f = a) = \begin{cases} \bar{n}(r, f), & a = \infty, \\ \bar{n}(r, \frac{1}{f-a}), & a \neq \infty \end{cases}$$

as the number of roots that equation $f(z) = a$ has in the disk $|z| < r$ $(0 < r \leq R)$ (each multiple root is only counted once), and denote

$$\bar{n}(0, a) = \bar{n}(0, f = a) = \begin{cases} \bar{n}(0, f), & a = \infty, \\ \bar{n}(0, \frac{1}{f-a}), & a \neq \infty \end{cases}$$

as the number of roots of the equation $f(z) = a$ at the origin $z = 0$ (counted once only) and also let

$$\bar{N}(r, a) = \bar{N}(r, f = a)$$
$$= \begin{cases} \bar{N}(r, f) = \int_0^r \frac{\bar{n}(t, a) - \bar{n}(0, a)}{t} dt + \bar{n}(0, a) \log r, & a = \infty, \\ \bar{N}(r, \frac{1}{f-a}) = \int_0^r \frac{\bar{n}(t, a) - \bar{n}(0, a)}{t} dt + \bar{n}(0, a) \log r, & a \neq \infty. \end{cases}$$

We further assume that when $r \to R$, $T(r, f) \to +\infty$. Hence, according to the First Fundamental Theorem, for any arbitrary complex number a we obtain

$$m(r, a) + N(r, a) = T(r, f) + O(1),$$
$$0 \leq \varliminf_{r \to R} \frac{m(r, f)}{T(r, f)} = 1 - \varlimsup_{r \to +\infty} \frac{N(r, f)}{T(r, f)} \leq 1. \qquad (1.49)$$

R. Nevanlinna [32a] defined

$$\delta(a) = \delta(a, f) = \lim_{r \to R} \frac{m(r, a)}{T(r, f)} = 1 - \overline{\lim_{r \to R}} \frac{N(r, a)}{T(r, f)},$$

$$\Theta(a) = \Theta(a, f) = 1 - \overline{\lim_{r \to R}} \frac{\overline{N}(r, a)}{T(r, f)},$$

$$\theta(a) = \lim_{r \to R} \frac{N(r, a) - \overline{N}(r, a)}{T(r, f)}.$$

Therefore, when $r \to R$, we get

$$N(r, a) - \overline{N}(r, a) > \{\theta(a) - o(1)\}T(r, f),$$
$$N(r, a) < \{1 - \delta(a) + o(1)\}T(r, f),$$
$$\overline{N}(r, a) < \{1 - \delta(a) + \theta(a) + o(1)\}T(r, f).$$

In turn we arrive at the inequality

$$\Theta(a) \geq \delta(a) + \theta(a). \tag{1.50}$$

R. Nevanlinna called $\theta(a)$ the index of multiplicity (Verzweigungsindex) and $\delta(a)$ the deficiency of value a. According to formula (1.49) it is always true that $0 \leq \delta(a) \leq 1$. When $\delta(a) > 0$ in particular, we call such values the deficient values or Nevanlinna exceptional values.

THEOREM 1.7. *Let $f(z)$ be a function meromorphic in the disk $|z| < R$ $(0 < R \leq +\infty)$, and that when $R = +\infty$, $f(z)$ is not a constant, or when $R < +\infty$, $f(z)$ satisfies the condition*

$$\overline{\lim_{r \to R}} \frac{T(r, f)}{\log \frac{1}{R-r}} = +\infty. \tag{1.51}$$

Then the set $\{a | \delta(a) > 0\}$ of all the deficient values of $f(z)$ forms a countable set, and

$$\sum_a \{\delta(a) + \theta(a)\} \leq \sum_a \Theta(a) \leq 2. \tag{1.52}$$

PROOF. First we consider $R = +\infty$. Based on the assumption that $f(z)$ is not a constant, then $f(z)$ is either a rational function or it is a transcendental meromorphic function. When $f(z)$ is a rational function, we should have according to formula (1.29) $T(r, f) = O(\log r)$. On the other hand, for any arbitrary finite complex number a, when $z \to \infty$ we should have $f'/(f - a) \to 0$, meaning that $m(r, f'/(f - a)) = 0$. Hence we have from formula (1.40) that $S(r, f) = O(1)$. It follows that

$$\lim_{r \to +\infty} \frac{S(r, f)}{T(r, f)} = 0. \tag{1.53}$$

When $f(z)$ is a transcendental meromorphic function, we obtain, according to formula (1.28),

$$\lim_{r \to +\infty} \frac{T(r, f)}{\log r} = +\infty.$$

On the other hand, from formulas (1.41) and (1.42) we get $S(r, f) = O\{\log[rT(r, f)]\}$, but probably excluding an exceptional set E_0 of value r which has linear measure $\text{meas}\, E_0 \leq 2$. Hence we derive

$$\lim_{\substack{r \to +\infty \\ \overline{r \varepsilon E_0}}} \frac{S(r, f)}{T(r, f)} = 0. \tag{1.54}$$

Based on formulas (1.53) and (1.54), we may select a sequence $\{r_n\}$, such that when $n \to +\infty$, $r_n \to +\infty$, and also $S(r_n, f) = o\{T(r_n, f)\}$.

Next we consider $0 < R < +\infty$. According to formula (1.51), there exists a sequence $\{\rho_n\}$, such that when $n \to +\infty$, $\rho_n \to R$, and also

$$\lim_{n \to +\infty} \frac{T(\rho_n, f)}{\log(1/(R - \rho_n))} = +\infty.$$

From Lemma 1.4 we may select outside the exceptional set E_0 a sequence $\{r_n\}$, such that $R - \rho_n > R - r_n > (R - \rho_n)/q$. Hence when $n \to +\infty$, $r_n \to R$, and also

$$\log \frac{1}{R - r_n} = \log \frac{1}{R - \rho_n} + O(1) = o\{T(\rho_n, f)\} = o\{T(r_n, f)\}.$$

Consequently we have, from formula (1.43), $S(r_n, f) = o\{T(r_n, f)\}$.

The above discussion illustrates that when $0 < R \leq +\infty$, we may select a sequence $\{r_n\}$, such that when $n \to +\infty$, $r_n \to R$, and also $S(r_n, f) = o\{T(r_n, F)\}$. An application of the First and Second Fundamental Theorems to be selected sequence $\{r_n\}$ derives

$$(q + 1)T(r_n, f) \leq \sum_{i=1}^{q} N(r_n, a_i) + N(r_n, \infty) - N_1(r_n)$$
$$+ (2 + o(1))T(r_n, f),$$

$$(q - 1 - o(1))T(r_n, f) \leq \sum_{i=1}^{q} \overline{N}(r_n, a_i) + \overline{N}(r_n, \infty).$$

And hence

$$\sum_{i=1}^{q} \overline{\lim_{r \to R}} \frac{\overline{N}(r_n, a_i)}{T(r_n, f)} + \overline{\lim_{r \to R}} \frac{\overline{N}(r_n, \infty)}{T(r_n, f)}$$
$$\geq \overline{\lim_{r \to R}} \frac{\sum_{i=1}^{q} \overline{N}(r_n, a_i) + \overline{N}(r_n, \infty)}{T(r_n, f)} \geq q - 1,$$

$$\sum_{i=1}^{q} \{1 - \Theta(a_i)\} + 1 - \Theta(\infty) \geq q - 1,$$

$$\sum_{i=1}^{q} \Theta(a_i) + \Theta(\infty) \leq 2, \tag{1.55}$$

$$\sum_{i=1}^{q} \delta(a_i) \leq 2. \tag{1.56}$$

Formula (1.56) illustrates that there are at most $2N - 1$ finite values a which lead to $\delta(a) > \frac{1}{N}$. Hence if by letting $N = 2, 3, \ldots$, we may form a sequence which will exhaust all deficient values $\{a | \delta(a) > 0)\}$. I.e., a countable set consisting of all deficient values is formed. Notice that since the upper bound 2 is unrelated to q in formula (1.55), formula (1.52) is valid, implying that Theorem 1.7 is proved.

(4) Finally we make an application to meromorphic functions in the unit disk $|z| < 1$. Let $T(r) \geq 0$ be a nondecreasing function on the interval $[0, 1)$. If

$$\varlimsup_{r \to 1} \frac{\log^+ T(r)}{\log 1/(1 - r)} \leq \lambda, \qquad \lambda < +\infty, \tag{1.57}$$

then for any arbitrarily small number $\varepsilon > 0$, we have the integral

$$\int_0^1 T(r)(1 - r)^{\lambda + \varepsilon - 1} \, dr < +\infty. \tag{1.58}$$

Indeed, according to formula (1.57), there exists $r_\varepsilon \geq 0$ such that when $r \geq r_\varepsilon$ we have

$$T(r) < \left\{ \frac{1}{1 - r} \right\}^{\lambda + (\varepsilon/2)}$$

$$T(r)(1 - r)^{\lambda + \varepsilon - 1} < (1 - \lambda)^{(\varepsilon/2) - 1}.$$

Hence

$$\int_0^1 T(r)(1 - r)^{\lambda + \varepsilon - 1} \, dr < \int_{r_\varepsilon}^1 (1 - r)^{(\varepsilon/2) - 1} \, dr + O(1)$$

$$= \frac{2}{\varepsilon} \left\{ -(1 - r)^{\varepsilon/2} \right\}_{r_\varepsilon}^1 + O(1) < +\infty.$$

On the contrary, if, for a value k $(k < +\infty)$, the integration:

$$\int_0^1 T(r)(1 - r)^{k - 1} \, dr < +\infty, \tag{1.59}$$

holds, then

$$\varlimsup_{r \to 1} \frac{\log^+ T(r)}{\log(1/(1 - r))} \leq k. \tag{1.60}$$

Indeed, since

$$+\infty > \int_r^1 T(r)(1-r)^{k-1}\,dr \geq T(r)\int_r^1 (1-r)^{k-1}\,dr$$

$$= T(r)\cdot\left\{-\frac{1}{k}(1-r)^k\right\}_r^1$$

$$= \frac{1}{k}T(r)(1-r)^k,$$

we conclude that formula (1.60) holds.

THEOREM 1.8. *Suppose* $f(z)$ *is a function meromorphic in the disk* $|z| < 1$. *If there exist three distinct complex numbers* a_1 , a_2 , *and* a_3 , *such that*

$$\varlimsup_{r\to 1} \frac{\log^+ n(r,X)}{\log(1/(1-r))} \leq \lambda, \qquad 1 < \lambda < +\infty, \ X = a_1, a_2, a_3, \qquad (1.61)$$

then the order of $f(z) \leq \lambda - 1$.

PROOF. Suppose $a_3 = \infty$, otherwise we simply consider the transformation $f_1(z) = 1/(f(z) - a_3)$. According to the First Fundamental Theorem, $f(z)$ and $f_1(z)$ have the same order.

An application of the Second Fundamental Theorem, where we let $R = 1$ and $q = 2$, yields

$$m(r,\infty) + \sum_{i=1}^2 m(r,a_i) \leq 2T(r,f) + S(r,f), \qquad (1.62)$$

where

$$S(r,f) = m\left(r,\frac{f'}{f}\right) + m\left(r,\sum_{i=1}^2 \frac{f'}{f-a_i}\right) + O(1).$$

According to formula (1.47), when $\frac{1}{2} < r < \rho < 1$, then

$$S(r,f) \leq O\left\{\log^+ T(\rho,f) + \log\frac{1}{\rho-r}\right\}. \qquad (1.63)$$

We first select arbitrarily a value r' , $\frac{1}{2} < r' < 1$, and then choose a value r , $\frac{1}{2} < r < r'$ and also let $\rho = (r'+r)/2$. Hence when $\eta \geq 1$, we obtain

$$(1-r)^{\eta-1} = (1+r'-2\rho)^{\eta-1} \leq 2^{\eta-1}(1-\rho)^{\eta-1}$$
$$(1-r)^{\eta-1} \leq 2^{\eta-1},$$

and when $0 < \eta < 1$, it follows that

$$(1-r)^{\eta-1} \leq (1-\rho)^{\eta-1}, \qquad (1-r)^{\eta-1} \leq (r'-r)^{\eta-1}.$$

Therefore, when $\eta > 0$, we conclude that

$$\int_{1/2}^{r'} \log^+ T(\rho, f)(1 - r)^{\eta-1}\, dr$$

$$\leq 2^{\eta+1} \int_{1/2}^{r'} \log^+ T(\rho, f)(1 - \rho)^{\eta-1}\, d\rho$$

$$\leq 2^{\eta+1} \int_{1/2}^{r'} \log^+ T(r, f)(1 - r)^{\eta-1}\, dr,$$

and

$$\int_{1/2}^{r'} \log \frac{1}{\rho - r}(1 - r)^{\eta-1}\, dr = \int_{1/2}^{r'} \log \frac{2}{r' - r}(1 - r)^{\eta-1}\, dr$$

$$\leq \begin{cases} 2^{\eta-1} \int_{1/2}^{r'} \log \frac{2}{r'-r}\, dr < +\infty, & \eta \geq 1, \\ \int_{1/2}^{r'} \log \frac{2}{r'-r}(r' - r)^{\eta-1}\, dr < +\infty, & 0 < \eta < 1. \end{cases}$$

Based on formula (1.63) we arrive at the following inequalities:

$$\int_{r_0}^{r} S(t, f)(1 - t)^{\eta-1}\, dt$$

$$\leq O\left\{ \int_{r_0}^{r} \log^+ T(t, f)(1 - t)^{\eta-1}\, dt \right\} + O(1), \qquad r_0 > 0. \tag{1.64}$$

On the other hand, according to formulas (1.57) and (1.58), we have for any arbitrarily small number $\varepsilon > 0$:

$$\int^{1} n(r, X)(1 - r)^{\lambda+\varepsilon-1}\, dr < +\infty, \qquad X = a_1, a_2, a_3.$$

Since

$$(\lambda - 1 + \varepsilon) \int_{r_0}^{r} N(t, X)(1 - t)^{\lambda+\varepsilon-2}\, dt$$

$$= -(1 - t)^{\lambda-1+\varepsilon} N(t, X) \big|_{r_0}^{r} + \int_{r_0}^{r} n(t, X)(1 - t)^{\lambda-1+\varepsilon} \cdot \frac{dt}{t}$$

$$\leq (1 - r_0)^{\lambda-1+\varepsilon} N(r_0, X) + \frac{1}{r_0} \int_{r_0}^{r} n(t, X)(1 - t)^{\lambda-1+\varepsilon}\, dt$$

$$< +\infty,$$

we obtain therefore

$$\int^{1} N(t, X)(1 - t)^{\lambda+\varepsilon-2}\, dt < +\infty, \qquad X = a_1, a_2, a_3. \tag{1.65}$$

Now applying formula (1.62) and the First Fundamental Theorem we derive

$$T(r, f) \leq N(r, a_1) + N(r, a_2) + N(r, a_3) + S(r, f).$$

Further, according to formulas (1.64) and (1.65), we get

$$\int^1 T(t, f)(1 - t)^{\lambda + \varepsilon - 2} \, dt < +\infty.$$

Consequently we deduce from formula (1.60)

$$\varlimsup_{r \to 1} \frac{\log^+ T(r, f)}{\log(1/(1 - r))} \leq \lambda + \varepsilon - 1.$$

Since ε is arbitrary, the order of $f(z) \leq \lambda - 1$, meaning that Theorem 1.8 is proved.

§1.7. Annotated notes

The Nevanlinna Theory formed in 1925 marked the beginning of the contemporary study concerning the value distribution theory of meromorphic functions. Throughout fifty years of development, the Nevanlinna Theory has made considerable progress and gained impressive achievement in various aspects. Here we will briefly introduce some major developments about the basic inequality (1.39).

1.7.1. Milloux's inequality. Nevanlinna's Second Fundamental Theorem involves merely how the meromorphic function $f(z)$ takes its values. If we consider simultaneously how the derivative $f^{(k)}(z)$ $(k \geq 1)$ takes its values, then the basic inequality (1.39) has some important extensions. H. Milloux [30c] and Xiong Qing-lai [22a] made a profound study in this area, and in the following we will illustrate the simplest form worked out by H. Milloux.

Suppose $f(z)$ is a transcendental meromorphic function on the open plane $|z| < +\infty$. Then for any arbitrary value r, $0 < r < +\infty$, we have

$$T(r, f) \leq \overline{N}(r, f) + N\left(r, \frac{1}{f}\right) + N\left(r, \frac{1}{f^{(k)} - 1}\right)$$
$$- N\left(r, \frac{1}{f^{(k+1)}}\right) + S(r, f),$$

where $S(r, f) = o\{T(r, f)\}$, $r \to +\infty$, but probably excluding an exceptional set E_0 of value r which has linear measure meas $E_0 < +\infty$.

According to Milloux inequality, we can easily conclude: If $f(z)$ has only finitely many zeros and poles, then $f^{(k)}(z)$ assumes the value of each finite complex number for infinite times, but probably with the exception of the value zero.

1.7.2. Hayman's inequality.[*] In 1959, W. Hayman found a striking result

---

[*] *Translator's note*: Concerning variations of this inequality, we refer the reader to some recent results obtained by Frank and Hennekemper [Resultate Math. **4** (1981), 39–54], Xin-hou Hua [Kodaĭ Math. J. **13**, no. 3, (1990), 386–390], and L. Yang [Sci. Sinica Ser. A **34** (1991), 157–167]. Yang's result is the best thus far, and he conjectured that the coefficient $(2 + \frac{1}{k})$ in the inequality can be replaced by 1.

and established the following inequality [21b]:

$$T(r, f) \leq \left(2 + \frac{1}{k}\right) N\left(r, \frac{1}{f}\right)$$

$$+ \left(2 + \frac{1}{k}\right) \overline{N}\left(r, \frac{1}{f^{(k)} - 1}\right) + S(r, f), \qquad k \geq 1,$$

where $S(r, f) = o(T(r, f))$, $r \to +\infty$. In fact, among the various inequalities mentioned before, in finding the bound for the characteristic function $T(r, f)$, they all involve at least three values on how these values are taken by the function itself. However, Hayman inequality involves merely how two values were taken. This is, in principle, a really great improvement. Moreover, through the illustration of simple example (such as function e^z), it shows that, in general, it is impossible to estimate the $T(r, f)$ by considering how $f(z)$ takes two values only. Thus Hayman inequality is much more powerful than Milloux inequality, and it has, therefore, a stronger corollary as follows:

Either a transcendental meromorphic function $f(z)$ takes the value of each complex number for infinite times, or its kth $(k \geq 1)$ derivatives $f^{(k)}(z)$ takes the value of each complex number for infinite times (with the exception of the value zero).

1.7.3. Zhuang Chi-tai's inequality. In 1929, R. Nevanlinna [32a] raised the following question: In the inequality

$$(q - 2)T(r, f) \leq \sum_{i=1}^{q} N(r, a_i) - N_1(r) + S(r, f)$$

can the q constants a_i $(i = 1, 2, \ldots, q)$ be replaced by q meromorphic functions $a_i(z)$ $(i = 1, 2, \ldots, q)$ satisfying the condition

$$T(r, a_i(z)) = o\{T(r, f)\} \qquad (i = 1, 2, \ldots, q)?$$

When $q = 3$, this question can easily be solved. In fact, using the auxiliary function

$$F(z) = \frac{f(z) - a_1(z)}{f(z) - a_2(a)} \cdot \frac{a_3(z) - a_2(z)}{a_3(z) - a_1(z)}$$

and inequality

$$T(r, F) \leq N(r, F) + N\left(r, \frac{1}{F}\right) + N\left(r, \frac{1}{F - 1}\right) + S(r, F)$$

one can derive

$$(1 - o(1))T(r, f) < \sum_{i=1}^{3} N\left(r, \frac{1}{f - a_i}\right) + S(r, f).$$

However, when $q > 3$, it will be very difficult to solve the question. For a long time, this question has been at a standstill except in the case when each

$a_i(z)$ $(i = 1, 2, \ldots, q)$ degenerates into a polynomial. In 1964 Zhuang Chi-tai solved the problem concerning the entire function by utilizing cleverly Wronski's determinant. As for the meromorphic function, it is necessary to impose a reasonable restriction on the number of its poles. To date, the problem is still unsolved, and is worth studying.* We only restate Zhuang Chi-tai's result as follows [10b]:

Suppose $f(z)$ is a nonconstant meromorphic function on the open plane $|z| < +\infty$, and that $a_i(z)$ $(i = 1, 2, \ldots, q)$ $(q \geq 2)$ are q mutually distinct meromorphic functions satisfying the condition $T(r, a_i) = o\{T(r, f)\}$ $(i = 1, 2, \ldots, q)$. Then for any arbitrary value r, $0 < r < +\infty$, we have

$$(q - 1 - o(1))T(r, f) < \sum_{i=1}^{q} \overline{N}\left(r, \frac{1}{f - a_i}\right)$$
$$+ q\overline{N}(r, f) + S(r, f),$$

where $S(r, f) = O\{\log[rT(r, f)]\}$ $(r \to +\infty)$, but probably excluding an exceptional set E_0 of r which has linear measure $\operatorname{meas} E_0 < +\infty$.

1.7.4. The Ahlfors Theory. R. Nevanlinna at the outset adopted a pure analytical technique to establish the two famous fundamental theorems. L. Ahlfors later developed a geometric method; that is, the so-called Theory of the Finite Covering Surfaces was obtained, with its result parallel to that of the Nevanlinna Theory [1d]. Here we will only state a very special case [21c].

Let $w = f(z)$ be a nonconstant meromorphic function on the disk $|z| < R$ $(0 < R \leq +\infty)$, and that $a_1, a_2, \ldots, a_q$ $(q \geq 3)$ are q distinct points on the w-Riemann sphere; then for any arbitrary value r, $0 < r < R$, we have

$$(q - 2)S(r) \leq \sum_{i=1}^{q} n(r, a_i) + hL(r), \qquad (1.66)$$

where h is a constant depending only on $a_1, a_2, \ldots, q_q$, and that

$$S(r) = \frac{1}{\pi} \iint_{|z| < r} \frac{|f'(re^{i\theta})|^2}{\{1 + |f(re^{i\theta})|^2\}^2} r\, dr\, d\theta, \qquad z = re^{i\theta},$$

$$L(r) = \int_0^{2\pi} \frac{|f'(re^{i\theta})|}{1 + |f(re^{i\theta})|^2} r\, d\theta.$$

We find that $L(r)$ plays a role similar to $S(r, f)$ of the Second Fundamental Theorem. Namely when $R = +\infty$, only if $f(z)$ does not degenerate into a constant; or when $R < +\infty$, only if $f(z)$ satisfies the condition $\overline{\lim}_{r \to R}(R - r)S(r) = +\infty$ or

$$\overline{\lim_{r \to R}} \frac{T_0(r, f)}{\log(R/(R - r))} = +\infty,$$

Translator's note: This problem has been completely and independently solved by C. Osgood [J. Number Theory **21** (1985), 347–389] and N. Steinmetz [J. Reine Angew. Math. **368** (1986), 134–141] by entirely different arguments.

then there exists definitely a sequence $\{r_n\}$, such that when $n \to +\infty$ and $r_n \to R$,

$$\frac{L(r_n)}{S(r_n)} \to 0. \tag{1.67}$$

Indeed, if formula (1.67) is invalid, then there exists a value r_0 $(r_0 < R)$ and a constant c $(0 < c < +\infty)$, such that when $r_0 \le r < R$, we obtain $S(r) \le cL(r)$. According to Cauchy's inequality we derive

$$L^2(r) \le \int_0^{2\pi} r\,d\theta \int_0^{2\pi} \frac{|f'(re^{i\theta})|^2}{\{1 + |f(re^{i\theta})|^2\}^2} r\,d\theta = 2\pi^2 rS'(r).$$

And hence $S^2(r) \le 2\pi^2 c^2 rS'(r)$. By integrating the above inequality we have

$$\frac{1}{S(r_0)} - \frac{1}{S(r)} = \int_{r_0}^r \frac{S'(t)}{S(t)}\,dt \ge \frac{1}{2\pi^2 c^2} \int_{r_0}^r \frac{dt}{t},$$

$$S(r_0) \le \frac{2\pi^2 c^2}{\log(r/r_0)}. \tag{1.68}$$

When $R = +\infty$, if we let $r \to +\infty$, then $S(r_0) = 0$. Hence $f(z)$ is a constant which contradicts our assumption. On the other hand, when $R < +\infty$, similar to formula (1.68) we have

$$S(r) \le \frac{2\pi^2 c^2}{\log(R/r)} \le \frac{2\pi^2 c^2 R}{R - r} \qquad (r_0 \le r < R)$$

$$T_0(r, f) = \int_0^r \frac{S(t)}{t}\,dt = T_0(r_0, f) + \int_{r_0}^r \frac{S(t)}{t}\,dt$$

$$\le T_0(r_0, f) + 2\pi^2 c^2 \log \frac{R^2}{r_0(R - r)} \qquad (r_0 \le r < R).$$

And therefore

$$\varlimsup_{r \to R}(R - r)T_0(r_0, f) < +\infty,$$

$$\varlimsup_{r \to R} \frac{T_0(r, f)}{\log(R/(R - r))} < +\infty,$$

which contradicts our assumption.

The above Ahlfors' result is a very special formula because in the Ahlfors theory, in general, not a set of q distinct points on the w-Riemann sphere is not considered, but a set of q unrelated simply-connected closed domains $\overline{D}_1, \overline{D}_2, \dots, \overline{D}_q$ on the sphere is taken into account; also $n(r, a_i)$ is replaced by $n(r, D_i)$ in the inequality. Here $n(r, D_i)$ denotes the total number of simply-connected covers that entirely cover domain D_i when $w = f(z)$ maps the disk $|z| < r$ to the w-Riemann sphere. Readers can find the detailed illustration of Ahlfors' result with reference to Ahlfors' original text or other relevant books. Finally we should note that Ahlfors theory is applicable for not only the meromorphic functions, but also for some other general functions, such as the quasiconformal mapping type.

CHAPTER 2

The Singular Directions

The value distribution theory of meromorphic functions includes the modulus and argument distribution. The modulus distribution theory studies how a meromorphic function attains its values on the open plane, with the Picard theorem, the Borel theorem and the Nevanlinna deficiency relation in Chapter 1 being its fundamental results. The argument distribution theory was first developed in 1919 by G. Julia [25a], who proved that for a transcendental entire function, there exists on the open plane at least one half straight line starting from the origin, such that in any arbitrary angular region bisected by this half straight line, the Picard theorem remains valid. We call any half straight lines with such a property the Julia direction. Hence, the argument distribution theory studies how a meromorphic function attains its values from regions around a half straight line starting from the origin.

Corresponding to the Borel theorem, G. Valiron proved the existence of the Borel direction in 1928 [39b]. This chapter not only introduces briefly the fundamental results of the argument distribution theory, such as the Julia direction and the Borel direction, but also, corresponding to the Nevanlinna deficiency relations, discusses preliminarily the existence of the Nevanlinna direction. Meanwhile, this chapter is also essentially a lead for the rest of the book.

§2.1. On some properties of monotonic functions

The maximum modulus of an entire function and the characteristic function of a meromorphic function are the main tools for studying the entire and meromorphic functions, respectively. Both of them are monotonic increasing functions. It is, hence, of great significance to study the properties of monotonic functions. The following lemmas on monotonic functions play an important role in this chapter as well as in the sequels.

LEMMA 2.1.([1])*Suppose that $T(r)$ defined on the interval $(r_0, +\infty)$ $(r_0 \geq 1)$ is a continuous, nondecreasing, positive function tending to $\to \infty$, and that*

Please see [10c] and [39d] for the reference books of this chapter.

([1]) Lemma 2.1 originates from the work of Hayman [21e]. When $h_1 = 0$, we immediately get Hayman's result.

there exists a monotonic sequence $\{r_n\}$ tending to ∞, such that

$$\lim_{n \to +\infty} \frac{\log T(r_n)}{\log r_n} \leq \nu < +\infty. \tag{2.1}$$

For any arbitrarily selected number h $(0 < h < +\infty)$, h_1 $(0 \leq h_1 < h)$ and H $(\nu < H)$, let

$$K = h\left(1 + \frac{h_1}{h}\right)H, \qquad K_1 = h_1\left(1 + \frac{h_1}{h}\right)H,$$

$$E = \{t \mid T(te^h) \leq e^K T(t), \ T(te^{h_1}) \leq e^{K_1} T(t), \ t \geq r_0\},$$

$$E[r_0, r_n] = E \cap [r_0, r_n].$$

Then it follows that

$$\lim_{n \to +\infty} \frac{1}{\log r_n} \int_{E[r_0, r_n]} \frac{dt}{t} \geq 1 - \frac{h(1 + (h_1/h))\nu}{K} = 1 - \frac{\nu}{H}.$$

PROOF. Select arbitrarily any fixed number $\eta > 0$. Then according to formula (2.1) there exists a positive integer $N = N(\eta)$, such that when $n \geq N$, we have $T(r_n) < r_n^{\nu + \eta}$. Consider the inequality

$$T(te^h) < e^K T(t). \tag{2.2}$$

Suppose that t_1 is the smallest value not satisfying formula (2.2) on the interval $[r_0, +\infty)$, t_2 is the smallest value not satisfying formula (2.2) on the interval $[t_1 e^h, +\infty)$, t_3 is the smallest value not satisfying formula (2.2) on the interval $[t_2 e^h, +\infty)$, ..., and t_{m+1} is the smallest value not satisfying formula (2.2) on the interval $[t_m e^h, +\infty)$, and that $t_m e^h < r_n$ and $t_{m+1} e^h \geq r_n$, then we deduce from formula (2.2) that

$$r_n^{\nu + \eta} \geq T(r_n) \geq T(t_m e^h) \geq e^K T(t_m) \geq e^K T(t_{m-1} e^h)$$

$$\geq e^{2K} T(t_{m-1}) \geq \cdots \geq e^{mK} T(t_1) \geq e^{mK} T(r_0).$$

And hence

$$mK \leq (\nu + \eta)\log r_n - \log T(r_0),$$

$$m \leq \frac{(\nu + \eta)}{K}\log r_n - \frac{1}{K}\log T(r_0). \tag{2.3}$$

Let $r_0 = t_0 e^h$ and $h_1 \neq 0$. We consider the interval $[t_k e^h, +\infty)$ $(0 \leq k \leq m - 1)$ and the inequality

$$T(te^{h_1}) < e^{K_1} T(t). \tag{2.4}$$

Suppose that τ_{k1} is the smallest value not satisfying formula (2.4) on the interval $[t_k e^{h_1}, +\infty)$, τ_{k2} is the smallest value not satisfying formula (2.4) on the interval $[\tau_{k1} e^{h_1}, +\infty)$, ..., τ_{kl_k+1} is the smallest value not satisfying formula (2.4) on the interval $[\tau_{kl_k} e^{h_1}, +\infty)$, and that $\tau_{kl_k} e^{h_1} < t_{k+1}$, $\tau_{kl_k+1} e^{h_1} \geq$

t_{k+1}. Finally we consider the interval $[t_m e^h, +\infty)$ and define correspondingly $\tau_{m1}, \tau_{m2}, \ldots, \tau_{ml_m}$ and τ_{ml_m+1}, such that $\tau_{ml_m} e^{h_1} < \min\{r_n, t_{m+1}\}$ and $\tau_{ml_m+1} e^{h_1} \geq \min\{r_n, t_{m+1}\}$. We therefore obtain

$$r_n^{\nu+\eta} \geq T(r_n) \geq T(\tau_{ml_m} e^{h_1}) \geq e^{K_1} T(\tau_{ml_m}) \geq \cdots \geq e^{l_m K_1} T(\tau_{m1})$$

$$\geq e^{l_m K_1} T(t_m e^h) \geq e^{l_m K_1 + K} T(t_m) \geq \cdots \geq e^{\sum_{k=0}^m l_k K_1 + mK} T(r_0).$$

And hence,

$$mK + \sum_{k=0}^m l_k K_1 \leq (\nu + \eta) \log r_n - \log T(r_0),$$

$$mh + \sum_{k=0}^m l_k \cdot \frac{K_1}{K} \cdot h \leq \frac{h}{K}(\nu + \eta) \log r_n - \frac{h}{K} \log T(r_0),$$

$$(m+1)h + \sum_{k=0}^m l_k h_1 + (m+1)h_1$$

$$\leq \frac{h}{K}(\nu + \eta) \log r_n - \frac{h}{K} \log T(r_0) + h + (m+1)h_1.$$

Furthermore we conclude according to formula (2.3) that

$$(m+1)h + \sum_{k=0}^m l_k h_1 + (m+1)h_1$$

$$\leq \frac{h + h_1}{K}(\nu + \eta) \log r_n - \frac{h + h_1}{K} \log T(r_0) + h + h_1$$

$$= \left\{ \frac{h(1 + (h_1/h))}{K}(\nu + \eta) - \frac{h + h_1}{K} \cdot \frac{\log T(r_0)}{\log r_n} + \frac{h + h_1}{\log r_n} \right\} \log r_n.$$

Notice that

$$\int_{E[r_0, r_n)} \frac{dt}{t} \geq \log \frac{r_n}{r_0} - \left\{ (m+1)h + \sum_{k=0}^m l_k h_1 + (m+1)h_1 \right\}$$

$$\geq \left\{ 1 - \frac{h(1 + (h_1/h))}{K}(\nu + \eta) + \frac{h + h_1}{K} \cdot \frac{\log T(r_0)}{\log r_n} \right.$$

$$\left. - \frac{h + h_1}{\log r_n} - \frac{\log r_0}{\log r_n} \right\} \log r_n.$$

It follows that

$$\lim_{n \to +\infty} \frac{1}{\log r_n} \int_{E[r_0, r_n]} \frac{dt}{t} \geq 1 - \frac{h(1 + (h_1/h))}{K}(\nu + \eta).$$

Taking the arbitrary nature of η into account, Lemma 2.1 is proved under the assumption that $h_1 \neq 0$.

Assume now that $h_1 = 0$, then it yields from formula (2.3) that

$$(m+1)h \leq \frac{h(\nu + \eta)}{K} \log r_n - \frac{h}{K} \log T(r_0) + h$$

$$\leq \left\{ \frac{h(\nu + \eta)}{K} - \frac{h}{K} \cdot \frac{\log T(r_0)}{\log r_n} + \frac{h}{\log r_n} \right\} \log r_n.$$

Notice also that

$$\int_{E[r_0, r_n]} \frac{dt}{t} \geq \log \frac{r_n}{r_0} - (m+1)h$$

$$\geq \left\{ 1 - \frac{h(\nu + \eta)}{K} + \frac{h}{K} \cdot \frac{\log T(r_0)}{\log r_n} - \frac{h}{\log r_n} - \frac{\log r_0}{\log r_n} \right\} \log r_n.$$

Then we get

$$\lim_{n \to +\infty} \frac{1}{\log r_n} \int_{E[r_0, r_n]} \frac{dt}{t} \geq 1 - \frac{h}{K}(\nu + \eta).$$

Since η is arbitrary, Lemma 2.1 is thus proved completely.

LEMMA 2.2. *Suppose that $T(r)$ defined on the positive real axis is a continuous, nondecreasing positive function tending to ∞, and satisfies the condition*

$$\lim_{r \to +\infty} \frac{\log T(r)}{\log r} = \mu < +\infty. \tag{2.5}$$

Then for any arbitrarily selected number h $(0 < h < +\infty)$ and h_1 $(0 \leq h_1 < h)$, there exists definitely a sequence $\{R_n\}$, such that

$$\lim_{n \to +\infty} \frac{\log T(R_n)}{\log R_n} = \mu \tag{2.6}$$

and

$$T(R_n e^h) < e^{h(1 + (h_1/h))\mu} T(R_n)(1 + o(1)), \qquad (n \to +\infty),$$

$$T(R_n e^{h_1}) < e^{h_1(1 + (h_1/h))\mu} T(R_n)(1 + o(1)), \qquad (n \to +\infty).$$

PROOF. Let $\varepsilon_n = \frac{1}{n}$, $\eta_n = \frac{1}{\sqrt{n}}$. Then, according to formula (2.5), a sequence $\{r_n\}$ exists such that

$$\left. \begin{array}{l} \displaystyle \lim_{n \to +\infty} \frac{\log T(r_n)}{\log r_n} = \mu, \\[2mm] r_n^{\mu - \varepsilon_n} \leq T(r_n) \leq r_n^{\mu + \varepsilon_n}, \\[2mm] T(r_n^{1 - \eta_n}) \geq \{r_n^{1 - \eta_n}\}^{\mu - \varepsilon_n}, \\[2mm] \log r_n \geq \dfrac{n}{\eta_n}(h + h_1)\left(\mu + \dfrac{2}{\sqrt{n}}\right). \end{array} \right\} \tag{2.7}$$

In the following, we need only to prove that the existence of value R_n in the closed interval $[r_n^{1 - \eta_n}, r_n]$, such that formula (2.6) and

$$T(R_n e^h) \leq e^K T(R_n), \qquad K = h\left(1 + \frac{h_1}{h}\right)\left(\mu + \frac{2}{\sqrt{n}}\right) \tag{2.8}$$

as well as

$$T(R_n e^{h_1}) \le e^{K_1} T(R_n), \qquad K_1 = h_1 \left(1 + \frac{h_1}{h}\right)\left(\mu + \frac{2}{\sqrt{n}}\right) \qquad (2.9)$$

hold. Consider the inequality

$$T(R_n e^h) < e^K T(R_n). \qquad (2.10)$$

Suppose t_1 is the smallest value that not satisfying formula (2.10) on the interval $[r_n^{1-\eta_n}, r_n]$, t_2 is the smallest value not satisfying formula (2.10) on the interval $[t_1 e^h, r_n]$, t_3 is the smallest value not satisfying formula (2.10) on the interval $[t_2 e^h, r_n]$, ..., t_{m+1} is the smallest value not satisfying formula (2.10) on the interval $[t_m e^h, r_n]$, and that $t_m e^h < r_n$, $t_{m+1} e^h \ge r_n$. Hence from formula (2.7) we obtain

$$r_n^{\mu + \varepsilon_n} \ge T(r_n) \ge T(t_m e^h) \ge e^K T(t_m) \ge e^K T(t_{m-1} e^h)$$
$$\ge e^{2K} T(t_{m-1}) \ge \cdots \ge e^{mK} T(r_n^{1-\eta_n}) \ge e^{mK} r_n^{(1-\eta_n)(\mu - \varepsilon_n)}$$

which implies that

$$mK \le \left\{\mu + \frac{2\varepsilon_n}{\eta_n} - \varepsilon_n\right\} \eta_n \log r_n,$$
$$m \le \frac{1}{K}\left\{\mu + \frac{2\varepsilon_n}{\eta_n} - \varepsilon_n\right\} \eta_n \log r_n. \qquad (2.11)$$

Let $r_n^{1-\eta_n} = t_0 e^h$ and assume that $h_1 \ne 0$. We consider the closed interval $[t_k e^h, +\infty)$ $(0 \le k \le m - 1)$ and the inequality

$$T(R_n e^{h_1}) < e^{K_1} T(R_n). \qquad (2.12)$$

Suppose that τ_{k1} is the smallest value not satisfying formula (2.12) on the interval $[t_k e^h, +\infty)$, τ_{k2} is the smallest value not satisfying formula (2.12) on the interval $[\tau_{k1} e^{h_1}, +\infty)$, ..., τ_{kl_k+1} is the smallest value not satisfying formula (2.12) on the interval $[\tau_{kl_k} e^{h_1}, +\infty)$, and that $\tau_{kl_k} e^{h_1} < t_{k+1}$, $\tau_{kl_k+1} e^{h_1} \ge t_{k+1}$. Finally we consider the interval $[t_m e^h, +\infty)$ and define correspondingly $\tau_{m1}, \tau_{m2}, \ldots, \tau_{ml_m}$ and τ_{ml_m+1} such that $\tau_{ml_m} e^{h_1} < \min\{r_n, t_{m+1}\}$, $\tau_{ml_m+1} e^{h_1} \ge \min\{r_n, t_{m+1}\}$. Hence, from formula (2.7) we get

$$r_n^{\mu + \varepsilon_n} \ge T(r_n) \ge T(\tau_{ml_m} e^{h_1}) \ge e^{K_1} T(\tau_{ml_m}) \ge \cdots$$
$$\ge e^{l_m K_1} T(\tau_{m1}) \ge e^{l_m K_1} T(t_m e^h) \ge e^{l_m K_1 + K} T(t_m) \ge \cdots$$
$$\ge e^{\sum_{-k=0}^m l_k K_1 + mK} T(r_n^{1-\eta_n})$$
$$\ge e^{\sum_{k=0}^m l_k K_1 + mK} (r_n)^{(1-\eta_n)(\mu - \varepsilon_n)}.$$

And therefore

$$mK + \sum_{k=0}^{m} l_k K_1 \le \left\{ \mu + \frac{2\varepsilon_n}{\eta_n} - \varepsilon_n \right\} \eta_n \log r_n,$$

$$mh + \sum_{k=0}^{m} l_k \frac{K_1}{K} h \le \frac{h}{K} \left\{ \mu + \frac{2\varepsilon_n}{\eta_n} - \varepsilon_n \right\} \eta_n \log r_n,$$

$$(m+1)h + \sum_{k=0}^{m} l_k h_1 + (m+1)h_1$$

$$\le \left\{ \frac{h}{K} \left(\mu + \frac{2\varepsilon_n}{\eta_n} - \varepsilon_n \right) + \frac{h}{\eta_n \log r_n} + \frac{(m+1)h_1}{\eta_n \log r_n} \right\} \eta_n \log r_n.$$

Moreover, combining this with formulas (2.8), (2.9) and (2.11) we deduce

$$(m+1)h + \sum_{k=0}^{m} l_k h_1 + (m+1)h_1$$

$$\le \left\{ 1 - \frac{1}{n(\mu + (2/\sqrt{n}))} + \frac{h+h_1}{\eta_n \log r_n} \right\} \eta_n \log r_n.$$

Further using formula (2.7) we have

$$(m+1)h + \sum_{k=0}^{m} l_k h_1 + (m+1)h_1 \le \eta_n \log r_n.$$

Therefore there exists R_n in the interval $[r_n^{1-\eta_n}, r_n]$, such that formulas (2.8) and (2.9) hold.

Suppose now $h_1 = 0$, then according to formula (2.12) we derive

$$(m+1)h < \eta_n \log r_n \left\{ \frac{h}{K} \left(\mu + \frac{2\varepsilon_n}{\eta_n} - \varepsilon_n \right) + \frac{h}{\eta_n \log r_n} \right\}.$$

It further yields from formulas (2.7) and (2.8) that $(m+1)h < \eta_n \log r_n$. Consequently there exists also R_n in the interval $[r_n^{1-\eta_n}, r_n]$, such that formulas (2.8) and (2.9) are valid. Finally from formula (2.7) and the inequality

$$(1 - \eta_n) \frac{\log T(r_n^{1-\eta_n})}{(1 - \eta_n) \log r_n} \le \frac{\log T(R_n)}{\log R_n} \le \frac{\log T(r_n)}{(1 - \eta_n) \log r_n}$$

we conclude that

$$\mu = \lim_{r \to \infty} \frac{\log T(r)}{\log r} \le \lim_{n \to +\infty} \frac{\log T(R_n)}{\log R_n} \le \overline{\lim}_{n \to +\infty} \frac{\log T(R_n)}{\log R_n}$$

$$\le \overline{\lim}_{n \to +\infty} \frac{\log T(r_n)}{(1 - \eta_n) \log r_n} = \lim_{n \to +\infty} \frac{\log T(r_n)}{(1 - \eta_n) \log r_n} = \mu.$$

So $\{R_n\}$ satisfies formula (2.6), and hence, Lemma 2.2 is proved completely.

Analogously, we have

LEMMA 2.3. *Suppose that* $T(r)$ *defined on the positive real axis is a continuous, nondecreasing positive function tending to* ∞, *and satisfies the condition*

$$\overline{\lim_{r \to +\infty}} \frac{\log T(r)}{\log r} = \lambda < +\infty.$$

Then for any arbitrarily selected h $(0 < h < +\infty)$ *and* h_1 $(0 \le h_1 < h)$, *there exists definitely a sequence* $\{R_n\}$, *such that*

$$\lim_{n \to +\infty} \frac{\log T(R_n)}{\log R_n} = \lambda,$$

and

$$T(R_n e^h) < e^{h(1+(h_1/h))\lambda} T(R_n)(1 + o(1)), \qquad (n \to +\infty),$$
$$T(R_n e^{h_1}) < e^{h_1(1+(h_1/h))\lambda} T(R_n)(1 + o(1)), \qquad (n \to +\infty).$$

LEMMA 2.4. *Suppose that* $T(r)$ *defined on the positive real axis is a continuous, nondecreasing positive function, and satisfies the condition*

$$\overline{\lim_{r \to +\infty}} \frac{T(r)}{(\log r)^2} = +\infty, \tag{2.13}$$

and

$$\lim_{r \to +\infty} \frac{\log T(r)}{\log r} = \mu < +\infty. \tag{2.14}$$

Then for any arbitrarily selected number h $(0 < h < +\infty)$, *there exists definitely a sequence* $\{R_n\}$, *such that*

$$\lim_{n \to +\infty} \frac{T(R_n)}{(\log R_n)^2} = +\infty \tag{2.15}$$

and $T(R_n e^h) \le e^{h\mu} T(R_n)(1 + o(1))$, $(n \to +\infty)$.

PROOF. Let $\varepsilon_n = \frac{1}{n}$ and $\eta_n = \frac{1}{\sqrt{n}}$ $(n = 2, 3, \ldots)$. Then according to formulas (2.13) and (2.14), there exist two sequences $\{t_n\}$ and $\{r_n'\}$, $t_n < t_{n+1} \to +\infty$, $r_n' < r_{n+1}' \to +\infty$ $(n \to +\infty)$ such that

$$\lim_{n \to +\infty} \frac{T(r_n')}{(\log r_n')^2} = +\infty,$$

$$(\log r_n')^{\eta_n} > \frac{n^2 h}{\eta_n} \left(\mu + \frac{2}{\sqrt{n}}\right), \tag{2.16}$$

$$t_n^{1-\eta_n} \ge r_n', \tag{2.17}$$

$$T(t_n^{1-\eta_n}) \ge \{t_n^{1-\eta_n}\}^{\mu-\varepsilon_n}, \tag{2.18}$$

$$t_n^{\mu-\varepsilon_n} \le T(t_n) \le t_n^{\mu+\varepsilon_n}. \tag{2.19}$$

Further from formulas (2.16) and (2.17), the inequality

$$\log t_n \geq \frac{nh}{\eta_n}\left(\mu + \frac{2}{\sqrt{n}}\right). \tag{2.20}$$

holds.

Now we notice that formulas (2.18), (2.19) and (2.20) correspond just to $(h_1 = 0)$ of formula (2.7) proved in Lemma 2.2. Hence, there exists in the interval $[t_n^{1-\eta_n}, t_n)$ $(n = 1, 2, \ldots)$ a value R_n', such that

$$T(R_n'e^h) \leq e^K T(R_n'),$$
$$K = h\left(\mu + \frac{2}{\sqrt{n}}\right) = h\mu + o(1) \qquad (n \to +\infty). \tag{2.21}$$

This implies that the values R satisfying the inequality

$$T(Re^h) \leq e^K T(R) \tag{2.22}$$

on the interval $[r_n', t_n]$ form a nonempty set. Suppose that R_n is the smallest value satisfying formula (2.22) on the interval $[r_n', t_n]$. If there exist infinitely many values n, such that $R_n = r_n'$, then we need only to define $r_n' = R_n$ to obtain a sequence $\{R_n\}$ that conforms to the requirement of Lemma 2.4. Therefore, we need only to consider the case for those sufficiently large n; it is always true that $r_n' < R_n$. When $r_n' \leq R < R_n$, then

$$T(Re^h) > e^K T(R). \tag{2.23}$$

In the following, we will distinguish two cases for discussion:
(1) We have

$$\varlimsup_{n \to +\infty} \frac{T(R_n)}{(\log R_n)^2} = +\infty.$$

Without any loss of generality, we may assume that

$$\lim_{n \to +\infty} \frac{T(R_n)}{(\log R_n)^2} = +\infty,$$

otherwise we need only to choose a suitable subsequence. Clearly, sequence $\{R_n\}$ satisfies the requirement of Lemma 2.4.
(2) We have

$$\varlimsup_{n \to +\infty} \frac{T(R_n)}{(\log R_n)^2} < +\infty.$$

Hence, when n is sufficiently large, we obtain $T(R_n) < c(\log R_n)^2$, where c $(0 < c < +\infty)$ is a constant. We define value k, such that $R_n e^{-kh} \geq r_n' \geq R_n e^{-(k+1)h}$. From formula (2.23) we get

$$e(\log R_n)^2 \geq T(R_n) \geq e^K T(R_n e^{-h}) \geq e^{2K} T(R_n e^{-2h})$$
$$\geq \cdots \geq e^{kK} T(R_n e^{-kh}) \geq e^{kK} T(r_n') \geq c e^{kK}.$$

And hence

$$kK \leq 2 \log\log R_n,$$

$$\log \frac{R_n}{r'_n} \leq (k+1)h \leq \left\{ \frac{2h}{K} \cdot \frac{\log\log R_n}{\log R_n} + \frac{h}{\log R_n} \right\} \qquad (2.24)$$

$$\cdot \frac{\log R_n}{\log R_n/r'_n} \cdot \log \frac{R_n}{r'_n}.$$

On the other hand, it yields

$$(\log r'_n)^{\eta_n} > \frac{2hn}{\eta_n^2}, \qquad K \geq \frac{2h}{\sqrt{n}},$$

from formulas (2.16) and (2.21), respectively. Consequently,

$$K \geq \frac{4h^2 n}{\sqrt{n}\,\eta_n^2} \frac{1}{(\log r'_n)^{\eta_n}} > \frac{4h^2}{\eta_n^3 (\log R_n)^{\eta_n}}.$$

Substituting this inequality into formula (2.24) we deduce

$$1 \leq \left\{ \frac{\eta_n^3}{2h} \cdot \frac{\log\log R_n}{(\log R_n)^{1-\eta_n}} + \frac{h}{\log R_n} \right\} \frac{\log R_n}{\log(R_n/r'_n)}.$$

And hence $R_n^{1-o(1)} \leq r'_n \ (n \to +\infty)$. Accordingly we conclude that

$$\lim_{n \to +\infty} \frac{T(R_n)}{(\log R_n)^2} \geq \lim_{n \to +\infty} \frac{T(r_n)}{(\log r'_n)^2} (1 - o(1))^2 = +\infty,$$

which implies the impossibility for case (2) to happen, and hence, Lemma 2.4 is proved completely.

LEMMA 2.5. *Suppose that* $T(r)$ *defined on the positive real axis is a continuous, nondecreasing positive function tending to* ∞, *and satisfies the condition*

$$\overline{\lim_{r \to +\infty}} \frac{\log T(r)}{\log r} = \lambda < +\infty. \qquad (2.25)$$

Then for any arbitrarily selected numbers η $(0 < \eta < 1)$, h *and* h_1 $(0 \leq h_1 < h < +\infty)$, *there exists definitely a value* r_0, *such that when* $r \geq r_0$, *values* R *on the closed interval* $[r, r^{1+\eta}]$ *exist, so that*

$$T(Re^h) \leq e^K T(R), \qquad K = 2h\left(1 + \frac{h_1}{h}\right)(\lambda + 1)(1 + \eta)\eta^{-1} \qquad (2.26)$$

and

$$T(Re^{h_1}) \leq e^{K_1} T(R), \qquad K_1 = 2h_1\left(1 + \frac{h_1}{h}\right)(\lambda + 1)(1 + \eta)\eta^{-1}. \qquad (2.27)$$

PROOF. First, according to formula (2.25), there exists a value r_0, such that when $r \geq r_0$, we have

$$1 \leq T(r) \leq r^{\lambda+1}, \qquad (2.28)$$

and

$$h_1 < h < \frac{1}{4}\eta \log r. \tag{2.29}$$

Next, we consider the inequality

$$T(Re^h) < e^K T(R). \tag{2.30}$$

Suppose that t_1 is the smallest value not satisfying formula (2.30) on the interval $[r, +\infty)$, t_2 is the smallest value not satisfying formula (2.30) on the interval $[t_1 e^h, +\infty)$, t_3 is the smallest value not satisfying formula (2.30) on the interval $[t_2 e^h, +\infty)$, ..., t_{m+1} is the smallest value not satisfying formula (2.30) on the interval $[t_m e^h, +\infty)$, and that $t_m e^h < r^{1+\eta}$ and $t_{m+1} e^h \geq r^{1+\eta}$. Hence, it follows from formula (2.30) that

$$(r^{1+\eta})^{\lambda+1} \geq T(r^{1+\eta}) \geq T(t_m e^h) \geq e^K T(t_m) \geq e^K T(t_{m-1} e^h)$$
$$\geq e^{2K} T(t_{m-1}) \geq \cdots \geq e^{mK} T(t_1) \geq e^{mK} T(r) \geq e^{mK}.$$

And hence

$$mK \leq (\lambda+1)(1+\eta) \log r,$$
$$m \leq \frac{(\lambda+1)(1+\eta)}{K} \log r. \tag{2.31}$$

Let $r = t_0 e^h$ and assume that $h_1 \neq 0$. We consider the interval $[t_k e^h, +\infty)$ $(0 \leq k \leq m-1)$ and the inequality

$$T(Re^{h_1}) < e^{K_1} T(R). \tag{2.32}$$

Suppose that τ_{k1} is the smallest value not satisfying formula (2.32) on the interval $[t_k e^h, +\infty)$, τ_{k2} is the smallest value not satisfying formula (2.32) on the interval $[\tau_{k1} e^{h_1}, +\infty)$, τ_{k3} is the smallest value not satisfying formula (2.32) on the interval $[\tau_{k2} e^{h_1}, +\infty)$, ..., τ_{kl_k+1} is the smallest value not satisfying formula (2.30) on the interval $[t_{kl_k} e^{h_1}, +\infty)$, and that $\tau_{kl_k} e^{h_1} < t_{k+1}$ and $\tau_{kl_k+1} e^{h_1} \geq t_{k+1}$. Finally we consider the interval $[t_m e^h, +\infty)$, and define correspondingly $\tau_{m1}, \tau_{m2}, \ldots, \tau_{ml_m}$ and τ_{ml_m+1}, such that $\tau_{ml_m} e^{h_1} < \min\{r^{1+\eta}, t_{m+1}\}$ and $\tau_{ml_m+1} e^{h_1} \geq \min\{r^{1+\eta}, t_{m+1}\}$. Hence we have

$$(r^{1+\eta})^{\lambda+1} \geq T(r^{1+\eta}) \geq T(\tau_m e^{h_1}) \geq e^{K_1} T(\tau_{ml_m}) \geq \cdots$$
$$\geq e^{lmK1} T(\tau_{m1}) \geq e^{l_m K_1} T(t_m e^h) \geq e^{l_m K_1 + K} T(t_m) \geq \cdots$$
$$\geq e^{\sum_{k=0}^m l_k K_1 + mK} T(\tau_{01}) \geq e^{\sum_{k=0}^m l_k K_1 + mK} T(r) \geq e^{\sum_{k=0}^m l_k K_1 + mK}.$$

And therefore

$$mK + \sum_{k=0}^{m} l_k K_1 \leq (\lambda + 1)(1 + \eta)\log r,$$

$$mh + \sum_{k=0}^{m} l_k \frac{K_1}{K} h \leq \frac{h}{K}(\lambda + 1)(1 + \eta)\log r,$$

$$(m+1)h + \sum_{k=0}^{m} l_k h_1 + (m+1)h_1$$

$$\leq \frac{h}{K}(\lambda + 1)(1 + \eta)\log r + h + (m+1)h_1 .$$

It then yields from formula (2.31) that

$$(m+1)h + \sum_{k=0}^{m} l_k h_1 + (m+1)h_1$$

$$\leq \left\{ \frac{h(1 + (h_1/h))(\lambda + 1)(1 + \eta)}{K} + \frac{h_1 + h}{\log r} \right\} \log r .$$

Further, based on formulas (2.26) and (2.29) we conclude that

$$(m+1)h + \sum_{k=0}^{m} l_k h_1 + (m+1)h_1 < \left\{ \frac{\eta}{2} + \frac{\eta}{4} + \frac{\eta}{4} \right\} \log r = \eta \log r .$$

Hence, there exist values R in the interval $[r, r^{1+\eta}]$, such that formulas (2.26) and (2.27) hold, meaning that when $h_1 \neq 0$, Lemma 2.5 is valid.

Let $h_1 = 0$; then it follows from formula (2.31) that

$$(m+1)h \leq \frac{h}{K}(\lambda + 1)(1 + \eta)\log r + h$$

$$= \left\{ \frac{h(\lambda + 1)(1 + \eta)}{K} + \frac{h}{\log r} \right\} \log r .$$

Further, based on formulas (2.26) and (2.29) we conclude that

$$(m+1)h \leq \left(\frac{\eta}{2} + \frac{\eta}{4} \right) \log r < \eta \log r .$$

Hence, there exist values R on the interval $[r, r^{1+\eta}]$, such that formulas (2.26) and (2.27) hold, implying that Lemma 2.5 is proved completely.

In the following, we will prove another kind of lemma, which originates from the efforts made by E. Borel [6a], F. Bureau [7a] and H. Milloux [30a].

LEMMA 2.6. *Suppose that $T(r)$ is a nondecreasing positive function on the interval $(0, R)$, a and b are two positive numbers, and that $b \geq 2a$ and $b \geq 8a^2$. If for any arbitrary value r and ρ, $0 < r < \rho < R$, the following inequality holds:*

$$T(r) < a \log^+ T(\rho) + a \log \frac{\rho}{\rho - r} + b . \tag{2.33}$$

Then we further have the inequality

$$T(r) < 2a \log \frac{\rho}{\rho - r} + 2b. \tag{2.34}$$

PROOF. First we prove the inequality

$$e^{b/a} x > 8a \log x + 8b \qquad (x \geq 2). \tag{2.35}$$

Let

$$\varphi(x) = e^{b/a} x - 8a \log x - 8b.$$

Then we need only to prove that $\varphi(2) > 0$, and that when $x \geq 2$, we have

$$\varphi'(x) = e^{b/a} - \frac{8a}{x} > 0.$$

Indeed, since we have

$$2e^{b/a} > 2\left(1 + \frac{b}{a} + \frac{1}{2} \cdot \frac{b^2}{a^2}\right) > \left(1 + \frac{b}{a}\right)^2,$$

and that $b > 8a^2$, it follows that $1 + \frac{b}{a} > 8a$, $2e^{b/a} > 8a(1 + \frac{b}{a}) = 8a + 8b > 8a \log 2 + 8b$. Hence $\varphi(2) > 0$ and when $x \geq 2$, we have $\varphi'(x) > 0$.

In the following we will prove the validity of formula (2.34). If formula (2.34) is not valid, then there exist two values r and ρ $(0 < r < \rho < R)$, such that

$$T(r) \geq 2a \log \frac{\rho}{\rho - r} + 2b. \tag{2.36}$$

Let $r' = \frac{1}{2}(r + \rho)$. Then according to formula (2.33), we get

$$T(r) < a \log^+ T(r') + a \log \frac{r'}{r' - r} + b$$
$$< a \log^+ T(r') + a \log \frac{\rho}{\rho - r'} + b.$$

On the other hand, formula (2.36) can be rewritten as

$$T(r) \geq 2a \log \frac{\rho}{\rho - r'} - 2a \log 2 + 2b.$$

And hence

$$\log^+ T(r') > \log \frac{\rho}{\rho - r'} - 2 \log 2 + \frac{b}{a} = \log\left(\frac{1}{4} e^{b/a} \frac{\rho}{\rho - r'}\right).$$

Since $b \geq 2a$, we conclude that

$$\frac{1}{4} e^{b/a} \frac{\rho}{\rho - r'} > \frac{1}{4} e^{b/a} > \frac{1}{4} e^2 > 1.$$

And therefore

$$T(r') > \frac{1}{4} e^{b/a} \frac{\rho}{\rho - r'}.$$

Further by applying formula (2.35), where we let $x = \rho/(\rho - r') \geq 2$, we have

$$e^{b/a} \frac{\rho}{\rho - r'} > 8a \log \frac{\rho}{\rho - r'} + 8b.$$

And therefore $T(r') > 2a \log(\rho/(\rho - r')) + 2b$. Analogously, by letting $r_n = \frac{1}{2}(r_{n-1} + \rho)$ $(n = 1, 2, \ldots, r_0 = r)$, we have

$$T(r_n) \geq 2a \log \frac{\rho}{\rho - r_n} + 2b, \qquad \rho - r_n = \frac{1}{2^n}(\rho - r), \qquad (2.37)$$

and

$$T(r_n) \leq T(\rho) \qquad (0 < r_n < \rho < R). \qquad (2.38)$$

Notice that when $n \to +\infty$, according to formula (2.37), $T(r_n)$ should tend to ∞. Based further on formula (2.38), $T(r_n)$ should be bounded. Hence we derive a contradiction, and consequently Lemma 2.6 is proved completely.

§2.2. The Boutroux-Cartan Theorem

2.2.1. The Boutroux-Cartan Theorem.

THEOREM 2.1. *Suppose that* $z_1, z_2, \ldots, z_n$ *is a finite sequence of* n *points on the open plane* $|z| < +\infty$; *then the set of points* z *satisfying the inequality*

$$\prod_{i=1}^{n} |z - z_i| < h^n$$

can be included only in a few circles (γ), *with the sum of their radii not exceeding* $2eh$.

Theorem 2.1 is called the Boutroux-Cartan Theorem [9a]. For simplicity, we will, for the rest of the book, denote (γ) the Euclidean exceptional circles that correspond to these n points and number h.

PROOF. Let $E = (z_1, z_2, \ldots, z_n)$ and C_{zK} be the open disk, with its respective center and radius being point z and $K\frac{eh}{n}$ (K is a positive integer). We also use P_{zK} to denote the number of points in E that belongs to C_{zK}. If $P_{zK} > K$, then there exists definitely a positive integer K', $K' > K$, such that $P_{zK'} = K'$. Otherwise, it follows that $P_{zK+1} > K + 1$, $P_{zK+2} > K + 2, \ldots, P_{zn} > n$. However, it is always true that $P_{zn} \leq n$, which results in a contradiction. We consider particularly $C_{z_1 1}$; then there exists a positive integer K' such that $P_{z_1 K'} = K'$. We prove, therefore, that a positive integer K $(1 \leq K \leq n)$ must exist, such that circle C_{zK} with center z and radius $K\frac{eh}{n}$ contains exactly K points of E, meaning that $P_{zK} = K$. We denote K_1 the largest positive integer with such properties, and C_1 the corresponding circle. We also let S_1 be the set of K_1 points in C_1, and $E_1 = E - S_1$. Applying similar discussion for E_1, we may obtain a positive integer K_2

and a corresponding circle C_2. By repeating the preceding steps, we get (K_j, C_j, S_j) $(j = 1, 2, \ldots, m)$, such that

$$m \leq n, \qquad \sum_{j=1}^{m} K_j = n, \qquad E = \bigcup_{j=1}^{m} S_j,$$

and $K_1 \geq K_2 \geq \cdots \geq K_m$.

In the following we prove that if $P_{zK} \geq K$, then there exist a circle C_{zK_j} a point z_i $(1 \leq i \leq n)$ and a positive integer j $(1 \leq j \leq m)$, such that

$$z_i \in C_{zK_j} \cap S_j, \qquad K_j \geq K. \tag{2.39}$$

When $K_m \geq K$, we need only to select a point $z_i \in E \cap C_{zK}$, and then identify the set S_j to which z_i belongs. When $K_m < K$, and suppose that $K_{P+1} < K \leq K_P \leq K_1$, if we denote S the set of P_{zK} points in C_{zK}, then it is impossible to have $S \subset E - \bigcup_{j=1}^{P} S_j$. Otherwise, it would lead to $K_{P+1} \geq K > K_{P+1}$, which is a contradiction. Hence we may get a point $z_i \in S \cap \{\bigcup_{j=1}^{P} S_j\}$ and then identify the set S_j $(1 \leq j \leq P)$ to which z_i belongs. The z_i and j derived in this way satisfy formula (2.39).

Now we denote Γ_j $(1 \leq j \leq m)$ a concentric circle of C_j, with radius $2K_j \frac{eh}{n}$. If the center z of C_{zK} does not belong to $\bigcup_{j=1}^{m} \Gamma_j \equiv (\gamma)$, then C_{zK} may contain at most $K - 1$ points of E. In fact, otherwise, there exist a point z_i and a positive integer j, such that formula (2.39) holds. Therefore, on the one hand since points $z_i \in C_{zK}$, we have $|z - z_i| < K\frac{eh}{n}$, while on the other hand, we get $|z - z_i| \geq K_j \frac{eh}{n}$ because the points $z_i \in C_j$ and z are outside the circle Γ_j. Furthermore, we derive

$$K_j < K \tag{2.40}$$

which, however, contradicts formula (2.39).

Finally we select arbitrarily any point z outside (γ), and arrange orderly the point sequence in E as $z_1', z_2', \ldots, z_n'$, such that $|z - z_1'| \leq |z - z_2'| \leq \cdots \leq |z - z_n'|$. Notice that C_{z1} does not contain any points of E, and therefore we have $|z - z_1| \geq \frac{eh}{n}$. Analogously, C_{z2} contains at most one point of E, and it follows that $|z - z_2| \geq 2\frac{eh}{n}$. Generally, we have $|z - z_K'| \geq K\frac{eh}{n}$, $(1 \leq K \leq n)$. And consequently

$$\prod_{i=1}^{n} |z - z_i| \geq n! \left(\frac{eh}{n}\right)^n > h^n.$$

Theorem 2.1 is thus proved.

We make a supplementary illustration for Theorem 2.1. Among the m exceptional circles Γ_j $(1 \leq j \leq m)$ denoted in (γ), we may assume that these m closed disks $\overline{\Gamma}_j$ $(j = 1, 2, \ldots, m)$ have no intersections among one another. In fact, otherwise, suppose that $\overline{\Gamma}_j$ intersects $\overline{\Gamma}_{j'}$ $(1 \leq j \neq j' \leq m)$. We may construct an open disk Γ containing Γ_j and $\Gamma_{j'}$. Apparently, the radius of Γ does not exceed the sum of radii of Γ_j and $\Gamma_{j'}$. Hence, if we replace Γ_j and $\Gamma_{j'}$ of (γ) by Γ, the result of Theorem 2.1 remains valid. In the following chapters, whenever Theorem 2.1 is applied, we always consider that the exceptional circles denoted by (γ) do not intersect or tangentially touch one another.

2.2.2. Extensions. Under the concept of spherical distance, it is obvious that the Boutroux-Cartan Theorem also holds.

THEOREM 2.2. *Suppose that* $z_1, z_2, \ldots, z_n$ *are* n *points on the Riemann sphere; then the set of points* z *satisfying the inequality*

$$\prod_{i=1}^{n} |z, z_i| < h^n$$

can only be included in at most a few circles (γ) *that do not intersect or touch tangentially among one another, and the sum of their spherical radii does not exceed* $2eh$.

H. Milloux introduced the concept of pseudo-non-Euclidean distance [30b], and extended further the Boutroux-Cartan Theorem. Let z and z' be two points in the unit disk $|z| < 1$, with their pseudo-non-Euclidean distance defined by H. Milloux as

$$|(z, z')| = \left| \frac{z - z'}{1 - \overline{z}'z} \right|.$$

Then obviously $0 \le |(z, z')| \le 1$ and that the pseudo-non-Euclidean distance remains invariant under the fractional linear transformation which maps the unit disk into itself. We select arbitrarily a point z_0, $|z_0| < 1$ and a value r, $0 < r < 1$. If we denote C the set of points z satisfying $|(z, z_0)| < r$, we call C the pseudo-non-Euclidean circle, with z_0 being the pseudo-non-Euclidean center and r the pseudo-non-Euclidean radius. Indeed, it is easy to prove that such a pseudo-non-Euclidean circle C is just the same as the Euclidean circle in general case. But, of course, the centers and radii of these two kinds of circles are different. The two circles will be the same only when both are centered at the origin $z = 0$.

Under this concept, H. Milloux extended the Boutroux-Cartan Theorem as follows [30b]:

THEOREM 2.3. *Suppose that* $z_1, z_2, \ldots, z_n$ *are* n *points in the unit circle* $|z| < 1$; *then the set of points* z *(in the unit circle* $|z| < 1$) *satisfying the inequality*

$$\prod_{i=1}^{n} |(z, z_i)| < h^n, \qquad \left(h < \frac{1}{2e} \right)$$

can only be included in at most n *circles* (γ) *(these circles are also in the disk* $|z| < 1$), *with the sum of their pseudo-non-Euclidean radii not exceeding* $2eh$.

For the rest of the book, we will denote (γ) the pseudo-non-Euclidean circles corresponding to these n points and number h.

Indeed, in order to prove Theorem 2.3, we need only to repeat the procedure of proving Theorem 2.1. But if we intend to derive the corresponding formula (2.40), we have to make some illustrations. This is because under the concept of pseudo-non-Euclidean distance, the following fact is not easy to note: When $z_i \in C_i$ and z is outside circle Γ_j, we have $|(z, z_j)| > K_j \frac{eh}{n}$. Therefore, we need to prove that when C is denoted a circle with its pseudo-non-Euclidean center and radius being z_0 ($|z_0| < 1$) and r ($r < \frac{1}{2}$), respectively, and C' is denoted a concentric circle with its pseudo-non-Euclidean center and radius being z_0 and $2r$, respectively. Then the pseudo-non-Euclidean distance between any point z outside C' and any point z' within the circle C or on the boundary of C is $|(z, z')| \geq r$. Hence we need only to prove that if Γ is denoted a circle with z being its pseudo-non-Euclidean center and r its pseudo-non-Euclidean radius, then circles Γ and C do not intersect or touch tangentially with each other. Furthermore, if we notice that $|(z, z_0)| \geq 2r$, then we need to prove that if Γ intersects C, then $|(z, z_0)| < 2r$. Now we prove a more general fact:

Assume that Γ' and Γ'' are two pseudo-non-Euclidean circles, with z' and z'' being the pseudo-non-Euclidean centers and r_1 and r_2 ($r_1 + r_2 < 1$) the pseudo-non-Euclidean radii, respectively. If Γ' intersects Γ'' or Γ' touches Γ'' tangentially, then it follows that $|(z', z'')| < r_1 + r_2$.

First we prove the case when Γ' touches Γ'' tangentially. Without any loss of generality, we may suppose that $z' = 0$ and that z'' is on the positive real axis. Otherwise, we only need to consider a suitable fractional linear transformation which preserves the status quo of the unit disk $|z| < 1$. Moreover, we assume that x is the point of tangency between Γ' and Γ''. Hence we conclude that

$$|(z', x)| = x = r_1, \qquad |(x, z'')| = \frac{z'' - x}{1 - xz''} = r_2,$$

$$|(z', z'')| = z'' = \frac{r_1 + r_2}{1 + r_1 r_2} < r_1 + r_2.$$

Next we prove the case when Γ' intersects Γ''. Analogously, we may assume that $z' = 0$ and z'' is on the positive real axis. We then denote x_1 the intersecting points between Γ' and the positive real axis, and x_2 the intersecting points between Γ'' and the positive real axis, and assume that x_2 is on the left-hand side of z''. Consequently, we conclude that

$$x_2 < x_1, \quad |(z', x_1)| = x_1 = r_1, \quad |(x_2, z'')| = \frac{z'' - x_2}{1 - x_2 z''} = r_2,$$

$$|(z', z'')| = z'' = \frac{r_2 + x_2}{1 + r_2 x_2} \leq \frac{r_2 + x_1}{1 + x_1 r_2} = \frac{r_1 + r_2}{1 + r_1 r_2} < r_1 + r_2.$$

The above discussion illustrates the validity of Theorem 2.3.

Similarly, we may always assume that the pseudo-non-Euclidean circles denoted by (γ) in Theorem 2.3 do not intersect and touch tangentially among one another. However, we need to make a supplementary illustration for the following fact:

Suppose that Γ' and Γ'' are two pseudo-non-Euclidean circles, with z' and z'' being the pseudo-non-Euclidean centers, and r_1 and r_2 $(r_1 + r_2 < 1)$ the pseudo-non-Euclidean radii. If Γ' intersects Γ'' or Γ' touches Γ'' tangentially, then there exists definitely a circle Γ''', such that Γ''' contains Γ' and Γ'' in its interior, with the pseudo-non-Euclidean radius $r_3 < r_1 + r_2$.

In fact, we know from the concept of elementary geometry that there exists in the unit disk $|z| < 1$ a disk Γ''' that touches internally both Γ' and Γ''. Let r_3 $(r_3 < 1)$ be the pseudo-non-Euclidean radius of Γ'''. In the following we need only to prove that $r_3 < r_1 + r_2$. We may assume that the pseudo-non-Euclidean center of Γ''' is at the origin $z = 0$, z' and z'' are on the real axis and that $z' < z''$.

Then we assume that Γ' intersects the real axis, and their intersecting point on the right-hand side of z' is x_1; Γ'' intersects the real axis and their intersecting point on the left-hand side of z'' is x_2. Under the hypothesis that Γ' intersects Γ'', or Γ'' touches Γ''' externally, it follows that $x_2 \leq x_1$. Hence we have

$$x_1 = \frac{r_1 + z'}{1 + r_1 z'}, \quad x_2 = \frac{z'' - r_2}{1 - r_2 z''}, \quad z' = \frac{r_1 - r_3}{1 - r_1 r_3} \quad z'' = \frac{r_3 - r_2}{1 - r_2 r_3}.$$

Accordingly,

$$\frac{z'' - r_2}{1 - r_2 z''} \leq \frac{r_1 + z'}{1 + r_1 z'},$$

$$(r_1 + r_2)(1 + r_1 r_2)r_3^2 - [(1 + r_1^2)(1 + r_2^2) + 4r_1 r_2]r_3$$
$$+ (r_1 + r_2)(1 + r_1 r_2) \geq 0.$$

Now we consider the following polynomial of R of degree two:

$$f(R) = (r_1 + r_2)(1 + r_1 r_2)R^2 - [(1 + r_1^2)(1 + r_2^2) + 4r_1 r_2]R$$
$$+ (r_1 + r_2)(1 + r_1 r_2).$$

Suppose that R_1 and R_2 are two roots of $f(R)$. It is apparent that $R_1 = 1/R_2$. We may, therefore, assume that $R_1 \leq 1 \leq R_2$. When $R \leq R_1$ or $R \geq R_2$, we have $f(R) \geq 0$). On the other hand, since $f(r_3) \geq 0$ and $r_3 < 1$, we conclude that $r_3 \leq R_1$. Also notice that when $f(r_1 + r_2) < 0$, we have $r_3 < r_1 + r_2$.

Finally we discuss a very important issue on application: If we rotate all the circles in (γ), such that all their centers lie on the positive real axis, then among such a group of circles covering the positive real axis, how long will the largest interval covered be? Suppose that (r', r'') $(r'' < 1)$ is the largest interval and that $|(r', r'')|$ is the pseudo-non-Euclidean diameter of the disk Γ, then the pseudo-non-Euclidean radius of Γ is

$$R = \frac{\frac{r''-r'}{1-r'r''}}{1 + \sqrt{1 - (\frac{r''-r'}{1-r'r''})^2}}.$$

Obviously $R < 2eh$. Hence, if r' and r'' satisfy the condition

$$\frac{r''-r'}{1 - r'r'' + \sqrt{(1-r'r'')^2 - (r''-r')^2}} \geq 2eh, \qquad (2.41)$$

then in the annulus $r' \leq |z| \leq r''$, there exists definitely a circumference $|z| = r$, $r' \leq r \leq r''$ that is disjoint with (γ). In fact, we can provide a much more practical sufficient condition:

$$(1 - r') \geq (1 - r'') \left(\frac{1 + 2eh}{1 - 2eh} \right)^2. \qquad (2.42)$$

For which, we have to prove that formula (2.42) implies formula (2.37). First we find that

$$(1 - 2eh)^2 (r'' - r') \geq (1 - r'')8eh,$$

$$(1 + 2eh)^2 (r'' - r') \geq (1 - r')8eh \geq (1 - r')r''8eh.$$

It follows that

$$[(1 - 2eh)^2 + (1 + 2eh)^2](r'' - r') \geq (1 - r'r'')8eh,$$

$$\frac{r'' - r'}{1 - r'r''} \geq \frac{4eh}{1 + 4e^2h^2}. \qquad (2.43)$$

And hence

$$\sqrt{1 - \left(\frac{r'' - r'}{1 - r'r''} \right)^2} \leq \sqrt{1 - \left(\frac{4eh}{1 + 4e^2h^2} \right)^2},$$

$$\frac{1}{1 + \sqrt{1 - \left(\frac{r''-r'}{1-r'r''} \right)^2}} \geq \frac{1}{1 + \sqrt{1 - \left(\frac{4eh}{1+4e^2h^2} \right)^2}} = \frac{1 + 4e^2h^2}{2}.$$

Furthermore, according to formula (2.43), we deduce

$$\frac{r'' - r''}{1 - r'r'' + \sqrt{(1-r'r'')^2 - (r''-r')^2}} \geq 2eh,$$

meaning that formula (2.41) holds.

§2.3. Fundamental theorem of value distribution of functions meromorphic in a disk

2.3.2. The theorem of bound. In 1928 G. Valiron proved the existence of the Borel direction by means of applying the Nevanlinna theory. To begin with, he established the following theorem of bound [39d]:

THEOREM 2.4. *Let $f(z)$ be a function meromorphic on the disk $|z| < R$ $(0 < R < +\infty)$, and that $f(0) \neq 0, 1, \infty$, $f'(0) \neq 0$. Then for any arbitrary value r $(0 \leq r < R)$ we have*

$$T(r, f) \leq 2\{N(R, 0) + N(R, 1) + N(R, \infty)\}$$
$$+ 4\log^+ |f(0)| + 2\log^+ \frac{1}{R|f'(0)|} \tag{2.44}$$
$$+ 36\log \frac{R}{R - r} + 5220.$$

PROOF. First we prove that if $a > e$ and $x > 0$, then

$$\log x + a\log^+ \log^+ \frac{1}{x} \leq a(\log a - 1) + \log^+ x. \tag{2.45}$$

In fact, when $x \geq \frac{1}{e}$, it follows apparently that $\log^+ \log^+ \frac{1}{x} = 0$, meaning that formula (2.45) holds. When $x < \frac{1}{e}$, formula (2.45) can be rewritten as

$$\log x + a\log\log \frac{1}{x} \leq a(\log a - 1). \tag{2.46}$$

Let $\varphi(y) = a\log y - y - a(\log a - 1)(y > 0)$. Then we have

$$\varphi'(y) = \frac{a}{y} - 1, \qquad \varphi''(y) = -\frac{a}{y^2}.$$

And hence, $\varphi'(a) = 0$, $\varphi''(a) < 0$. We therefore conclude that

$$\varphi(y) = a\log y - y - a(\log a - 1) \leq \varphi(a) = 0.$$

Replacing y by $\log \frac{1}{x}$, we obtain immediately formula (2.46).

Next we prove the validity of formula (2.44). According to Theorem 1.4, where we let $a_1 = 0$, $a_2 = 1$, then for any arbitrary value r $(0 < r < R)$ we have

$$m(r, 0) + m(r, 1) + m(r, \infty) \leq 2T(r, f) + S(r, f),$$

where

$$S(r, f) = m\left(r, \frac{f'}{f}\right) + m\left(r, \frac{f'}{f} + \frac{f'}{f - 1}\right)$$
$$+ 2\log 6 + \log 2 + \log \frac{1}{|f'(0)|}.$$

An application of formulas (1.11) and (1.12) yields

$$T(r, f - 1) \leq N(r, 0) + N(r, 1) + N(r, \infty)$$
$$+ \log |f(0)(f(0) - 1)| + S(r, f),$$
$$T(r, f) \leq N(r, 0) + N(r, 1) + N(r, \infty) \tag{2.47}$$
$$+ \log |f(0)(f(0) - 1)| + S(r, f) + \log 2.$$

Further based on Lemma 1.3, for any arbitrary value ρ $(0 < r < \rho < R)$ we obtain

$$S(r, f) \leq 2m\left(r, \frac{f'}{f}\right) + m\left(r, \frac{f'}{f-1}\right) + 8\log 2 + \log \frac{1}{|f'(0)|}$$

$$\leq 8\log^+ T(\rho, f) + 8\log^+\log^+ \frac{1}{|f(0)|} + 10\log^+ \rho$$

$$+ 12\log^+ \frac{1}{\rho - r} + 2\log^+ \frac{1}{r} + 28 + 4\log^+ T(\rho, f - 1)$$

$$+ 4\log^+\log^+ \frac{1}{|f(0) - 1|} + 5\log^+ \rho$$

$$+ 6\log^+ \frac{1}{\rho - r} + \log^+ \frac{1}{r} + 14 + 8\log 2 + \log \frac{1}{|f'(0)|}$$

$$\leq 12\log^+ T(\rho, f) + 8\log^+\log^+ \frac{1}{|f(0)|}$$

$$+ 4\log^+\log^+ \frac{1}{|f(0) - 1|} + 15\log^+ \rho + 18\log^+ \frac{1}{\rho - r}$$

$$+ 3\log^+ \frac{1}{r} + 42 + 15\log 2 + \log \frac{1}{|f'(0)|}.$$

Therefore formula (2.47) gives

$$T(r, f) \leq N(r, 0) + N(r, 1) + N(r, \infty)$$

$$+ \log|f(0)(f(0) - 1)| + 12\log^+ T(\rho, f)$$

$$+ 8\log^+\log^+ \frac{1}{|f(0)|} + 4\log^+\log^+ \frac{1}{|f(0) - 1|}$$

$$+ 15\log^+ \rho + 18\log^+ \frac{1}{\rho - r} + 3\log^+ \frac{1}{r}$$

$$+ 42 + 16\log 2 + \log \frac{1}{|f'(0)|}.$$

Taking formula (2.46) into consideration the following inequalities hold:

$$\log|f(0)| + 8\log^+\log^+ \frac{1}{|f(0)|}$$

$$\leq 8(\log 8 - 1) + \log^+ |f(0)|$$

$$\leq 24\log 2 + \log^+ |f(0)|,$$

$$\log|f(0) - 1| + 4\log^+\log^+ \frac{1}{|f(0) - 1|}$$

$$\leq 4(\log 4 - 1) + \log^+ |f(0) - 1|$$

$$\leq 9\log 2 + \log^+ |f(0)|.$$

Hence we conclude that

$$T(r, f) \leq N(r, 0) + N(r, 1) + N(r, \infty) + 2 \log^+ |f(0)|$$
$$+ \log \frac{1}{|f'(0)|} + 91 + 3 \log^+ \frac{1}{r} + 18 \log^+ \frac{1}{\rho}$$
$$+ 15 \log^+ \rho + 18 \log^+ \frac{\rho}{\rho - r} + 12 \log^+ T(\rho, f).$$

Particularly when $\frac{R}{2} \leq r < \rho < R$, we have

$$T(r, f) < 12 \log^+ T(\rho, f) + 18 \log \frac{\rho}{\rho - r} + H,$$

where

$$H = N(R, 0) + N(R, 1) + N(R, \infty) + 21 \log^+ \frac{1}{R}$$
$$+ 15 \log^+ R + 112 + 2 \log^+ |f(0)| + \log^+ \frac{1}{|f'(0)|}.$$

Now in order to apply Lemma 2.6, we set

$$T(r) = \begin{cases} T(r, f), & \frac{R}{2} \leq r < R, \\ 0, & 0 < r < \frac{R}{2}, \end{cases}$$

$a = 18$ and $b = H + 2480$. Then when $\frac{R}{2} \leq r < \rho < R$, it follows that

$$T(r) < 36 \log \frac{\rho}{\rho - r} + 2H + 4960.$$

Let $\rho \to R$. We find that

$$T(r) \leq 36 \log \frac{R}{R - r} + 2H + 4960.$$

Then letting $r = \frac{R}{2}$ in particular we deduce $T(\frac{R}{2}, f) \leq 36 \log 2 + 2H + 4960$. Therefore, when $0 < r < R$, we assure that

$$\begin{aligned} T(r, f) &\leq 36 \log \frac{R}{R - r} + 2H + 4996 \\ &\leq 2\{N(R, 0) + N(R, 1) + N(R, \infty)\} \\ &\quad + 42 \log^+ \frac{1}{R} + 30 \log^+ R + 4 \log^+ |f(0)| \\ &\quad + 2 \log^+ \frac{1}{|f'(0)|} + 36 \log \frac{R}{R - r} + 5220. \end{aligned} \tag{2.48}$$

In order to eliminate items $\log^+ \frac{1}{R}$ and $\log^+ R$, we let $g(z) = f(Rz)$. Then $g(z)$ is meromorphic in the disk $|z| < 1$, and also

$$g(0) = f(0), \quad g'(0) = Rf'(0), \quad T\left(\frac{r}{R}, g\right) = T(r, f)$$
$$(0 < r < R).$$

Hence, an application of formula (2.48) to $g(z)$ derives

$$T(r, f) = T\left(\frac{r}{R}, g\right) \leq 2\{N(R, 0) + N(R, 1) + N(R, \infty)\}$$
$$+ 36 \log \frac{R}{R - r} + 4 \log^+ |f(0)|$$
$$+ 2 \log^+ \frac{1}{|Rf'(0)|} + 5220,$$

implying that formula (2.44) is valid.

Examining the proof of Theorem 2.4, we may find that Theorem 2.4 will have a more general form without the assumption that $f'(0) \neq 0$.

THEOREM 2.5. *Let $f(z)$ be a function meromorphic on the disk $|z| < R$ $(0 < R < +\infty)$ and that it has a neighborhood around $z = 0$ the following expansion:*

$$f(z) = f(0) + c_s z^s + \cdots, \quad f(0) \neq 0, 1, \infty; \quad c_s \neq 0, \ s \geq 1.$$

Then for any arbitrary value r $(0 \leq r < R)$ we have

$$T(r, f) \leq 2\{N(R, 0) + N(R, 1) + N(R, \infty)\} + 36 \log \frac{R}{R - r}$$
$$+ 4 \log^+ |f(0)| + 2 \log^+ \frac{1}{R|sc_s|} + 5220.$$

2.3.2. The fundamental theorem.   G. Valiron further applied Theorem 2.4 to prove the following fundamental theorem on the value distribution of meromorphic functions in the disk [39d]. This theorem plays a crucial role in proving the existence of the singular directions.

THEOREM 2.6. *Let $f(z)$ be a nonconstant meromorphic function in the disk $|z| < 1$, and that $n(1, 0) \leq n$, $n(1, 1) \leq n$, $n(1, \infty) \leq n$. Then for any complex value X and arbitrary value r $(0 < r < 1)$*

$$n(r, X) \leq \frac{1}{(1 - r)^2} \left\{ An \log \frac{1}{h} + B \log \frac{2}{1 - r} C \log^+ \frac{1}{|X, X(r)|} \right\}, \quad (2.49)$$

where $0 < h \leq 0.01$, and A, B and C are constants. Also $X(r)$ is a complex value depending on $f(z)$ and value r.

PROOF. First we prove that if z_1 and z_2 are two finite complex numbers, then

$$\log^+ |z_1| + \log^+ |z_2| + \log \frac{1}{|z_1 - z_2|} \leq \log \frac{1}{|z_1, z_2|}. \quad (2.50)$$

Indeed, on the one hand, we have according to the spherical distance:

$$\log \frac{1}{|z_1, z_2|} = \log \frac{1}{|z_1 - z_2|} + \frac{1}{2} \log(1 + |z_1|^2) + \frac{1}{2} \log(1 + |z_2|^2).$$

On the other hand, when $x \geq 0$, we have, in general, the inequality

$$\log(1 + x^2) \geq 2 \log^+ x.$$

Hence, formula (2.50) is valid.

Next, we prove the validity of formula (2.49). Suppose (γ) are the pseudo-non-Euclidean exceptional circles that correspond to these $n(1, 0) + n(1, 1) + n(1, \infty)$ points and number h $(0 < h \leq 0.01)$. From formula (2.42), we conclude that there exists a point z_0 in the circle $|z| \leq 8eh$ as well as outside circles (γ). Then consider the transformation $\zeta = (z - z_0)/(1 - \overline{z}_0 z)$, with its inverse transformation as

$$z = (\zeta + z_0)/(1 + \overline{z}_0 \zeta).$$

We find that function

$$F(\zeta) = f\left(\frac{\zeta + z_0}{1 + \overline{z}_0 \zeta}\right)$$

is meromorphic in the disk $|\zeta| < 1$, and it follows that

$$n(1, F = 0) \leq n, \qquad n(1, F = 1) \leq n, \qquad n(1, F = \infty) \leq n. \qquad (2.51)$$

Meanwhile, circles (γ) are transformed into pseudo-non-Euclidean exceptional circles $(\gamma)_\zeta$ in the ζ-plane that correspond to these

$$n(1, F = 0) + n(1, F = 1) + n(1, F = \infty) \qquad (2.52)$$

points and number h. Notice that point $\zeta = 0 \notin (\gamma)_\zeta$. According to the inequality

$$|\zeta| \leq \frac{|z| + |z_0|}{1 + |z_0||z|},$$

we may conclude that the image domain of the disk $|z| < r$ on the ζ-plane is contained in the disk $|\zeta| \leq \tau$, and that

$$\tau = \frac{r + |z_0|}{1 + |z_0|r}. \qquad (2.53)$$

Further based on formula (2.42), we may conclude that there exists value ρ, $\frac{1+\tau}{2} \leq \rho \leq \frac{3+\tau}{4}$, such that circumference $\Gamma: |\zeta| = \rho$ does not intersect $(\gamma)_\zeta$.

In the following we will distinguish three cases for discussion:

(1) It is always true that $|F(\zeta)| > 1$ on Γ. Hence, obviously $m(\rho, 1/F) = 0$. On the other hand, notice that point $\zeta = 0 \notin (\gamma)_\zeta$; then we derive from formulas (1.5), (2.51) and (2.52) that

$$N\left(\rho, \frac{1}{F}\right) \leq N\left(1, \frac{1}{F}\right)$$

$$\leq N\left(1, \frac{1}{F}\right) + N\left(1, \frac{1}{F - 1}\right) + N(1, F)$$

$$\leq \{n(1, F = 0) + n(1, F = 1) + n(1, F = \infty)\} \log\frac{1}{h}$$

$$\leq 3n \log\frac{1}{h}.$$

And hence

$$T(\rho, F) = T\left(\rho, \frac{1}{F}\right) + \log|F(0)| \le 3n \log \frac{1}{h} + \log^+ |F(0)|.$$

Therefore, when $X \ne F(0) = f(z_0)$, we have the following estimation:

$$n(r, f = X) \le n(\tau, F = X) \le \frac{\rho}{\rho - \tau} N\left(\rho, \frac{1}{F - X}\right)$$

$$\le \frac{\rho}{\rho - \tau} T\left(\rho, \frac{1}{F - X}\right)$$

$$= \frac{\rho}{\rho - \tau} \left\{T(\rho, F - X) + \log \frac{1}{|F(0) - X|}\right\}$$

$$\le \frac{2}{1 - \tau} \left\{T(\rho, F) + \log^+ |X| + \log \frac{1}{|F(0) - X|} + \log 2\right\}$$

$$\le \frac{2}{1 - \tau} \left\{3n \log \frac{1}{h} + \log^+ |f(z_0)| + \log^+ |X|\right.$$

$$\left. + \log \frac{1}{|f(z_0) - X|} + \log 2\right\}.$$

Furthermore, from formulas (2.50) and (2.53) and $|z_0| \le 8eh$ ($h \le 0.01$), we conclude that

$$n(r, f = X) \le \frac{6}{1 - r} \left\{3n \log \frac{1}{h} + \log^+ \frac{1}{|f(z_0), X|} + \log 2\right\}. \tag{2.54}$$

When $X = f(z_0)$, it is apparent that formula (2.50) is valid.

(2) There exists a point ζ_1 on Γ, such that $|F(\zeta_1)| \le 1$, and that $|F'(\zeta)| \le 1$ on Γ. Hence, starting from point ζ_1 and integrating along Γ it yields

$$F(\zeta) - F(\zeta_1) = \int_\Gamma F'(\zeta)\, d\zeta,$$

$$|F(\zeta)| \le |F(\zeta_1)| + \int_\Gamma |F'(\zeta)||d\zeta| \le 1 + 2\pi,$$

$$m(\rho, F) \le \log(1 + 2\pi).$$

On the other hand, notice that point $\zeta = 0 \notin (\gamma)_\zeta$; then according to formulas (1.5), (2.51) and (2.52) we have $N(\rho, F) \le 3n \log \frac{1}{h}$. And hence $T(\rho, F) \le 3n \log \frac{1}{h} + \log(1 + 2\pi)$. Analogously, we conclude that

$$n(r, f = X) \le \frac{6}{1 - r} \left\{3n \log \frac{1}{h} + \log^+ \frac{1}{|f(z_0), X|} + \log 2(1 + 2\pi)\right\}. \tag{2.55}$$

(3) There exist two points ζ_1 and ζ_2 on Γ, such that $|F(\zeta_1)| \le 1$ and $|F'(\zeta_2)| \ge 1$. Also starting from point ζ_1 along Γ to point ζ_2, it is always true that $|F'(\zeta)| \le 1$. Hence we have $|F(\zeta_2)| \le 1 + 2\pi$. Considering the transformation $\xi = (\zeta - \zeta_2)/(1 - \overline{\zeta_2}\zeta)$, with its inverse transformation being

$\zeta = (\xi + \zeta_2)/(1 + \overline{\zeta_2}\xi)$, we find that function

$$G(\xi) = F\left(\frac{\xi + \zeta_2}{1 + \overline{\zeta_2}\xi}\right)$$

is meromorphic in the disk $|\xi| < 1$, and that

$$n(1, G = 0) \leq n, \quad n(1, G = 1) \leq n, \quad n(1, G = \infty) \leq n.$$

Simultaneously, circles (γ) are transformed into pseudo-non-Euclidean exceptional circles $(\gamma)_\xi$ in the ξ-plane that correspond to these

$$n(1, G = 0) + n(1, G = 1) + n(1, G = \infty)$$

points and number h. Notice that $\xi = 0 \notin (\gamma)_\xi$, and that

$$|G(0)| = |F(\zeta_2)| \leq 1 + 2\pi,$$
$$|G'(0)| = |F'(\zeta_2)| \left\{\frac{1 - |\zeta_2|^2}{|1 + \overline{\zeta_2}\xi|^2}\right\}_{\xi=0}$$
$$\geq |F'(\zeta_2)|(1 - \rho^2) \geq 1 - \rho.$$

According to the inequality

$$|\xi| \leq \frac{|\zeta| + |\zeta_2|}{1 + |\zeta_2||\zeta|},$$

we may conclude that the image domain of the disk $|\zeta| < \tau$ on the ξ-plane is contained in the disk $|\xi| < s$, and

$$s = \frac{\tau + \rho}{1 + \tau\rho}. \tag{2.56}$$

Further based on formula (2.42), we may arrive at the conclusion that there exists ρ', $\frac{1+s}{2} \leq \rho' \leq \frac{3+s}{4}$, such that circumference Γ': $|\xi| = \rho'$ does not intersect $(\gamma)_\xi$. Now applying Theorem 2.4 to $G(\xi)$, we obtain

$$T(\rho', G) \leq 2\{N(1, G = 0) + N(1, G = 1) + N(1, G = \infty)\}$$
$$+ 4\log^+ |G(0)| + 2\log^+ \frac{1}{|G'(0)|} + 36\log\frac{1}{1 - \rho'} + 5220$$
$$\leq 6n\log\frac{1}{h} + 4\log(1 + 2\pi) + 36\log\frac{1}{1 - \rho'} + 2\log\frac{1}{1 - \rho}.$$

And hence

$$n(r, f = X)$$

$$\leq n(s, G = X) \leq \frac{\rho'}{\rho' - s} \int_s^{\rho'} \frac{n(t, G = X)}{t} \, dt$$

$$\leq \frac{\rho'}{\rho' - s} \left(\rho', \frac{1}{G - X} \right) \leq \frac{1}{\rho' - s} T \left(\rho', \frac{1}{G - X} \right)$$

$$\leq \frac{1}{\rho' - s} \left\{ T(\rho', G) + \log^+ |G(0)| + \log \frac{1}{|G(0) - X|} + \log 2 \right\}$$

$$\leq \frac{2}{1 - s} \left\{ 6n \log \frac{1}{h} + 36 \log \frac{1}{1 - \rho'} + 2 \log \frac{1}{1 - \rho} + \log^+ |G(0)| \right.$$

$$\left. + \log \frac{1}{|G(0) - X|} + 4 \log(1 + 2\pi) + \log 2 + 5220 \right\}$$

$$\leq \frac{2}{1 - s} \left\{ 6n \log \frac{1}{h} + 36 \log \frac{1}{1 - \rho'} + 2 \log \frac{1}{1 - \rho} \right.$$

$$\left. + \log^+ \frac{1}{|G(0), X|} + 4 \log(1 + 2\pi) + \log 2 + 5220 \right\}.$$

Taking formulas (2.53) and (2.56) into account, it follows that

$$1 - \rho \geq 1 - \frac{3 + \tau}{4} = \frac{1 - \tau}{4} \geq \frac{1}{4} \cdot \frac{(1 - |z_0|)(1 - r)}{2} \geq \frac{1 - r}{16},$$

$$1 - s = \frac{(1 - \tau)(1 - \rho)}{1 + \tau\rho} \geq \frac{1}{2} \cdot \frac{1 - r}{16} \cdot (1 - \tau)$$

$$\geq \frac{1}{2} \cdot \frac{1 - r}{16} \cdot \frac{1 - r}{4} = \frac{(1 - r)^2}{8 \times 16}.$$

$$1 - \rho' \geq 1 - \frac{3 + s}{4} = \frac{1 - s}{4} \geq \frac{(1 - r)^2}{32 \times 16}.$$

And consequently we conclude further that

$$n(r, f = X)$$

$$\leq \frac{16^2}{(1 - r)^2} \left\{ 6n \log \frac{1}{h} + 36 \log \frac{32 \times 16}{(1 - r)^2} + 2 \log \frac{16}{1 - r} \right.$$

$$\left. + \log^+ \frac{1}{|f(z_2), X|} + 4 \log(1 + 2\pi) + \log 2 + 5220 \right\}$$

$$= \frac{16^2}{(1 - r)^2} \left\{ 6n \log \frac{1}{h} + 74 \log \frac{2}{1 - r} + \log^+ \frac{1}{|f(z_2), X|} \right. \tag{2.57}$$

$$\left. + 72 \log 16 + 2 \log 8 + 4 \log(1 + 2\pi) + \log 2 + 5220 \right\},$$

where z_2 is a corresponding point of ζ_2 on the z-plane.

Notice that when $0 \leq r < 1$, it is always true that $\log \frac{2}{1-r} > \log 2 > 0$. Hence we may enlarge the coefficient of item $\log \frac{2}{1-r}$ to eliminate the constant item in the inequality. Consequently it derives formula (2.49), from formulas (2.54), (2.55) and (2.57). Theorem 2.6 is thus proved completely.

By considering the transformation

$$F(z) = \frac{(f(z) - a)(c - b)}{(f(z) - b)(c - a)}, \tag{2.58}$$

we may derive a more general form:

THEOREM 2.7. *Suppose that $f(z)$ is a nonconstant meromorphic function in the disk $|z| < 1$, and a, b, and c are three distinct complex numbers, with their spherical distances among one another $\geq d$ $(0 < d \leq \frac{1}{2})$. Also assume that*

$$n(1, a) \leq n, \quad n(1, b) \leq n, \quad n(1, c) \leq n.$$

Then for any arbitrary complex value X and value r $(0 < r < 1)$ we have

$$n(r, X) \leq \frac{1}{(1 - r)^2} \left\{ An \log \frac{1}{h} + B \log \frac{2}{1 - r} + C \log \frac{1}{d} \right.$$
$$\left. + D \log^+ \frac{1}{|X, X(r)|} \right\},$$

where $0 < h \leq 0.01$ and A, B, C and D are constants. Also $X(r)$ is a complex value depending on $f(z)$ and value r.

In fact, among the three values a, b, and c, there exists at most one value, which may assume to be the value c, such that $|c, \infty| < \frac{d}{3}$. Otherwise, for both a and c we have $|a, \infty| < \frac{d}{3}$ and $|c, \infty| < \frac{d}{3}$, respectively. Hence

$$|a, c| = \frac{|a - c|}{\sqrt{1 + |a|^2}\sqrt{1 + |c|^2}} \leq \left| \frac{1}{a} - \frac{1}{c} \right| \leq \frac{1}{|a|} + \frac{1}{|c|}$$

$$= \frac{|a, \infty|}{\sqrt{1 - |a, \infty|^2}} + \frac{|c, \infty|}{\sqrt{1 - |c, \infty|^2}} \leq 2 \cdot \frac{d/3}{\sqrt{1 - d^2/9}} \leq \frac{4}{\sqrt{35}} d,$$

which contradicts the assumption that $|a, c| \geq d$. Therefore, we may assume that $|a, \infty| \geq \frac{d}{3}$ and $|b, \infty| \geq \frac{d}{3}$, and hence we have the estimation: $|a| < \frac{3}{d}$ and $|b| < \frac{3}{d}$. Furthermore we may assume $|\frac{c-b}{c-a}| \leq 1$, otherwise we need only to exchange the positions of a and b in formula (2.58). Now an application of Theorem 2.6 to $F(z)$ yields

$$n(r, f = X) = n(r, F = Y)$$
$$\leq \frac{1}{(1 - r)^2} \left\{ An \log \frac{1}{h} + B \log \frac{2}{1 - r} + C' \log^+ \frac{1}{|Y, Y(r)|} \right\},$$

where

$$Y = \frac{X - a}{X - b} \cdot \frac{c - b}{c - a}, \qquad Y(r) = \frac{X(r) - a}{X(r) - b} \cdot \frac{c - b}{c - a}.$$

In the following we need only to illustrate that

$$C' \log^+ \frac{1}{|Y, Y(r)|} \leq C \log \frac{1}{d} + D \log^+ \frac{1}{|X, X(r)|}.$$

Indeed we have

$$|Y, Y(r)| = \frac{\left|\frac{X-a}{X-b} - \frac{X(r)-a}{X(r)-b}\right| \left|\frac{c-b}{c-a}\right|}{\sqrt{1 + \left|\frac{c-b}{c-a}\right|^2 \left|\frac{X-a}{X-b}\right|^2} \sqrt{1 + \left|\frac{c-b}{c-a}\right|^2 \left|\frac{X(r)-a}{X(r)-b}\right|^2}}$$

$$\geq \frac{|X - X(r)| \, |a - b| \cdot \left|\frac{c-b}{c-a}\right|}{\sqrt{|X-b|^2 + |X-a|^2} \sqrt{|X(r)-b|^2 + |X(r)-a|^2}} \, .$$

Notice that

$$|X - b|^2 \leq |X|^2 + |b|^2 + 2|X| \, |b|$$
$$\leq 2 + |X|^2 + |b|^2 + 2|X|^2 |b|^2$$
$$\leq 2(1 + |X|^2)(1 + |b|^2)$$

$$|X - b|^2 + |X - a|^2 \leq 2(1 + |X|^2)\{1 + |a|^2 + 1 + |b|^2\}$$
$$\leq 4(1 + |X|^2)(1 + |a|^2)(1 + |b|^2),$$

and $|X(r) - b|^2 + |X(r) - a|^2 \leq 4(1 + |X(r)|^2)(1 + |a|^2)(1 + |b|^2)$. It follows that

$$|Y, Y(r)| \geq \tfrac{1}{4} \cdot |X, X(r)| \, |a, b| \cdot \frac{\left|\frac{c-b}{c-a}\right|}{\sqrt{1 + |a|^2} \sqrt{1 + |b|^2}} \, .$$

And hence

$$C' \log^+ \frac{1}{|Y, Y(r)|} \leq C' \left\{ \log^+ \frac{1}{|X, X(r)|} + \log^+ \frac{1}{d} + \log^+ \left|\frac{c-a}{c-b}\right| \right.$$
$$\left. + 2\log 2 + \log^+ \sqrt{(1 + |a|^2)(1 + |b|^2)} \right\} \, .$$

Further notice that

$$\log^+ \left|\frac{c-a}{c-b}\right| \leq \log^+ \frac{1}{|c-b|} + \log^+ |a| + \log^+ |b| + 2\log 2$$
$$\leq \log^+ \frac{1}{d} + 2\log^+ \frac{3}{d} + 2\log 2$$
$$\leq 3\log \frac{1}{d} + 2\log 6,$$

and

$$\log \sqrt{(1 + |a|^2)(1 + |b|^2)} \leq \log^+ |a| + \log^+ |b| + \log 2$$
$$\leq 2\log \frac{1}{d} + 2\log 6 \, .$$

Then we derive

$$C' \log \frac{1}{|Y, Y(r)|} \leq C' \log^+ \frac{1}{|X, X(r)|} + 6C' \log \frac{1}{d} + 6C' \log 6 \, .$$

Since $\log \frac{1}{d} \geq \log 2$, we need only to enlarge suitably the coefficient of item $\log \frac{1}{d}$ to eliminate the constant item and, in turn, conclude that

$$C' \log^+ \frac{1}{|Y, Y(r)|} \leq C \log \frac{1}{d} + D \log^+ \frac{1}{|X, X(r)|},$$

where C and D are constants. Hence Theorem 2.7 is proved.

2.3.3. Schottky-type theorem. First we utilize the method of proving the fundamental theorem to derive two theorems of bound concerning the characteristic functions. These two results will be applied in Chapters 3 and 5.

THEOREM 2.8 [43a]. *Let $f(z)$ be a function meromorphic in the disk $|z| < 1$, and*

$$n(1, f = 0) \leq n, \quad n(1, f = 1) \leq n, \quad n(1, f = \infty) \leq n.$$

We also denote (γ) the pseudo-non-Euclidean circles that correspond to these $n(1, f = 0) + n(1, f = 1) + n(1, f = \infty)$ points and number h $(0 < h \leq 0.01)$, and assume that origin $z = 0 \notin (\gamma)$ and $|f(0)| \leq 1$; then for an arbitrary value r $(0 \leq r < 1)$ we have

$$T(r, f) \leq \frac{1}{(1-r)^2} \left\{ An \log \frac{1}{h} + B \log \frac{2}{1-r} \right\}, \tag{2.59}$$

where A and B are two constants.

PROOF. Selecting arbitrarily a value r $(0 < r < 1)$, we may conclude, according to formula (2.42) and $0 < h \leq 0.01$, that there exists a value ρ, $r \leq \rho \leq \frac{1}{2}(1+r)$, such that circumference $\Gamma: |z| = \rho$ does not intersect (γ). Obviously we have

$$T(r, f) \leq T(\rho, f). \tag{2.60}$$

Suppose there exists a point z_1 on the circumference Γ satisfying $|f(z_1)| = \max_{|z|=\rho} |f(z)|$. We then use a straight line to connect point z_1 and the origin $z = 0$. If we come across circles (γ), then we use those relatively smaller arcs of (γ) to replace the lines included in circles (γ). Hence a curve L is derived, with its length $\leq 2 + 2\pi eh < 4$.

In the following we try to find an upper bound of $T(\rho, f)$. To this end, we distinguish two cases for discussion:

(1) It is always true that $|f'(z)| \leq 1$ on Γ. Then start from the origin $z = 0$ and integrate along L to point z_1, and we get

$$|f(z_1)| \leq |f(0)| + \left| \int_L f'(z) \, dz \right| \leq 5.$$

It follows that $m(\rho, f) \leq \log 5$. On the other hand, notice that point $z = 0 \notin (\gamma)$, we obtain $N(\rho, f) \leq N(1, f) \leq 3n \log \frac{1}{h}$. And hence

$$T(\rho, f) \leq 3n \log \frac{1}{h} + \log 5. \tag{2.61}$$

(2) There exists a point z_2 on L, such that $|f'(z_2)| \geq 1$, and that $|f'(z)| \leq 1$ on the path starting from the origin $z = 0$ and along L to point z_2. Therefore we can conclude that $|f(z_2)| \leq 5$. Considering the transformation $\zeta = (z - z_2)/(1 - \overline{z}_2 z)$, with its inverse transformation being $z = (\zeta + z_2)/(1 - \overline{z}_2 \zeta)$, then function

$$F(\zeta) = f\left(\frac{\zeta + z_2}{1 + \overline{z}_2 \zeta}\right)$$

is meromorphic on the disk $|\zeta| \leq 1$, and that

$$n(1, F = 0) \leq n, \quad n(1, F = 1) \leq n, \quad n(1, F = \infty) \leq n.$$

Meanwhile circles (γ) are transformed into pseudo-non-Euclidean exceptional circles $(\gamma)_\zeta$ on the ζ-plane that correspond to these $n(1, F = 0) + n(1, F = 1) + n(1, F = \infty)$ points and number h. Notice that point $\zeta = 0 \notin (\gamma)_\zeta$ and also

$$|F(0)| = |f(z_2)| \leq 5, \qquad |F'(0)| = |f'(z_2)|(1 - |z_2|^2) \geq 1 - \rho.$$

According to the inequality

$$|\zeta| \leq \frac{|z| + |z_2|}{1 + |z_2||z|},$$

we may conclude that the image domain of the disk $|z| < \rho$ on the ζ-plane is contained in the disk $|\zeta| \leq t$, and that

$$t = \frac{2\rho}{1 + \rho^2}. \tag{2.62}$$

Now, applying Theorem 2.4 we obtain

$$T\left(\frac{1 + t}{2}, F\right) \leq 2\left\{N\left(1, \frac{1}{F}\right) + N\left(1, \frac{1}{F - 1}\right) + N(1, F)\right\}$$
$$+ 4\log^+ |F(0)| + 2\log^+ \frac{1}{|F'(0)|} + 36\log\frac{2}{1 - t} + 5220 \tag{2.63}$$

$$\leq 6n\log\frac{1}{h} + 74\log\frac{1}{1 - \rho} + 4\log 5 + 5220.$$

In the following we denote, respectively, a_i $(i = 1, 2, \ldots, p; p = n(1, F = 0))$, b_j $(j = 1, 2, \ldots, q; q = n(1, F = 1))$ and c_k $(k = 1, 2, \ldots, l; l = n(1, F = \infty))$ the zero points, the 1-value points and the poles of $F(\zeta)$ in the disk $|\zeta| < 1$. Let

$$I(\zeta) = \prod_{i=1}^{p} \frac{\zeta - a_i}{1 - \overline{a}_i \zeta}, \qquad J(\zeta) = \prod_{j=1}^{q} \frac{\zeta - b_j}{1 - \overline{b}_j \zeta},$$

$$K(\zeta) = \prod_{k=1}^{l} \frac{\zeta - c_k}{1 - \overline{c}_k \zeta}.$$

Then function $\Phi(\zeta) = K(\zeta)F(\zeta)$ is regular in the disk $|\zeta| < 1$, and that

$$T\left(\frac{1+t}{2}, \Phi\right) \leq T\left(\frac{1+t}{2}, F\right).$$

Hence, it follows from formulas (1.21) and (2.63) that

$$\log M(t, \Phi) \leq \frac{\frac{1+t}{2}+t}{\frac{1+t}{2}-t}T\left(\frac{1+t}{2}, \Phi\right) \leq \frac{4}{1-t}T\left(\frac{1+t}{2}, F\right)$$

$$\leq \frac{4}{1-t}\left\{6n\log\frac{1}{h} + 74\log\frac{2}{1-\rho} + 4\log 5 + 5220\right\}.$$

Further based on formulas (2.62) and $\rho \leq \frac{1}{2}(1+r)$, we derive

$$\log M(t, \Phi) \leq \frac{32}{(1-r)^2}\left\{6n\log\frac{1}{h} + 74\log\frac{2}{1-r}\right.$$

$$\left. + 74\log 2 + 4\log 5 + 5220\right\}. \tag{2.64}$$

Suppose ζ_1 is a corresponding point of z_1 on the ζ-plane, such that $\zeta_1 \notin (\gamma)_\zeta$. Then we find that

$$\log\frac{1}{|k(\zeta_1)|} \leq \log\frac{1}{|I(\zeta_1)J(\zeta_1)k(\zeta_1)|} \leq 3n\log\frac{1}{h}.$$

It follows further from formula (2.64) that

$$\log|f(z_1)| = \log|F(\zeta_1)| = \log|\Phi(\zeta_1)| + \log\frac{1}{|K(\zeta_1)|}$$

$$\leq \log M(t, \Phi) + 3n\log\frac{1}{h}$$

$$\leq \frac{32}{(1-r)^2}\left\{6n\log\frac{1}{h} + 74\log\frac{2}{1-r} + 74\log 2\right.$$

$$\left. + 4\log 5 + 5220\right\} + 3n\log\frac{1}{h}.$$

Notice that $m(\rho, f) \leq \log^+|f(z_1)|$, and $N(\rho, f) \leq 3n\log\frac{1}{h}$; then we conclude that

$$T(\rho, f) \leq \frac{1}{(1-r)^2}\left\{An\log\frac{1}{h} + B\log\frac{2}{1-r}\right\}, \tag{2.65}$$

where A and B are two constants.

Finally, from formulas (2.60), (2.61) and (2.65) we obtain formula (2.59). Theorem 2.8 is thus proved completely.

THEOREM 2.9 [42a]. *Suppose $f(z)$ is a function meromorphic in the disk $|z| < 1$ and that the distances between origin $z = 0$ and zero-points, 1-value points and poles $f(z)$ are $\geq d$ $(0 < d < \frac{1}{2})$. Let*

$$n = n(1, f = 0) + n(1, f = 1) + n(1, f = \infty).$$

Then for any arbitrary value r, $0 < r < 1$ *we have*

$$T(r, f) < \frac{c(n+1)}{1-r} \left\{ \log \frac{n+1}{d} + \log \frac{2}{1-r} \right\} + \log^+ |f(0)|,$$

where c *is a constant.*

PROOF. We construct the exceptional circles, with each value-point of $f(z) = 0, 1, \infty$ in the disk $|z| < 1$ being the center, and $\frac{d(1-r)}{12(n+1)}$ the radius. We then denote all these exceptional circles by (γ). Obviously, the origin $z = 0 \notin (\gamma)$, and there exists a value ρ, $r \leq \rho \leq \frac{1}{2}$, such that circumference $\Gamma: |z| = \rho$ has no intersection with circles (γ). Also $T(r, f) \leq T(\rho, f)$.

Let point z_1 be a point on the circumference Γ, such that $|f(z_1)| = \text{Max}_{|z|=\rho} |f(z)|$, and that a straight line is used to connect point z_1 and the origin $z = 0$. If it comes across circles (γ), we replace it by arcs of (γ). Hence, a curve L is derived, with its length ≤ 2.

In the following we try to find an upper bound of $T(\rho, f)$. To this end, we distinguish two cases for discussion:

(1) It is always true that $|f'(z)| \leq 1$ on L. Then start from the origin $z = 0$ and integrate along L to point z_1 and get

$$|f(z_1)| \leq |f(0)| + \left| \int_L f'(z)\,dz \right| \leq |f(0)| + 2.$$

Hence $m(\rho, f) \leq \log^+ |f(0)| + 2\log 2$. On the other hand, notice that the distances between origin $z = 0$ and the value-points of $f(z) = 0, 1, \infty$ are $\geq d$, and it follows that $N(\rho, f) \leq n \log \frac{1}{d}$. And hence

$$T(r, f) \leq T(\rho, f) \leq n \log \frac{1}{d} + 2\log \frac{2}{1-r} + \log^+ |f(0)|,$$

meaning that for case (1) Theorem 2.9 holds.

(2) There exists a point z_2 on L, such that $|f'(z_2)| > 1$, and $|f'(z)| \leq 1$ on the path starting from the origin $z = 0$ along L to point z_2.

We construct a circle Γ', with z_2 being the center and $R = \frac{3+r}{4} - |z_2|$ the radius. Notice that

$$|z_1 - z_2| < (\rho - |z|) + \frac{d(1-r)}{6(n+1)}.$$

We let $m(z_2, R, f)$ be the logarithmic mean-value of $f(z)$ on the disk $|z - z_2| = R$, and b_ν be the poles of $f(z)$ in the disk $|z - z_2| < R$; then according to the Poisson-Jensen Formula we have

$$m(\rho, f) \leq \log^+ |f(z_1)|$$

$$\leq \frac{2}{R - (\rho - |z_2| + \frac{d(1-r)}{6(n+1)})} m(z_2, R, f)$$

$$+ \sum_\nu \log \left| \frac{R^2 - \overline{(b_\nu - z_2)}(z_1 - z_2)}{R(z_1 - b_\nu)} \right|$$

$$< \frac{12}{1-r} m(z_2, R, f) + n(z_2, R, f = \infty) \log \frac{24(n+1)}{d(1-r)}.$$

We then apply Theorem 2.4 to estimate $m(z_2, R, f)$. Let $R = 1 - |z_2|$ in Theorem 2.4 and notice that $|f'(z_2)| > 1$ and that the distances between point z_2 and the value-points of $f(z) = 0, 1, \infty \geq \frac{d(1-r)}{12(n+1)}$, and $|f(z_2)| \leq |f(0)| + 2$. Then we get

$$m(z_2, R, f) \leq c(n+1) \left\{ \log \frac{n+1}{d} + \log \frac{2}{1-r} + \log^+ |f(0)| \right\},$$

where c is a suitably large constant. Hence

$$m(\rho, f) \leq \frac{c(n+1)}{1-r} \left\{ \log \frac{n+1}{d} + \log \frac{2}{1-r} + \log^+ |f(0)| \right\}.$$

Further notice that $N(\rho, f) \leq n \log \frac{1}{d}$. It follows that

$$T(r, f) \leq T(\rho, f)$$
$$\leq \frac{c(n+1)}{1-r} \left\{ \log \frac{n+1}{d} + \log \frac{1}{1-r} + \log^+ |f(0)| \right\},$$

where c is a constant. When $|f(0)| < 1$, it is apparent that Theorem 2.9 holds. When $|f(0)| > 1$, we replace $f(z)$ by $1/f(z)$ and note that $T(r, f) = T(r, \frac{1}{f}) + \log |f(0)|$; then we conclude that Theorem 2.9 is valid.

Finally, in order to prove the existence of the Julia direction, we prove the following Schottky-type theorem.

THEOREM 2.10 [39d]. *Let $f(z)$ be a nonconstant function meromorphic in the disk $|z| < 1$, and a, b and c are three distinct complex numbers, with their spherical distances among one another $\geq d$, $0 < d \leq \frac{1}{2}$ and also*

$$n(1, a) \leq n, \quad n(1, b) \leq n, \quad n(1, c) \leq n.$$

Further assume that the set of points satisfying the inequality $|f(z)| \leq M < +\infty$ in the disk $|z| < \tau < 1$ cannot be included in $3n$ disks, with the sum of their radii not exceeding $2eh$ $(0 < h < 0.01, 8eh < \tau)$. Then for the point z in the disk $|z| \leq r$ $(\tau < r < 1)$ we have

$$\log |f(z)| \leq \frac{1}{(1-r)^2} \left\{ An \log \frac{1}{h} + B \log \frac{2}{1-r} + C \log^+ M + D \log \frac{1}{d} \right\}$$
$$\times \left\{ 1 + \log \frac{2}{H} \right\},$$

excluding at most some exceptional disks, with the sum of their radii not exceeding $2eH < 1$ $(h < H)$, and where A, B, C and D are positive constants.

PROOF. Assume that (γ) are the pseudo-non-Euclidean exceptional circles that correspond to these $n(1, f = a) + n(1, f = b) + n(1, f = c)$ points and number h, and (γ) contain at most $3n$ circles. Also notice that a pseudo-non-Euclidean circle is itself an Euclidean circle, and that the Euclidean radius of this circle does not exceed the pseudo-non-Euclidean radius. Then it follows that the sum of the Euclidean radii of circles (γ) does not exceed $2eh$. Hence, by assumption, there exists point z_0 in the disk $|z| \leq \tau$ but outside circles (γ), such that $|f(z_0)| \leq M$. In the following, we distinguish two cases for discussion:

(1) It is always true that $|f'(z)| \leq 1$ in the disk $|z| \leq r$ but outside (γ). We use a straight line to connect point z_0 and an arbitrary point z in the disk $|z| \leq r$ but outside circles (γ). If we come across circles (γ), we use those relatively smaller arcs in (γ) to replace the straight line in the circles (γ). Consequently a curve L is derived, with its length ≤ 4. Hence for point z in the disk $|z| \leq r$ but outside circles (γ) we have

$$|f(z)| \leq |f(z_0)| + \left| \int_L f'(z)\,dz \right| \leq f(z_0)| + 4,$$

$$\log|f(z)| \leq \log^+|f(z_0)| + 3\log 2.$$

Since $H > h$, thus, for case (1), Theorem 2.10 holds.

(2) There exists a point z_1 in the disk $|z| \leq r$ but outside circles (γ), such that $|f'(z_1)| \geq 1$, and that it is true that $|f'(z)| \leq 1$ on the above founded curve L which connects between point z_0 and point z_1. Hence we conclude that $|f(z_1)| \leq M+4$. Considering the transformation $\zeta = (z-z_1)/(1-\overline{z}_1 z)$, with its inverse transformation being $z = (\zeta + z_1)/(1 + \overline{z}_1\zeta)$, then function

$$\varphi(\zeta) = f\left(\frac{\zeta + z_1}{1 + \overline{z}_1\zeta} \right)$$

is meromorphic in the disk $|z| < 1$, and that

$$n(1, \varphi = a) \leq n, \quad n(1, \varphi = b) \leq n, \quad n(1, \varphi = c) \leq n.$$

Meanwhile, circles (γ) are transformed into pseudo-non-Euclidean exceptional circles $(\gamma)_\zeta$ in the ζ-plane that correspond to these $n(1, \varphi = a) + n(1, \varphi = b) + n(1, \varphi = c)$ points and number h. Notice that point $\zeta = 0 \notin (\gamma)$, and also

$$|\varphi(0)| = |f(z_1)| \leq M + 4, \qquad |\varphi'(0)| = |f'(z_1)|(1 - |z_1|^2) \geq 1 - r. \quad (2.66)$$

According to the inequality

$$|\zeta| \leq \frac{|z| + |z_1|}{1 + |z_1||z|},$$

we may conclude that the image domain of disk $|z| \leq r$ in the ζ-plane is contained in the disk $|\zeta| \leq t$, and

$$t = \frac{2r}{1 + r^2}. \quad (2.67)$$

Based on the fact that the spherical distances $\geq d$ among the three values a, b and c, we may assume that a and b are finite, and that $|a| < \frac{3}{d}$ and

$|b| < \frac{3}{d}$.(^{2}) Considering the transformation

$$\Phi(\zeta) = \frac{\varphi(\zeta) - a}{\varphi(\zeta) - b} \cdot \frac{c - b}{c - a},$$

then we have

$$n(1, \Phi = 0) \leq n, \quad n(1, \Phi = 1) \leq n, \quad n(1, \Phi = \infty) \leq n.$$

We may further assume that $|\Phi(0)| \leq 1$, otherwise we need only to exchange the positions of a and b. Besides, we have according to formula (2.64),

$$|\Phi'(0)| = \left| \frac{c - b}{c - a} \right| |a - b| \frac{|\varphi'(0)|}{|\varphi(0) - b|^2} \geq \left| \frac{c - b}{c - a} \right| |a - b| \frac{1 - r}{(M + 4 + |b|)^2}.$$

Applying Theorem 2.4 to $\Phi(\zeta)$, then for any arbitrary value ρ, $0 < \rho < 1$, we get

$$T(\rho, \Phi) \leq 6n \log \frac{1}{h} + 2 \log^+ \frac{(M + 4 + |b|)^2 |c - a|}{(1 - r)|a - b||c - b|}$$

$$+ 36 \log \frac{1}{1 - \rho} + 5220$$

$$\leq 6n \log \frac{1}{h} + 2 \log \frac{1}{1 - r} + 2 \log^+ \frac{1}{|a - b|} + 2 \log^+ \left| \frac{c - a}{c - b} \right|$$

$$+ 4 \log^+ M + 4 \log^+ |b| + 16 \log 2 + 36 \log \frac{1}{1 - \rho} + 5220.$$

On the other hand, from formulas (1.19), (1.20) and the equality

$$\frac{1}{\varphi(\zeta) - b} = \frac{1}{b - a} \left\{ \frac{c - a}{c - b} \Phi(\zeta) - 1 \right\},$$

we derive

$$T\left(\rho, \frac{1}{\varphi - b} \right) \leq T(\rho, \Phi) + \log^+ \frac{1}{|b - a|} + \log^+ \left| \frac{c - a}{c - b} \right| + \log 2,$$

$$T(\rho, \varphi - b) \leq T(\rho, \Phi) + \log^+ \frac{1}{|b - a|} + \log^+ \left| \frac{c - a}{c - b} \right|$$

$$+ \log^+ M + 2 \log^+ |b| + 6 \log 2$$

$$\leq 6n \log \frac{1}{h} + 2 \log \frac{1}{1 - r} + 3 \log^+ \left| \frac{c - a}{c - b} \right|$$

$$+ 3 \log^+ \frac{1}{|b - a|} + 6 \log^+ |b| + 5 \log^+ M$$

$$+ 22 \log 2 + 36 \log \frac{1}{1 - \rho} + 5220.$$

Further taking into account the conditions

$$|a| \leq \frac{3}{d}, \quad |b| \leq \frac{3}{d}, \quad |a - b| \geq |a, b| \geq d, \quad |c - b| \geq |c, b| \geq d,$$

(^{2}) See the proof of Theorem 2.7 for the corresponding illustration.

and

$$\log^+ \left| \frac{c-a}{c-b} \right| = \log^+ \left| 1 + \frac{b-a}{c-b} \right| \le \log^+ |a| + \log^+ |b| + \log^+ \frac{1}{|c-b|} + 2\log 2,$$

we get

$$T(\rho,\varphi) \le 6n\log\frac{1}{h} + 2\log\frac{1}{1-r} + 36\log\frac{1}{1-\rho} + 5\log M \tag{2.68}$$
$$+ 18\log\frac{1}{d} + 52\log 2 + 5220.$$

where $0 < \rho < 1$.

Suppose C_k $(k = 1, 2, \ldots, l;\ l = n(\frac{1+t}{2}, \varphi = \infty))$ are the poles of $\varphi(\zeta)$ in the disk $|\zeta| < \frac{1+t}{2}$, $(\gamma)'_\zeta$ are the pseudo-non-Euclidean exceptional circles that correspond to these $n(\frac{1+t}{2}, \varphi = \infty)$ points and number H. Let

$$K(\zeta) = \prod_{k=1}^{l} \frac{\zeta - c_k}{1 - \bar{c}_k \zeta}.$$

Then function $F(\zeta) = \varphi(\zeta)K(\zeta)$ is regular in the disk $|\zeta| < \frac{1+t}{2}$, and

$$T\left(\frac{1+t}{2}, F\right) \le T\left(\frac{1+t}{2}, \varphi\right),$$
$$\log M(t, F) \le \frac{\frac{1+t}{2} + t}{\frac{1+t}{2} - t} T\left(\frac{1+t}{2}, F\right) \le \frac{4}{1-t} T\left(\frac{1+t}{2}, \varphi\right).$$

Also for point ζ in the disk $|\zeta| \le t$ but outside $(\gamma)_\zeta$ we have

$$\log|\varphi(\zeta)| = \log|F(\zeta)| + \log\frac{1}{|K(\zeta)|}$$
$$\le \log M(t, F) + n\left(\frac{1+t}{2}, \varphi = \infty\right)\log\frac{2}{H}$$
$$\le \frac{4}{1-t} T\left(\frac{1+t}{2}, \varphi\right)$$
$$+ \log\frac{2}{H} \cdot \frac{\frac{1}{4}(3+t)}{\frac{1}{4}(3+t) - \frac{1}{2}(1+t)} \int_{(1+t)/2}^{(3+t)/4} \frac{n(t, \varphi = \infty)}{t} dt$$
$$\le \frac{4}{1-t} T\left(\frac{1+t}{2}, \varphi\right) + \log\frac{2}{H}\frac{4}{1-t} T\left(\frac{3+t}{4}, \varphi\right)$$
$$\le \frac{4}{1-t} T\left(\frac{3+t}{4}, \varphi\right)\left(\log\frac{2}{H} + 1\right).$$

Furthermore, by applying formula (2.68), where we let $\rho = \frac{3+t}{4}$, and based on formula (2.67) we get

$$\log|\varphi(\zeta)| \leq \frac{16}{(1-r)^2}\left\{6n\log\frac{1}{h} + 2\log\frac{1}{1-r} + 36\log\frac{8}{(1-r)^2}\right.$$

$$\left. + 5\log^+ M + 18\log\frac{1}{d} + 52\log 2 + 5220\right\}\left\{1 + \log\frac{2}{H}\right\}$$

$$\leq \frac{1}{(1-r)^2}\left\{An\log\frac{1}{h} + B\log\frac{2}{1-r}\right. \tag{2.69}$$

$$\left. + c\log^+ M + D\log\frac{1}{d}\right\}\left\{1 + \log\frac{2}{H}\right\},$$

where A, B, C and D are constants.

Suppose $(\gamma)'$ is the pseudo-non-Euclidean exceptional circles corresponding to $(\gamma)'_\zeta$ on the z-plane; then the sum of the pseudo-non-Euclidean radii of $(\gamma)'$ does not exceed $2eH$. According to formula (2.69), when point z lies in the disk $|z| \leq r$ but outside circles $(\gamma)'$, it follows that

$$\log|f(z)| = \log|\varphi(\zeta)|$$

$$\leq \frac{1}{(1-r)^2}\left\{An\log\frac{1}{h} + B\log\frac{2}{1-r} + C\log^+ M + D\log\frac{1}{d}\right\}$$

$$\cdot\left\{1 + \log\frac{2}{H}\right\}.$$

Thus Theorem 2.10 is proved completely.

§2.4. The Julia and Borel directions

In this section, we shall prove two main results, namely the existences of the Julia and Borel directions. However, we shall first introduce some notations to facilitate descriptions for the sequels: We denote, on the open plane $|z| < +\infty$, $\Delta(\theta)$ a half straight line: $\arg z = \theta$, $0 < |z| + \infty$ starting from the origin; $\Delta(\theta; R_1, R_2)$ $(0 \leq R_1 < R_2 \leq +\infty)$ the area where this half straight line lies between point $R_1 e^{i\theta}$ and $R_2 e^{i\theta}$; $\Omega(\theta_1, \theta_2)$ $(\theta_1 < \theta_2)$ the set: $\theta_1 < \arg z < \theta_2$; $\Omega(\theta_1, \theta_2; R)$ $(\theta_1 < \theta_2; 0 < R < +\infty)$ the set: $\theta_1 < \arg z < \theta_2$, $|z| < R$; $\Omega(\theta_1, \theta_2; R_1, R_2)$ $(\theta_1 < \theta_2; 0 < R_1 < R_2 \leq +\infty)$ the set: $\theta_1 < \arg z < \theta_2$, $R_1| < |z| < R_2$, and $\Gamma(\theta_1, \theta_2; R)$ $(\theta_1 < \theta_2; 0 < R < +\infty)$ the set: $\theta_1 < \arg z < \theta_2$, $|z| = R$. Generally, we let $\overline{E}$ denotes the closure of the set E with respect to the open plane $|z| < +\infty$.

We also denote $n(E, f = a) = n(E, a)$ the number of roots of equation $f(z) = a$ in the set E (multiple roots being counted with their multiplicities), $M(\overline{E}, f)$ the $\max_{z\in\overline{E}}|f(z)|$, and $[x]$ the integral part of x.

2.4.1. The filling circle. Let $f(z)$ be a nonconstant meromorphic function in domain D. A circle $\Gamma: |z - z_0| < \delta$ in this domain is called a filling circle of $f(z)$ with an index $m(\geq 1)$, if for each complex value X, we have

$n(\Gamma, X) \geq m$, excluding at most some values X, which can be included in two spheres with their radii being e^{-m}.

The concept of filling circle was introduced by A. Ostrowski [33a] and H. Milloux [30a].

LEMMA 2.7. *Suppose that* $f(z)$ *is a nonconstant meromorphic function in the disk* $|z| < 2R$, $m \geq 1$ *is an integer and* $0 < \varepsilon < \frac{1}{2}$. *If point* z_0 *does not exist in annulus* $K: r < |z| < R$, *such that circle:* $|z - z_0| < \varepsilon|z_0|$ *is a filling circle of* $f(z)$ *with an index* m, *then for each complex value* X *we have that*

$$n(\overline{K}, f = X) \leq 4 \left\{ \frac{10\pi}{\varepsilon} \cdot \frac{\log \frac{R}{r}}{\log(1 + \frac{\varepsilon}{5})} + \frac{10\pi}{\varepsilon} \right\}$$

$$\cdot \left\{ An \log 100 + B \log 4 + Cm + D \log \frac{1}{H} \right\},$$

(2.70)

but probably excluding some values X, *which may be including in a group of spheres* (γ), *with the sum of their radii not exceeding* $2eH$ ($2eH < 1$), *where* A, B, C *and* D *are the constants as defined in Theorem 2.7.*

PROOF. Let $P = [\frac{8\pi}{\varepsilon}] + 1$ and $\theta_i = \frac{2\pi}{P} i$ $(i = 0, 1, \ldots, P - 1)$. On the other hand we assume that integer Q satisfies the inequality

$$r\left(1 + \frac{2\pi}{P}\right)^{Q-1} < R \leq r\left(1 + \frac{2\pi}{P}\right)^{Q}.$$

And hence

$$Q \leq \frac{\log \frac{R}{r}}{\log(1 + \frac{2\pi}{P})} + 1 \leq \frac{\log \frac{R}{r}}{\log(1 + \frac{\varepsilon}{5})} + 1.$$

(2.71)

Now we consider a group of circumferences $\Gamma_j: |z| = r(1 + \frac{2\pi}{P})^j$ $(j = 0, 1, 2, \ldots, Q - 1)$ and circumference $\Gamma_Q: |z| = R$. Hence, these P half straight lines $\Delta(\theta_i)$ $(i = 0, 1, \ldots, P - 1)$ and $Q + 1$ circumferences Γ_j $(j = 0, 1, \ldots, Q)$ divide the circular ring K into PQ sectors:

$$\Omega\left(\theta_i, \theta_{i+1}; r\left(1 + \frac{2\pi}{P}\right)^j, r\left(1 + \frac{2\pi}{P}\right)^{j+1}\right)$$

$$(i = 0, 1, \ldots, P - 1; j = 0, 1, \ldots, Q - 1).$$

We then construct circles $c_{ij}: |z - z_{ij}| < \frac{\varepsilon}{2}|z_{ij}|$, with point

$$z_{ij} = r\left(1 + \frac{2\pi}{P}\right)^j \left(1 + \frac{\pi}{P}\right) e^{i(\theta_i + \theta_{i+1})/2}$$

being the centers and $\frac{\varepsilon}{2}|z_{ij}|$ the radii; then the whole of

$$\Omega\left(\theta_i, \theta_{i+1}; r\left(1 + \frac{2\pi}{P}\right), r\left(1 + \frac{2\pi}{P}\right)^{j+1}\right)$$

lies entirely in the circle C_{ij}. We use C'_{ij} to denote the concentric circle: $|z - z_{ij}| < \varepsilon|z_{ij}|$ of C_{ij}. Under the assumption that all C'_{ij} $(i = 0, 1, \ldots, P-$

1; $j = 0, 1, \ldots Q - 1$) are not the filling circles of $f(z)$ with an index m, then there exist three complex values a_{ij}, b_{ij} and c_{ij} that correspond to each C'_{ij}, with their spherical distances among one another $\geq e^{-m}$, and also

$$n\{C'_{ij}, f = X\} < m, \qquad X_{ij} = a_{ij}, b_{ij}, c_{ij}.$$

In the following, we consider the transformation

$$\zeta = \frac{z - z_{ij}}{\varepsilon |z_{ij}|}$$

and let $\varphi(\zeta) = f(z_{ij} + \varepsilon |z_{ij}| \zeta)$. An application of Theorem 2.7 to $\varphi(\zeta)$, where we let $h = 0.01$, may yield, for each complex value X, the following expressions:

$$n(c_{ij}, f = x) = n\left(\frac{1}{2}, \varphi = x\right)$$

$$\leq 4\left\{ An \log 100 + B \log 4 + cm + D \log \frac{1}{|X, X_{ij}(\frac{1}{2})|} \right\}$$

$$n(K, f = X)$$

$$\leq \sum_{ij} n(c_{ij}, f = X)$$

$$\leq 4P \cdot Q\{An \log 100 + B \log 4 + cm\} + \sum_{ij} 4D \log \frac{1}{|X, X_{ij}(\frac{1}{2})|}$$

$$= 4P \cdot Q\{An \log 100 + B \log 4 + cm\} + 4D \log \frac{1}{\prod_{ij} |X, X_{ij}(\frac{1}{2})|}.$$

A further application of Theorem 2.2 results in a point set that satisfies the inequality

$$\prod_{ij} \left| X, X_{ij}\left(\frac{1}{2}\right) \right| < H^{p \cdot Q}$$

and that the point set may be included in at most PQ spheres (γ), with the sum of their radii not exceeding $2eH$ $(2eH < 1)$. Hence for each complex value X not belonging to spheres (γ) it follows that

$$n(K, f = X) \leq 4P \cdot Q\left\{ Am \log 100 + B \log 4 + Cm + D \log \frac{1}{H} \right\}.$$

Finally, according to $P = [\frac{8\pi}{\varepsilon}] + 1 \leq \frac{10\pi}{\varepsilon}$ and formula (2.71) we may derive formula (2.70). Lemma 2.7 is thus proved.

LEMMA 2.8. *Let $f(z)$ be a function meromorphic on the open plane $|z| < +\infty$, $f(0) \neq \infty$. Suppose that there exists a sequence $r_n : r_n < r_{n+1} \to +\infty$ $(n \to +\infty)$, such that*

$$\lim_{n \to +\infty} \frac{T(r_n, f)}{(\log r_n)^2} = +\infty.$$

Then for any arbitrarily selected value r, $r \geq 1$, and value ε, $0 < \varepsilon \leq \frac{1}{2}$, provided that r_n is suitably large, point z_n must exist in annulus: K_n: $r < |z| < 2r_n$, such that circle Γ_n: $|z - z_n| < \varepsilon|z_n|$ is a filling circle of $f(z)$, with its index being

$$m_n \geq C \cdot \frac{\varepsilon^2 T(r_n, f)}{(\log r_n)^2} \geq 1, \tag{2.72}$$

where $C > 0$ is a suitably small constant.

PROOF. Indeed, suppose that Lemma 2.8 is not valid, then for a sufficiently large value r_n, there exists no point z_n in annulus K_n, such that circle Γ_n is a filling circle of $f(z)$ with an index m_n. Therefore, for each complex value X, we have according to Lemma 2.7,

$$n(\overline{K}_n, f = X) \leq 4 \left\{ \frac{10\pi}{\varepsilon} \frac{\log \frac{2r_n}{r_n}}{\log(1 + \frac{\varepsilon}{5})} + \frac{10\pi}{\varepsilon} \right\}$$
$$\cdot \left\{ Am_n \log 100 + B \log 4 + Cm_n + D \log \frac{1}{H} \right\},$$

but probably excluding some values X, which may be included in a group of spheres (γ), with the sum of their spherical radii not exceeding $2eH$. We assume $H = \frac{1}{8(e+2)}$, according to $m_n \geq 1$, and $\log(1 + \frac{\varepsilon}{5}) \geq \frac{\varepsilon}{10}$, we further conclude that

$$n(\overline{K}_n, f = X) \leq \frac{A \log r_n}{\varepsilon^2} m_n, \tag{2.73}$$

where $A > 0$ is a suitably large constant.

We construct spheres D_1: $|X, \infty| < H$ and D_2: $|X, f(0)| < H$. Based on the condition that $2eH + 4H \leq \frac{1}{4}$, we arrive at a conclusion that there exist three points a_n, b_n and c_n outside circles D_1, D_2 and (γ), such that the spherical distances among $f(0)$, a_n, b_n, and c_n are $\geq H$, and also none of $|a_n|$, $|b_n|$ and $|c_n|$ exceeds $\frac{1}{H}$. Hence formula (2.73) yields particularly

$$n(\overline{K}_n, f = X) < \frac{A \log r_n}{\varepsilon^2} \cdot m_n, \qquad X = a_n, b_n, c_n.$$

It follows when $X = a_n$, b_n and c_n we have further that

$$n(2r_n, f = X) \leq n(r, f = X) + n(\overline{K}_n, f = X)$$
$$\leq \frac{1}{\log 2} \left\{ T(2r_n, f) + \log^+ \frac{1}{|f(0), X|} + \log 2 \right\}$$
$$+ \frac{A \log r_n}{\varepsilon^2} \cdot m_n.$$

Since r is a given value and $|f(0), X| \geq H = \frac{1}{8(e+2)}$, thus provided that r_n is suitably large, we may conclude that

$$n(2r_n, f = X) \leq \frac{A \log r_n}{\varepsilon^2} m_n, \qquad X = a_n, b_n, c_n, \tag{2.74}$$

where $A > 0$ is a suitably large constant.

Let

$$F(z) = \frac{f(z) - a_n}{f(z) - b_n} \cdot \frac{c_n - b_n}{c_n - a_n}.$$

Then

$$n(2r_n, F = 0) = n(2r_n, f = a_n),$$
$$n(2r_n, F = 1) = n(2r_n, f = c_n),$$
$$n(2r_n, F = \infty) = n(2r_n, f = b_n).$$

We may also assume that $|F(0)| \leq 1$, otherwise we need only to exchange the positions of a_n and b_n. Besides, according to $f(0) \neq a_n, b_n, c_n$, we conclude that $F(0) \neq 0, 1, \infty$. Now suppose that at a neighborhood around the origin $z = 0$, $f(z)$ has the following expansion:

$$f(z) = f(0) + c_s z^s + \cdots, \qquad f(0) \neq \infty, \ c_s \neq 0, \ s \geq 1.$$

Then the corresponding expansion of $F(z)$ is

$$F(z) = F(0) + \frac{(c_n - b_n)(a_n - b_n)}{(c_n - a_n)(f(0) - b_n)^2} c_s z^s + \cdots.$$

An application of Theorem 2.5, therefore, yields

$$T(r_n, F) \leq 2\left\{ N\left(2r_n, \frac{1}{F}\right) + N\left(2r_n, \frac{1}{F-1}\right) + N(2r_n, F)\right\}$$
$$+ 4\log^+ |F(0)| + 2\log^+ \frac{|c_n - b_n||f(0) - b_n|^2}{2r_n s |c_s| |c_n - b_n| |a_n - b_n|}$$
$$+ 36\log 2 + 5220 \qquad (2.75)$$
$$\leq 2\left\{ N\left(2r_n, \frac{1}{f - a_n}\right) + N\left(2r_n, \frac{1}{f - b_n}\right) + N\left(2r_n, \frac{1}{f - c_n}\right)\right\}$$
$$+ 2\log^+ \frac{1}{|c_s|} + 12\log \frac{1}{H} + 4\log^+ |f(0)|$$
$$+ 48\log 2 + 5220.$$

On the other hand, based on the equality

$$\frac{1}{f(z) - b_n} = \frac{1}{b_n - a_n}\left\{\frac{c_n - a_n}{c_n - b_n} F(z) - 1\right\},$$

we derive

$$T\left(r_n, \frac{1}{f - b_n}\right) \leq T(r_n, F) + 4\log \frac{1}{H} + 2\log 2,$$

$$T(r_n, f - b_n) \leq T(r_n, F) + 4\log \frac{1}{H} + 2\log 2 + \log|f(0) - b_n|,$$

$$T(r_n, f) \leq T(r_n, F) + 6\log \frac{1}{H} + \log^+ |f(0)| + 4\log 2.$$

And further from formula (2.75) we deduce

$$T(r_n, f) \le 2\left\{N\left(2r_n, \frac{1}{f - a_n}\right) + N\left(2r_n, \frac{1}{f - b_n}\right) + N\left(2r_n, \frac{1}{f - c_n}\right)\right\}$$
$$+ 2\log\frac{1}{|C_s|} + 18\log\frac{1}{H}$$
$$+ 5\log^+|f(0)| + 52\log 2 + 5220.$$

Notice that when $X = a_n$, b_n and c_n, we have

$$N\left(2r_n, \frac{1}{f - X}\right) \le N\left(r, \frac{1}{f - X}\right) + n(2r_n f = X)\log\frac{2r_n}{r}$$
$$\le T(r, f) + \log\frac{1}{H} + n(2r_n, f = X)\log(2r_n) + \log 2.$$

And hence

$$T(r_n, f) \le 2\{n(2r_n, f = a_n) + n(2r_n, f = b_n) + n(2r_n, f = c_n)\}\log 2r_n$$
$$+ 6T(r, f) + 2\log\frac{1}{|c_s|}$$
$$+ 24\log\frac{1}{H} + 5\log^+|f(0)| + 58\log 2 + 5220.$$

Further based on formula (2.74), when r_n is suitably large, we conclude that

$$T(r_n, f) \le \frac{A(\log r_n)^2}{\varepsilon^2}m_n,$$

where $A > 0$ is a suitably large constant. Therefore, by selecting a suitably small constant C in formula (2.72), such that $A \cdot C < 1$, we will arrive at a contradiction, implying that Lemma 2.8 holds.

2.4.2. The Borel direction. Suppose that $f(z)$ is a meromorphic function of order λ $(\lambda \ge 0)$ on the open plane $|z| < +\infty$. If for any arbitrarily small number $\varepsilon > 0$ and complex value X we have

$$\varlimsup_{r \to +\infty} \frac{\log^+ n\{\Omega(\theta - \varepsilon, \theta + \varepsilon; r), f = X\}}{\log r} = \lambda,$$

but probably excluding at most two values X, then we call $\Delta(\theta)$ a Borel direction of $f(z)$ of order λ.

G. Valiron first proved the existence of the Borel direction in 1928 [39b].

THEOREM 2.11. *Let $f(z)$ be a meromorphic function of order λ, $0 < \lambda \le +\infty$ on the open plane $|z| < +\infty$. Then $f(z)$ has at least one Borel direction of order λ.*

PROOF. According to the definition of the Borel direction, $f(z)$ and $\frac{1}{f(z)}$ have the same Borel directions. We may, therefore, assume that $f(0) \ne \infty$, otherwise we need only to consider $\frac{1}{f(z)}$. Besides, based on the definition of order, there exists a sequence r_n, $1 \leq r_n < r_{n+1} \to +\infty$ $(n \to +\infty)$, such that

$$\lim_{n \to +\infty} \frac{\log T(r_n, f)}{\log r_n} = \lambda > 0.$$

Hence it follows that

$$\lim_{n \to +\infty} \frac{T(r_n, f)}{(\log r_n)^4} = +\infty.$$

Now from $\{r_n\}$ we select suitably a subsequence $r_{n_k} : e^2 < r_{n_k} < r_{n_{k+1}}$, $r_{n_k} \to +\infty$ $(k \to +\infty)$. Concretely speaking, after r_{n_k} has been selected, we let

$$\varepsilon_{n_k} = \frac{1}{\log r_{n_k}} < \frac{1}{2}$$

and then choose $r_{n_{k+1}}$ sufficiently large, such that according to Lemma 2.8, there exists a point $z_{n_{k+1}}$ in the annulus $K_{n_{k+1}} : r_{n_k} < |z| < 2r_{n_{k+1}}$, such that circle $\Gamma_{n_{k+1}} : |z - z_{n_{k+1}}| < \varepsilon_{n_k}|z_{n_{k+1}}|$ is a filling circle of $f(z)$ with its index being

$$m_{n_{k+1}} = C \cdot \frac{1}{(\log r_{n_k})^2} \cdot \frac{T(r_{n_{k+1}}, f)}{(\log r_{n_{k+1}})^2} \geq C \cdot \frac{T(r_{n_{k+1}}, f)}{(\log r_{n_{k+1}})^4} \geq 1.$$

Considering the set $E = \{\theta_{n_k} = \arg z_{n_k} | k = 1, 2, \dots, 0 \leq \theta_{n_k} \leq 2\pi\}$, then we find that E contains at least one cluster point θ, $0 \leq \theta \leq 2\pi$. We may also assume that $\theta_{n_k} \to \theta$ $(k \to +\infty)$, otherwise we need only to select a suitable subsequence. In the following we will prove that $\Delta(\theta)$ is a Borel direction of $f(z)$ of order λ. Indeed, otherwise, there exist certain value $\varepsilon > 0$ and three corresponding complex numbers a, b and c, such that

$$\overline{\lim_{r \to +\infty}} \frac{\log^+ n\{\Omega(\theta - \varepsilon, \theta + \varepsilon; r), f = X\}}{\log r} < \lambda, \qquad X = a, b, c. \quad (2.76)$$

On the other hand, when k is sufficiently large, we have $\Gamma_{n_k} \subset \Omega(\theta - \varepsilon, \theta + \varepsilon)$ and $e^{-m_{n_k}} \leq \frac{1}{4} \min\{|a, b|, |a, c|, |b, c|\}$. Therefore, among the three values a, b and c, there exists at least one value, for example a, such that at an infinite sequence on Γ_{n_k}, we have the following estimation:

$$n(\Gamma_{n_k}, f = a) \geq m_{n_k} \geq C \cdot \frac{T(r_{n_k}, f)}{(\log r_{n_k})^4}.$$

Consequently

$$\varlimsup_{r\to+\infty} \frac{\log^+ n\{\Omega(\theta-\varepsilon,\theta+\varepsilon;r),f=a\}}{\log r}$$

$$\geq \varlimsup_{k\to+\infty} \frac{\log^+ n\{\Omega(\theta-\varepsilon,\theta+\varepsilon;2r_{n_k}(1+\varepsilon_{n_{k-1}})),f=a\}}{\log 2r_{n_k}(1+\varepsilon_{n_{k-1}})}$$

$$\geq \varlimsup_{k\to+\infty} \frac{\log^+ n(\Gamma_{n_k},f=a)}{\log 2r_{n_k}(1+\varepsilon_{n_{k-1}})}$$

$$\geq \varlimsup_{k\to+\infty} \frac{\log T(r_{n_k},f) - 4\log\log r_{n_k} + \log C}{\log r_{n_k} + \log 2(1+\varepsilon_{n_{k-1}})} = \lambda,$$

but this contradicts formula (2.76), thus Theorem 2.11 holds.

From the proof of Theorem 2.11 we find that a series of filling circles determines a Borel direction. A. Rauch, however, proved conversely that a Borel direction determines a series of filling circles [35a].

THEOREM 2.12. *Let* $f(z)$ *be a meromorphic function of order* λ, $0 < \lambda < +\infty$ *on the open plan* $|z| < +\infty$, *and* $\Delta(\theta)$ *be a Borel direction of* f *of order* λ. *Then there exists an infinite point sequence* z_n, $z_n = |z_n|e^{i\theta}$, $|z_n| < |z_{n+1}| \to \infty$ $(n \to +\infty)$ *on* $\Delta(\theta)$, *such that each circle* $\Gamma_n\colon |z - z_n| < \varepsilon_n|z_n|$, $0 < \varepsilon_n \to 0$ $(n \to +\infty)$ *is a filling circle of* $f(z)$, *with its index* m_n *satisfying the condition: for any arbitrarily selected* λ', $0 < \lambda' < \lambda$, *provided that* n *is sufficiently large, we have* $m_n \geq |z_n|^{\lambda'}$.

For the sequels we will call a series of filling circles with the above properties the sequence of filling circles of order λ.

PROOF. Without any loss of generality, we may assume that $\Delta(\theta) = \Delta(0)$. In the following, we need only to prove that for any arbitrarily given value R, $R > 1$, value ε, $0 < \varepsilon < \frac{1}{2}$, and value λ', $0 < \lambda' < \lambda$, there must exist a point z_0 in $\Delta(0)$, such that the circle $\Gamma\colon |z - z_0| < \varepsilon|z_0|$ is a filling circle of $f(z)$ with an index $m \geq |z_0|^{\lambda'}$. In fact, otherwise, for any point z_0 in $\Delta(0; R, +\infty)$, all circles $\Gamma\colon |z - z_0| < \varepsilon|z_0|$ will not be the filling circles of $f(z)$ with an index m, $m \geq |z_0|^{\lambda'}$. We shall derive a contradiction from this.

First we construct a sequence $r_n = (1+\frac{\varepsilon}{4})^n$ $(n = 1, 2, \dots)$. For the given value R there exists a positive integer n_0, such that when $n \geq n_0$, $r_n > R$. If we construct a circle $c_n\colon |z - z_n| < \frac{\varepsilon}{2}|z_n|$ centered at z_n, $z_n = (1+\frac{\varepsilon}{4})^n(1+\frac{\varepsilon}{8})$ with radius $\frac{\varepsilon}{2}|z_n|$, then the whole of $\overline{\Omega}(-\frac{\varepsilon}{4},\frac{\varepsilon}{4};r_n,r_{n+1})$ lies entirely in circle c_n. We also construct a concentric circle $c_n'\colon |z - z_n| < \varepsilon_n|z_n|$, then c_n' is not a filling circle of $f(z)$ with an index m. Hence, there exist three distinct complex numbers a_n, b_n and c_n, with their spherical distances among one another $\geq d$, $d = e^{-|z_n|^{\lambda'}}$, and also

$$n(c'_n, f = X) < |z_n|^{\lambda'}, \qquad X = a_n, b_n, c_n.$$

In the following we consider the transformation

$$\zeta = \frac{z - z_n}{\varepsilon |z_n|}$$

and let $\varphi(\zeta) = f(z_n + \varepsilon |z_n|\zeta)$. Then we apply Theorem 2.7 to $\varphi(\zeta)$, where we let $h = 0.01$. Hence we may conclude that for each complex value X we have

$$n(c_n, f = X) \leq 4\left\{ A|z_n|^{\lambda'} \log 100 + B \log 4 + C|z_n|^{\lambda'} + D \log^+ \frac{1}{|X, X_n(\frac{1}{2})|} \right\}.$$

We construct a sphere D_n: $|X, X_n(\frac{1}{2})| < e^{-|z_n|^{\lambda'}}$. When X is outside circle D_n, we further have

$$n(c_n, f = X) \leq 4\{A|z_n|^{\lambda'} \log 100 + B \log 4 + C|z_n|^{\lambda'} + D|z_n|^{\lambda'}\} \tag{2.77}$$
$$\leq A|z_n|^{\lambda'},$$

where $A > 0$ is a constant independent of n.

Let

$$(\gamma)_{n_1} = \bigcup_{n=n_1}^{\infty} D_n,$$

then the sum of radii of spheres $(\gamma)_{n_1}$ is

$$H_{n_1} = \sum_{n=n_1}^{\infty} e^{-|z_n|^{\lambda'}} \leq \sum_{n=n_1}^{\infty} \frac{1}{|z_n|^{\lambda'}} \leq \sum_{n=n_1}^{\infty} \frac{1}{r_n^{\lambda'}}$$

$$= \frac{1}{r_{n_1}^{\lambda'}} \left(1 + \frac{1}{(1 + \frac{\varepsilon}{4})^{\lambda'}} + \frac{1}{(1 + \frac{\varepsilon}{4})^{2\lambda'}} + \cdots \right)$$

$$= \frac{(1 + \frac{\varepsilon}{4})^{2'}}{(1 + \frac{\varepsilon}{4})^{\lambda'} - 1} \cdot \frac{1}{r_{n_1}^{\lambda'}}.$$

Hence, we need only to choose a sufficiently large n_1, $n_1 \geq n_0$, such that $H_{n_1} < \frac{1}{4}$. We may then find three distinct complex values a, b and c outside spheres $(\gamma)_{n_1}$ and it follows from formula (2.77) that when $n \geq n_1$,

$$n(c_n, f = X) \leq A|z_n|^{\lambda'}, \qquad X = a, b, c.$$

For any arbitrarily selected value r, $r \geq r_{n_1}$, there exists a positive integer N, $N \geq n_1$, such that $r_N \leq r < r_{N+1}$. Therefore, when $X = a$, b and c,

we have

$$n\left\{\Omega\left(-\frac{\varepsilon}{4},\frac{\varepsilon}{4};r\right),f=X\right\}$$

$$\leq n\left\{Q\left(-\frac{\varepsilon}{4},\frac{\varepsilon}{4};r_{n_1}\right),f=X\right\}+\sum_{n=n_1}^{N}n(c_n,f=X)$$

$$\leq n\left\{\Omega\left(-\frac{\varepsilon}{4},\frac{\varepsilon}{4};r_{n_1}\right),f=X\right\}+A\sum_{n=n_1}^{N}|z_n|^{\lambda'}.$$

Notice that

$$\sum_{n=n_1}^{N}|z_n|^{\lambda'}\leq\sum_{n=n_1}^{N}r_{n+1}^{\lambda'}\leq r_{N+1}^{\lambda'}\sum_{n=0}^{\infty}\frac{1}{r_n^{\lambda'}}$$

$$\leq\frac{(1+\frac{\varepsilon}{4})^{\lambda'}}{(1+\frac{\varepsilon}{4})^{\lambda'}-1}(1+\frac{\varepsilon}{4})^{\lambda'}\cdot r^{\lambda'}.$$

Then

$$n\left\{\Omega\left(-\frac{\varepsilon}{4},\frac{\varepsilon}{4};r\right),f=X\right\}\leq n\left\{\Omega\left(-\frac{\varepsilon}{4},\frac{\varepsilon}{4};r_n\right),f=X\right\}+Ar^{\lambda'},$$

where $A>0$ is a constant independent of r. Thus, when $X=a$, b and c, we conclude that

$$\varlimsup_{r\to+\infty}\frac{\log^+n\{\Omega(-\frac{\varepsilon}{4},\frac{\varepsilon}{4};r),f=X\}}{\log r}\leq\lambda'<\lambda,$$

but this contradicts the assumption that $\Delta(0)$ is a Borel direction of $f(z)$ of order λ. Theorem 2.12 is thus proved.

We need the following result for the sequels [39d]:

THEOREM 2.13. *Let $f(z)$ be meromorphic on $\Omega(\theta-\varepsilon,\theta+\varepsilon)$ $(0<\varepsilon<\frac{\pi}{2})$ and a, b, and c be three distinct complex numbers such that*

$$\varlimsup_{r\to+\infty}\frac{\log^+n\{\Omega(\theta-\varepsilon,\theta+\varepsilon;r),f=X\}}{\log r}\leq\lambda,\qquad(2.78)$$

$$0\leq\lambda<+\infty,\quad X=a,b,c.$$

Then for each complex value X

$$\varlimsup_{r\to+\infty}\frac{\log^+n\{\Omega(\theta-\frac{\varepsilon}{2\pi},\theta+\frac{\varepsilon}{2\pi};r),f=X\}}{\log r}\leq\lambda,$$

but probably excluding some values X, with their linear measure on the Riemann sphere being zero.

PROOF. Without any loss of generality, we may assume that $\theta=0$. We form a sequence $r_n=(1+\frac{\varepsilon}{2\pi})^n$ $(n=1,2,\ldots)$ and, to each n, we construct a circle $C_n\colon|z-z_n|<\frac{\varepsilon}{\pi}|z_n|$, centered at $z_n=(1+\frac{\varepsilon}{2\pi})^n(1+\frac{\varepsilon}{4\pi})$ with radius $\frac{\varepsilon}{\pi}|z_n|$. Then the whole of $\overline{\Omega}(-\frac{\varepsilon}{2\pi},\frac{\varepsilon}{2\pi};r_n,r_{n+1})$ lies entirely in circle C_n.

We then construct a concentric circle C_n': $|z - z_n| < \frac{2\varepsilon}{\pi}|z_n|$, hence the whole of C_n' lies entirely within $\Omega(-\varepsilon, \varepsilon; r_{n+5})$. According to formula (2.78), for any arbitrarily selected number η $(\eta > 0)$, there exists a value R_0, such that when $r > R_0$, it follows that

$$n\{\Omega(-\varepsilon, \varepsilon; r), f = X\} \leq r^{\lambda+\eta}, \qquad X = a, b, c.$$

Therefore, there exists a positive integer n_0, such that when $n \geq n_0$, we have $r_n > R_0$. Accordingly, the following inequalities hold:

$$n(C_n', f = X) \leq n\{\Omega(-\varepsilon, \varepsilon; r_{n+5}), f = X\} \leq r_{n+5}^{\lambda+\eta}, \qquad X = a, b, c.$$

Suppose that the spherical distances among a, b and c are $\geq d$ $(0 < d < \frac{1}{2})$. In the following we consider the transformation

$$\zeta = \frac{z - z_n}{2\varepsilon|z_n|/\pi}$$

and let $\varphi(\zeta) = f(z_n + \frac{2\varepsilon}{\pi}|z_n|\zeta)$. Then we apply Theorem 2.7 to $\varphi(\zeta)$, where we let $h = 0.01$. Consequently we may conclude that for each complex value X, when $n \geq n_0$,

$$n(C_n, f = X) \leq 4\left\{ Ar_{n+5}^{\lambda+\eta}\log 100 + B\log 4 + C\log\frac{1}{d} \right.$$
$$\left. + D\log^+\frac{1}{|X, X_n(\frac{1}{2})|} \right\}.$$

We construct a sphere D_n: $|X, X_n(\frac{1}{2})| < e^{-r_{n+1}^{\lambda+\eta}}$; then, when X is outside sphere D_n,

$$n(C_n, f = X) \leq 4\left\{ Ar_{n+5}^{\lambda+\eta}\log 100 + B\log 4 + C\log\frac{1}{d} + Dr_{n+5}^{\lambda+\eta} \right\} \tag{2.79}$$
$$\leq Ar_{n+5}^{\lambda+\eta},$$

where $A > 0$ is a constant unrelated to n.

Let

$$(\gamma)_{n_1} = \bigcup_{n=n_1}^{\infty} D_n, \qquad n_1 \geq n_0.$$

Then the sum of radii of spheres $(\gamma)_{n_1}$ is

$$H_{n_1} = \sum_{n=n_1}^{\infty} e^{-r_{n+5}^{\lambda+\eta}} \leq \sum_{n=n_1}^{\infty} \frac{1}{r_{n+5}^{\lambda+\eta}} = \frac{1}{r_{n_1+5}^{\lambda+\eta}}\sum_{n=n_1}^{\infty} \frac{1}{r_{n-n_1}^{\lambda+\eta}}$$
$$= \frac{(1 + \frac{\varepsilon}{2\pi})^{\lambda+\eta}}{(1 + \frac{\varepsilon}{2\pi})^{\lambda+\eta} - 1} \cdot \frac{1}{r_{n_1+5}^{\lambda+\eta}}. \tag{2.80}$$

For any arbitrarily assumed value r $(r \geq r_{n_1})$, there always exists a positive integer $N(\geq n_1)$, such that $r_N \leq r < r_{N+1}$. Hence when $X \notin (\gamma)_{n_1}$, it follows from formula (2.79) that

$$n\left\{\Omega\left(-\frac{\varepsilon}{2\pi}, \frac{\varepsilon}{2\pi}; r\right), f = X\right\}$$

$$\leq n\left\{\Omega\left(-\frac{\varepsilon}{2\pi}, \frac{\varepsilon}{2\pi}; r_{n_1}\right), f = X\right\} + \sum_{n=n_1}^{N} n(C_n, f = X)$$

$$\leq n\left\{\Omega\left(-\frac{\varepsilon}{2\pi}, \frac{\varepsilon}{2\pi}; r_{n_1}\right), f = X\right\} + A\sum_{n=n_1}^{N} r_{n+5}^{\lambda+\eta}.$$

Notice that

$$\sum_{n=n_1}^{N} r_{n+5}^{\lambda+\eta} \leq r_{n+5}^{\lambda+\eta} \sum_{n=0}^{\infty} \frac{1}{r_n^{\lambda+\eta}}$$

$$\leq \frac{(1 + \frac{\varepsilon}{2\pi})^{\lambda+\eta}}{(1 + \frac{\varepsilon}{2\pi})^{\lambda+\eta} - 1} \left(1 + \frac{\varepsilon}{2\pi}\right)^{5(\lambda+\eta)} \cdot r^{\lambda+\eta}.$$

Then

$$n\left\{\Omega\left(-\frac{\varepsilon}{2\pi}, \frac{\varepsilon}{2\pi}; r\right), f = X\right\}$$

$$\leq n\left\{\Omega\left(-\frac{\varepsilon}{2\pi}, \frac{\varepsilon}{2\pi}; r_{n_1}\right), f = X\right\} + Ar^{\lambda+\eta},$$

where $A > 0$ is a constant independent of r. Therefore, we conclude that

$$\varlimsup_{r \to +\infty} \frac{\log^+ n\{\Omega(-\frac{\varepsilon}{2\pi}, \frac{\varepsilon}{2\pi}; r), f = X\}}{\log r} \leq \lambda + \eta.$$

Let $\eta \to 0$. Then for each complex value X not belonging to spheres $(\gamma)_{n_1}$, we have

$$\varlimsup_{r \to +\infty} \frac{\log^+ n\{\Omega(-\frac{\varepsilon}{2\pi}, \frac{\varepsilon}{2\pi}; r), f = X\}}{\log r} \leq \lambda. \qquad (2.81)$$

The above discussion shows that those complex values X not satisfying formula (2.81) belong definitely to the set $\bigcap_{n=n_1}^{\infty} (\gamma)_n$. On the other hand, it is apparent that

$$(\gamma)_{n_1} \supset (\gamma)_{n_1+1} \supset (\gamma)_{n_1+2} \supset \cdots.$$

Also according to formula (2.80), when n_1 is sufficiently large, the linear measure of spheres $(\gamma)_{n_1}$ may be sufficiently small. Hence, $\bigcap_{n=n_1}^{\infty} (\gamma)_n$ is a set of zero linear measure on the Riemann sphere, which also completes the proof of Theorem 2.13.

THEOREM 2.14. *Suppose that a meromorphic function* $f(z)$ *of order* λ $(\lambda > 0)$ *on* $\Omega(-\theta, \theta)$ $(0 < \theta < \pi)$ *has no Borel direction of order* λ. *Then for any arbitrarily assumed number* ε $(0 < \varepsilon < \theta)$, *there exist definitely three distinct complex numbers* a, b *and* c, *such that*

$$\varlimsup_{r \to +\infty} \frac{\log^+ n\{\overline{\Omega}(-\theta + \varepsilon, \theta - \varepsilon; r), f = X\}}{\log r} \le \lambda' < \lambda, \qquad X = a, b, c.$$

PROOF. Under the assumption that $f(z)$ has no Borel direction of order λ on $\Omega(-\theta, \theta)$, then for any arbitrary value φ, $-\theta + \varepsilon \le \varphi \le \theta - \varepsilon$, $\Delta(\varphi)$ is not a Borel direction of order λ of $f(z)$, meaning that there exist a particular number δ $(0 < \delta < \varepsilon)$ and three corresponding distinct complex numbers α, β and γ, such that

$$\varlimsup_{r \to +\infty} \frac{\log^+ n\{\Omega(\varphi - \delta, \varphi + \delta; r), f = X\}}{\log r} \le \lambda' < \lambda, \qquad X = \alpha, \beta, \gamma.$$

According to Theorem 2.13, for each complex value X we have

$$\varlimsup_{r \to +\infty} \frac{\log^+ n\{\Omega(\varphi - \frac{\delta}{2\pi}, \varphi + \frac{\delta}{2\pi}; r), f = X\}}{\log r} \le \lambda';$$

but probably excluding some values X, with their linear measure on the Riemann sphere being zero.

The set $\{\Omega(\varphi - \frac{\delta}{2\pi}, \varphi + \frac{\delta}{2\pi}) | \varphi \in [-\theta + \varepsilon, \theta - \varepsilon]\}$ constitutes an open covering of $\overline{\Omega}(-\theta + \varepsilon, \theta - \varepsilon)$. According to the finite covering theorem, we may select a finite covering. Since the exceptional values X of each member of the finite covering on the Riemann sphere has linear measure zero, the linear measure of all exceptional values X of finite coverings on the Riemann sphere is also equal to zero. Hence three distinct complex numbers a, b and c exist, such that

$$\varlimsup_{r \to +\infty} \frac{\log^+ n\{\Omega(-\theta + \varepsilon, \theta - \varepsilon; r), f = X\}}{\log r} \le \lambda', \qquad X = a, b, c.$$

Theorem 2.14 is thus proved.

2.4.3. The Julia direction. Let $f(z)$ be a transcendental meromorphic function on the open plane $|z| < +\infty$. If for any arbitrarily small number $\varepsilon > 0$, $f(z)$ takes each complex value X in $\Omega(\theta - \varepsilon, \theta + \varepsilon)$ for infinitely many times, probably with the exception of at most two values X, then we call $\Delta(\theta)$ a Julia direction of $f(z)$.

In 1919 G. Julia proved the existence of the Julia direction [25a], and thus initiating the study of the argument distribution theory.

THEOREM 2.15 [30a, 43c]. *Let* $f(z)$ *be a transcendental meromorphic function on the open plane* $|z| < +\infty$, *which satisfies one of the following conditions:*

(1)
$$\varlimsup_{r \to \infty} \frac{T(r, f)}{(\log r)^2} = +\infty,$$

(2) $f(z)$ *has an asymptotic value* α,

(3) *for a certain complex value* α, $\lim_{r\to\infty} m(r, \alpha) = +\infty$.
Then $f(z)$ *has at least one Julia direction.*

PROOF. When $f(z)$ satisfies condition (1), there exists a sequence r_n, $1 < r_n < r_{n+1} \to +\infty$ $(n \to +\infty)$, such that

$$\lim_{n\to+\infty} \frac{T(r_n, f)}{(\log r_n)^2} = +\infty.$$

Now we select suitably a subsequence $r_{n_k}: 1 < r_{n_k} < r_{n_{k+1}} \to \infty$ $(k \to +\infty)$ of $\{r_n\}$. Concretely speaking, after r_{n_k} has been selected, we let $\varepsilon_{n_k} = \frac{1}{4k} < 1$ and choose $r_{n_{k+1}}$ sufficiently large. According to Lemma 2.8, there exists a point $z_{n_{k+1}}$ in annulus $K_{n_{k+1}}: r_{n_k} < |z| < 2r_{n_{k+1}}$, such that circle $\Gamma_{n_{k+1}}: |z - z_{n_{k+1}}| < \varepsilon_{n_k}|z_{n_{k+1}}|$ is a filling circle of $f(z)$, with the index

$$m_{n_{k+1}} = c\frac{1}{16k^2}\frac{T(r_{n_{k+1}}, f)}{(\log r_{n_{k+1}})^2} \geq k+1.$$

Let θ be a cluster point of the set $E = E\{\theta_{n_k} = \arg z_{n_k} | k = 1, 2, \ldots, 0 \leq \theta_{n_k} \leq 2\pi\}$. We prove that $\Delta(\theta)$ is a Julia direction of $f(z)$. In fact, otherwise, there exist a certain value $\varepsilon > 0$ and three corresponding distinct complex numbers a, b and c, such that $f(z)$ takes values a, b and c for at most finite times in $\Omega(\theta - \varepsilon, \theta + \varepsilon)$, implying that value R_0 exists, such that $f(z)$ does not take values a, b and c in $\Omega(\theta - \varepsilon, \theta + \varepsilon; R_0, +\infty)$. On the other hand, when k is sufficiently large, there exists circle $\Gamma_{n_k} \subset \Omega(\theta - \varepsilon, \theta + \varepsilon; R_0, +\infty)$, and also $e^{-m_{n_k}} \leq \frac{1}{4}\min\{|a, b|, |a, c|, |b, c|\}$. Among the three values a, b and c, there is, therefore, at least one value, for example a, such that on Γ_{n_k} we obtain

$$n(\Gamma_{n_k}, f = a) \geq m_{n_k} \geq k,$$

which derives a contradiction. Hence when $f(z)$ satisfies condition (1), Theorem 2.15 holds.

Now we prove that when $f(z)$ satisfies condition (2) or (3), it has at least one Julia direction, Without any loss of generality, we may assume that $\alpha = \infty$. In the following we prove by contradiction:

Suppose $f(z)$ does not possess any Julia direction. Then for each direction $\Delta(\theta)$ $(0 \leq \theta \leq 2\pi)$ there exist definitely an angular domain $\Omega(\theta - \varepsilon, \theta + \varepsilon)$ $(\varepsilon > 0)$ and three corresponding distinct complex numbers a, b, and c, such that in $\Omega(\theta - \varepsilon, \theta + \varepsilon)$ $f(z)$ takes values a, b, and c for at most finite times. Since the set $\{\Omega(\theta - \frac{\varepsilon}{4}, \theta + \frac{\varepsilon}{4}) | \theta \in [0, 2\pi]\}$ forms a covering of the open plane $|z| < +\infty$, we may select within the set a finite covering $\Omega(\theta_i - \frac{\varepsilon_i}{4}, \theta_i + \frac{\varepsilon_i}{4})$ $(i = 1, 2, \ldots, N)$, with a_i, b_i and c_i being three corresponding values. Clearly $R_0 > 1$, such that in $\Omega(\theta_i - \frac{\varepsilon_i}{4}, \theta_i + \frac{\varepsilon_i}{4}; R_0, +\infty)$ $(1 \leq i \leq N)$ $f(z)$ does not take values a_i, b_i, and c_i. Besides, numbers

d $(0 < d < \frac{1}{2})$ exist, such that the spherical distances among a_i, b_i and c_i $(1 \le i \le N)$ are $\ge d$.

Since $f(z)$ is a transcendental meromorphic function, there exists a point sequence z_n: $|z_n| < |z_{n+1}| \to +\infty$ $(n \to +\infty)$, such that $|f(z_n)| \le 1$ $(n = 1, 2, \dots)$. Let $\varepsilon_0 = \min_{1 \le i \le N}\{\frac{\varepsilon_i}{4}\}$ and construct circle Γ_{n0}: $|z - z_n| < \varepsilon_0|z_n|$; then the whole of Γ_{n0} lies entirely in an angular domain $\Omega(\theta_i - \varepsilon_i, \theta_i + \varepsilon_i)$, meaning that a positive integer n exists, such that when $n \ge n_0$, we have $R_0 < |z_n|$. Hence $f(z)$ on Γ_{n0} does not take a_i, b_i, and c_i.

We consider the transformation

$$\zeta = \frac{z - z_n}{\varepsilon_0|z_n|}$$

and let $\varphi(\zeta) = f(z_n + \varepsilon_0|z_n|\zeta)$; then $\varphi(\zeta)$ does not take values a_i, b_i, and c_i in the disk $|\zeta| < 1$, and that $|\varphi(0)| = |f(z_n)| \le 1$. We then apply Theorem 2.10, where we let $n = 0$, $d = d$, $M = 1$ and $r = \frac{1}{2}$. It follows that $\varphi(\zeta)$ satisfies

$$\log|\varphi(\zeta)| \le 4\left\{B\log 4 + D\log\frac{1}{d}\right\}\left(1 + \log\frac{2}{H}\right),$$

in the disk $|\zeta| < \frac{1}{2}$, excluding at most some circles $(\gamma)_H$, with the sum of their radii not exceeding $2eH$. Hence, $f(z)$ in the disk Γ'_{n0}: $|z - z_n| < \frac{1}{2}\varepsilon_0|z_n|$ satisfies

$$\log|f(z)| \le A_0\left(1 + \log\frac{2}{H}\right),$$

excluding at most some circles $(\gamma)_H^0$, with the sum of their radii not more than $2e\varepsilon_0|z_n|H$, where A_0 is a constant independent of n.

We now construct a sequence of circles Γ_{nj}: $|z - z_n e^{i((j/4)\varepsilon_0)}| < \varepsilon_0|z_n|$ $(j = 1, 2, \dots, K; K = [\frac{8\pi}{\varepsilon_0}] + 1)$ and then apply Theorem 2.10 to each circle orderly. Then analogously, we may prove that $f(z)$, in the disk Γ'_{nj}: $|z - z_n e^{i((j/4)\varepsilon_0)}| \le \frac{1}{2}\varepsilon_0|z_n|$, satisfies

$$\log|f(z)| \le A_j\left(1 + \frac{2}{H}\right)^{j+1},$$

excluding at most some circles $(\gamma)_H^j$, with the sum of their radii not more than $2e\varepsilon_0|z_n|H$, where A_j is a constant independent of n. Let

$$(\gamma) = \bigcup_{j=0}^{K}(\gamma)_H^j,$$

then the sum of radii of (γ) does not exceed $2e\varepsilon_0|z_n|KH$. Assume $H = 7/(64eH)$; then to each n there exists circumference Γ_n: $|z| = R_n$ in the annulus $|z_n|(1 - \frac{7}{8}\cdot\frac{\varepsilon_0}{2}) \le |z| \le |z_n|(1 + \frac{7}{8}\cdot\frac{\varepsilon_0}{2})$, such that Γ_n does not intersect (γ), and that Γ_n is covered by the set $\{\Gamma'_{nj}|j = 0, 1, \dots, K - 1\}$. Hence when $z \in \Gamma_n$, we have

$$\log |f(z)| \le A \left(1 + \frac{2}{H}\right)^K, \qquad A = \max_{0 \le j \le K-1}\{A_j\},$$

where A is a constant independent of n. Therefore, we may assure on the one hand that

$$\lim_{r \to +\infty} m(r, \infty) \le \lim_{n \to +\infty} m(R_n, \infty) < +\infty,$$

while on the other hand we may conclude that the value ∞ cannot be an asymptotic value of $f(z)$. Both cases, however, contradict conditions (2) and (3), respectively. Thus Theorem 2.15 is proved completely.

COROLLARY. *Suppose that $f(z)$ is a transcendental entire function or a transcendental meromorphic function having some deficient values. Then $f(z)$ possesses at least one Julia direction.*

A. Ostrowski [33a] once exhibited an example to show that there exists a meromorphic function having no Julia direction with its growth satisfying $T(r, f) = O\{(\log r)^2\}$ $(r \to +\infty)$.

Finally corresponding to Theorem 2.14 we have

THEOREM 2.16. *Suppose that meromorphic function $f(z)$ on $\Omega(-\theta, \theta)$ $(0 < \theta < \pi)$ has no Julia direction. Then for any arbitrarily selected number ε, $0 < \varepsilon < \theta$, there must exist three distinct complex numbers a, b and c, such that*

$$\overline{\lim_{r \to +\infty}} \frac{\log^+ n\{\overline{\Omega}(-\theta + \varepsilon, \theta - \varepsilon; r), f = X\}}{\log r} = 0, \qquad X = a, b, c.$$

§2.5. On the growth of the entire function

2.5.1. Some lemmas.
We consider the transformation:

$$\zeta = \frac{z^{\pi/(2\theta)} - R_1^{\pi/(2\theta)}}{z^{\pi/(2\theta)} + R_1^{\pi/(2\theta)}}, \qquad 0 < \theta \le \pi, \ 1 \le R_1 < +\infty \qquad (2.82)$$

which maps $\Omega(-\theta, \theta)$ into the unit disk $|\zeta| < 1$ of the ζ-plane, and point $z = R_1$ into point $\zeta = 0$.

LEMMA 2.9. *Under the transformation (2.82), the image of domain of $\overline{\Omega}(-\theta + \varepsilon, \theta - \varepsilon; R_1, R)$ $(0 < \varepsilon < \theta, \ 1 \le R_1 < R < +\infty)$ on the ζ-plane lies entirely in the disk $|\zeta| \le \rho$, and that*

$$\rho = 1 - \frac{\varepsilon}{2\theta} R^{-\pi/(2\theta)},$$

while the preimage domain of circle $|\zeta| \le r$ $(r < 1)$ on the z-plane lies entirely in domain $\overline{\Omega}(-\theta, \theta; R_2)$, where

$$R_2 = R_1 \left(\frac{2}{1-r}\right)^{2\theta/\pi}.$$

Furthermore, when $|\zeta| \leq r$, *the following inequalities hold:*

$$R_1 \cdot \frac{2\theta}{\pi} \left(\frac{1-r}{2}\right)^{2\theta/\pi} \leq |z'(\zeta)| \leq R_1 \cdot \frac{2\theta}{\pi} \left(\frac{2}{1-r}\right)^{(2\theta/\pi)+1}.$$

PROOF. Let $z = re^{i\varphi} \in \overline{\Omega}(-\theta + \varepsilon, \theta - \varepsilon; R_1, R)$. Then we have

$$|\zeta| = \left| \frac{r^{\pi/(2\theta)} e^{i\pi\varphi/(2\theta)} - R_1^{\pi/(2\theta)}}{r^{\pi/(2\theta)} e^{i\pi\varphi/(2\theta)} + R_1^{\pi/(2\theta)}} \right|$$

$$= \sqrt{1 - \frac{4r^{\pi/(2\theta)} R_1^{\pi/(2\theta)} \cos \frac{\pi\varphi}{2\theta}}{r^{\pi/\theta} + R_1^{\pi/\theta} + 2r^{\pi/(2\theta)} R_1^{\pi/(2\theta)} \cos \frac{\pi\varphi}{2\theta}}}.$$

Notice that

$$r^{\pi/\theta} + R_1^{\pi/\theta} + 2r^{\pi/(2\theta)} R_1^{\pi/(2\theta)} \cos \frac{\pi\varphi}{2\theta} \leq 4r^{\pi/\theta},$$

$$4r^{\pi/(2\theta)} R_1^{\pi/(2\theta)} \cos \frac{\pi\varphi}{2\theta} \geq \frac{4\varepsilon}{\theta} r^{\pi/(2\theta)}.$$

Then we obtain

$$|\zeta| \leq \sqrt{1 - \tfrac{\varepsilon}{\theta} r^{-\pi/(2\theta)}} < 1 - \tfrac{\varepsilon}{2\theta} R^{-\pi/(2\theta)} = \rho.$$

On the other hand, the inverse transformation of formula (2.82) is

$$z = R_1 \left(\frac{1+\zeta}{1-\zeta}\right)^{2\theta/\pi};$$

then when $|\zeta| \leq r(< 1)$, we get

$$|z| \leq R_1 \left(\frac{2}{1-r}\right)^{2\theta/\pi}.$$

Finally, according to the expression

$$z'(\zeta) = \frac{4\theta R_1}{\pi} \left(\frac{1+\zeta}{1-\zeta}\right)^{2\theta/\pi} \cdot \frac{1}{1-\zeta^2},$$

when $|\zeta| \leq r$, it follows that

$$\frac{2\theta R_1}{\pi} \left(\frac{1-r}{2}\right)^{2\theta/\pi} \leq |z'(\zeta)| \leq \frac{2\theta R_1}{\pi} \left(\frac{2}{1-r}\right)^{1+(2\theta/\pi)}.$$

Hence Lemma 2.9 is proved completely.

LEMMA 2.10. *If we denote* $u(z; \theta_1, \theta_2; R)$ *the harmonic measure of* $\Gamma(\theta_1, \theta_2; R)$ *corresponding to point* z *and domain* $\Omega(\theta_1, \theta_2; R)$. *Then when* $0 < \theta_2 - \theta_1 \leq 2\pi$, *it follows that*

$$u(z; \theta_1, \theta_2, R) \leq \frac{4(\frac{|z|}{R})^{\pi/(\theta_2-\theta_1)}}{\pi[1 - (\frac{|z|}{R})^{2\pi/(\theta_2-\theta_1)}]}.$$

PROOF. We consider the transformation

$$\zeta = \xi + i\eta = \left\{ \frac{r}{R} e^{-i(\theta_1 + \theta_2)/2} \right\}^{\pi/(\theta_2 - \theta_1)} .$$

Then $\Omega(\theta_1, \theta_2, R)$ transforms into a semi-circle $\Omega_\zeta(-\frac{\pi}{2}, \frac{\pi}{2}; 1)$ on the ζ-plane and $\Gamma(\theta_1, \theta_2; R)$ into a semi-circumference $\Gamma_\zeta(-\frac{\pi}{2}, \frac{\pi}{2}, 1)$. Hence we have

$$u(z; \theta_1, \theta_2, R) = u_\zeta \left(\zeta; -\frac{\pi}{2}, \frac{\pi}{2}, 1 \right) .$$

But

$$u_\zeta \left(\zeta; -\frac{\pi}{2}, \frac{\pi}{2}, 1 \right) = \frac{2\varphi}{\pi} ,$$

where φ is the complement of the extended angle of point ζ with respect to the diameter $(\xi = 0) \cap (-1 < \eta < 1)$. Therefore, it follows that

$$u(z; \theta_1, \theta_2, R) \leq \frac{2}{\pi} \left\{ \arctan \frac{\xi}{1 + \eta} + \arctan \frac{\xi}{1 - \eta} \right\}$$

$$\leq \frac{2}{\pi} \left\{ \frac{\xi}{1 + \eta} + \frac{\xi}{1 - \eta} \right\} = \frac{4\xi}{\pi(1 - \eta^2)}$$

$$\leq \frac{4|\zeta|}{\pi(1 - |\zeta|^2)} = \frac{4(\frac{|z|}{R})^{\pi/(\theta_2 - \theta_1)}}{\pi(1 - (\frac{|z|}{R})^{2\pi/(\theta_2 - \theta_1)})} .$$

Consequently Lemma 2.10 is proved.

LEMMA 2.11 [43c]. *Suppose that the order of an entire function $f(z)$ is $\lambda > \frac{\pi}{2\theta}$ $(0 < \theta \leq \pi)$, and $f(z)$ has no Julia direction on $\Omega(-\theta, \theta)$. Selecting arbitrarily a number ε, such that $0 < \varepsilon < \theta$, $\frac{\pi}{2\theta - \varepsilon} < \lambda$, we then have*

$$\varlimsup_{R \to +\infty} \frac{\log \log^+ M\{\overline{\Omega}(-\theta + \varepsilon, \theta - \varepsilon; R), f\}}{\log R} \leq \frac{\pi}{2\theta - \varepsilon} . \tag{2.83}$$

PROOF. Under the hypothesis that $f(z)$ has no Julia direction on $\Omega(-\theta, \theta)$, it follows according to Theorem 2.16 that for any arbitrary number $\varepsilon > 0$, two distinct finite values a and b exist, such that

$$\varlimsup_{R \to +\infty} \frac{\log^+ n\{\overline{\Omega}(-\theta + \frac{\varepsilon}{2}, \theta - \frac{\varepsilon}{2}; R), f = X\}}{\log R} = 0, \qquad X = a, b. \tag{2.84}$$

We construct the transformation

$$\zeta = \frac{z^{\pi/(2\theta - \varepsilon)} - 1}{z^{\pi/(2\theta - \varepsilon)} + 1} ,$$

which maps $\Omega(-\theta + \frac{\varepsilon}{2}, \theta - \frac{\varepsilon}{2})$ into the unit disk $|\zeta| < 1$ on the ζ-plane. On the other hand, its inverse transformation is

$$z = z(\zeta) = \left(\frac{1 + \zeta}{1 - \zeta} \right)^{(2\theta - \varepsilon)/\pi} .$$

According to Lemma 2.9 (where we let $R_1 = 1$), we conclude that the image domain of disk $|\zeta| \leq r(< 1)$ on the z-plane must be contained in domain $\overline{\Omega}(-\theta + \frac{\varepsilon}{2}, \theta - \frac{\varepsilon}{2}; R)$, and that

$$R = \left(\frac{2}{1-r}\right)^{(2\theta-\varepsilon)/\pi} .$$

Let $\varphi(\zeta) = f(z(\zeta))$; then it follows from formula (2.84) that for any arbitrarily small number $\eta > 0$, provided that r is sufficiently close to 1, the following inequalities hold:

$$n(r, \varphi = X) \leq n\left\{\overline{\Omega}\left(-\theta + \frac{\varepsilon}{2}, \theta - \frac{\varepsilon}{2}; R\right), f = X\right\} \leq R^\eta$$

$$= 2^{(2\theta-\varepsilon)\eta/\pi} \left(\frac{1}{1-r}\right)^{(2\theta-\varepsilon)\eta/\pi} , \qquad X = a, b,$$

$$\varlimsup_{r \to 1} \frac{\log^+ n(r, \varphi = X)}{\log \frac{1}{1-r}} \leq \frac{2\theta - \varepsilon}{\pi} \cdot \eta, \qquad X = a, b.$$

Further based on Theorem 1.8, we conclude that the order of $\varphi(\zeta)$ does not exceed $\frac{2\theta-\varepsilon}{\pi}\eta$. However, since η can be arbitrarily small, the order of $\varphi(\zeta)$ must be zero. Hence, when r is sufficiently close to 1, for any arbitrarily small number $\eta > 0$, we obtain

$$T(r, \varphi) < \left(\frac{1}{1-r}\right)^\eta .$$

Further according to formula (1.21) we get the following estimation:

$$\log^+ M(r, \varphi) \leq \frac{4}{1-r} T\left(\frac{1+r}{2}, \varphi\right) \leq 2^{2+\eta} \left(\frac{1}{1-r}\right)^{1+\eta} . \tag{2.85}$$

On the other hand, an application of Lemma 2.9 yields

$$\log^+ M\{\overline{\Omega}(-\theta + \varepsilon, \theta - \varepsilon; R), f\}$$

$$\leq \log^+ M(\rho, \varphi) + \log^+ M(1, f) + \log 2,$$

$$\rho = 1 - \frac{\varepsilon}{2(2\theta - \varepsilon)} R^{-\pi/(2\theta-\varepsilon)} .$$

Notice that when $R \to +\infty$, $\rho \to 1$, we conclude from formula (2.85) that

$$\varlimsup_{R \to +\infty} \frac{\log^+ \log^+ M\{\overline{\Omega}(-\theta + \varepsilon, \theta - \varepsilon; R), f\}}{\log R} \leq \frac{\pi}{2\theta - \varepsilon}(1 + \eta).$$

Since η may be selected arbitrarily small, formula (2.83) holds. This completes the proof of Lemma 2.11.

If it is assumed that $f(z)$ on $\Omega(-\theta, \theta)$ has no Borel direction of order λ $(\lambda > \frac{\pi}{2\theta})$, then instead of formula (2.84), we have according to Theorem 2.14 that

$$\varlimsup_{R \to +\infty} \frac{\log^+ n\{\overline{\Omega}(-\theta + \frac{\varepsilon}{2}, \theta - \frac{\varepsilon}{2}; R), f = X\}}{\log R} \leq \lambda' < \lambda, \qquad X = a, b.$$

And we deduce further that

$$\overline{\lim_{r \to 1}} \frac{\log^+ n(r, \varphi = X)}{\log \frac{1}{1-r}} \leq \frac{2\theta - \varepsilon}{\pi}(\lambda' + \eta), \qquad X = a, b.$$

Moreover, we may assume that $\frac{2\theta - \varepsilon}{\pi}\lambda' > 1$. According to Theorem 1.8, we conclude, therefore, that the order of $\varphi(\zeta)$ does not exceed $\frac{2\theta - \varepsilon}{\pi}\lambda' - 1$. Hence instead of formula (2.85) we have

$$\log^+ M(r, \varphi) \leq 2^{2+(2\theta-\varepsilon)/\pi\lambda'-1+\eta} \left(\frac{1}{1-r}\right)^{(2\theta-\varepsilon)/\pi\lambda'+\eta}.$$

We derive further that

$$\overline{\lim_{R \to +\infty}} \frac{\log^+ \log^+ M\{\overline{\Omega}(-\theta + \varepsilon, \theta - \varepsilon; R), f\}}{\log R} \leq \lambda' + \frac{\pi}{2\theta - \varepsilon}\eta.$$

Since η may be selected arbitrarily small, we arrive at the conclusion that

$$\overline{\lim_{R \to +\infty}} \frac{\log^+ \log^+ M\{\overline{\Omega}(-\theta + \varepsilon, \theta - \varepsilon; R), f\}}{\log R} \leq \lambda' < \lambda.$$

Consequently we prove the following result:

LEMMA 2.12. *Suppose that the order* λ *of an entire function* $f(z)$ *is* $> \frac{\pi}{2\theta}$ $(0 < \theta \leq \pi)$, *and* $f(z)$ *has no Borel direction of order* λ *on* $\Omega(-\theta, \theta)$. *Selecting arbitrarily a number* ε, *such that* $0 < \varepsilon < \theta$, $\frac{\pi}{2\theta-\varepsilon} < \lambda$, *we have*

$$\overline{\lim_{R \to +\infty}} \frac{\log^+ \log^+ M\{\overline{\Omega}(-\theta + \varepsilon, \theta - \varepsilon; R), f\}}{\log R} < \lambda.$$

2.5.2. Distribution of the Julia directions. Let E be the set of the intersecting points between all Julia directions of a transcendental entire function $f(z)$ and the unit circumferences $\overline{\Gamma}(0, 2\pi; 1)$. According to the corollary of Theorem 2.15 and the definition of the Julia direction, E is a nonempty closed set. hence, the complementary set $\Omega = \{\overline{\Gamma}(0, 2\pi; 1) - E)\}$ is an open set, with each connected component being an open arc and called simply the maximal open arc. Ω is made up of at most countable maximal open arcs ω_i $(i = 1, 2, \ldots, p; 0 \leq p \leq +\infty)$: $\Omega = \bigcup_i \omega_i$. Let $\omega = \min_i\{\text{meas}\, \omega_i\}$. If Ω is an empty set, then we assign $\omega = 0$.

Each connected branch of E is a closed arc (may be degenerated into a point), and is called simply the maximal closed arc. E is made up of at most countable maximal closed arcs I_j $(j = 1, 2, \ldots, p)$: $E = \bigcup_j I_j$. Let $I = \max_j\{\text{meas}\, I_j\}$. Then we have the following result [43c]:

THEOREM 2.17. *When the order* λ *of an entire function* $f(z)$ *is* $> \frac{\pi}{\omega}$, *then we have* $I \geq \min\{\frac{\pi}{\mu}, \omega\}$, *where* μ *is the lower order of* $f(z)$.

PROOF. When $\mu = +\infty$, it is obvious that Theorem 2.17 holds. Hence, we need only to consider the case when $\mu < +\infty$. Under the assumption that $\lambda > \frac{\pi}{\omega}$, Ω cannot be an empty set, and it only contains finitely many maximal open arcs ω_i $(i = 1, 2, \ldots, p; 1 \le p < +\infty)$. Correspondingly, E also contains p maximal closed arcs I_j $(j = 1, 2, \ldots, p)$. We take on $\overline{\Gamma}(0, 2\pi; 1)$ $2p$ points $e^{i\theta_1}, e^{i\theta_2}, \ldots, e^{i\theta_{2p}}$; $0 \le \theta_1 < \theta_2 \le \theta_3 < \theta_4 \le \cdots \le \theta_{2p-1} < \theta_{2p} \le \theta_{2p+1}$, $\theta_{2p+1} = \theta_1 + 2\pi$, such that $\overline{\Gamma}(\theta_{2k}, \theta_{2k+1}; 1)$ $(k = 1, 2, \ldots, p)$ refer to all maximal closed arcs of E, and $\overline{\Gamma}(\theta_{2k-1}, \theta_{2k}; 1)$ $(k = 1, 2, \ldots, p)$ refer to all maximal open arcs of Ω.

In the following, we distinguish two cases for the proof:

(1) $\mu > \frac{\pi}{\omega}$.

We prove by contradiction. Suppose Theorem 2.17 is invalid, then $I < \frac{\pi}{\mu} < \omega$. We select arbitrarily a number ε, such that

$$0 < \varepsilon < \frac{\omega}{2}, \qquad \frac{\pi}{\omega - 2\varepsilon} < \mu, \qquad I + 2\varepsilon < \frac{\pi}{\mu}.$$

Then we choose a number η, such that

$$0 < \eta$$
$$< \min\left\{ \frac{1}{3}\left(\mu - \frac{\pi}{\omega - 2\varepsilon} \right) \right.$$
$$\left(\mu - \sqrt{\frac{\mu\pi}{\omega - 2\varepsilon}} \right)\left[3 + \sqrt{\mu\pi^{-1}(\omega - 2\varepsilon)} \right]^{-1},$$
$$\left. \left(\frac{\pi}{I + 2\varepsilon} - \mu \right)\left[3 + \sqrt{\mu\pi^{-1}(\omega - 2\varepsilon)} \right]^{-1}\left[\sqrt[3]{\mu\pi^{-1}(\omega - 2\varepsilon)} - 1 \right] \right\}. \tag{2.86}$$

According to Lemma 2.11 there exists a value r_0, such that

$$r_0^\eta \ge (1 + 8\pi^{-1}), \qquad r_0^{\pi/(I+2\varepsilon)\cdot(\sqrt[3]{\mu_\pi^{-1}(\omega - 2_\varepsilon)} - 1)} \ge 2,$$

and when $r \ge r_0$, the following formula holds:

$$\log M\{\overline{\Omega}(\theta_{2k-1} + \varepsilon, \theta_{2k} - \varepsilon; r), f\} \le r^{(\pi/(\omega - 2\varepsilon) + \eta)}, \tag{2.87}$$
$$k = 1, 2, \ldots, p.$$

On the other hand, under the definition of the lower order, there exists an infinite sequence r_n $(n = 1, 2, \ldots)$, such that

$$r_n^{\mu - \eta} \le \log M(r_n, f) \le r_n^{\mu + \eta}, \qquad n = 1, 2, \ldots, \tag{2.88}$$

and

$$r_{n+1} \ge r_n^{\mu(\omega - 2\varepsilon)/\pi}, \qquad r_n \ge r_0, \qquad n = 1, 2, \ldots. \tag{2.89}$$

Now for each value n, we define a positive integer m_n, such that

$$\left\{ \frac{\log r_{n+1}}{\log r_n} \right\}^{1/m_n} < \left\{ \frac{\mu(\omega - 2\varepsilon)}{\pi} \right\}^{1/2}, \tag{2.90}$$

and

$$\left\{ \frac{\log r_{n+1}}{\log r_n} \right\}^{1/(m_n - 1)} \geq \left\{ \frac{\mu(\omega - 2\varepsilon)}{\pi} \right\}^{1/2} . \qquad (2.91)$$

According to formulas (2.89) and (2.90), we conclude that $m_n \geq 3$. Further based on formula (2.91) we assure that $m_n < +\infty$. Let

$$\nu_n = \left\{ \frac{\log r_{n+1}}{\log r_n} \right\}^{1/m_n} - 1 ;$$

then

$$r_{n+1} = r_n^{(1 + \nu_n)^{m_n}} . \qquad (2.92)$$

We have from formula (2.91) that

$$\left\{ \frac{\log r_{n+1}}{\log r_n} \right\}^{1/m_n} \geq \left\{ \frac{\mu(\omega - 2\varepsilon)}{\pi} \right\}^{(m_n - 1)/(2 m_n)} \geq \left\{ \frac{\mu(\omega - 2\varepsilon)}{\pi} \right\}^{1/3} .$$

And hence

$$\nu_n \geq \left\{ \frac{\mu(\omega - 2\varepsilon)}{\pi} \right\}^{1/3} - 1 > 0 . \qquad (2.93)$$

On the other hand, we obtain according to formula (2.90) that

$$\nu_n < \left\{ \frac{\mu(\omega - 2\varepsilon)}{\pi} \right\}^{1/2} - 1 . \qquad (2.94)$$

Applying Lemma 2.10, we derive from formulas (2.87) and (2.88) that

$$\log M\{\overline{\Gamma}(\theta_{2k} - \varepsilon, \theta_{2k+1} + \varepsilon ; r_n^{(1 + \nu_n)^{m_n - 1}}), f\}$$

$$\leq r_{n+1}^{\pi/(\omega - 2\varepsilon) + \eta}$$

$$+ \frac{4 \left(\dfrac{r_n^{(1 + \nu_n)^{m_n - 1}}}{r_{n+1}} \right)^{\pi/(I + 2\varepsilon)}}{\pi \left\{ 1 - \left(\dfrac{r_n^{(1 + \nu_n)^{m_n - 1}}}{r_{n+1}} \right)^{2\pi/(I + 2\varepsilon)} \right\}} \cdot r_{n+1}^{\mu + \eta}, \qquad k = 1, 2, \ldots, p .$$

Further based on formula (2.92) we get

$$\log M\{\overline{\Gamma}(\theta_{2k} - \varepsilon, \theta_{2k+1} + \varepsilon ; r_n^{(1 + \nu_n)^{m_n - 1}}), f\},$$

$$\leq r_n^{(1 + \nu_n)^{m_n - 1}(1 + \nu_n)(\pi/(\omega - 2\varepsilon) + \eta)}$$

$$+ \frac{4}{\pi \left\{ 1 - r_n^{(1 + \nu_n)^{m_n - 1} \cdot (-2\pi\nu_n)/(I + 2\varepsilon)} \right\}} r_n^{(1 + \nu_n)^{m_n - 1}\{(1 + \nu_n)(\mu + \eta) - (\pi\nu_n)/(I + 2\varepsilon)\}} ,$$

$$k = 1, 2, \ldots, p .$$

Notice that

$$r_n^{(1 + \nu_n)^{m_n - 1} \cdot (2\pi\nu_n)/(I + 2\varepsilon)} \geq r_0^{\pi/(I + 2\varepsilon) \cdot (\sqrt{\mu\pi^{-1}(\omega - 2\varepsilon) - 1})} \geq 2 .$$

Then

$$\log M\{\overline{\Gamma}(\theta_{2k} - \varepsilon, \theta_{2k+1} + \varepsilon; r_n^{(1+\nu_n)^{m_n-1}}), f\}$$
$$\leq r_n^{(1+\nu_n)^{m_n-1}(1+\nu_n)(\pi/(\omega-2\varepsilon)+\eta)}$$
$$+ \frac{8}{\pi} r_n^{(1+\nu_n)^{m_n-1}\{(1+\nu_n)(\mu+\eta)-(\pi\nu_n)/(I+2\varepsilon)\}}, \qquad k = 1, 2, \ldots, p.$$

It yields from formulas (2.86) and (2.94) that

$$(1 + \nu_n)\left(\frac{\pi}{\omega - 2\varepsilon} + \eta\right) \leq \sqrt{\mu\pi^{-1}(\omega - 2\varepsilon)}\left(\frac{\pi}{\omega - 2\varepsilon} + \eta\right)$$
$$= \sqrt{\mu\pi(\omega - 2\varepsilon)^{-1}} + \sqrt{\mu\pi^{-1}(\omega - 2\varepsilon)} \cdot \eta$$
$$\leq \mu - 3\eta.$$

Besides, according to formulas (2.86), (2.93) and (2.94) we deduce

$$(1 + \nu_n)(\mu + \eta) - \frac{\pi\nu_n}{I + 2\varepsilon}$$
$$= \mu + (1 + \nu_n)\eta - \left(\frac{\pi}{I + 2\varepsilon} - \mu\right)\nu_n$$
$$\leq \mu + \sqrt{\mu\pi^{-1}(\omega - 2\varepsilon)} \cdot \eta - \left(\frac{\pi}{I + 2\varepsilon} - \mu\right)$$
$$\times \left\{\sqrt[3]{\mu\pi^{-1}(\omega - 2\varepsilon)} - 1\right\} \leq \mu - 3\eta.$$

Hence we find that

$$\log M\{\overline{\Gamma}(\theta_{2k} - \varepsilon, \theta_{2k+1} + \varepsilon; r_n^{(1+\nu_n)^{m_n-1}}), f\}$$
$$\leq \left(1 + \frac{8}{\pi}\right) r_n^{(1+\nu_n)^{m_n-1}(\mu-3\eta)}, \qquad k = 1, 2, \ldots, p.$$

Notice that

$$\left(1 + \frac{8}{\pi}\right) r_n^{-\eta(1+\nu_n)^{m_n-1}} \leq \left(1 + \frac{8}{\pi}\right) r_0^{-\eta} \leq 1;$$

then it follows that

$$\log M\{\overline{\Gamma}(\theta_{2k} - \varepsilon, \theta_{2k+1} + \varepsilon; r_n^{(1+\nu_n)^{m_n-1}}), f\} \tag{2.95}$$
$$\leq r_n^{(1+\nu_n)^{m_n-1}(\mu-2\eta)}, \qquad k = 1, 2, \ldots, p.$$

On the other hand, according to formula (2.86) we obtain $\frac{\pi}{\omega-2\varepsilon} + \eta \leq \mu - 2\eta$.
Consequently, we derive from formula (2.87) that

$$\log M\{\overline{\Omega}(\theta_{2k-1} + \varepsilon, \theta_{2k} - \varepsilon, r_n^{(1+\nu_n)^{m_n-1}}), f\}$$
$$\leq r_n^{(1+\nu_n)^{m_n-1}(\mu-2\eta)}, \qquad k = 1, 2, \ldots, p.$$

Combining this with formula (2.95) we arrive at the conclusion that

$$\log M(r_n^{(1+\nu_n)^{m_n-1}}, f) \leq r_n^{(1+\nu_n)^{m_n-1}(\mu-2\eta)}.$$

In the following, by repeating the above discussion $m_n - 1$ times we may obtain successively the following expressions:

$$\log M(r_n^{(1+\nu_n)^{m_n-2}}, f) \leq r_n^{(1+\nu_n)^{m_n-2}(\mu-2\eta)},$$

$$\cdots\cdots$$

$$\log M(r_n, f) \leq r_n^{\mu-2\eta},$$

which, however, contradict formula (2.88).

(2) $\mu \leq \frac{\pi}{\omega}$.

Assume that Theorem 2.17 is not valid, then $I < \omega \leq \frac{\pi}{\mu}$. We select arbitrarily a fixed number ε, such that

$$0 < \varepsilon < \frac{\omega}{2}, \quad I + 2\varepsilon < \omega - 2\varepsilon, \quad \frac{\pi}{\omega - 2\varepsilon} < \lambda.$$

And then we choose a finite value K, $\frac{\pi}{\omega-2\varepsilon} < K < \lambda$. According to the definition of order, there exists an infinite sequence r_n $(n = 1, 2, \ldots)$, such that

$$\log M(r_n, f) \geq r_n^K. \tag{2.96}$$

Finally we select a number η, such that

$$0 < \eta < (1 + P)^{1/2} - 1, \tag{2.97}$$

and

$$P = \min\left\{1, \left(3 + \frac{\pi}{\omega - 2\varepsilon}\right)^{-1}\left(\frac{\pi}{I + 2\varepsilon} - \frac{\pi}{\omega - 2\varepsilon}\right), \right.$$
$$\left.\left(4 + \frac{\pi}{\omega - 2\varepsilon}\right)^{-1}\left(K - \frac{\pi}{\omega - 2\varepsilon}\right)\right\}. \tag{2.98}$$

Taking Lemma 2.11 into account, we find that a value r_0 exists, such that

$$r_0^\eta \geq 2, \quad r_0^{[(1+p)^{1/2}-1]\eta} \geq \frac{16}{\pi}, \quad r_0^{(\pi/(I+2\varepsilon)[(1+p)^{1/2}-1]} \geq 2, \tag{2.99}$$

and when $r \geq r_0$, we have

$$\log M\{\overline{\Omega}(\theta_{2k-1} + \varepsilon, \theta_{2k} - \varepsilon; r), f\}$$
$$\leq r^{\pi/(\omega-2\varepsilon)+\eta}, \quad k = 1, 2, \ldots, P. \tag{2.100}$$

On the other hand, again according to the definition of the lower order, there exists an infinite sequence t_n $(n = 1, 2, \ldots)$, such that

$$\log M(t_n, f) \leq t_n^{\mu+\eta}, \tag{2.101}$$

and

$$t_n \geq r_n^{1+P}. \tag{2.102}$$

From formulas (2.96) and (2.102), we find there must exist a positive integer n_0, such that when $n \geq n_0$, we have $t_n \geq r_n \geq r_0$. Now, for each number n $(n > n_0)$, we denote a positive integer m_n, such that

$$\left\{ \frac{\log t_n}{\log r_n} \right\}^{1/m_n} < 1 + P, \tag{2.103}$$

and

$$\left\{ \frac{\log t_n}{\log r_n} \right\}^{1/(m_n-1)} \geq 1 + P. \tag{2.104}$$

We therefore conclude from formulas (2.102) and (2.103) that $m_n \geq 2$. Besides, according to formulas (2.104) we conclude that $m_n < +\infty$. Let

$$\nu_n = \left\{ \frac{\log t_n}{\log r_n} \right\}^{1/m_n} - 1;$$

then $t_n = r_n^{(1+\nu_n)m_n}$. According to (2.104) we have

$$\left\{ \frac{\log t_n}{\log r_n} \right\}^{1/m_n} \geq (1+P)^{(m_n-1)/m_n} \geq (1+P)^{1/2}.$$

And hence

$$\nu_n \geq (1+P)^{1/2} - 1. \tag{2.105}$$

On the other hand, according to (2.103) we have

$$\nu_n < P. \tag{2.106}$$

Notice that $\mu \leq \frac{\pi}{\omega}$. Therefore, $\mu < \frac{\pi}{\omega-2\varepsilon}$. It follows from formula (2.101) that

$$\log M(t_n, f) \leq t_n^{\pi/(\omega-2\varepsilon)+\eta}.$$

Furthermore, when $n > n_0$, we have

$$\log M(r_n^{(1+\nu_n)^{m_n-1}}, f) \leq 2r_n^{(1+\nu_n)^{m_n-1}(1+\nu_n)(\pi/(\omega-2\varepsilon)+\eta)}. \tag{2.107}$$

Applying Lemma 2.10, we derive from formulas (2.100) and (2.107) that

$$\log M\{\overline{\Gamma}(\theta_{2k} - \varepsilon, \theta_{2k+1} + \varepsilon; r_n^{(1+\nu_n)^{m_n-2}}), f\}$$
$$\leq r_n^{(1+\nu_n)^{m_n-2}(1+\nu_n)\{\pi/(\omega-2\varepsilon)+\eta\}}$$
$$+ \frac{4\left(\frac{r_n^{(1+\nu_n)^{m_n-2}}}{r_n^{(1+\nu_n)^{m_n-1}}}\right)^{\pi/(I+2\varepsilon)}}{\pi\left\{1 - \left(\frac{r_n^{(1+\nu_n)^{m_n-2}}}{r_n^{(1+\nu_n)^{m_n-1}}}\right)^{2\pi/(I+2\varepsilon)}\right\}} 2r_n^{(1+\nu_n)^{m_n-1}(1+\nu_n)(\pi/(\omega-2\varepsilon)+\eta)},$$

$$k = 1, 2, \ldots, P.$$

Notice that

$$r_n^{(1+\nu_n)^{m_n-2} \cdot \frac{2\pi\nu_n}{I+2\varepsilon}} \geq r_0^{\frac{2\pi}{I+2\varepsilon}\{(1+P)^{1/2}-1\}} \geq 2.$$

Then

$$\log M\{\overline{\Gamma}(\theta_{2k} - \varepsilon, \, \theta_{2k+1} + \varepsilon; \, r_n^{(1+\nu_n)^{m_n-2}}), f\}$$

$$\leq r_n^{(1+\nu_n)^{m_n-2}(1+\nu_n)(\pi/(\omega-2\varepsilon)+\eta)} \tag{2.108}$$

$$+ \frac{16}{\pi} r_n^{(1+\nu_n)^{m_n-2}\{(1+\nu_n)^2(\pi/(\omega-2\varepsilon)+\eta)-(\pi\nu_n)/(I+2\varepsilon)\}},$$

$$k = 1, 2, \ldots, P.$$

Further note that

$$(1 + \nu_n)^2 \left(\frac{\pi}{\omega - 2\varepsilon} + \eta \right) - \frac{\pi\nu_n}{I + 2\varepsilon}$$

$$= (1 + \nu_n)^2 \left(\frac{\pi}{\omega - 2\varepsilon} + \eta \right) - (1 + \nu_n) \left(\frac{\pi}{\omega - 2\varepsilon} + \eta \right)$$

$$+ (1 + \nu_n) \left(\frac{\pi}{\omega - 2\varepsilon} + \eta \right) - \frac{\pi\nu_n}{I + 2\varepsilon} + \nu_n\eta - \nu_n\eta$$

$$= (1 + \nu_n)\nu_n \left(\frac{\pi}{\omega - 2\varepsilon} + \eta \right) - \frac{\pi\nu_n}{I + 2\varepsilon} + \nu_n\eta$$

$$+ (1 + \nu_n) \left(\frac{\pi}{\omega - 2\varepsilon} + \eta \right) - \nu_n\eta$$

$$= \nu_n \left\{ \frac{\pi}{\omega - 2\varepsilon} - \frac{\pi}{I + 2\varepsilon} + \frac{\pi\nu_n}{\omega - 2\varepsilon} + 2\eta + \nu_n\eta \right\}$$

$$+ (1 + \nu_n) \left(\frac{\pi}{\omega - 2\varepsilon} + \eta \right) - \nu_n\eta.$$

It follows from formulas (2.97) and (2.105) that

$$(1 + \nu_n)^2 \left(\frac{\pi}{\omega - 2\varepsilon} + \eta \right) - \frac{\pi\nu_n}{I + 2\varepsilon}$$

$$\leq \nu_n \left\{ \frac{\pi}{\omega - 2\varepsilon} - \frac{\pi}{I + 2\varepsilon} + \left(\frac{\pi}{\omega - 2\varepsilon} + 2 + \eta \right) \nu_n \right\}$$

$$+ (1 + \nu_n) \left(\frac{\pi}{\omega - 2\varepsilon} + \eta \right) - \nu_n\eta.$$

By formulas (2.98) and (2.106) and $\eta \leq 1$ we obtain

$$(1 + \nu_n)^2 \left(\frac{\pi}{\omega - 2\varepsilon} + \eta \right) - \frac{\pi\nu_n}{I + 2\varepsilon} \leq (1 + \nu_n) \left(\frac{\pi}{\omega - 2\varepsilon} + \eta \right) - \nu_n\eta.$$

Hence formula (2.108) yields

$$\log M\{\overline{\Gamma}(\theta_{2k} - \varepsilon, \, \theta_{2k+1} + \varepsilon; \, r_n^{(1+\nu_n)^{m_n-2}}), f\}$$

$$\leq r_n^{(1+\nu_n)^{m_n-2}(1+\nu_n)(\pi/(\omega-2\varepsilon)+\eta)}$$

$$+ \frac{16}{\pi} r_n^{(1+\nu_n)^{m_n-2}(1+\nu_n)(\pi/(\omega-2\varepsilon)+\eta)-(1+\nu_n)^{m_n-2}\nu_n\eta}, \qquad k = 1, 2, \ldots, p.$$

Further based on formulas (2.99) and (2.105) we get

$$\frac{16}{\pi} r_n^{-\eta\nu_n(1+\nu_n)^{m_n-2}} \leq \frac{16}{\pi} r_0^{-\eta\{(1+p)^{1/2}-1\}} \leq 1.$$

And hence

$$\log M\{\overline{\Gamma}(\theta_{2k} - \varepsilon, \theta_{2k+1} + \varepsilon; r_n^{(1+\nu_n)^{m_n-2}}), f\}$$
$$\leq 2r_n^{(1+\nu_n)^{m_n-2}(1+\nu_n)(\pi/(\omega-2\varepsilon)+\eta)}, \qquad k = 1, 2, \ldots, p.$$

Combining this with formula (2.100), we deduce

$$\log M(r_n^{(1+\nu_n)^{m_n-2}}, f) \leq 2r_n^{(1+\nu_n)^{m_n-2}(1+\nu_n)(\pi/(\omega-2\varepsilon)+\eta)}.$$

In the following, by repeating the above discussion $m_n - 2$ times, we may derive successively:

$$\log M(r_n^{(1+\nu_n)^{m_n-3}}, f) \leq 2r_n^{(1+\nu_n)^{m_n-3}(1+\nu_n)(\pi/(\omega-2\varepsilon)+\eta)},$$
$$\cdots\cdots \tag{2.109}$$
$$\log M(r_n, f) \leq 2r_n^{(1+\nu_n)(\pi/(\omega-2\varepsilon)+\eta)}.$$

From formulas (2.97), (2.105) and (2.106) we obtain $\eta < \nu_n < P$. Therefore

$$(1 + \nu)\left(\frac{\pi}{\omega - 2\varepsilon} + \eta\right) \leq (1 + P)\left(\frac{\pi}{\omega - 2\varepsilon} + P\right)$$
$$= \frac{\pi}{\omega - 2\varepsilon} + \frac{\pi P}{\omega - 2\varepsilon} + P + P^2.$$

Again based on formula (2.98) we get

$$(1 + \nu_n)\left(\frac{\pi}{\omega - 2\varepsilon} + \eta\right) \leq K - 2P.$$

Notice that $\eta < P$; then

$$(1 + \nu_n)\left(\frac{\pi}{\omega - 2\varepsilon} + \eta\right) \leq K - 2\eta.$$

Hence, formula (2.109) yields $\log M(r_n, f) \leq 2r_n^{K-2\eta}$. Hence it follows from formula (2.99) that $2r_n^{-\eta} \leq 2r_0^{-\eta} \leq 1$. Finally we derive $\log M(r_n, f) \leq r_n^{K-\eta}$, which, however, contradicts formula (2.96). Theorem 2.17 is thus proved.

Theorem 2.17 has the following corollaries:

COROLLARY 1. *Let $f(z)$ be an entire function of lower order $\mu < +\infty$, and has only finitely many Julia directions. Then the order λ of $f(z)$ must be finite.*

Indeed, when $f(z)$ merely has finitely many Julia directions, it follows that $\omega > 0$ and $I = 0$. Besides, according to $\mu < +\infty$, we have $\frac{\pi}{\mu} > 0$. Hence $I < \min\{\frac{\pi}{\mu}, \omega\}$. Consequently from Theorem 2.17 we arrive at a conclusion that $\lambda \leq \frac{\pi}{\omega} < +\infty$.

COROLLARY 2. *Suppose that $f(z)$ is an entire function of lower order $\mu < +\infty$, and has merely one Julie direction; then the order λ of $f(z)$ is $\leq \frac{1}{2}$.*

Indeed, according to the proof of Corollary 1, first we have $\lambda \leq \frac{\pi}{\omega}$. Next, since $f(z)$ has only one Julia direction, we have $\omega = 2\pi$. And hence $\lambda \leq \frac{1}{2}$.

COROLLARY 3. *Suppose that the entire function $f(z)$ has merely one maximal closed arc, and that $I < \min\{\frac{\pi}{\mu}, \omega\}$; then the order λ of $f(z)$ is < 1, and also $I < \frac{\pi}{\lambda}$.*

Indeed, according to $I < \min\{\frac{\pi}{\mu}, \omega\}$, then $\mu < +\infty$ and $\lambda \leq \frac{\pi}{\omega}$. Hence, $\omega \leq \frac{\pi}{\mu}$ and, in turn, $I < \omega \leq \frac{\pi}{\lambda}$. On the other hand, from $\lambda \leq \frac{\pi}{\omega} = \frac{\pi}{2\pi - I}$ and $I < \frac{\pi}{\lambda}$, it follows that $\lambda < 1$.

§2.6. On the Nevanlinna direction

2.6.1. Definition of the Nevanlinna direction. The Julia direction and the Borel direction correspond to the Picard Theorem and the Borel Theorem on the open plane, respectively. People may ask naturally: Is there the so-called Nevanlinna direction that corresponds to the Nevanlinna deficiency relation? This is a natural question but it is quite difficult to describe the answer precisely from a mathematical point of view. For instance, how to define a deficient value of a meromorphic function for a particular direction is itself a problem worth studying. Recently, Li Yi-nian and the author worked out a method to define the Nevanlinna direction, and under this meaning, they proved its existence [29a].

We recall some notations and introduce the new ones. Let $w = w(z)$ be a nonconstant meromorphic function on the open plane $|z| < +\infty$, and let

$$S(r) = \frac{1}{\pi} \iint_{|z|<r} \frac{|w'(z)|^2}{(1 + |w(z)|^2)^2} \, d\omega, \qquad d\omega = r \, dr \, d\theta, \quad z = re^{i\theta},$$

$$S(E) = \frac{1}{\pi} \iint_E \frac{|w'(z)|^2 \, d\omega}{(1 + |w(z)|^2)^2},$$

$$T_0(r, w) = \int_0^r \frac{S(t)}{t} \, dt,$$

$$T_0(\Omega(\theta_1, \theta_2; r)) = \int_0^r \frac{S\{\Omega(\theta_1, \theta_2; t)\}}{t} \, dt,$$

as well as

$$N\{\Omega(\theta_1, \theta_2; r), a\} = N\{\Omega(\theta_1, \theta_2; r), w = a\}$$
$$= \int_0^r \frac{n\{\Omega(\theta_1, \theta_2; t), a\}}{t} \, dt.$$

Now we introduce the definition of the Nevanlinna direction: For any arbitrary complex value a, let

$$\delta(a, \varphi) = 1 - \limsup_{\varepsilon \to 0} \varlimsup_{\substack{\Omega_\varphi \\ r \to +\infty}} \frac{N\{\Omega(\theta_1, \theta_2; r), a\}}{T_0\{\Omega(\theta_1, \theta_2; r)\}},$$

where

$$\Omega_\varphi = \{\Omega(\varphi_1, \varphi_2) | \varphi - \varepsilon < \varphi_1 < \varphi_2 < \varphi + \varepsilon, \ \varepsilon > 0\}.$$

Then we call $\delta(a, \varphi)$ the deficiencies of value a regarding the direction $\Delta(\varphi)$. If $\delta(a, \varphi) > 0$, we call value a the deficient value regarding the direction $\Delta(\varphi)$. Apparently we have $\delta(a, \varphi) \le 1$. When $w(z)$ does not take value a in $\Omega(\varphi - \varepsilon, \varphi + \varepsilon)$ $(\varepsilon > 0)$, we obtain $\delta(a, \varphi) = 1$.

Let

$$\delta\{a, \Omega(\varphi_1, \varphi_2)\} = 1 - \varlimsup_{r \to +\infty} \frac{N\{\Omega(\varphi_1, \varphi_2; r), a\}}{T_0\{\Omega(\varphi_1, \varphi_2; r)\}};$$

then we call $\delta\{a, \Omega(\varphi_1, \varphi_2)\}$ the deficiency of value a regarding the angular domain $\Omega(\varphi_1, \varphi_2)$. If $\delta\{a, \Omega(\varphi_1, \varphi_2)\} > 0$, then we call value a the deficient value regarding the angular domain $\Omega(\varphi_1, \varphi_2)$.

If for the arbitrary finite deficient values $a_1, a_2, \ldots, a_q$ $(q < +\infty)$ regarding the direction $\Delta(\varphi)$, it is always true that $\sum_{\nu=1}^{q} \delta(a_\nu, \varphi) \le 2$, then we call $\Delta(\varphi)$ a Nevanlinna direction of $w(z)$.

Under the definition, we conclude that if $\Delta(\varphi)$ is a Nevanlinna direction of $w(z)$, then the set of all deficient values of the direction $\Delta(\varphi)$ constitutes a countable set, with the sum of all deficiencies of direction $\Delta(\varphi) \le 2$. Besides, a Nevanlinna direction must also be a Julia direction.

2.6.2. Some lemmas.

LEMMA 2.13. *Let* $w = w(z)$ *be a meromorphic function in the disk* $|z| \le 1$ *and* $a_1, a_2, \ldots, q_q$ $(3 \le q < +\infty)$ *be* q *distinct complex numbers with*

$$\sum_{\nu=1}^{q} n(1, a_\nu) < +\infty.$$

Then for any arbitrary value r $(0 < r < 1)$ *we have*

$$(q - 2)S(r) \le \sum_{\nu=1}^{q} n(1, a_\nu) + \frac{A}{1 - r}, \qquad (2.110)$$

where $A > 0$ *is a constant depending merely on* $a_1, a_2, \ldots, a_q$.

PROOF. According to formula (1.66) we have

$$(q - 2)S(r) \le \sum_{\nu=1}^{q} n(r, a_\nu) + hL(r) \le \sum_{\nu=1}^{q} n(1, a_\nu) + hL(r),$$

where h is a constant depending merely on $a_1, a_2, \ldots, a_q$, and

$$L(r) = \int_0^{2\pi} \frac{|w'(re^{i\theta})|}{1 + |w(re^{i\theta})|^2} r \, d\theta.$$

If for any arbitrary value r', $r \le r' < 1$, we have $(q-2)S(r') - \sum_{\nu=1}^{q} n(1, a_\nu) > 0$, then from the inequality $[L(r)]^2 \le 2\pi^2 r \frac{dS(r)}{dr}$, we conclude that

$$\left\{(q - 2)S(r') - \sum_{\nu=1}^{q} n(1, a_\nu)\right\}^2 \le h^2 (L(r'))^2 \le 2\pi^2 h^2 r' \frac{dS(r')}{dr'}.$$

And hence

$$1 - r < \int_r^1 \frac{dr'}{r'} \le 2\pi^2 h^2 \int_r^1 \frac{dS(r')}{\{(q-2)S(r') - \sum_{\nu=1}^q n(1, a_\nu)\}^2}$$

$$\le \frac{2\pi^2 h^2}{q-2} \cdot \frac{1}{\{(q-2)S(r) - \sum_{\nu=1}^q n(1, a_\nu)\}},$$

$$(q-2)S(r) \le \sum_{\nu=1}^q n(1, a_\nu) + \frac{2\pi^2 h^2}{(q-2)(1-r)}.$$

If for a certain value r', $r \le r' < 1$, $S(r')(q-2) - \sum_{\nu=1}^q n(1, a_\nu) \le 0$, then $(q-2)S(r) \le (q-2)S(r') \le \sum_{\nu=1}^q n(1, a_\nu)$. Therefore, formula (2.110) holds, implying that Lemma 2.13 is proved.

LEMMA 2.14. *Let $w = w(z)$ be a meromorphic function on $\overline{\Omega}(-\theta, \theta)$ $(0 < \theta < \pi)$ and $a_1, a_2, \ldots, a_q$ $(3 \le q \le +\infty)$ be q distinct complex numbers. Then for any arbitrary value θ' $(0 < \theta' < \theta)$, value σ $(\sigma > 1)$, and positive integer m, provided that r is sufficiently large we have*

$$(q-2)T_0\{\Omega(-\theta', \theta'; r)\}$$

$$\le \left(1 + \frac{1}{m}\right) \sum_{\nu=1}^q N(\Omega(-\theta, \theta; r\sigma^{2m}), a_\nu) + O((\log r)^2).$$

PROOF. Let $r_i = \sigma^{m_i}$, $i = 0, 1, 2, \ldots$; $r_{ij} = \sigma^{mi+j}$, $j = 0, 1, \ldots, m-1$. It is obvious that $r_{i0} = r_i$, $r_{im} = r_{i+1}$. For any arbitrarily selected number t $(t \ge r_1)$, there exists a positive integer k such that $r_k \le t < r_{k+1}$. Clearly, there is an integer j_0, $0 \le j_0 \le m-1$, such that

$$\sum_{\nu=1}^q n\left\{\bigcup_{i=0}^{k+1} \Omega(-\theta, \theta; r_{ij_0}, r_{ij_0+1}), a_\nu\right\}$$

$$\le \frac{1}{m} \sum_{\nu=1}^q n\{\Omega(-\theta, \theta; r_{k+2}), a_\nu\}.$$

We construct the transformation

$$\zeta = \frac{z}{r_{i+1j_0+1}} \qquad (0 \le i \le k);$$

then $\Omega(-\theta, \theta; r_{ij_0}, r_{i+1j_0+1})$ and $\Omega(-\theta', \theta'; r_i', r_{i+1}')$ $(r_i' = \sqrt{r_{ij_0} \cdot r_{ij_0+1}}$, $r_{i+1}' = \sqrt{r_{i+1j_0} \cdot r_{i+1j_0+1})}$ are mapped into $\Omega(-\theta, \theta; \frac{1}{\sigma^{m+1}}, 1)$ and $\Omega(-\theta', \theta'; \frac{1}{\sigma^{m+(1/2)}}, \frac{1}{\sigma^{1/2}})$ on the ζ-plane, respectively, and also point $\sqrt{r_{ij_0} \cdot r_{i+1j_0+1}}$ changes into point $\sigma^{-(m+1)/2}$ on the ζ-plane. Further, $\Omega(-\theta, \theta, \frac{1}{\sigma^{m+1}}, 1)$ is mapped conformally into the disk $|\xi| < 1$ on the ξ-plane, with point $\zeta = \sigma^{-(m+1)/2}$ corresponding to the origin $\xi = 0$. Obviously, there exists a constant χ depending merely on m, σ, θ, and θ', such that the image domain of $\Omega(-\theta', \theta'; \frac{1}{\sigma^{m+(1/2)}}, \frac{1}{\sigma^{1/2}})$ on the ζ-plane is contained in the disk $|\zeta| < \chi$. Applying Lemma 2.13 we get

$$(q-2)S\{\Omega(-\theta', \theta'; r_i', r_{i+1}')\}$$

$$\leq \sum_{\nu=1}^{q} n\{\Omega(-\theta, \theta; r_{ij_0}, r_{i+1j_0+1}), a_\nu\} + \frac{A}{1-\chi}, \qquad i = 0, 1, \ldots, k,$$

where $A > 0$ is a constant merely depending on $a_1, a_2, \ldots, a_q$. And hence we have

$$(q-2)\sum_{i=0}^{k} S\{\Omega(-\theta', \theta'; r_i', r_{i+1}')\}$$

$$\leq \sum_{\nu=1}^{q}\sum_{i=0}^{k} n\{\Omega(-\theta, \theta; r_{ij_0}, r_{i+1j_0+1}), a_\nu\} + \frac{k+1}{1-\chi}A.$$

Notice that $r_k \leq t < r_{k+1}$ and $r_{k+2} \leq t\sigma^{2m}$. It follows that

$$(q-2)S\{\Omega(-\theta', \theta'; t)\}$$

$$\leq \left(1 + \frac{1}{m}\right)\sum_{\nu=1}^{q} n\{\Omega(-\theta, \theta; t\sigma^{2m}), a_\nu\}$$

$$+ (q-2)S(\Omega(-\theta', \theta'; \sigma^m)) + \frac{2A\log t}{m(1-x)\log\sigma}.$$

Therefore

$$(q-2)\int_{\sigma^m}^{r} \frac{S\{\Omega(-\theta', \theta'; t)\}}{t}\,dt$$

$$\leq \left(1 + \frac{1}{m}\right)\sum_{\nu=1}^{q}\int_{0}^{r} \frac{n\{\Omega(-\theta, \theta'; t\sigma^{2m}), a_\nu\}}{t}\,dt$$

$$+ (q-2)S\{\Omega(-\theta', \theta'; \sigma^m)\}\int_{\sigma^m}^{r} \frac{dt}{t}$$

$$+ \frac{2A}{m(1-\chi)\log\sigma}\int_{\sigma^m}^{r} \frac{\log t}{t}\,dt.$$

Consequently

$$(q-2)T_0\{\Omega(-\theta', \sigma'; r)\}$$

$$\leq \left(1 + \frac{1}{m}\right)\sum_{\nu=1}^{2} N\{\Omega(-\theta, \theta; re^{2\sigma}), a_\nu\}$$

$$+ (q-2)T_0\{\Omega(-\theta', \theta'; \sigma^m)\} + (q-2)S\{\Omega(-\theta', \theta'; \sigma^m)\}\log r$$

$$+ \frac{A}{m(1-x)\log\sigma}(\log r)^2$$

$$= \left(1 + \frac{1}{m}\right)\sum_{\nu=1}^{q} N\{\Omega(-\theta, \theta; r\sigma^{2m}), a_\nu\} + O((\log r)^2),$$

and Lemma 2.14 is thus proved.

For any arbitrarily given positive integer l and even positive integer k, we denote

$$\theta_{ij} = \frac{2\pi}{l}i + \frac{2\pi}{lk}j, \qquad i = 0, 1, \ldots, l-1; \; j = 0, 1, \ldots, k-1,$$

$$\theta_{ik} = \theta_{i+1o},$$

and

$$\Omega = \left\{ \Omega\left(\theta_{ij}, \theta_{ij} + \frac{2\pi}{l}\right) \mid i = 0, 1, \ldots, l-1; \; j = 0, 1, \ldots, k-1 \right\}.$$

Clearly there exist kl distinct angular domains in Ω.

LEMMA 2.15. *Let $w = w(z)$ be a meromorphic function on the open plane, which satisfies the following condition:*

$$\overline{\lim_{r \to +\infty}} \frac{T_0(r, w)}{(\log r)^2} = +\infty,$$

$$\underline{\lim_{r \to +\infty}} \frac{\log T_0(r, w)}{\log r} = \mu < +\infty.$$

Then for any arbitrarily selected number $\delta > 0$, positive integer N, $N > 3$ and positive integer l, there exists at least one angular domain $\Omega(\theta, \theta+\frac{2\pi}{l}) = \Omega_{Nl}$, such that the sum of the deficiencies of any q $(3 \leq q \leq N)$ deficient values regarding Ω_{Nl} is $\leq 2 + \delta$, and that

$$\overline{\lim_{r \to +\infty}} \frac{T_0\{\Omega(\theta, \theta + \frac{2\pi}{l}; r)\}}{(\log)^2} = +\infty.$$

PROOF. Indeed, otherwise, for any arbitrary index pair (i, j) that corresponds to $\Omega(\theta_{ij}, \theta_{ij} + \frac{2\pi}{l}) \in \Omega$, we have

$$\overline{\lim_{r \to +\infty}} \frac{T_0\{\Omega(\theta_{ij}, \theta_{ij} + \frac{2\pi}{l}; r)\}}{(\log r)^2} < +\infty, \tag{2.111}$$

or for the q_{ij} $(3 \leq q_{ij} \leq N)$ deficient values $a_1^{(i,j)}, a_2^{(i,j)}, \ldots, a_{q_{ij}}^{(i,j)}$ regarding $\Omega(\theta_{ij}, \theta_{ij} + \frac{2\pi}{l})$, it follows that

$$\sum_{\nu=1}^{q_{ij}} \delta\left\{a_\nu^{(ij)}, \Omega\left(\theta_{ij}, \theta_{ij} + \frac{2\pi}{l}\right)\right\} > 2 + \delta. \tag{2.112}$$

Let

$$K = \max_{ij} \left\{ \frac{q_{ij} - \sum_{\nu=1}^{q_{ij}} \delta\{a_\nu^{(ij)}, \Omega(\theta_{ij}, \theta_{ij} + \frac{2\pi}{l})\}}{q_{ij} - 2} \right\}.$$

Then from formula (2.112) we obtain

$$K \leq 1 - \frac{\delta}{N-2} < 1 - \frac{\delta}{N}.$$

On the other hand, according to Lemma 2.4, for any arbitrarily given number $\sigma > 1$ and positive integer m, there exists a sequence $\{R_n\}$, such that

$$R_n < R_{n+1} \to +\infty \qquad (n \to +\infty),$$

$$\lim_{n \to +\infty} \frac{T_0(R_n, w)}{(\log R_n)^2} = +\infty \tag{2.113}$$

and

$$T_0(R_n \sigma^{2m}, w) < \sigma^{2m\mu} T_0(R_n, w)(1 + o(1)) \qquad (n \to +\infty). \tag{2.114}$$

Obviously, there is an even integer j_n, $0 \le j_n \le k - 2$, such that

$$\sum_{i=0}^{i-1} T_0\{\Omega(\theta_{ij_n}, \theta_{ij_n+2}; R_n), w\} \le \frac{2}{k} T_0(R_n, w). \tag{2.115}$$

For the index pair $(i, j_n + 1)$, we find that if formula (2.111) is valid; then the following inequality holds:

$$T_0\left\{\Omega\left(\theta_{ij_n+1}, \theta_{ij_n+1} + \frac{2\pi}{l}; R_n\right), w\right\} = O\{(\log R_n)^2\}. \tag{2.116}$$

If formula (2.112) holds, then according to Lemma 2.14 we have

$$(q_{ij} - 2)T_0\{\Omega(\theta_{ij_n+2}, \theta_{i+1j_n}, R_n), w\}$$

$$\le \left(1 + \frac{1}{m}\right)\left\{q_{ij} - \sum_{\nu=1}^{q_{ij}} \delta(a_\nu^{(ij_n+1)}, \Omega(\theta_{ij_n+1}, \theta_{i+1j_n+1})) + o(1)\right\} .$$

$$T_0\{\Omega(\theta_{ij_n+1}, \theta_{i+1j_n+1}; R_n\sigma^{2m}), w\} + O\{\log R_n\}^2\}.$$

And hence

$$T_0\{\Omega(\theta_{ij_n+2}, \theta_{i+1j_n}; R_n), w\}$$

$$\le \left(1 + \frac{1}{m}\right)\{K + o(1)\} \cdot T_0\{\Omega(\theta_{ij_n+1}, \theta_{i+1j_n+1}; R_n\sigma^{2m}), w\} \tag{2.117}$$

$$+ o\{(\log R_n)^2\}.$$

Then it follows from formulas (2.115), (2.116) and (2.117) that

$$T_0(R_n, w) \le \left(1 + \frac{1}{m}\right)(K + o(1))T_0(R_n\sigma^{2m}, w)$$

$$+ \frac{2}{k} T_0(R_n, w) + o\{(\log R_n)^2\}.$$

Let $n \to +\infty$. Then according to formulas (2.114) and (2.113) we get

$$1 \le \left(1 + \frac{1}{m}\right)K\sigma^{2m\mu} + \frac{2}{k}. \tag{2.118}$$

Notice that $K < 1 - \frac{\delta}{N}$, if we select orderly m first, such that

$$\left(1 + \frac{1}{m}\right)K < 1 - \frac{\delta}{2N},$$

and then select $\sigma > 1$, such that

$$\left(1 + \frac{1}{m}\right) K \sigma^{2m\mu} < 1 - \frac{\delta}{4N}.$$

And finally we select k, such that $\frac{2}{k} < \frac{\delta}{\delta N}$; then formula (2.118) yields $1 < 1 - \frac{\delta}{\delta N}$, which leads to a contradiction, and hence Lemma 2.15 is proved.

2.6.3. Theorem on the existence of the Nevanlinna direction.

THEOREM 2.18. *Let* $w = w(z)$ *be a meromorphic function on the open plane* $|z| < +\infty$, *which satisfies the following condition*:

$$\varlimsup_{r \to +\infty} \frac{T_0(r, w)}{(\log r)^2} = +\infty,$$

$$\varliminf_{r \to +\infty} \frac{\log T_0(r, w)}{\log r} = \mu < +\infty.$$

Then $w(z)$ *has at least one Nevanlinna direction* $\Delta(\varphi)$, *and also for any arbitrary given number* $\varepsilon > 0$,

$$\varlimsup_{r \to +\infty} \frac{T_0\{\Omega(\varphi - \varepsilon, \varphi + \varepsilon; r), w\}}{(\log r)^2} = +\infty.$$

PROOF. Apply Lemma 2.15, where by letting $\delta = \delta_N = \frac{1}{N}$, $N = l$, $\Omega_{Nl} = \Omega_N = \Omega(\theta_N, \theta_N + \frac{2\pi}{N})$, we conclude that the sum of the deficiencies of any q $(3 \leq q \leq N)$ deficient values regarding Ω is $\leq 2 + \delta_N$, and also

$$\varlimsup_{r \to +\infty} \frac{T_0\{\Omega(\theta_N, \theta_N + \frac{2\pi}{N}; r), w\}}{(\log r)^2} = +\infty. \qquad (2.119)$$

Without any loss of generality, we may assume that $\theta_N \to \varphi$ $(N \to +\infty)$, otherwise, we need only to select a convergent subsequence. In the following, we prove that $\Delta(\varphi)$ is a Nevanlinna direction. In fact, otherwise, there exist q $(3 \leq q < +\infty)$ deficient values $a_1, a_2, \ldots, a_q$ regarding $\Delta(\varphi)$, such that $\sum_{\nu=1}^{q} \delta(a_\nu, \varphi) > 2$. We select a sufficiently small number $\alpha > 0$, such that $\delta(a_\nu, \varphi) > \frac{\alpha}{q}$, $\nu = 1, 2, \ldots, q$, and $\sum_{\nu+1}^{q} \delta(a_\nu, \varphi) \geq 2 + 2\alpha$. According to the definition of $\delta(a_\nu, \varphi)$, value $\varepsilon_0 > 0$ exists, such that for any arbitrary value ε $(0 < \varepsilon < \varepsilon_0)$ and angular domain $\Omega(\varphi_1, \varphi_2) \subset \Omega(\varphi - \varepsilon, \varphi + \varepsilon)$ it follows that

$$\varlimsup_{r \to +\infty} \frac{N\{\Omega(\varphi_1, \varphi_2; r), a_\nu\}}{T_0\{\Omega(\varphi_1, \varphi_2; r), w\}} \leq 1 - \delta(a_\nu, \varphi) + \frac{\alpha}{q},$$

$$1 - \varlimsup_{r \to +\infty} \frac{N\{\Omega(\varphi_1, \varphi_2; r), a_\nu\}}{T_0\{\Omega(\varphi_1, \varphi_2; r), w\}} \geq \delta(a_\nu, \varphi) - \frac{\alpha}{q}.$$

And hence

$$\sum_{\nu=1}^{q} \left\{ 1 - \varlimsup_{r \to +\infty} \frac{N\{\Omega(\varphi_1, \varphi_2; r), a_\nu\}}{T_0\{\Omega(\varphi_1, \varphi_2; r), w\}} \right\} > 2 + \alpha. \qquad (2.120)$$

On the other hand, we select a sufficiently large positive integer N, such that $\delta_n < \alpha$, $q \leq N$, and also $\Omega_N \subset \Omega(\varphi - \varepsilon, \varphi + \varepsilon)$. Let $\Omega_N = \Omega(\varphi_1, \varphi_2)$; then

$$\sum_{\nu=1}^{q} \delta(a_\nu, \Omega_N) \leq 2 + \delta_N < 2 + \alpha.$$

This, however, contradicts formula (2.120). Hence, $\Delta(\varphi)$ is a Nevanlinna direction of $w(z)$.

For any arbitrary value $\varepsilon > 0$, there exists a sufficiently large positive integer N, such that $\Omega_N \subset \Omega(\varphi - \varepsilon, \varphi + \varepsilon)$. Hence, from formula (2.119) we obtain

$$\varlimsup_{r \to +\infty} \frac{T_0\{\Omega(\varphi - \varepsilon, \varphi + \varepsilon; r), w\}}{(\log r)^2} = +\infty.$$

Theorem 2.18 is thus proved completely.

§2.7. Annotated notes

2.7.1. The common Borel direction. In 1928, G. Valiron proved the existence of the Borel direction, and raised, at the same time, a critical but difficult question: Is there a common Borel direction between a meromorphic function and its derivative? By adding different conditions, G. Valiron himself [39a], A. Rauch [35b], Zhuang Chi-Tai [10a] and others, later worked to prove the existence of the common Borel direction. It was in 1951 that H. Milloux made great progress and proved the following result [30d]:

Let $f(z)$ be an entire function of order λ, $0 < \lambda < +\infty$ on the open plane $|z| < +\infty$. Then each Borel direction of order λ of its derivative $f'(z)$ is also a Borel direction of order λ of $f(z)$.

We may prove that the entire function $f(z)$ and its derivative $f'(z)$ have the same order. Hence, according to G. Valiron's result (Theorem 2.11), $f'(z)$ has at least one Borel direction of order λ. Based further on the above H. Milloux's result, we conclude that there exists at least one common Borel direction between $f(z)$ and $f'(z)$. In fact, if we make further discussion, we may derive H. Milloux's important result concerning the common Borel direction [30d]:

Let $f(z)$ be an entire function of order λ, $0 < \lambda < +\infty$ on the open plane $|z| < +\infty$. Then there is at least one common Borel direction among $f(z)$ and all its derivatives as well as all its antiderivatives.

The author had once extended respectively H. Milloux's and Zhuang Chi-Tai's results as follows [43a]:

Let $f(z)$ be a meromorphic function of order λ, $0 < \lambda < +\infty$ on the open plane $|z| < +\infty$. If $f(z)$ has ∞ as its Borel exceptional value, then each Borel direction of order λ of its derivative $f'(z)$ is also a Borel direction of order λ of $f(z)$. If $f(z)$ has a finite complex number as its Borel exceptional value,([3]) then each Borel direction of order λ of $f(z)$ is also a

---

([3]) Zhuang Chi-tai requires, in the meantime, ∞ to be a Borel exceptional value.

Borel direction of order λ of its nth derivative $f^{(n)}(z)$ $(n \geq 1)$. Hence, if $f(z)$ has a complex number as its Borel exceptional value, then there exists at least one common Borel direction between $f(z)$ and of its nth derivative $f^{(n)}(z)$ $(n \geq 1)$.

To date Valiron's question has not been resolved completely and is subject to further study.

2.7.2. The distribution regularity of the Borel direction. Let $f(z)$ be a meromorphic function of order λ, $0 < \lambda < +\infty$ on the open plane $|z| < +\infty$. We denote E to be the set consisting of the intersecting points between all the Borel directions of order λ of $f(z)$ and the unit circle. Clearly, according to G. Valiron's result (Theorem 2.11), we conclude that the set E is not empty. Further based on the definition of the Borel direction, we may easily prove that the set E is closed. Hence, the intersecting points between all the Borel directions of order λ of $f(z)$ and the unit circle constitute a nonempty closed set. Do the properties of such a nonempty closed set illustrate completely the distribution regularity of the Borel direction of order λ of $f(z)$? In 1976 Yang Le and the author gave an affirmative answer to this question as follows [42d]:

Let λ be any arbitrarily given finite positive number and E be an arbitrarily given nonempty closed set on the unit circle. Then there exists definitely a meromorphic function $f(z)$ of order λ, such that the set composed of the intersecting points between all the Borel directions of order λ of $f(z)$ and the unit circle is exactly E.

Earlier, D. Drasin and A. Weitsman had solved the distribution regularity [13b] of the Borel directions of entire functions of finite positive order. They first introduced a concept:

If for a particular number λ $(> \frac{1}{2})$, a finite sequence $\theta_1 < \theta_2 < \cdots < \theta_n \leq \theta_1 + 2\pi$ satisfies the following conditions:
  (1) $\theta_{j+1} - \theta_j \leq \frac{\pi}{\lambda}$ $(j = 1, 2, \ldots, n - 1)$,
  (2) $\theta_n - \theta_1 \geq \frac{\pi}{\lambda}$,
  (3) only when $n = 2$, (1) an (2) take the "$=$" sign,
then it is called a chain of real numbers of order λ.

Then they proved the following result:

Given arbitrarily value λ, $0 < \lambda < +\infty$, and a nonempty closed set E on $[0, 2\pi]$, when $\lambda \leq \frac{1}{2}$, without imposing on E any additional conditions, E is just a set composed of the intersecting points between all the Borel directions of a certain entire function of order λ and the unit circle. When $\lambda > \frac{1}{2}$, in order to let E be the set composed of the intersecting points between all the Borel directions of an entire function of order λ and the unit circle, a necessary and sufficient condition is to have any one value of E be an element of a chain of order λ, and also all the elements θ_j (mod 2π) $(j = 1, 2, \ldots, n)$ in this chain of order λ belong to E.

CHAPTER 3

The Deficient Value Theory

Ever since the establishment of the Nevanlinna Theory, particularly over the recent twenty years or more, the deficient value theory has become the central issue of studying the value distribution theory. Some fruitful and impressive results have already been obtained. W. H. J. Fuchs in his recent outstanding paper sums up systematically the development of deficient values since R. Nevanlinna [16b]. In this chapter it is impossible for us to introduce the deficient value theory comprehensively. We will focus only on the discussion of the following special topic: according to Theorem 1.7, for a transcendental meromorphic function, all its deficient values constitute a countable set. On the other hand, there exhibit examples of meromorphic [18a] and entire [3a] functions having infinitely many deficient values. Then, under what circumstances will the number of deficient values be finite? Many scholars have concentrated on the study of this problem and have made estimations on its upper bound.

§3.1. The harmonic measure and the Lindelöf-type theorem

3.1.1. An estimation on the harmonic measure.([1]) Let D be a bounded domain on the open plane $|z| < +\infty$, with its boundary B formed by finitely many Jordan curves that do not intersect among one another. B is divided into B' and B'', which are made up of finitely many arcs and closed curves, respectively. By solving the Dirichlet problem, we conclude that there exists a unique bounded harmonic function $u_D(z, B')$ defined on D, such that when z tends to the interior point of B', $u_D(z, B') \to 1$, and when z tends to the interior point of B'', $u_D(z, B') \to 0$. Also when z is within D, $0 < u_D(z, B') < 1$. We call such a harmonic function $u_D(z, B')$ the harmonic measure on B' with respect to domain D that corresponds to point z. Clearly, we have that

$$u_D(z, B') + u_D(z, B'') = 1.$$

For the sequels we will often make use of the following fact:

([1]) The content of this section is extracted mainly from the book [38a].

Let $f(z)$ be a function regular on $\overline{D}$ with $|f(z)| \leq M_1$ on B' and $|f(z)| \leq M_2$ on B''; then, according to the maximum modulus principle for a sub-harmonic function, we have

$$\log|f(z)| \leq u_D(z, B') \log M_1 + U_D(z, B'') \log M_2$$
$$\leq \log M_2 + U_D(z, B') \log M_1.$$

In the following we will make an estimation on a typical harmonic measure, which plays an important role for the rest of the book.

Let D be a domain in the disk $|z| < r$ $(0 < r < +\infty)$ and satisfy the following conditions:

(1) The origin $z = 0$ belongs to domain D.

(2) $\overline{D} \cap (|z| = r)$ is a nonempty set containing at least one interior point.

(3) Assume that B is the boundary of D, the part Γ_r of B in the disk $|z| < r$ is analytical and consists of only finitely many connected branches, and also $\theta_r = B - \Gamma_r$ is made up of finitely many arcs.

For any arbitrary value t, $0 < t \leq r$, let D_t denote the connected branch of D in the disk $|z| < t$ containing the point $z = 0$, Γ_t denote the part of the boundary B_t of D_t that lies in the disk $|z| < t$, and $\theta_t = B_t - \overline{\Gamma}_t$ is composed of finitely many arcs with linear measure $t\theta(t)$. Clearly, we have $D_r = D$. If circumference $|z| = t$ intersects B, then we define $\theta^*(t) = \theta(t)$. Otherwise, we define $\theta^*(t) = +\infty$.

LEMMA 3.1. *We have, for the harmonic measure* $u_D(z, \theta_r)$,

$$U_D(0, \theta_r) \leq \frac{\sqrt{2e}}{\sqrt{1 - \kappa}} \exp\left\{ -\pi \int_0^{\kappa r} \frac{dr}{r\theta^*(r)} \right\}, \qquad 0 < \kappa < 1. \tag{3.1}$$

PROOF. First we prove the Wirtinger's inequality: Suppose that $f(x)$ and $f'(x)$ are continuous on the interval $[a, b]$ with $f(a) = f(b) = 0$; then

$$\int_a^b [f'(x)]^2 \, dx \geq \frac{\pi^2}{(b-a)^2} \int_a^b [f(x)]^2 \, dx.$$

Indeed, we can simply assume $a = 0$, $b = \pi$. Hence, we need only to prove

$$\int_0^\pi [f'(x)]^2 \, dx \geq \int_0^\pi [f(x)]^2 \, dx. \tag{3.2}$$

When $-\pi \leq x \leq 0$, we define $f(x) = -f(-x)$. Then, $f(x)$ becomes a continuous function on the closed interval $[-\pi, \pi]$ with $f(-\pi) = f(0) = f(\pi) = 0$. Hence, $f(x)$ has the following expression:

$$f(x) = \sum_{n=1}^\infty a_n \sin nx, \qquad f'(x) = \sum_{n=1}^\infty na_n \cos nx.$$

We apply the Parseval Theorem and conclude consequently that

$$\frac{2}{\pi} \int_0^\pi [f'(x)]^2\, dx = \frac{1}{\pi} \int_{-\pi}^\pi [f'(x)]^2\, dx = \sum_{n=1}^\infty n^2 a_n^2 \geq \sum_{n=1}^\infty a_n^2$$

$$= \frac{1}{\pi} \int_{-\pi}^\pi [f(x)]^2\, dx = \frac{2}{\pi} \int_0^\pi [f(x)]^2\, dx,$$

meaning that formula (3.2) holds.

In the following we prove the validity of formula (3.1). Since Γ_r is analytic, there can exist finitely many intersecting points between circumference $|z| = t$ $(0 < t < r)$ and Γ_r. Therefore, θ_t is made up of finitely many arcs θ_t^i $(i = 1, 2, \ldots, n_t)$. Let $t\theta_i(t)$ denote the linear measure of θ_t^i; then $\theta(t) = \sum_{i=1}^{n_t} \theta_i(t)$. Obviously, according to the definition, $\theta(t)$ is continuous on the interval $[0, r]$, but probably excluding at most finite points $0 < t_1 < t_2 < \cdots < t_n < r$, with $\theta(t_i - 0) = \theta(t_i) < \theta(t_i + 0)$ $(^2)$ at the exceptional point t_i.

Let $u(z) = u_D(z, \theta_r)$ and consider

$$m(t) = \frac{1}{2\pi} \int_{\theta_t} u^2(te^{i\theta})\, d\theta, \qquad 0 < t < r. \tag{3.3}$$

Notice that if circumference $|z| = t$ intersects B, then $u(z)$, is zero at the two end points of θ_t. Therefore, when $t \neq t_i$, by formula (3.3), we have

$$\frac{dm(t)}{d\log t} = \frac{1}{\pi} \int_{\theta_t} u\frac{\partial u}{\partial \log t}\, d\theta, \tag{3.4}$$

$$\frac{d^2 m(t)}{d\log t^2} = \frac{1}{\pi} \int_{\theta_t} \left\{ \left(\frac{\partial u}{\partial \log t}\right)^2 + u\frac{\partial^2 u}{\partial \log t^2} \right\}\, d\theta$$

$$= \frac{1}{\pi} \int_{\theta_t} \left\{ \left(\frac{\partial u}{\partial \log t}\right)^2 - u\frac{\partial^2 u}{\partial \theta^2} \right\}\, d\theta \tag{3.5}$$

$$= \frac{1}{\pi} \int_{\theta_t} \left\{ \left(\frac{\partial u}{\partial \log t}\right)^2 + \left(\frac{\partial u}{\partial \theta}\right)^2 \right\}\, d\theta > 0.$$

Accordingly, $m(t)$ is a convex function of $\log t$ on the interval (t_i, t_{i+1}), and discontinues at $t = t_i$. It satisfies, however, the condition $m(t_i - 0) = m(t_i) < m(t_i + 0)$. Let ν be the exterior normal line of B with respect to D. According to the condition that $u(z)$ is zero on Γ_t, an application of the Green formula to D_t yields

$$tm'(t) = \frac{dm(t)}{d\log t} = \frac{1}{\pi} \int_{\theta_t} u\frac{\partial u}{\partial t} t\, d\theta = \frac{1}{\pi} \int_{\Gamma_t + \theta_t} u\frac{\partial u}{\partial \nu}\, ds$$

$$= \frac{1}{\pi} \iint_{D_t} \left\{ \left(\frac{\partial u}{\partial x}\right)^2 + \left(\frac{\partial u}{\partial y}\right)^2 \right\}\, dx\, dy, \qquad z = x + iy.$$

$(^2)$ $\theta(t_i - 0) = \lim_{\varepsilon \to 0,\ \varepsilon > 0} \theta(t_i - \varepsilon)$, $\theta(t_i + 0) = \lim_{\varepsilon \to 0,\ \varepsilon > 0} \theta(t_i + \varepsilon)$.

Hence, if we let

$$s(t) = \frac{1}{\pi} \iint_{D_t} \left\{ \left(\frac{\partial u}{\partial x} \right)^2 + \left(\frac{\partial u}{\partial y} \right)^2 \right\} dx \, dy,$$

then

$$m'(t) = \frac{s(t)}{t} > 0, \qquad t \neq t_i. \tag{3.6}$$

Therefore, $m(t)$ is a monotonic increasing function of t on the interval (t_i, t_{i+1}). Notice that $m(t_i - 0) = m(t_i) < m(t_i + 0)$; then $m(t)$ is a monotonic increasing function of t on the interval $[0, r]$, and it follows that

$$m(r) - m(t) \geq \int_t^r m'(t) \, dt, \qquad 0 \leq t < r. \tag{3.7}$$

By formula (3.4) we find that

$$\left(\frac{dm(t)}{d \log t} \right)^2 \leq \frac{1}{\pi^2} \int_{\theta_t} u^2 \, d\theta \cdot \int_{\theta_t} \left(\frac{\partial u}{\partial \log t} \right)^2 d\theta$$

$$= \frac{2m(t)}{\pi} \int_{\theta_t} \left(\frac{\partial u}{\partial \log t} \right)^2 d\theta.$$

And hence

$$\frac{1}{\pi} \int_{\theta_t} \left(\frac{\partial u}{\partial \log t} \right)^2 d\theta \geq \frac{1}{2m(t)} \left(\frac{dm(t)}{d \log t} \right)^2. \tag{3.8}$$

Suppose that circumference $|z| = t$ intersects B. Notice the definition of $\theta^*(t)$ and the values of $u(z)$ at the two end points of θ_i^i are zero. Then by an application of the Wirtinger's inequality, we obtain

$$\int_{\theta_t^i} \left(\frac{\partial u}{\partial \theta} \right)^2 d\theta \geq \frac{\pi^2}{[\theta_i(t)]^2} \int_{\theta_t^i} u^2 \, d\theta \geq \frac{\pi^2}{[\theta^*(t)]^2} \int_{\theta_t^i} u^2 \, d\theta,$$

$$\frac{1}{\pi} \int_{\theta_1} \left(\frac{\partial u}{\partial \theta} \right)^2 = \frac{1}{\pi} \sum_{i=1}^{n_t} \int_{\theta_t^i} \left(\frac{\partial u}{\partial \theta} \right)^2 \geq \frac{\pi}{[\theta^*(t)]^2} \int_{\theta_t} u^2 \, d\theta$$

$$= \frac{2\pi^2}{[\theta^*(t)]^2} m(t). \tag{3.9}$$

Furthermore, by formulas (3.5), (3.8) and (3.9) we get

$$\frac{d^2 m(t)}{d \log t^2} \geq \frac{1}{2m(t)} \left(\frac{dm(t)}{d \log t} \right)^2 + \frac{1}{2} \left(\frac{2\pi}{\theta^*(t)} \right)^2 m(t). \tag{3.10}$$

Suppose that circumference $|z| = t$ does not intersect B, meaning that the whole of circumference $|z| = t$ lies in domain D, then $\theta^*(t) = +\infty$. Further based on formulas (3.5) and (3.8), we find that

$$\frac{d^2 m(t)}{d \log t^2} \geq \frac{1}{2m(t)} \left(\frac{dm(t)}{d \log t} \right)^2$$

$$= \frac{1}{2m(t)} \left(\frac{dm(t)}{d \log t} \right)^2 + \frac{1}{2} \left(\frac{2\pi}{\theta^*(t)} \right)^2 m(t).$$

Hence, formula (3.10) holds under general circumstances. Let

$$\rho = \log t, \quad \varphi(\rho) = m(t), \quad F(\rho) = \frac{2\pi}{\theta^*(t)}, \quad \rho_i = \log t_i.$$

Then formula (3.10) yields

$$\varphi''(\rho) \geq \frac{\varphi'(\rho)^2}{2\varphi(\rho)} + \frac{1}{2}F^2(\rho)\varphi(\rho). \tag{3.11}$$

This is the noted Carleman's differential inequality. Furthermore, we let $\psi(\rho) = \log \varphi(\rho)$; then $\psi'(\rho)^2 + 2\psi''(\rho) \geq F^2(\rho)$. Accordingly, we conclude that

$$\left(\frac{\varphi''(\rho)}{\varphi'(\rho)}\right)^2 = \left(\psi'(\rho) + \frac{\psi''(\rho)}{\psi'(\rho)}\right)^2 \geq \psi'(\rho)^2 + 2\psi''(\rho) \geq F^2(\rho).$$

By formula (3.6), we have $\varphi'(\rho) > 0$ in the interval (ρ_i, ρ_{i+1}). Further based on formula (3.11), we have $\varphi''(\rho) > 0$. Hence, on the interval (ρ_i, ρ_{i+1}), we have $\varphi''(\rho)/\varphi'(\rho) > 0$. And consequently $\varphi''(\rho)/\varphi'(\rho) \geq F(\rho)$.

When $t \neq t_i$, formula (3.6) yields

$$\varphi'(\rho) = \frac{dm(t)}{d\log t} = s(t).$$

Hence, it follows that $\varphi'(\rho_i - 0) < \varphi'(\rho_i + 0)$. If $\rho < \tau$, then

$$\log \varphi'(\tau) - \log \varphi'(\rho) \geq \int_\rho^\tau \frac{\varphi''(\rho)}{\varphi'(\rho)} d\rho \geq \int_\rho^\tau F(\rho) d\rho,$$

$$\varphi'(\tau) \geq \varphi'(\rho) \exp\left\{\int_\rho^\tau F(\rho) d\rho\right\}, \quad \rho < \tau. \tag{3.12}$$

Next, when $t \leq \sigma \leq r$, let $\rho = \log t$, $\tau = \log \sigma$, $\rho_0 = \log r$. If $t \leq \kappa r$ ($0 < \kappa < 1$), then we conclude from formulas (3.3) and (3.12) that

$$1 \geq \varphi(\rho_0) \geq \varphi(\rho_0) - \varphi(\rho)$$

$$\geq \int_\rho^{\rho_0} \varphi'(\tau) d\tau \geq \varphi'(\rho) \int_\rho^{\rho_0} \exp\left\{\int_\rho^\tau F(\rho) d\rho\right\} d\tau$$

$$= \varphi'(\rho) \int_t^r \exp\left\{2\pi \int_t^\sigma \frac{dt}{t\theta^*(t)}\right\} \frac{d\sigma}{\sigma}$$

$$\geq \varphi'(\rho) \int_{\kappa r}^r \exp\left\{2\pi \int_t^\sigma \frac{dt}{t\theta^*(t)}\right\} \frac{d\sigma}{\sigma} \tag{3.13}$$

$$\geq (1 - \kappa)\varphi'(\rho) \exp\left\{2\pi \int_t^{\kappa r} \frac{dt}{t\theta^*(t)}\right\},$$

$$\varphi'(\rho) \leq \frac{1}{1 - \kappa} \exp\left\{-2\pi \int_t^{\kappa r} \frac{dt}{t\theta^*(t)}\right\}, \quad 0 < \kappa < 1.$$

According to formula (3.11) we obtain

$$\varphi''(\rho) \geq \frac{1}{2} F^2(\rho)\varphi(\rho) = \frac{1}{2}\left[\frac{2\pi}{\theta^*(t)}\right]^2 \varphi(\rho) \geq \frac{1}{2}\frac{2\pi}{\theta^*(t)} \cdot \varphi(\rho),$$

$$\varphi''(\rho) \geq \frac{\pi m(t)}{\theta^*(t)}.$$

Further notice that $m(t)$ is a monotonic increasing function of t. Then

$$\varphi'(\tau) \geq \varphi'(\tau) - \varphi'(\rho) \geq \int_\rho^\tau \varphi''(\rho)\,d\rho$$

$$\geq \pi \int_t^\sigma \frac{m(t)}{t\theta^*(t)}\,dt \geq \pi m(t)\int_t^\sigma \frac{dt}{t\theta^*(t)}.$$

Accordingly, when $\sigma < \kappa r$, we obtain from formula (3.13) that

$$\frac{1}{1-\kappa}\exp\left\{-2\pi\int_\sigma^{\kappa r}\frac{dt}{t\theta^*(t)}\right\} \geq \varphi'(\tau) \geq \pi m(t)\int_t^\sigma \frac{dt}{t\theta^*(t)},$$

or

$$\frac{1}{1-\kappa}\exp\left\{2\pi\int_t^\sigma \frac{dt}{t\theta^*(t)} - 2\pi\int_t^{\kappa r}\frac{dt}{t\theta^*(t)}\right\}$$

$$\geq \frac{m(t)}{2}\cdot 2\pi\int_t^\sigma \frac{dt}{t\theta^*(t)}.$$

When $2\pi\int_t^{\kappa r} 1/(t\theta^*(t))\,dt > 1$, we assume σ $(t < \sigma < \kappa r)$ such that $2\pi\int_t^\sigma 1/(t\theta^*(t))\,dt = 1$. Hence

$$m(t) \leq \frac{2e}{1-\kappa}\exp\left\{-2\pi\int_t^{\kappa r}\frac{dt}{t\theta^*(t)}\right\}. \tag{3.14}$$

When $2\pi\int_t^{\kappa r} 1/(t\theta^*(t))\,dt \leq 1$, clearly formula (3.14) is valid if $m(t) \leq 1$ and $2/(1-\kappa) > 1$. Hence, formula (3.14) holds under general circumstances. Consequently,

$$u_D(0, \theta_r) = u(0) = \sqrt{m(0)}$$

$$\leq \sqrt{\frac{2e}{1-\kappa}}\exp\left\{-\pi\int_0^{\kappa r}\frac{dt}{t\theta^*(t)}\right\}, \qquad 0 < \kappa < 1.$$

Lemma 3.1 is thus proved.

COROLLARY. *When $z \in D$ with $|z| < \kappa r/2$, we have*

$$u_D(z, \theta_r) < \frac{9}{\sqrt{1-\kappa}}\exp\left\{-\pi\int_{2|z|}^{\kappa r}\frac{dt}{t\theta^*(t)}\right\}. \tag{3.15}$$

PROOF. We select arbitrarily a number ε, $0 < \varepsilon < \frac{3-\sqrt{2e}}{9-\sqrt{2e}}|z|$, and then construct a circle $\Delta: |\zeta - \varepsilon| < 2|z| - \varepsilon$, centered at ε with radius $2|z| - \varepsilon$. Let $\tilde{D} = D \cup \Delta$ and define correspondingly $\tilde{\theta}^*(t)$ and $\tilde{\theta}(t)$. Since the discontinued points t_i $(i = 1, 2, \ldots, n_t)$ of $\theta(t)$ is a finite set, once a suitable ε is selected, we may assert that $\tilde{\theta}(t)$, with the exception of points

t_i $(i = 1, 2, \ldots, n_t)$, is continuous. Even though the corresponding boundary of $\widetilde{D}$ in the disk $|z| < r$ is only piecewise analytic, the proof of Lemma 3.1 remains valid. On the other hand, obviously, when $0 \leq t \leq 2|z| - 2\varepsilon$ we have $\tilde{\theta}^*(t) = +\infty$, and when $t \geq 2|z|$ it follows that $\tilde{\theta}^*(t) = \theta^*(t)$. Hence, by Lemma 3.1 we conclude that

$$
\begin{aligned}
u_{\widetilde{D}}(0, \theta_r) &\leq \sqrt{\frac{2e}{1-\kappa}} \exp\left\{ -\pi \int_0^{\kappa r} \frac{dt}{t\tilde{\theta}^*(t)} \right\} \\
&\leq \sqrt{\frac{2e}{1-\kappa}} \exp\left\{ -\pi \int_{2|z|}^{\kappa r} \frac{dt}{t\tilde{\theta}^*(t)} \right\}.
\end{aligned}
\tag{3.16}
$$

$u_{\widetilde{D}}(\zeta, \theta_r)$ is harmonic in the disk $|\zeta| < 2|z| - 2\varepsilon$. Let $v_{\widetilde{D}}(\zeta, \theta_r)$ be the conjugate harmonic function of $u_{\widetilde{D}}(\zeta, \theta_r)$, then $f(\zeta) = e^{u_{\widetilde{D}}(\zeta, \theta_r) + i v_{\widetilde{D}}(\zeta, \theta_r)}$ is a regular function in the disk $|\zeta| \leq 2|z| - 2\varepsilon$ and has no zeros. Also $u_{\widetilde{D}}(\zeta, \theta_r) = \log|f(\zeta)| \geq 0$. Hence, applying the Poisson-Jensen formula, we obtain

$$
\begin{aligned}
u_{\widetilde{D}}(z, \theta_r) = \log|f(z)| &\leq \frac{2|z| - 3\varepsilon + |z|}{2|z| - 3\varepsilon - |z|} \cdot \frac{1}{2\pi} \int_0^{2\pi} u_{\widetilde{D}}((2|z| - 3\varepsilon)e^{i\theta}) \, d\theta \\
&= \frac{3(|z| - \varepsilon)}{|z| - 3\varepsilon} u_{\widetilde{D}}(0, \theta_r).
\end{aligned}
$$

Based further on the maximum modulus principle and formula (3.16), we conclude that

$$
\begin{aligned}
u_D(z, \theta_r) \leq u_{\widetilde{D}}(z, \theta_r) &\leq \frac{3(|z| - \varepsilon)}{|z| - 3\varepsilon} \cdot \frac{\sqrt{2e}}{\sqrt{1-\kappa}} \cdot \exp\left\{ -\pi \int_{2|z|}^{\kappa r} \frac{dt}{t\theta^*(t)} \right\} \\
&\leq \frac{9}{\sqrt{1-\kappa}} \exp\left\{ -\pi \int_{2|z|}^{\kappa r} \frac{dt}{t\theta^*(t)} \right\},
\end{aligned}
$$

meaning that formula (3.15) is proved.

From the above proof, we find that if the origin $z = 0 \notin D$, we need only to construct a suitable circle Δ, and then consider $\widetilde{D} = D \cup \Delta$, we may immediately prove that the corollary of Lemma 3.1 remains valid. Correspondingly, we have the following result:

THEOREM 3.1. *Let D be a domain in the disk $|z| < r$ $(0 < r < +\infty)$ satisfying conditions (2) and (3); then the harmonic measure of θ_r with respect to domain D that corresponds to point z, $|z| < \frac{\kappa r}{2}$ $(0 < \kappa < 1)$ is*

$$
u_D(z, \theta_r) \leq \frac{9}{\sqrt{1-\kappa}} \exp\left\{ -\pi \int_{2|z|}^{\kappa r} \frac{dt}{t\theta^*(t)} \right\}.
$$

3.1.2. A local version of the Lindelöf Theorem. First we illustrate a simple result:

LEMMA 3.2. *Let Ω be the upper semi-circle: $|z| < 1$, $\mathrm{Im}\, z > 0$, and Γ be the straight line starting from -1 to $+1$ on the real axis. Then the harmonic measure of Γ corresponding to point $z \in \Omega$ is*

$$u_\Omega(z, \Gamma) = \frac{2\varphi}{\pi} - 1,$$

and φ is the degree of the extended angle observing Γ from point z. Particularly, when $z = re^{i\pi/2}$, $r \in [0, \frac{1}{\sqrt{3}}]$, it follows that $u_\Omega(z, \Gamma) \geq \frac{1}{3}$.

Next, we prove a special form of the Grötzsch principle [43c].

LEMMA 3.3. *Suppose that two simple, continuous curves L_i $(i = 1, 2)$ divide annulus $r < |z| < R$, $0 < r < R < +\infty$ into two simply connected domains, with one denoted Ω. Also L_i $(i = 1, 2)$ connects the point A_i $(i = 1, 2)$ on circumference $|z| = r$ and point B_i $(i = 1, 2)$ on circumference $|z| = R$, and there is no coincidence between A_1 and A_2, and B_1 and B_2. We further assume that Ω is mapped conformally onto rectangle $C_1 C_2 D_2 D_1$, where point A_i $(i = 1, 2)$ corresponds to point C_i, and point B_i $(i = 1, 2)$ corresponds to point D_i. Then we have*

$$\frac{\overline{C_1 C_2}}{\overline{C_1 D_1}} \leq \frac{2\pi}{\log(R/r)},$$

where $\overline{C_1 C_2}$ and $\overline{C_1 D_1}$ indicate the length of straight lines $C_1 C_2$ and $C_1 D_1$, respectively.

PROOF. First we divide circular ring $r < |z| < R$ along L_1 into a simply connected domain, and then we consider the transformation $t = \log z = u + iv$, which maps this simply connected domain onto t-plane; and Ω onto a stripe-shaped region Ω_t with curved boundary consisting of two straight lines $u = \log r$ and $u = \log R$ on the t-plane. We also let $z = \varphi(\zeta)$ that maps rectangle $C_1 C_2 D_2 D_1$ conformally onto Ω, then $t = \log \varphi(\zeta) = f(\zeta)$ maps $C_1 C_2 D_2 D_1$ conformally onto Ω_t. Hence we obtain

$$\log \frac{R}{r} \leq \int_{\xi_1}^{\xi_1 + \overline{C_1 D_1}} |f'(\zeta)|\, d\xi, \qquad \zeta = \xi + i\eta,$$

where $C_i = \xi_1 + i\eta_1$. An application of the Schwarz's inequality yields

$$\left(\log \frac{R}{r}\right)^2 \leq \overline{C_1 D_1} \int_{\xi_1}^{\xi_1 + \overline{C_1 D_1}} |f'(\zeta)|^2\, d\xi,$$

$$\left(\log \frac{R}{r}\right)^2 \overline{C_1 C_2} \leq \overline{C_1 D_1} \int_{\eta_1}^{\eta_1 + \overline{C_1 C_2}} d\eta \int_{\xi_1}^{\xi_1 + \overline{C_1 D_1}} |f'(\zeta)|^2\, d\xi$$

$$\leq \overline{C_1 D_1}\, 2\pi \log \frac{R}{r}.$$

Consequently, we conclude that

$$\frac{\overline{C_1 C_2}}{\overline{C_1 D_1}} \leq \frac{2\pi}{\log(R/r)},$$

implying that Lemma 3.3 is proved.

Analogously, we may prove the following result:

LEMMA 3.4. *Suppose there exist two simple, continuous curves* L_i ($i = 1, 2$) *in* $\overline{\Omega}(\theta_1, \theta_2; R_1, R_2)$ ($0 \leq \theta_1 < \theta_2 < 2\pi + \theta_1$; $0 < R_1 < R_2 < +\infty$) *linking point* A_i ($i = 1, 2$) *on* $\Gamma(\theta_1, \theta_2; R_1)$ *and point* B_i ($i = 1, 2$) *on* $\Gamma(\theta_1, \theta_2; R_2)$, *respectively.* L_1 *does not intersect* L_2 *in* $\overline{\Omega}(\theta_1, \theta_2; R_1, R_2)$. *Then a simply connected domain* $\Omega' \subset \Omega(\theta_1, \theta_2; R_1, R_2)$ *is bounded by* L_1 *and* L_2, *as well as some arcs on* $\Gamma(\theta_1, \theta_2; R)$ *and* $\Gamma(\theta_1, \theta_2; R_2)$. *We further assume that the conformal mapping* $\zeta = \varphi(z)$ *maps* Ω' *onto a domain* $\Omega_\zeta(\theta_1, \theta_2; R_1, R_2')$, *with point* A_1 *mapped into point* $R_1 e^{i\theta_1}$, *point* A_2 *into point* $R_1 e^{i\theta_2}$, *point* B_1 *into point* $R_2^1 e^{i\theta_1}$ *and point* B_2 *into point* $R_2^1 e^{i\theta_2}$. *Then* $R_2^1 \geq R_2$.

Now, we prove a Lindelöf-type theorem.

THEOREM 3.2. *Suppose that two simple, continuous curves* L_i ($i = 1, 2$) *divide annulus* $\Gamma: 1 < |z| < R$ ($R > e^{4\pi}$) *into two simple connected domains, where one domain denotes* Ω. *Also* L_i ($i = 1, 2$) *connected point* A_i ($i = 1, 2$) *on circumference* $|z| = 1$ *and point* B_i ($i = 1, 2$) *on circumference* $|z| = R$, *where there is no coincidence between* A_1 *and* A_2, *and* B_1 *and* B_2. *Further assume that function* $f(z)$ *is regular in* Ω, *and is continuous on* $\overline{\Omega}$, *and also*

$$|f(z)| \leq N < +\infty, \qquad z \in \Omega,$$
$$|f(z) - a_i| \leq \varepsilon_i < 1, \qquad z \in L_i, \ i = 1, 2,$$

where a_1 *and* a_2 *are two finite complex numbers. Moreover,* $|a_i| \leq M$, $M < +\infty$, $i = 1, 2$.

Under the above assumption, we may have $a_1 = a_2$, *and a curve* l *linking both* L_1 *and* L_2 *on* $\overline{\Omega}$, *such that when* $z \in l$, *we have* $|f(z) - a| \leq \varepsilon_3$, $a = a_1 = a_2$, $\varepsilon_3 = (M + N) \max\{\varepsilon_1^{1/3}, \varepsilon_2^{1/3}\}$; *or we may have* $a_1 \neq a_2$, *and*

$$(M + N)(\varepsilon_1^{1/3} + \varepsilon_2^{1/3}) \geq |a_1 - a_2|.$$

PROOF. First there exists the conformal mapping $\zeta = \zeta(z)$, which maps Ω onto rectangle $C_1 C_2 D_2 D_1$ on the ζ-plane, with point A_i ($i = 1, 2$) corresponding to C_i ($i = 1, 2$), point B_i ($i = 1, 2$) corresponding to point D_i ($i = 1, 2$), and that $\overline{C_1 C_2} = \overline{D_1 D_2} = 1$. Then we obtain by Lemma 3.3

$$\frac{1}{\overline{C_1 D_1}} \leq \frac{2\pi}{\log(R/r)}.$$

And hence

$$\overline{C_1 D_1} = \overline{C_2 D_2} \geq \frac{1}{2\pi} \log R > 2. \tag{3.17}$$

Let $z = z(\zeta)$ be the inverse transformation of $\zeta = \zeta(z)$; then function $F(\zeta) = f(z(\zeta))$ is regular in rectangle $C_1 C_2 D_2 D_1$, and is bounded above by N. Moreover, when $\zeta \in C_1 D_1$, $|F(\zeta) - a_1| \leq \varepsilon_1$ holds, and when $\zeta \in C_2 D_2$, $|F(\zeta) - a_2| \leq \varepsilon_2$. Let E_2 be the midpoint of $C_1 D_1$ and E_2 be the midpoint of $C_2 D_2$. Then we choose point F_2 on line $E_1 E_2$, such that $\overline{E_1 F_2} = \frac{1}{\sqrt{3}}$, and select point F_1, such that $\overline{E_2 F_1} = \frac{1}{\sqrt{3}}$. Finally, we construct a unit circle, with center E_1. According to formula (3.17), half of the unit circle locates within rectangle $C_1 C_2 D_2 D_1$. Then by applying Lemma 3.2, when $\zeta \in E_1 F_2$, we obtain

$$\log |F(\zeta) - a_1| < \frac{1}{3} \log \varepsilon_1 + \log(M + N).$$

Analogously, when $\zeta \in E_2 F_1$, we have:

$$\log |F(\zeta) - a_2| < \frac{1}{3} \log \varepsilon_2 + \log(M + N).$$

Hence, when $a_1 \neq a_2$, by choosing a point on the common part of $E_1 F_2$ and $E_2 F_1$, we may derive $|a_2 - a_1| \leq (M + N)(\varepsilon_1^{1/3} + \varepsilon_2^{1/3})$. When $a_1 = a_2$, if we let $\varepsilon_3 = (M + N) \max\{\varepsilon_1^{1/3}, \varepsilon_2^{1/3}\}$, then when $\delta \in E_1 E_2$, we have $|F(\zeta) - a| \leq \varepsilon_3$, $a = a_1 = a_2$. Let l be the image of $E_1 E_2$ on the z-plane; then l is a continuous curve which connects L_1 and L_2, and that, when $z \in l$, $|f(z) - a| \leq \varepsilon_3$, meaning that Theorem 3.2 is proved.

§3.2. The Length-Area Principle

3.2.1. The Length-Area Principle.([3])
The following Theorem 3.3 is the so-called Length-Area Principle.

THEOREM 3.3. *Assume that $f(z)$ is a meromorphic function in the open set Δ, $l(t) = l(t, \Delta)$ is the total length of the level curve $|f(z)| = t$ in Δ, and A ($A < +\infty$) is the area of Δ. Let*

$$p(t) = p(t, \Delta) = \frac{1}{2\pi} \int_0^{2\pi} n(\Delta, f = t e^{i\theta}) \, d\theta.$$

Then

$$\int_0^{+\infty} \frac{l(t)^2}{t p(t)} \, dt \leq 2\pi A, \tag{3.18}$$

where we define the integrand to be zero if $p(t) = +\infty$. Particularly, for almost all values t satisfying condition $p(t) < +\infty$, it is true that $l(t) < +\infty$.

PROOF. First we consider that Δ is an open rectangle, and $f(z)$ is univalent and has no zeros or poles in $\overline{\Delta}$ (i.e., a domain that includes $\overline{\Delta}$). Selecting arbitrarily a branch of the function

$$s(z) = \log f(z) = \sigma(z) + i\tau(z),$$

([3]) The content of this section is extracted from [21a].

we find that $s(z)$ is also univalent on $\overline{\Delta}$, and it maps Δ into a domain Ω on the open plane $|s| < +\infty$. Obviously, the boundary of Ω is a piecewise analytic Jordan curve. Hence, if θ_σ denotes a set of intersecting points between straight line $\sigma = $ const. and Ω, then θ_σ is made up of finitely many straight lines

$$\tau_1 < \tau < \tau_1', \quad \tau_2 < \tau < \tau_2', \quad \ldots .$$

Besides, if γ_σ denotes the pre-image of θ_σ in Δ, then $|f(z)| = e^\sigma$ on γ_σ. Taking into account the univalency of $f(z) = e^{s(z)}$, we find that

$$p(e^\sigma) = \frac{1}{2\pi} \sum (\tau_\nu' - \tau_\nu) = \frac{\theta(\sigma)}{2\pi},$$

where $\theta(\sigma)$ represents the linear measure of θ_σ.

On the other hand, according to the Schwarz's inequality, we obtain

$$l(e^\sigma)^2 = \left\{ \int_{\theta_\sigma} \left| \frac{dz}{ds} \right| d\tau \right\}^2 \leq \int_{\theta_\sigma} d\tau \int_{\theta_\sigma} \left| \frac{dz}{ds} \right|^2 d\tau$$

$$= \theta(\sigma) \int_{\theta_\sigma} \left| \frac{dz}{ds} \right|^2 d\tau = 2\pi p(e^\sigma) \int_{\theta_\sigma} \left| \frac{dz}{ds} \right|^2 d\tau. \tag{3.19}$$

Let

$$\sigma_1 = \inf\{\sigma | \sigma + i\tau \in \Omega\}, \qquad \sigma_2 = \sup\{\sigma | \sigma + i\tau \in \Omega\}.$$

Then we derive by formula (3.19)

$$\int_{-+\infty}^{+\infty} \frac{l(e^\sigma)^2}{p(e^\sigma)} d\sigma = \int_{\sigma_1}^{\sigma_2} \frac{l(e^\sigma)^2}{p(e^\sigma)} d\sigma \leq 2\pi \int_{\sigma_1}^{\sigma_2} d\sigma \int_{\theta_\sigma} \left| \frac{dz}{ds} \right|^2 d\tau = 2\pi A,$$

where A represents the area of Δ. Let $t = e^\sigma$; then

$$\int_0^{+\infty} \frac{l(t)^2}{tp(t)} dt \leq 2\pi A \tag{3.20}$$

holds, meaning that formula (3.18) is valid.

Next, we consider, in the general case, that Δ is an open set. Without any loss of generality, we may assume that $f(z) \neq 0, \infty$ and $f'(z) \neq 0$ in Δ. Otherwise, we may use Δ_0 to replace Δ, where Δ_0 is an open set excluding the zeros and poles of $f(z)$, as well as the zeros of $f'(z)$. Obviously, $p(t)$, $l(t)$, and A remain invariant. Now, by applying the method commonly used in the Lebesgue Integration Theory, we construct a net

$$G_m: x = \frac{\pm n}{2^{m-1}}, y = \frac{\pm n}{2^{m-1}} : n = 0, 1, 2, \ldots; \quad m = 1, 2, \ldots; \quad z = x + iy$$

on an open plane $|z| < +\infty$, such that Δ may be represented as the union of the countable closed rectangles: $\Delta = \bigcup_{\nu=1}^{\infty} \overline{\Delta}_\nu$ that do not intersect among one another internally. Moreover, under the assumption that $f'(z)$ has no zeros on $\overline{\Delta}_\nu$, we may assume that $f(z)$ is univalent on Δ_ν. Otherwise, we need only to divide Δ_ν repeatedly to arrive at the result.

Let γ_t denote the level curve $|f(z)| = t$; then there may be no intersections between a particular side of rectangle Δ_ν $(\nu \geq 1)$ and γ_t or the side of this rectangle intersects γ_t for finitely many times, or the side belongs entirely to γ_t. Hence, it is obvious that there are at most countable values t which satisfy the last condition. We denote these values as t_i $(i = 1, 2, \ldots)$. When $t \neq t_i$ $(i = 1, 2, \ldots)$, we have

$$p(t, \Delta) = \sum_{\nu=1}^{+\infty} p(t, \Delta_\nu), \qquad l(t, \Delta) = \sum_{\nu=1}^{+\infty} l(t, \Delta_\nu), \qquad A = \sum_{\nu=1}^{+\infty} A_\nu,$$

where A_ν represents the area of Δ_ν. When $p(t, \Delta_\nu) = 0$, it follows that $l(t, \Delta_\nu) = 0$. For this case, we assert that

$$\frac{l^2(t, \Delta_\nu)}{p(t, \Delta_\nu)} = 0.$$

Then based on the Schwarz's inequality, we conclude that

$$l(t, \Delta)^2 = \left\{ \sum_{\nu=1}^{+\infty} \frac{l(t, \Delta_\nu)}{\sqrt{p(t, \Delta_\nu)}} \cdot \sqrt{p(t, \Delta_\nu)} \right\}^2 \leq \sum_{\nu=1}^{+\infty} \frac{l(t, \Delta_\nu)^2}{p(t, \Delta_\nu)} \cdot \sum_{\nu=1}^{+\infty} p(t, \Delta_\nu)$$

$$= \sum_{\nu=1}^{+\infty} \frac{l(t, \Delta_\nu)^2}{p(t, \Delta_\nu)} \cdot p(t, \Delta),$$

$$\frac{l(t, \Delta)^2}{p(t, \Delta)} \leq \sum_{\nu=1}^{+\infty} \frac{l(t, \Delta_\nu)^2}{p(t, \Delta_\nu)}.$$

Finally, by taking the Lebesgue's integration as well as applying formula (3.20), we get

$$\int_0^{+\infty} \frac{l(t)^2}{tp(t)} \, dt \leq \sum_{\nu=1}^{\infty} \int_0^{+\infty} \frac{l(t, \Delta_\nu)^2}{tp(t, \Delta_\nu)} \, dt \leq 2\pi \sum_{\nu=1}^{+\infty} A_\nu = 2\pi A.$$

Hence Theorem 3.3 is proved completely.

3.2.2. Applications.   As an application of Theorem 3.3, we prove an important result which is basically attributed to A. Weitsman's efforts [40a].

LEMMA 3.5. *Let $f(z)$ be regular in the disk $|z| \leq R$ $(0 < R < +\infty)$, $|f(0)| = 1$. Suppose that there is a point z_0 in the disk $|z| < r$ $(0 < r < R)$, such that $|f(z_0)| \geq A$, $A \geq 16$. Then there must exist a number A' in the interval $I = [\sqrt[4]{A}, \sqrt{A}]$, such that derivative $f'(z)$ has no zeros on the level curve $|f(z)| = A'$, meaning that the level curve is analytic and the following fact holds: We consider the set*

$$\Omega(A') = E\{z | |f(z)| > A', \ |z| < R\},$$

and denote $\Omega_r(A')$ as the connected branch of $\Omega(A')$ in the disk $|z| < r$ which contains point z_0. Then for any two arbitrary points z_1 and z_2 on *the closure $\overline{\Omega}_r(A')$, surely we will find a piecewise analytic curve $L \subset \overline{\Omega}_r(A')$ which links these two points, with the length being*

$$\text{meas } L \leq 2r + 2\sqrt{2}\pi r\sqrt{\left(\log \frac{R}{r}\right)^{-1} T(R, f)}.$$

Also for point z on L it follows that $|f(z)| \geq \sqrt[4]{A}$, $|z| \leq r$.

Proof. When $t \in I$ and $0 \leq \varphi < 2\pi$, we have

$$|f(0) - te^{i\varphi}| \geq t - |f(0)| \geq \sqrt[4]{A} - 1 \geq 1 \neq 0,$$

and hence

$$n\left(r, \frac{1}{f(z) - te^{i\varphi}}\right) \leq \left(\log \frac{R}{r}\right)^{-1} N\left(R, \frac{1}{f(z) - te^{i\varphi}}\right)$$

$$= \left(\log \frac{R}{r}\right)^{-1} N\left(R, \frac{1}{(f(z)/t) - e^{i\varphi}}\right).$$

According to Cartan's identical relation, we obtain

$$\frac{1}{2\pi} \int_0^{2\pi} n\left(r, \frac{1}{f(z) - te^{i\varphi}}\right) d\varphi$$

$$\leq \left(\log \frac{R}{r}\right)^{-1} \left\{T\left(R, \frac{f}{t}\right) - \log^+ \frac{|f(0)|}{t}\right\}$$

$$\leq \left(\log \frac{R}{r}\right)^{-1} T(R, f).$$

Based further on the Length-Area Principle (Theorem 3.3), we derive

$$\int_{\sqrt[4]{A}}^{\sqrt{A}} \frac{l^2(t)}{t} dt \leq 2\pi \cdot \pi r^2 \left(\log \frac{R}{r}\right)^{-1} T(R, f)$$

$$= 2\pi^2 r^2 \left(\log \frac{R}{r}\right)^{-1} T(R, f) = K_0,$$

where $l(t)$ represents the total length of the level curve $|f(z)| = t$ in the disk $|z| < r$.

We let J be the set of values t satisfying the condition

$$\frac{l^2(t)}{t} \geq \frac{2K_0}{\sqrt{A} - \sqrt[4]{A}}$$

in I. Then it follows that

$$K_0 \geq \int_J \frac{l^2(t)}{t} dt \geq \frac{2K_0}{\sqrt{A} - \sqrt[4]{A}} \text{ meas } J,$$

meaning that the following inequality holds:

$$\text{meas } J \leq \frac{\sqrt{A} - \sqrt[4]{A}}{2}.$$

Let $I^* = I - J$; then

$$\text{meas } I^* \geq \frac{\sqrt{A} - \sqrt[4]{A}}{2}.$$

When $t \in I^*$, we conclude that

$$\frac{l^2(t)}{t} \leq \frac{2K_0}{\sqrt{A} - \sqrt[4]{A}},$$

$$l^2(t) \leq \frac{2\sqrt{A}}{\sqrt{A} - \sqrt[4]{A}} K_0 = \frac{\sqrt{A}}{\sqrt{A} - \sqrt[4]{A}} \cdot 4\pi^2 r^2 \left(\log \frac{R}{r}\right)^{-1} T(R, f),$$

$$l(t) \leq 2\sqrt{2}\pi r \sqrt{\left(\log \frac{R}{r}\right)^{-1} T(R, f)}.$$

On the other hand, since the derivative $f'(z)$ has at most finitely many zeros in the disk $|z| \leq R$, there exists value A' in I^*, such that the level curve $|f(z)| = A'$ has no zeros of $f'(z)$, meaning that the level curve is analytic. Hence, the composition of the boundary of domain $\Omega_r(A')$ may either be the level curves or the arcs on circumference $|z| = r$. Also the intersecting points between circumference $|z| = r$ and the level curve are finite. Otherwise, by analytic continuation successively, we may conclude that the whole of circumference $|z| = r$ must be the level curve. So that, according to the maximum modulus principle, when point z lies in the disk $|z| < r$, it follows that $|f(z)| \leq A'$ which, however, contradicts the assumption that $|f(z_0)| \geq A$. Therefore, the intersecting points between circumference $|z| = r$ and the level curve can only be finite, implying that the boundary branches of $\Omega_r(A')$ are all simple, closed curves that are piecewise analytic.

We consider the complementary set of $\overline{\Omega}_r(A')$ with respect to the closed plane $|z| \leq +\infty$. Obviously, each connected component of this complementary set is a simply connected domain, with its boundary being the boundary of $\Omega_r(A')$, and is, therefore, a simple, closed curve that is piecewise analytic. Next, the number of the connected components of this complementary set must be finite. Otherwise, we assume an interior point from each component, with at least one cluster point existing among all these interior points, and each cluster point can only be on the boundary of $\Omega_r(A')$. Since the boundaries of $\Omega_r(A')$ are piecewise analytic curves, they can be partially divided, meaning that there exists a neighborhood of this cluster point which is divided by the boundary curve into two connected components. One component belongs to $\Omega_r(A')$ while the other to the complementary set of $\overline{\Omega}_r(A')$. In this way, among the selected interior points, there exist at least two points belonging to the same component. This however, contradicts the way that the interior points were selected.

For any two arbitrary points z_1 and z_2 on $\overline{\Omega}_r(A')$, we first use a straight line to connect these two points. Notice that this straight line may intersect finitely many components of the complementary set of $\overline{\Omega}_r(A')$. If we suppose that this straight line intersects a particular component E of the complementary set of $\overline{\Omega}_r(A')$, then we let z_1' and z_2' denote the first and final intersecting points between boundary E and this straight line starting from z_1 to z_2. For the segment between point z_1' and z_2', we replace it by the part of the boundary of E that lies between z_1' and z_2'. Hence, according to this method, we derive a piecewise analytic curve $L \subset \overline{\Omega}_r(A')$ that connects point z_1 and z_2, with the length being

$$\operatorname{meas} L \leq 2r + 2\sqrt{2}\pi r \sqrt{\left(\log \frac{R}{r}\right)^{-1} T(R, f)},$$

and for z on L it follows that

$$|f(z)| \geq A' \geq \sqrt[4]{A}, \qquad |z| \leq r,$$

implying that Lemma 3.5 is proved.

§3.3. On the growth of meromorphic functions with deficient values

3.3.1. Growth of a meromorphic function and its deficient values.[4] Let $f(z)$ be a meromorphic function of order λ and lower order μ on the open plane $|z| < +\infty$. If a sequence $\{\rho_m\}$, $\rho_m \to +\infty$ $(m \to +\infty)$ satisfies the condition

$$\varlimsup_{m \to +\infty} \frac{\rho_m T'(\rho_m, f)}{T(\rho_m, f)} \leq \gamma,$$

then we call sequence $\{\rho_m\}$ a γ-sequence. By the Cartan's identical relation, we assure the existence of derivative $T'(r, f)$ of $T(r, f)$. Hence,

$$\frac{d \log T(r, f)}{d \log r} = \frac{rT'(r, f)}{T(r, f)}.$$

Therefore, according to the definition of order λ and lower order μ, for any arbitrary value γ, $\mu \leq \gamma \leq \lambda$, there must exist a γ-sequence. Indeed, there exists, first of all, the sequence $\{\rho_m\}$ such that

$$\lim_{m \to +\infty} \frac{\log T(\rho_m, f)}{\log \rho_m} \leq \gamma.$$

Then, by the mean-value theorem, for any arbitrarily selected value $N > 0$ we obtain

$$\begin{aligned}
\gamma &\geq \lim_{m \to +\infty} \frac{\log T(\rho_m, f)}{\log \rho_m} \\
&= \lim_{m \to +\infty} \frac{\log T(\rho_m, f) - \log T(N, f)}{\log \rho_m - \log N} \\
&\geq \lim_{m \to +\infty} \left. \frac{d \log T(r, f)}{d \log r} \right|_{r = \tilde{\rho}_m} \\
&= \lim_{m \to +\infty} \frac{\tilde{\rho}_m T'(\tilde{\rho}_m, f)}{T(\tilde{\rho}_m, f)},
\end{aligned}$$

---

[4] The content of this section is extracted from [40b].

where $N \leq \tilde{\rho}_m \leq \rho_m \ (m \geq m_N)$. Hence, we may select a subsequence $\{\tilde{\rho}_{m_k}\} \to +\infty \ (k \to +\infty)$, such that

$$\gamma \geq \lim_{k \to +\infty} \frac{\log T(\rho_{m_k}, f)}{\log \rho_{m_k}} \geq \lim_{k \to +\infty} \frac{\tilde{\rho}_{m_k} T'(\tilde{\rho}_{m_k}, f)}{T(\tilde{\rho}_{m_k}, f)},$$

meaning that $\{\tilde{\rho}_{m_k}\}$ is a γ-sequence.

LEMMA 3.6. *Let $f(z)$ be a nonconstant meromorphic function on the open plane $|z| < +\infty$ and $\{\rho_m\}$ be a γ-sequence. Then*

$$\pi\gamma \geq \delta(\infty, f) \cot\left(\frac{s_\infty}{4}\right) + \delta(0, f) \cot\left(\frac{2\pi - s_\infty}{4}\right), \qquad (3.21)$$

where

$$s_\infty = \lim_{m \to +\infty} \text{meas } E(\rho_m),$$
$$E(\rho_m) = \{\theta \mid |f(\rho_m e^{i\theta})| > 1, \ 0 \leq \theta < 2\pi\}. \qquad (3.22)$$

PROOF. Let

$$P(R, r, \theta, \Phi) = \frac{1}{2\pi} \cdot \frac{R^2 - r^2}{R^2 - 2rR\cos(\theta - \Phi) + r^2}, \qquad r < R.$$

Then

$$\int_0^{2\pi} P(R, r, \theta, \Phi) \, d\theta = 1,$$
$$P(R, r, \theta, \Phi) > 0, \qquad P(R, r, \theta, \Phi) = P(R, r, \Phi, \theta).$$

Assume further that

$$g(R, r, \psi, \omega) = \log\left|\frac{R^2 - \overline{\omega}re^{i\Psi}}{R(re^{i\Psi} - \omega)}\right|.$$

Besides, provided that ε_m is sufficiently small, we may assume that $f(z)$ has no zeros and poles on the annulus: $\rho_m - \varepsilon_m \leq |z| \leq \rho_m.$[5] Hence, when

---

[5] We may assume that there are no zeros and poles of $f(z)$ on circumference $|z| = \rho_m$. Otherwise, we need only to change value ρ_m slightly.

$\rho_m - \varepsilon_m \leq \rho < \rho_m$, by the Poisson-Jensen formula, we have

$$\frac{\rho_m(m(\rho_m, f) - m(\rho, f))}{\rho_m - \rho}$$

$$= \frac{\rho_m}{\rho_m - \rho}\left\{ \frac{1}{2\pi}\int_{E(\rho_m)} \log|f(\rho_m e^{i\theta})| \int_0^{2\pi} P(\rho_m, \rho, \Phi, \theta)\, d\Phi\, d\theta \right.$$

$$- \frac{1}{2\pi}\int_{E(\rho)}\int_0^{2\pi} \log|f(\rho_m e^{i\theta})| P(\rho_m, \rho, \theta, \Phi)\, d\theta\, d\Phi$$

$$+ \frac{1}{2\pi}\sum_{f(a_\nu)=0,\ |a_\nu|<\rho_m} \int_{E(\rho)} g(\rho_m, \rho, \theta, a_\nu)\, d\theta$$

$$\left. - \frac{1}{2\pi}\sum_{f(b_\nu)=\infty,\ |b_\nu|<\rho_m} \int_{E(\rho)} g(\rho_m, \rho, \theta, b_\nu)\, d\theta \right\}.$$

Let

$$\sum_0 = \sum_{f(a_\nu)=0,\ |a_\nu|<\rho_m} \int_{E(\rho)} g(\rho_m, \rho, \theta, a_\nu)\, d\theta,$$

$$\sum_\infty = \sum_{f(b_\nu)=\infty,\ |b_\nu|<\rho_m} \int_{E(\rho)} g(\rho_m, \rho, \theta, b_\nu)\, d\theta.$$

Then it follows that

$$\frac{\rho_m(m(\rho_m, f) - m(\rho, f))}{\rho_m - \rho}$$

$$= \frac{\rho_m}{\rho_m - \rho}\left\{ \frac{1}{2\pi}\int_{E(\rho_m)} \log|f(\rho_m e^{i\theta})| \int_0^{2\pi} P(\rho_m, \rho, \Phi, \theta)\, d\Phi\, d\theta \right.$$

$$- \frac{1}{2\pi}\int_0^{2\pi} \log|f(\rho_m e^{i\theta})| \int_{E(\rho)} P(\rho_m, \rho, \Phi, \theta)\, d\Phi\, d\theta$$

$$\left. + \frac{1}{2\pi}\sum_0 - \frac{1}{2\pi}\sum_\infty \right\}$$

$$= \frac{\rho_m}{2\pi}\int_{E(\rho_m)} \log|f(\rho_m e^{i\theta})| \int_{CE(\rho)} \frac{P(\rho_m, \rho, \Phi, \theta)\, d\Phi}{\rho_m - \rho}\, d\theta$$

$$+ \frac{\rho_m}{2\pi}\int_{CE(\rho_m)} \log\frac{1}{|f(\rho_m e^{i\theta})|} \int_{E(\rho)} \frac{P(\rho_m, \rho, \Phi, \theta)\, d\Phi}{\rho_m - \rho}\, d\theta$$

$$+ \frac{\rho_m}{2\pi(\rho_m - \rho)}\sum_0 - \frac{\rho_m}{2\pi(\rho_m - \rho)}\sum_\infty, \tag{3.23}$$

where $CE(\rho)$ and $CE(\rho_m)$ denote respectively the complementary sets of $E(\rho)$ and $E(\rho_m)$ with respect to $[0, 2\pi]$. Notice that when $\rho \to \rho_m$ we have

$$\frac{\rho_m}{2\pi} \sum \int_{E(\rho)} \frac{g(\rho_m, \rho, \theta, a_\nu)}{\rho_m - \rho} d\theta$$

$$\to \sum \int_{E(\rho)} P(\rho_m, |a_\nu|, \theta, \arg a_\nu) d\theta \geq 0, \tag{3.24}$$

$$n(\rho_m, f) - \frac{\rho_m}{2\pi} \sum \int_{E(\rho)} \frac{g(\rho_m, \rho, \theta, b_\nu)}{\rho_m - \rho} d\theta$$

$$\to \sum \int_{CE(\rho)} P(\rho_m, |b_\nu|, \theta, \arg b_\nu) d\theta \geq 0, \tag{3.25}$$

as well as

$$\frac{\rho_m}{2\pi(\rho_m - \rho)} \int_{E(\rho)} \frac{(\rho_m - \rho)(\rho_m + \rho)}{\rho_m^2 - 2\rho_m \rho \cos(\theta_\Phi) + \rho^2} d\Phi$$

$$\geq \frac{\rho_m}{\pi} \int_{\frac{1}{2} \text{meas } CE(\rho)}^{\pi} \frac{(\rho_m + \rho) d\beta}{\rho_m^2 - 2\rho\rho_m \cos \beta + \rho^2} \tag{3.26}$$

$$\to \frac{1}{\pi} \int_{\frac{1}{2} \text{meas } CE(\rho_m)}^{\pi} \frac{d\beta}{1 - \cos \beta}$$

$$= \frac{1}{\pi} \cot \frac{\text{meas } CE(\rho_m)}{4}$$

and

$$\frac{\rho_m}{2\pi(\rho_m - \rho)} \int_{CE(\rho)} \frac{(\rho_m - \rho)(\rho_m + \rho) d\Phi}{\rho_m^2 - 2\rho\rho_m \cos(\theta - \Phi) + \rho^2}$$

$$\geq \frac{1}{\pi} \int_{\frac{1}{2} \text{meas } E(\rho)}^{\pi} \frac{(\rho_m + \rho) d\beta}{\rho_m^2 - 2\rho\rho_m \cos \beta + \rho^2} \tag{3.27}$$

$$\to \frac{1}{\pi} \int_{\frac{1}{2} \text{meas } E(\rho_m)}^{\pi} \frac{d\beta}{1 - \cos \beta} = \frac{1}{\pi} \cot \frac{\text{meas } E(\rho_m)}{4}.$$

Accordingly, if we add simultaneously item $\rho_m N'(\rho_m, f) = n(\rho_m, f)$ to both sides of formula (3.23), and let $\rho \to \rho_m$, and divide both sides by $T(\rho_m, f)$ at the same time, then by formulas (3.22), (3.24), (3.25), (3.26), and (3.27), we conclude that formula (3.21) holds, meaning that Lemma 3.6 is proved.

Let $f(z)$ be a function meromorphic on the open plane $|z| < +\infty$, with lower order μ and having at least two deficient values a_1 and a_2, with their corresponding deficiencies being $\delta(a_1, f) = \delta_1 > 0$ and $\delta(a_2, f) = \delta_2 > 0$, respectively. We consider the transformation

$$F(z) = \frac{f(z) - a_1}{f(z) - a_2};$$

then $F(z)$ has the deficient values 0 and ∞, with their corresponding deficiencies being $\delta(0, F) = \delta(a_1, f) = \delta_1 > 0$ and $\delta(\infty, F) = \delta(a_2, f) = \delta_2 > 0$. Moreover, $F(z)$ and $f(z)$ have the same lower order μ. Hence, applying Lemma 3.6, where we let $\gamma = \mu$, we arrive at Edrei–Fuchs' result [15b]:

THEOREM 3.4. *Let $f(z)$ be a function meromorphic on the open plane $|z| < +\infty$, having two deficient values. Then its lower order μ is > 0.*

COROLLARY. *Any meromorphic functions of lower order zero has at most one deficient value.*

Earlier G. Valiron proved that a meromorphic function of order zero has at most one deficient value [39e].

When $f(z)$ is an entire function, we may obtain the following stronger result [18b].

THEOREM 3.5. *Let $f(z)$ be an entire function on the open plane $|z| < +\infty$, having a finite deficient value. Then its lower order is $\mu > \frac{1}{2}$.*

We will prove this theorem in §4.2.

Now we continue to discuss $F(z)$. First of all, by procedures analogous to the proof of Lemma 3.6, we may conclude there exists a positive number $\eta = \eta(\delta_1, \delta_2) > 0$ depending on δ_1 and δ_2, such that

$$\lim_{R \to +\infty} \frac{RT'(R, F)}{T(R, F)} \geq \eta. \tag{3.28}$$

Moreover, for any arbitrary value $\sigma \geq 1$, we have according to the mean-value theorem

$$\frac{\log T(\sigma r, F) - \log T(r, F)}{\log \sigma r - \log r} = \frac{RT'(R, F)}{T(R, F)}, \qquad r \leq R \leq \sigma r.$$

Then it follows that

$$\lim_{r \to +\infty} \frac{\log T(\sigma r, F) - \log T(r, F)}{\log \sigma} \geq \lim_{R \to +\infty} \frac{RT'(R, F)}{T(R, F)} \geq \eta. \tag{3.29}$$

On the other hand, by the First Fundamental Theorem we obtain

$$T(\sigma r, F) \leq T(\sigma r, f) + O(1), \tag{3.30}$$

and

$$T(r, F) \geq T(r, f) - O(1). \tag{3.31}$$

Hence, by formulas (3.29), (3.30), and (3.31), we get

$$\lim_{r \to +\infty} \frac{\log T(\sigma r, f) - \log T(r, f)}{\log \sigma} \geq \eta,$$

$$\lim_{r \to +\infty} \frac{T(\sigma r, f)}{T(r, f)} \geq \sigma^\eta,$$

which, in turn, prove the following result [40b]:

LEMMA 3.7. *Let $f(z)$ be a function meromorphic on the open plane $|z| < +\infty$, having two deficient values a_1 and a_2, with their corresponding deficiencies $\delta(a_1, f) = \delta_1 > 0$ and $\delta(a_2, f) = \delta_2 > 0$, respectively. Then there* *exists a positive number* $\eta = \eta(\delta_1, \delta_2)$ *depending on* δ_1 *and* δ_2, *such that, for any arbitrary value* $\sigma \geq 1$,

$$\varliminf_{r \to +\infty} \frac{T(\sigma r, f)}{T(r, f)} \geq \sigma^{\eta}.$$

3.3.2. A lemma about the deficient values.   First of all we prove the following lemma:

LEMMA 3.8. *Let* $f(z)$ *be a function meromorphic on the disk* $|z| \leq R$ $(1 < R < +\infty)$, α_i $(i = 1, 2, \ldots, n(R', f = \infty))$ *be the poles in the disk* $|z| < R'$ $(1 < R' < R)$, *and* (γ) *be the Euclidean exceptional circles corresponding to these* $n(R', f = \infty)$ *points and positive number* H. *Then for point* z *in the disk* $|z| < r$ $(4eH < r < R')$ *and outside circles* (γ), *we have*

$$\log|f(z)| \leq \left\{ \frac{R' + r}{R' - r} + \frac{\log(2R'/H)}{\log(R/R')} \right\} T(R, f). \tag{3.32}$$

PROOF. By applying the Poisson-Jensen formula, when $|z| < r$, we have that

$$\log|f(z)| \leq \frac{R' + r}{R' - r} m(R', f) + \sum_{i=1}^{n(R', f=\infty)} \log \left| \frac{R'^2 - \overline{\alpha}_i z}{R'(z - \alpha_i)} \right|$$

$$\leq \frac{R' + r}{R' - r} m(R', f) + n(R', f = \infty) \log(2R')$$

$$+ \log \frac{1}{\prod_{i=1}^{n(R', f=\infty)} |z - \alpha_i|}.$$

Moreover, when $z \notin (\gamma)$, we derive

$$\log|f(z)| \leq \frac{R' + r}{R' - r} m(R', f) + n(R', f = \infty) \log \frac{2R'}{H}.$$

Notice that

$$n(R', f = \infty) \leq \frac{1}{\log(R/R')} \left\{ \int_0^R \frac{n(t, \infty) - n(0, \infty)}{t} \, dt \right.$$

$$\left. + n(0, \infty) \log R - n(0, \infty) \log R' \right\}$$

$$\leq \frac{1}{\log(R/R')} N(R, \infty).$$

Then we obtain formula (3.32), meaning that Lemma 3.8 holds.

In the following, we prove an important lemma concerning deficient values.

LEMMA 3.9. *Let* $f(z)$ *be a function meromorphic on the open plane* $|z| < +\infty$. *Let it have* p *deficient values* a_{ν} $(\nu = 1, 2, \ldots, p;\ 1 \leq p < +\infty)$, *with their corresponding deficiencies* $\delta(a_\nu, f) = \delta_\nu > 0$. *When* $a_\nu \neq \infty$ $(1 \leq \nu \leq p)$, *we let*

$$f(z) - a_\nu = c_\nu z^{s_\nu} + c_{\nu+1} z^{s_\nu+1} + \cdots, \qquad c_\nu \neq 0,$$

and assume that $\delta = \min_{1 < \nu < p}\{\delta_\nu\}$, *and*

$$|a| = \max_{1 \leq \nu \leq p}\{|a_\nu|, a_\nu \neq \infty\}, \tag{3.33}$$

$$|c| = \min_{1 \leq \nu \leq p}\{|c_\nu|, a_\nu \neq \infty\}. \tag{3.34}$$

We also choose arbitrary numbers h $(h > 0)$, H $(H > 0)$, *and* $K > 0$ $(hH \leq K \leq 2hK)$, *and number* σ, $0 < \sigma < \frac{1}{2}h$. *If for a certain sufficiently large value* r, $r \geq 1$, *the following condition is satisfied*:

$$T(re^h, f) \leq e^K T(r, f)(1 + o(1)) \leq 2e^K T(r, f), \tag{3.35}$$

and when $R \geq r$, *the following formulas hold*:

$$\frac{\delta}{2}T(R, f) \leq m(R, a_\nu), \qquad \nu = 1, 2, \ldots, p, \tag{3.36}$$

and

$$T\left(R, \frac{1}{f - a_\nu}\right) < 2T(R, f), \qquad \nu = 1, 2, \ldots, p, \tag{3.37}$$

as well as

$$\frac{1}{T(R, f)}\left\{8\log 2 + \log^+\log^+|a| + \log^+\log^+\frac{1}{|c|} + \frac{1}{2}h\right.$$
$$+ 3K + 3\log\frac{1}{e^{h/2} - e^\sigma} + \log^+\frac{4(p+1)e^\sigma}{e^\sigma - 1} \tag{3.38}$$
$$\left.+ 2\log^+\frac{2(2p+1)}{h} + 2\log R + 3\log^+ T(R, f)\right\} < \frac{\delta}{8},$$

then there must exist a value R *in the interval* $[r, e^\sigma r]$ *and a corresponding set* $E_\nu(R)$ $(1 \leq \nu \leq p)$ *of values* $\theta(0 \leq \theta < 2\pi)$, *such that, when* $\theta \in E_\nu(R)$ *and* $a_\nu \neq \infty$,

$$\log\frac{1}{|f(Re^{i\theta}) - a_\nu|} \geq \frac{\delta}{4}T(R, f), \tag{3.39}$$

and

$$\log\frac{1}{|f'(Re^{i\theta})|} \geq \frac{\delta}{8}T(R, f); \tag{3.40}$$

while $a_\nu = \infty$, *we have*

$$\log|f(Re^{i\theta})| \geq \frac{\delta}{4}T(R, f)$$

and

$$\operatorname{meas} E_\nu(R) \geq \frac{\pi\delta}{8e^{2hH}\left\{\frac{e^{h/2}+e^\sigma}{e^{h/2}-e^\sigma} + \frac{2}{h}\log\frac{16(p+1)e^{1+h/2}}{e^\sigma-1}\right\}} \tag{3.41}$$
$$= M(\delta, h, H, \sigma) > 0.$$

PROOF. We may assume without any loss of generality that $a_\nu \neq \infty$ ($1 \leq \nu \leq p$). Otherwise, we need only to make an obvious modification of the proof. Let $\alpha_{\nu j}$ ($j = 1, 2, \ldots, n(re^{h/2}, a_\nu)$) be the a_ν-value points and β_m ($m = 1, 2, \ldots, n(re^{h/2}, \infty)$) be the poles of $f(z)$ in the disk $|z| < re^{h/2}$. We also assume $(\gamma)_\nu$ the Euclidean exceptional circles corresponding to these $n(re^{h/2}, a_\nu)$ points and positive number H_1. Let

$$N = n(re^{h/2}, \infty) + \sum_{\nu=1}^{p} n(re^{h/2}, a_\nu),$$

$$(\gamma)' = \bigcup_{\nu=1}^{p} \left\{ \bigcup_{j=1}^{n(re^{h/2}, a_\nu)} \left(|z - a_{\nu j}| < \frac{2eH_1}{N} \right) \right\}$$

$$\cup \left\{ \bigcup_{m=1}^{n(re^{h/2}, \infty)} \left(|z - \beta_m| < \frac{2eH_1}{N} \right) \right\},$$

$$(\gamma) = \bigcup_{\nu=1}^{p} (\gamma)_\nu \cup (\gamma)';$$

then the sum of the radii of (γ) does not exist $2e(p+1)H_1$. If we choose

$$H_1 = \frac{e^\sigma - 1}{8e(p+1)} r,$$

then there exists value R in the interval $[r, re^\sigma]$, such that circumference $|z| = R$ does not intersect (γ). Furthermore, we let

$$E_\nu(R) = \left\{ \theta \,|\, 0 \leq \theta < 2\pi, \, \log \frac{1}{|f(Re^{i\theta}) - a_\nu|} \geq \frac{\delta}{4} T(R, f) \right\}. \qquad (3.42)$$

In the following we prove

$$\operatorname{meas} E_\nu(R) \geq M(\delta, h, H, \sigma). \qquad (3.43)$$

First of all, by formulas (3.36) and (3.39) we get

$$\frac{\delta}{2} T(R, f) \leq m(R, a_\nu) = \frac{1}{2\pi} \int_0^{2\pi} \log^+ \frac{1}{|f(Re^{i\theta}) - a_\nu|} \, d\theta$$

$$\leq \frac{1}{2\pi} \int_{E_\nu(R)} \log^+ \frac{1}{|f(Re^{i\theta}) - a_\nu|} \, d\theta + \frac{\delta}{4} T(R, f), \qquad (3.44)$$

$$\frac{\delta}{4} T(R, f) \leq \frac{1}{2\pi} \int_{E_\nu(R)} \log^+ \frac{1}{|f(Re^{i\theta}) - a_\nu|} \, d\theta.$$

Next, we apply Lemma 3.8, where we let $r = re^\sigma$, $R' = re^{h/2}$, $R = re^h$ and $(\gamma) = (\gamma)_\nu$. Then it yields from formula (3.32) that

$$
\log^+ \frac{1}{|f(Re^{i\theta}) - a|}
$$
$$
\leq \left\{ \frac{e^{h/2} + e^\sigma}{e^{h/2} - e^\sigma} + \frac{2}{h} \log \frac{16(p+1)e^{1+h/2}}{e^\sigma - 1} \right\} T\left(re^h, \frac{1}{f - a_\nu}\right).
$$

Substituting this result into formula (3.44), and considering $hH \leq K \leq 2hH$, as well as formulas (3.37) and (3.35), we obtain formula (3.43).

Now we prove formula (3.40). First of all, by formula (1.35), we conclude that when $z = Re^{i\theta}$, $\theta \in E_\nu(R)$, $z \notin (\gamma)'$, the following inequality holds:

$$
\log\left|\frac{f'(z)}{f(z) - a_\nu}\right| \leq \log^+(re^{h/2}) + 2\log^+ \frac{1}{re^{h/2} - R}
$$
$$
+ 2\log 2 + \log^+ T(re^{h/2}, f - a_\nu)
$$
$$
+ \log^+ \log^+ \frac{1}{|c_\nu|} \quad (^6) + \log 2
$$
$$
+ \log^+ \frac{n(re^{h/2}, a_\nu) + n(re^{h/2}, \infty)}{R}
$$
$$
+ \log^+ \left(\frac{4(p+1)RN}{(e^\sigma - 1)r}\right)
$$
$$
+ \log^+ \frac{R}{re^{h/2} - R} + 2\log 2.
$$

Moreover, by formulas (3.33) and (3.34) as well as $1 \leq r \leq R \leq re^\sigma$, we derive

$$
\log\left|\frac{f'(z)}{f(z) - a_\nu}\right| \leq \log(re^{h/2}) + 3\log^+ \frac{1}{e^{h/2} - e^\sigma}
$$
$$
+ 3\log 2 + \log^+ T(re^{h/2}, f) + \log^+ \log^+ |a|
$$
$$
+ \log 2 + \log^+ \log^+ \frac{1}{|c|} + 2\log^+ N + \log R
$$
$$
+ \log^+ \frac{4e^\sigma(p+1)}{e^\sigma - 1} + 2\log 2.
$$

Furthermore, according to formula (3.37), we get

(^{6}) Here we need to replace $\log^+ \log^+ \frac{1}{|f(0)-a_\nu|}$ by $\log^+ \log^+ \frac{1}{|c_\nu|}$ in formula (1.35).

$$N = n(re^{h/2}, \infty) + \sum_{\nu=1}^{p} n(re^{h/2}, a_\nu)$$

$$\leq \frac{2}{h} \left\{ N(re^h, \infty) + \sum_{\nu=1}^{p} N(re^h, a_\nu) \right\}$$

$$\leq \frac{2}{h} \left\{ T(re^h, f) + \sum_{\nu=1}^{p} T\left(re^h, \frac{1}{f(z) - a_\nu}\right) \right\}$$

$$\leq \frac{2}{h}(2p + 1)T(re^h, f).$$

And hence

$$\log \left| \frac{f'(z)}{f(z) - a_\nu} \right| \leq 6 \log 2 + \frac{1}{2}h + \log^+ \log^+ |a|$$

$$+ \log^+ \log^+ \frac{1}{|c|} + 3 \log \frac{1}{e^{h/2} - e^\sigma}$$

$$+ \log^+ \frac{4e^\sigma(p+1)}{e^\sigma - 1} + 2 \log^+ \frac{2(2p+1)}{h}$$

$$+ \log r + 3 \log^+ T(re^h, f) + \log R.$$

Finally, by formulas (3.39), (3.35) and (3.38), we conclude that

$$\log^+ \frac{1}{|f'(Re^{i\theta})|} \geq \frac{\delta}{8} T(R, f),$$

meaning that formula (3.40) holds. Thus, Lemma 3.9 is proved completely.

3.3.3. On the growth and distribution of zeros and poles of meromorphic functions. We consider the relationship between the growth of a meromorphic function with deficient values and the distribution of its zeros and poles, or the distribution of the Julia directions, and prove the following result [44a]:

THEOREM 3.6. *Let $f(z)$ be a function meromorphic on the open plane $|z| < +\infty$, having a nonzero finite deficient value. Let $\Delta(\theta_k)\,(k = 1, 2, \ldots, q;\ 0 \leq \theta_1 < \theta_2 < \cdots < \theta_q;\ \theta_{q+1} = \theta_1 + 2\pi)$ be $q\ (1 \leq q < +\infty)$ half straight lines on the z-plane. Moreover, for any arbitrary given small number $\varepsilon > 0$,*

$$\varlimsup_{r \to +\infty} \frac{\log^+ n\{\bigcup_{k=1}^q \overline{\Omega}(\theta_k + \varepsilon, \theta_{k+1} - \varepsilon; r), f = X\}}{\log r} = 0, \qquad X = 0, \infty.$$

$$(3.45)$$

Then, when the lower order of $f(z)$, $\mu < +\infty$, surely the order of $f(z)$, $\lambda \leq \frac{\pi}{\omega}$, where $\omega = \min_{1 \leq k \leq q}\{\theta_{k+1} - \theta_k\}$.

PROOF. (1) First of all, according to the definition of lower order, there exists a sequence $\{r_n\}$, $r_n < r_{n+1} \to +\infty\ (n \to +\infty)$, such that

$$\lim_{n \to +\infty} \frac{\log T(r_n, f)}{\log r_n} = \mu < +\infty.$$

Then we select arbitrary fixed numbers h $(0 < h < +\infty)$, h_1 $(0 < h_1 < h)$, and H $(\max\{\mu, \frac{\pi}{\omega}\} < H)$ and let

$$K_1 = h_1 \left(1 + \frac{h_1}{h}\right) H, \qquad K = h \left(1 + \frac{h_1}{h}\right) H,$$

$$
\begin{aligned}
E = \{t | T(te^{h_1}, f) \leq e^{K_1} T(t, f), \\
T(e^h t, f) \leq e^K T(t, f), \ t \geq r_0 \geq 1\}, \\
E[r_0, r_n] = E \cap [r_0, r_n],
\end{aligned}
\tag{3.46}
$$

then by formula Lemma 2.1 we obtain

$$\lim_{n \to +\infty} \frac{1}{\log r_n} \int_{E[r_0, r_n]} \frac{dt}{t} \geq 1 - \frac{\mu}{H}.$$

Let a be a nonzero finite deficient value of $f(z)$, with its corresponding deficiency being $\delta(a, f) = \delta > 0$. By Lemma 3.9, we conclude that for any arbitrarily selected value σ $(0 < \sigma < \frac{1}{2} h_1)$ and sufficiently large value $t \in E$, there must exist in the interval $[t, e^\sigma t]$ a value R_t and its corresponding $E(R_t)$ comprising values θ $(0 \leq \theta < 2\pi)$, such that when $\theta \in E(R_t)$, we have

$$\log \frac{1}{|f(R_t e^{i\theta}) - a|} \geq \frac{\delta}{4} T(R_t, f),\tag{3.47}$$

and

$$\operatorname{meas} E(R_t) \geq M = M(\delta, h_1, H, \sigma) > 0.$$

Now, we select an arbitrary fixed number ε,

$$0 < \varepsilon < \min\left\{\frac{\omega}{8}, \frac{M}{8q}\right\};$$

then it follows from formula (3.45) that

$$\overline{\lim_{r \to +\infty}} \frac{\log^+ n\{\bigcup_{k=1}^q \overline{\Omega}(\theta_k + \varepsilon, \theta_{k+1} - \varepsilon; r), f = X\}}{\log r} = 0, \qquad X = 0, \infty.\tag{3.48}$$

On the other hand, there exists at least one set $E(R_t) \cap [\theta_{k_t} + 2\varepsilon, \theta_{k_t+1} - 2\varepsilon]$ $(1 \leq k_t \leq q)$ within the q sets $E(R_t) \cap [\theta_k + 2\varepsilon, \theta_{k+1} - 2\varepsilon]$ $(k = 1, 2, \ldots, q)$, such that

$$\operatorname{meas}\{E(R_t) \cap [\theta_{k_t} + 2\varepsilon, \theta_{k_t+1} - 2\varepsilon]\} \geq \frac{M}{2q}.\tag{3.49}$$

(2) By a rotating transformation $ze^{-i(\theta_{k_t} - \theta_{k_t+1})/2}$, we may transform $\Omega(\theta_{k_t} + \varepsilon, \theta_{k_t+1} - \varepsilon)$ into $\Omega(-\theta, \theta)$, $\theta = \frac{1}{2}(\theta_{k_t+1} - \theta_{k_t}) - \varepsilon$. Hence, we may assume that $\Omega(\theta_{k_t} + \varepsilon, \theta_{k_t+1} - \varepsilon) \equiv \Omega(-\theta, \theta)$. Clearly, there exists a value α $(1 \leq \alpha \leq 2)$, such that $f'(\alpha)/f(\alpha) \neq 0, \infty$. Consider the transformation

$$\zeta = \zeta(z) = \frac{z^{\pi/2\theta} - \alpha^{\pi/2\theta}}{z^{\pi/2\theta} + \alpha^{\pi/2\theta}}.\tag{3.50}$$

Then $\Omega(-\theta, \theta)$ is mapped into a unit circle $|\zeta| < 1$ and point $z = \alpha$ into the origin $\zeta = 0$ on the ζ-plane. Moreover, by Lemma 2.9, we conclude that image domain of $\overline{\Omega}(-\theta+\varepsilon, \theta-\varepsilon; \alpha, R_t)$ on the ζ-plane must be contained in the disk $|\zeta| \leq \rho$, with

$$\rho = 1 - \frac{\varepsilon}{2\theta} R_t^{-\pi/2\theta}, \tag{3.51}$$

while the pre-image domain on the z-plane of disk $|\zeta| \leq \frac{1}{2}(1+\rho)$ is contained in domain $\Omega(-\theta, \theta; R_1)$, with

$$R_1 = \alpha \left(\frac{8\theta}{\varepsilon}\right)^{2\theta/\pi} \cdot R_t. \tag{3.52}$$

Furthermore, when $|\zeta| \leq \frac{1}{2}(1 + \rho)$, we have

$$\frac{2\alpha\theta}{\pi} \left(\frac{\varepsilon}{8\theta}\right)^{2\theta/\pi} \frac{1}{R_t} \leq |z'(\zeta)| \leq \frac{2\alpha\theta}{\pi} \left(\frac{8\theta}{\varepsilon}\right)^{1+2\theta/\pi} R_t^{1+\pi/2\theta}, \tag{3.53}$$

where $z(\zeta)$ is the inverse transformation of formula (3.50). We continue to consider the transformation

$$\xi = \xi(\zeta) = \frac{2}{1+\rho}\zeta. \tag{3.54}$$

Then disk $|\zeta| \leq \frac{1}{2}(1 + \rho)$ on the ζ-plane is mapped into a unit circle $|\xi| \leq 1$ on the ξ-plane, and disk $|\zeta| \leq \rho$ is mapped into $|\xi| \leq z$, with

$$\tau = \frac{2\rho}{1+\rho}. \tag{3.55}$$

Let $\zeta(\xi)$ be the inverse transformation of formula (3.54); then $\frac{1}{2} \leq |\zeta'(\xi)| \leq 1$. By denoting $z(\xi) = z(\zeta(\xi))$, then according to formula (3.53) we conclude that when $|\xi| \leq 1$

$$\frac{\alpha\theta}{\pi} \left(\frac{\varepsilon}{8\theta}\right)^{2\theta/\pi} \frac{1}{R_t} \leq |z'(\xi)| \leq \frac{2\alpha\theta}{\pi} \left(\frac{8\theta}{\varepsilon}\right)^{1+2\theta/\pi} R_t^{1+\pi/2\theta}. \tag{3.56}$$

Obviously, the image $\Gamma_\xi(-\theta, \theta, R_t)$ on the ξ-plane of arc $\Gamma(-\theta, \theta, R_t)$ on the z-plane is orthogonal to unit circumference $|\xi| = 1$, and image $\Gamma_\xi(-\theta+\varepsilon, \theta-\varepsilon; R_t)$ of arc $\Gamma(-\theta+\varepsilon, \theta+\varepsilon; R_t)$ on the ξ-plane is contained in the disk $|\xi| \leq \tau$. Let

$$\xi_0 = \frac{2}{1+\rho} \cdot \frac{R_t^{\pi/2\theta} - \alpha^{\pi/2\theta}}{R_t^{\pi/2\theta} + \alpha^{\pi/2\theta}}. \tag{3.57}$$

It follows that $\xi_0 \in \Gamma_\xi(-\theta + \varepsilon, \theta - \varepsilon; R_t)$. We consider the transformation

$$x = x(\xi) = \frac{\xi - \xi_0}{1 - \xi_0 \xi}; \tag{3.58}$$

then $\Gamma_\xi(-\theta, \theta; R_t)$ changes into a straight line $\Gamma_x(-\theta, \theta; R_t)$ passing through origin $x = 0$ on the x-plane. On the other hand, the inverse transformation of formula (3.58) is

$$\xi = \xi(x) = \frac{x + \xi_0}{1 + \xi_0 x}.$$

Hence, when $|x| < 1$, we have

$$\frac{1 - \xi_0}{2} \le |\xi'(x)| \le \frac{2}{1 - \xi_0}.$$

Moreover, according to formula (3.57), we get

$$1 - \xi_0 = 1 - \frac{2}{1 + \rho} \cdot \frac{R_t^{\pi/2\theta} - \alpha^{\pi/2\theta}}{R_t^{\pi/2\theta} + \alpha^{\pi/2\theta}}$$

$$= \frac{\rho(R_t^{\pi/2\theta} + \alpha^{\pi/2\theta}) - R_t^{\pi/2\theta} + 3\alpha^{\pi/2\theta}}{(1 + \rho)(R_t^{\pi/2\theta} + \alpha^{\pi/2\theta})}.$$

Furthermore, noting that $1 \le \alpha \le 2$, $0 < \varepsilon < \frac{\omega}{8}$ and $0 < \omega \le 2\pi$, we conclude from formula (3.51) that

$$1 - \xi_0 \ge R_t^{-\pi/2\theta}.$$

And hence,

$$\frac{1}{2} R_t^{-\pi/2\theta} \le |\xi'(x)| \le 2 R_t^{\pi/2\theta}. \tag{3.59}$$

Let $z(x) = z(\delta(\xi(\chi)))$, by formulas (3.56) and (3.59), we then arrive at a conclusion that when $|x| < 1$, the following inequality holds:

$$\frac{\alpha\theta}{2\pi}\left(\frac{\varepsilon}{8\theta}\right)^{2\theta/\pi}\left(\frac{1}{R_t}\right)^{1+\pi/2\theta} \le |z'(x)| \le \frac{4\alpha\theta}{\pi}\left(\frac{8\theta}{\varepsilon}\right)^{1+2\theta/\pi} R_t^{i+\pi/\theta}. \tag{3.60}$$

(3) Let $F(\xi) = f(z(\zeta(\xi)))$. Clearly, $F'(0)/F(0) \neq 0, \infty$ Hence, according to the Jensen-Nevanlinna formula, for any arbitrary value r, $0 < r < 1$, we have

$$\log\left|\frac{F'(0)}{F(0)}\right| = m\left(r, \frac{F'}{F}\right) - m\left(r, \frac{F}{F'}\right)$$

$$+ N\left(r, \frac{F'}{F}\right) - N\left(r, \frac{F}{F'}\right)$$

$$\le m\left(r, \frac{F'}{F}\right) - m\left(r, \frac{F}{F'}\right) + N(r, F)$$

$$+ N\left(r, \frac{1}{F}\right) - N\left(r, \frac{1}{F'}\right). \tag{3.61}$$

On the other hand, we construct circles using each zeros and poles of $F(\xi)$ as centers, and

$$\frac{1}{N} \cdot \frac{1 - \tau}{32 \times 2} \qquad (N = n(1, F = 0) + n(1, F = \infty))$$

as the radii. We then denote all these circles as $(\gamma)_\xi$, with the sum of their radii not exceeding $\frac{1-\tau}{32\times2}$. Hence, there exist values τ_1 and τ_2 in intervals $[\frac{1+\tau}{2}, \frac{3+\tau}{4}]$ and $[\frac{7+\tau}{8}, \frac{15+\tau}{16}]$, respectively, such that there are no intersections between circumference $|\xi| = \tau_i$ $(i = 1, 2)$ and circles $(\gamma)_\xi$. In the following, we make an estimation on $m(\tau_1, \frac{F}{F'})$ and $n(\tau_1, F' = 0)$ by using formula (3.61).

(1) First, we estimate $m(\tau', \frac{F'}{F})$, where $0 < \tau' \leq \frac{15+\tau}{16}$, and there is no intersection between circumference $|\xi| = \tau'$ and circles $(\gamma)_\xi$. Let Γ' be the image of circumference $|\xi| = \tau'$ on the z-plane, β_ξ be a zero or pole of $F(\xi)$, with its image on the z-plane being β. We also let l' be the shortest path linking β and Γ', and l'_ξ be the image of l' on the ξ-plane. Under the hypotheses, circumference $|\xi| = \tau'$ is outside circles $(\gamma)_\xi$. Consequently, we conclude that the length of l' is

$$\begin{aligned}
\operatorname{meas} l' &= \int_{l'} |dz| = \int_{l'_\xi} |z'(\xi)|\, |d\xi| \\
&\geq \min_{|\xi|\leq1} |z'(\xi)| \frac{1-\tau}{32\times2} \cdot \frac{1}{N}.
\end{aligned} \tag{3.62}$$

Moreover, according to formulas (3.48) and (3.52), for any arbitrarily selected number $n > 0$, provided that $t \in E$ is sufficiently large, we obtain

$$\begin{aligned}
n(1, F = 0) &+ n(1, F = \infty) \\
&\leq n\{\Omega(-\theta + \varepsilon, \theta - \varepsilon; R_1), f = 0\} \\
&\quad + n\{\Omega(-\theta_1 + \varepsilon, \theta - \varepsilon; R_1), f = \infty\} \\
&\leq 2R_1^\eta = 2\alpha^\eta \left(\frac{8\theta}{\varepsilon}\right)^{2\theta^\eta/\pi} R_t^\eta.
\end{aligned} \tag{3.63}$$

Furthermore, by formula (3.51), we get

$$1 - \tau = 1 - \frac{2\rho}{1+\rho} \geq \frac{1}{2}(1 - \rho) = \frac{\varepsilon}{4\theta} R_t^{-\pi/2\theta}. \tag{3.64}$$

Hence, it follows from formulas (3.56), (3.62), (3.63) and (3.64) that

$$\begin{aligned}
\operatorname{meas} l' &\geq \frac{\alpha\theta}{32\times8} \cdot \frac{1}{\pi\alpha^\eta} \cdot \left(\frac{1}{8\theta}\right)^{(2\theta/\pi)(1+\eta)} \left(\frac{1}{R_t}\right)^{1+(\pi/2\theta)+\eta} \\
&= AR_t^{-(1+(\pi/2\theta)+\eta)},
\end{aligned}$$

where $A > 0$ is a constant unrelated to t. On the other hand, let Γ be the image of circumference $|\xi| = 1$ on the z-plane, $d(\Gamma', \Gamma)$ be the distance between Γ' and Γ, l be the shortest straight line linking Γ and Γ', and l_ξ

be the image of l on the ξ-plane. Hence, we have

$$d(\Gamma'; \Gamma) = \operatorname{meas} l = \int_l |dz| = \int_{l_\xi} |z'(\xi)||d\xi|$$

$$\geq \min_{|\xi| \leq 1} |z'(\xi)|(1 - \tau')$$

$$\geq \min_{|\xi| \leq 1} |z'(\xi)| \left(\frac{1 - \tau}{16}\right) \geq A R_t^{-(1+(\pi/2\theta)+\eta)}.$$

Consequently, we conclude that the distance between each zero and pole of $f(z)$ on the z-plane and Γ' is $\geq A R_t^{-(1+(\pi/2\theta)+\eta)}$.

Now, we apply formula (1.35), where we let $r = R_1$, $\rho' = 2R_1$; then, when $|z| > R_1$, we have

$$\log\left|\frac{f'(z)}{f(z)}\right| \leq \log^+ 2R_1 + \log^+ T(2R_1, f)$$

$$+ \log^+ \frac{n(2R_1)}{R_1} + \log^+ \frac{R_1}{\delta(z)} + O(1),$$

where $\delta(z)$ denotes the shortest distance between point z and the zeros and poles of $f(z)$, and also

$$n(2R_1) = n\left(2R_1, \frac{1}{f}\right) + n(2R_1, f).$$

Notice that

$$n(2R_1, f) \leq \frac{1}{\log 2} \int_{2R_1}^{4R_1} \frac{n(t, f)}{t} dt$$

$$\leq \frac{1}{\log 2} \left\{N(4R_1, f) + n(0, f) \log \frac{1}{2R_1}\right\}$$

$$\leq \frac{1}{\log 2} T(4R_1, f)$$

and

$$n\left(2R_1, \frac{1}{f}\right) \leq \frac{1}{\log 2} T\left(4R_1, \frac{1}{f}\right).$$

Then it follows that

$$n(2R_1) \leq \frac{2}{\log 2} T(4R_1, f) + O(1).$$

Hence, when $z \in \Gamma'$, in particular, the following inequality holds:

$$\log\left|\frac{f'(z)}{f(z)}\right| \leq A \log[R_1 T(4R_1, f)] + O(1), \tag{3.65}$$

where $A < +\infty$ is a constant unrelated to t. Furthermore, according to the equality

$$\frac{F'}{F} = \frac{f'}{f} z'(\xi),$$

as well as formulas (3.56) and (3.65), when $|\xi| = \tau'$, we obtain

$$\log^+ \left| \frac{F'}{F} \right| \leq \log^+ \left| \frac{f'}{f} \right| + \log^+ |z'(\xi)|$$

$$\leq A \log[R_1 T(4R_1, f)] + \log R^{1+\pi/2\theta} + O(1)$$
$$\leq A \log[R_1 T(4R_1, f)] + O(1)$$

and, in turn, we conclude that when $0 < \tau' \leq \frac{15+\tau}{16}$, and when there is no intersection between circumference $|\xi| = \tau'$ and circles $(\gamma)_\xi$, the following inequality holds:

$$m\left(\tau', \frac{F'}{F}\right) \leq A \log[R_1 T(4R_1, f)] + O(1), \qquad (3.66)$$

where $A < +\infty$ is a constant unrelated to t.

(2) Next, we estimate $m(\tau_1, \frac{F}{F'})$. By formulas (3.61) and (3.66), where we let $r = \tau' = \tau_1$, we find that

$$\log \left| \frac{F'(0)}{F(0)} \right| \leq A \log[R_1 T(4R_1, f)] - m\left(\tau_1, \frac{F}{F'}\right)$$
$$+ N(\tau_1, F) + N\left(\tau_1, \frac{1}{F}\right).$$

Notice that when $F'(0)/F(0) \neq 0, \infty$, $F(0) \neq 0, \infty$. We assume arbitrarily a number σ_0, $0 < \sigma_0 < \tau_1$; then it follows that

$$N(\tau_1, F) + N\left(\tau_1, \frac{1}{F}\right) = \int_0^{\tau_1} \frac{n(t, F = \infty)}{t} dt + \int_0^{\tau_1} \frac{n(t, F = 0)}{t} dt$$

$$= \int_0^{\sigma_0} \frac{n(t, F = \infty)}{t} dt + \int_{\sigma_0}^{\tau_1} \frac{n(t, F = \infty)}{t} dt$$

$$+ \int_0^{\sigma_0} \frac{n(t, F = 0)}{t} dt + \int_{\sigma_0}^{\tau_1} \frac{n(t, F = 0)}{t} dt$$

$$\leq \int_0^{\sigma_0} \frac{n(t, F = \infty)}{t} dt + \int_0^{\sigma_0} \frac{n(t, F = 0)}{t} dt$$

$$+ \{n(1, F = 0) + n(1, F = \infty)\} \log \frac{\tau_1}{\sigma_0}.$$

Furthermore, by formula (3.63), we assure that for any sufficiently large $t \in E$, $N(\tau_1, F) + N(\tau_1, \frac{1}{F}) \leq BR_1^\eta$, holds, where $B < +\infty$ is a constant unrelated to t. Hence,

$$\log \left| \frac{F'(0)}{F(0)} \right| \leq A \log[R_1 T(4R_1, f)] + BR_1^\eta - m\left(\tau_1, \frac{F}{F'}\right),$$

$$m\left(\tau_1, \frac{F}{F'}\right) \leq A \log[R_1 T(4R_1, f)] + BR_1^\eta + \log \left| \frac{F(0)}{F'(0)} \right|.$$

Therefore, when $t \in E$ is sufficiently large, we conclude that

$$m\left(\tau_1, \frac{F}{F'}\right) \leq R_1^{2\eta} \log T(4R_1, f). \tag{3.67}$$

(3) Finally, we make an estimation on $n(\tau_1, F' = 0)$. By formulas (3.61) and (3.66), where we let $r = \tau' = \tau_2$, we get

$$\log\left|\frac{F'(0)}{F(0)}\right| \leq A \log[R_1 T(4R_1, f)] + BR_1^\eta - N\left(\tau_2, \frac{1}{F'}\right).$$

Notice that $F'(0) \neq 0$, then

$$N\left(\tau_2, \frac{1}{F'}\right) \geq \int_{\tau_1}^{\tau_2} \frac{n(t, F' = 0)}{t} dt \geq \frac{\tau_2 - \tau_1}{\tau_2} n(\tau_1, F' = 0).$$

Moreover,

$$\tau_2 - \tau_1 \geq \frac{7+\tau}{8} - \frac{3+\tau}{4} \geq \frac{1}{16}(1 - \rho)$$
$$= \frac{\varepsilon}{32\theta} R_t^{-\pi/2\theta}, \qquad 0 < \tau_2 < 1.$$

And hence

$$N\left(\tau_2, \frac{1}{F'}\right) \geq C \cdot R_t^{-\pi/2\theta} n(\tau_1, F' = 0),$$

where $C > 0$ is a constant unrelated to $t \in E$, and hence, we obtain

$$\log\left|\frac{F'(0)}{F(0)}\right| \leq A \log[R_1 T(4R_1, f)]$$
$$+ BR_1^\eta - CR_1^{-\pi/2\theta} n(\tau_1, F' = 0).$$

Therefore, for sufficiently large $t \in E$, we conclude that

$$n(\tau_1, F' = 0) \leq R_1^{(\pi/2\theta)+2\eta} \log T(4R_1, f). \tag{3.68}$$

(4) Let $E_z(R_t) = E\{R_t e^{i\theta} | \theta \in E(R_t) \cap [\theta_{k_t} + 2\varepsilon, \theta_{k_t+1} - 2\varepsilon]\}$. According to formula (3.47), when $z \in E_z(R_t)$, we have

$$\frac{\delta}{4} T(R_t, f) \leq \log \frac{1}{|f(z) - a|}.$$

Moreover, by the identical relation,

$$\frac{1}{f(z) - a} = \frac{1}{a} \cdot \frac{f(z)}{f'(z)} \left\{ \frac{f'(z)}{f(z) - a} - \frac{f'(z)}{f(z)} \right\}.$$

We then conclude that when $z \in E_z(R_t)$, the following inequality holds:

$$\frac{\delta}{4} T(R_t, f) \leq \log^+\left|\frac{f'(z)}{f(z) - a}\right| + \log^+\left|\frac{f'(z)}{f(z)}\right|$$
$$+ \log^+\left|\frac{f(z)}{f'(z)}\right| + O(1). \tag{3.69}$$

In the following, we estimate each of the items at the right-hand side of formula (3.69).

(1) First of all, from formula (3.49) we have $\operatorname{meas} E_z(R_t) \geq (M/2q)R_t$. Then, we construct circles using each zero and pole of $f(z)$ and a-value points as centers, and $\frac{M}{8q} \cdot \frac{R_t}{N'}$ ($N' = n(2R_1, f = 0) + n(2R_1, f = \infty) + n(2R_1, f = a)$) as radii. We denote all these circles as $(\gamma)_z$, with the sum of their radii not exceeding $\frac{MR_t}{8q}$. Hence, if we let $\widetilde{E}_z(R_t) = E_z(R_t) - [E_z(R_t) \cap (\gamma)_z]$, then it follows that $\operatorname{meas} \widetilde{E}_z(R_t) \geq (M/4q)R_t > 0$. In the following, when $z \in \widetilde{E}_z(R_t)$, analogous to the proof of formula (3.65), we may conclude that

$$\log^+ \left| \frac{f'(z)}{f(z)} \right| \leq A \log[R_1 T(4R_1, f)] + O(1) \tag{3.70}$$

and

$$\log^+ \left| \frac{f'(z)}{f(z) - a} \right| \leq A \log[R_1 T(4R_1, f)] + O(1), \tag{3.71}$$

where $A > 0$ is a constant unrelated to t.

(2) Let a_ν ($\nu = 1, 2, \ldots, n(\tau_1, F' = 0)$) be the zeros of $F'(\xi)$ in the disk $|\xi| < \tau_1$. Applying the Poisson-Jensen formula, where we let $\gamma = \tau$, $\rho = \tau_1$, we obtain

$$\log \left| \frac{F(\xi)}{F'(\xi)} \right| \leq \frac{\tau_1 + \tau}{\tau_1 - \tau} m\left(\tau_1, \frac{F}{F'}\right) + \sum_{\nu=1}^{n(\tau_1, F'=0)} \log \left| \frac{\tau_1^2 - \overline{a}_\nu \xi}{\tau_1(\xi - a_\nu)} \right|,$$

when $|\xi| \leq \tau$. Notice that

$$\left| \frac{\tau_1^2 - \overline{a}_\nu \xi}{\tau_1(\xi - a_\nu)} \right| = \left| \frac{1 - \overline{a}_\nu \xi}{\xi - a_\nu} \right| \left| \frac{\tau_1^2 - \overline{a}_\nu \xi}{\tau_1(1 - \overline{a}_\nu \xi)} \right|$$

$$\leq \left| \frac{1 - \overline{a}_\nu \xi}{\xi - a_\nu} \right| \left| 1 + \frac{\overline{a}_\nu \xi - \frac{\overline{a}_\nu \xi}{\tau_1^2}}{1 - \overline{a}_\nu \xi} \right|$$

$$\leq \left| \frac{1 - \overline{a}_\nu \xi}{\xi - a_\nu} \right| \left(1 + \frac{|\overline{a}_\nu \xi|}{\tau_1^2} \right)$$

$$\leq 2 \left| \frac{1 - \overline{a}_\nu \xi}{\xi - a_\nu} \right|.$$

Then it follows that

$$\log \left| \frac{F(\xi)}{F'(\xi)} \right| \leq \frac{\tau_1 + \tau}{\tau_1 - \tau} m\left(\tau_1, \frac{F}{F'}\right) + \sum_{\nu=1}^{n(\tau_1, F'=0)} \log \left| \frac{1 - \overline{a}_\nu \xi}{\xi - a_\nu} \right| \tag{3.72}$$
$$+ n(\tau_1, F' = 0) \log 2,$$

when $|\xi| < \tau$.

Let $\widetilde{E}_\xi(R_t)$ and $\widetilde{E}_\chi(R_t)$ be the images of $\widetilde{E}_z(R_t)$ on the ξ-plane and x-plane, respectively. Obviously, $\widetilde{E}_\xi(R_t) \subset \Gamma_\xi(-\theta+\varepsilon, \theta-\varepsilon; R_t)$ and $\widetilde{E}_x(R_t) \subset \Gamma_x(-\theta, \theta; R_t)$. Moreover, by formula (3.60), we get

$$\frac{M}{4q}R_t \leq \int_{\widetilde{E}_z(R_t)} |dz| = \int_{\widetilde{E}_\xi(R_t)} |z'(\xi)| \, |d\xi| = \int_{\widetilde{E}_x(R_t)} |z'(x)| \, |dx|$$

$$\leq \frac{4\alpha\theta}{\pi} \left(\frac{8\theta}{\varepsilon}\right)^{1+2\theta/\pi} R_t^{1+\pi/\theta} \, \text{meas}\, \widetilde{E}_x(R_t); \tag{3.73}$$

$$\text{meas}\, \widetilde{E}_x(R_t) \geq \frac{M\pi}{16q\alpha\theta} \left(\frac{\varepsilon}{8\theta}\right)^{1+2\theta/\pi} R_t^{-\pi/\theta}.$$

We then let $(\gamma)'_\xi$ be the pseudo-non-Euclidean exceptional circles which correspond to these $n(\tau_1, F' = 0)$ points and number H_1, with the sum of radii of these circles not exceeding $2eH_1$. We also assume that $(\gamma)'_\xi$ is the image of $(\gamma)'_\xi$ on the χ-plane. Hence, the sum of the pseudo-non-Euclidean radii of $(\gamma)'_x$ does not exceed $2eH_1$. Notice that the pseudo-non-Euclidean radius of a pseudo-non-Euclidean exceptional circle is no less than its Euclidean radius, and also $\Gamma_x(-\theta, \theta; R_t)$ is a straight line. Therefore, if we assume

$$H_1 = \frac{M\pi}{8e \times 16q\alpha\theta} \left(\frac{\varepsilon}{8\theta}\right)^{1+2\theta/\pi} \left(\frac{1}{R_1}\right)^{\pi/\theta}, \tag{3.74}$$

then there must exist a point $x_1 \notin (\gamma)'_x$ on $\widetilde{E}_x(R_t)$. Let ξ_1 be the image of x_1 on the ξ-plane; then $\xi_1 \in \widetilde{E}_\xi(R_t)$ and $\xi_1 \notin (\gamma)'_\xi$. Hence, by formula (2.72), when $\xi = \xi_1$, we get

$$\log\left|\frac{F(\xi_1)}{F'(\xi_1)}\right| \leq \frac{2}{\tau_1 - \tau} m\left(\tau_1, \frac{F}{F'}\right) + n(\tau_1, F' = 0)\log\frac{2}{H_1}.$$

Notice that

$$\tau_1 - \tau \geq \frac{1+\tau}{2} - \tau \geq \frac{1}{4}(1 - \rho) = \frac{\varepsilon}{8\theta} R_t^{-\pi/2\theta},$$

and by formulas (3.67), (3.68) and (3.74), we conclude that

$$\log\left|\frac{F(\xi_1)}{F'(\xi_1)}\right| \leq AR_1^{\pi/2\theta+2\eta} \log T(4R_1, f)$$

$$+ R_1^{\pi/2\theta+2\eta} \log R_1^{\pi/\theta} + O(1) \tag{3.75}$$

$$\leq AR_1^{\pi/2\theta+2\eta} \log T(4R_1, f) + O(1),$$

where $A > 0$ is a constant unrelated to t. Let z_1 be the image of ξ_1 on the z-plane; then $z_1 \in \widetilde{E}_z(R_t)$. Hence, according to the equality

$$\frac{f(z_1)}{f'(z_1)} = \frac{F(\xi_1)}{F'(\xi_1)} z'(\xi_1),$$

we have

$$\log^+ \left| \frac{f(z_1)}{f'(z_1)} \right| \leq \log^+ \left| \frac{F(\xi_1)}{F'(\xi_1)} \right| + \log^+ |z'(\xi_1)|.$$

Furthermore, by formulas (3.56) and (3.75), we obtain

$$\begin{aligned}
\log^+ \left| \frac{f(z_1)}{f'(z_1)} \right| &\leq AR_1^{\pi/2\theta+2\eta} \log T(4R_1, f) \\
&\quad + \log R_1^{1+\pi/2\theta} + O(1) \\
&\leq AR_1^{\pi/2\theta+2\eta} \log T(4R_1, f) + O(1),
\end{aligned} \tag{3.76}$$

where $A < +\infty$ is a constant unrelated to t. Finally, it follows from formulas (3.69), (3.70), (3.71) and (3.76) that

$$T(R_t, f) \leq AR_1^{\pi/2\theta+2\eta} \log T(4R_1 f) + O(1),$$

where $A < +\infty$ is a constant unrelated to t. Furthermore, due to the arbitrariness of h, we may assume h in such a way that $8e^{h_1}(\frac{8\theta}{\varepsilon})^2 < e^h$. This means that $4R_1 < e^h t$. Hence, according to formula (3.46) we get

$$T(R_t, f) \leq AR_t^{\pi/2\theta+2\eta} \log T(t, f) + O(1).$$

It follows that, for sufficiently large value $t \in E$,

$$\begin{aligned}
T(t, f) &\leq At^{\pi/2\theta+2\eta} \log T(t, f) + O(1) \\
&\leq AT^{\pi/(\omega-2\varepsilon)+2\eta} \log T(t, f) + O(1),
\end{aligned}$$

where $A < +\infty$ is a constant unrelated to t. Therefore,

$$\varlimsup_{\substack{t \to +\infty \\ t \in E}} \frac{\log T(t, f)}{\log t} \leq \frac{\pi}{\omega - 2\varepsilon} + 2\eta.$$

Let $\varepsilon \to 0$ and $\eta \to 0$, we conclude that

$$\varlimsup_{\substack{t \to +\infty \\ t \in E}} \frac{\log T(t, f)}{\log t} \leq \frac{\pi}{\omega}. \tag{3.77}$$

Moreover, by Lemma 2.1, we obtain

$$\varliminf_{\substack{t \to +\infty \\ t \in E}} \frac{1}{\log t} \int_{E[r_0, t]} \frac{dt}{t} \geq 1 - \frac{\pi}{H\omega}.$$

And it follows that

$$\varlimsup_{\substack{t \to +\infty \\ t \in E}} \frac{1}{\log t} \int_{CE[r_0, t]} \frac{dt}{t} \leq \frac{\pi}{H\omega}, \tag{3.78}$$

where $CE[r_0, t] = [r_0, t] - E[r_0, t]$. Hence, for any sufficiently large value $t \notin E$, when $H > \frac{6\pi}{\omega}$, we conclude that there must exist a value $t' \in E$ in the interval $[t, t^{1+2\pi/H\omega})$. In fact, otherwise, we have $[t, t^{1+2\pi/H\omega}] \subset$

$CE[r_0, t^{1+2\pi/H\omega}]$. We assume value t'', such that $t'' \in E$ and $[t, t'') \subset CE[r_0, t'']$. Obviously, we have $t'' \geq t^{1+2\pi/H\omega}$. Therefore, on the one hand, we obtain

$$\int_t^{t''} \frac{dt}{t} = \log t'' - \log t,$$

while on the other hand, we have from formula (3.78) that

$$\int_t^{t''} \frac{dt}{t} \leq \frac{3\pi}{2H\omega} \log t'',$$

meaning that

$$\log t'' - \log t \leq \frac{3\pi}{2H\omega} \log t'',$$

$$\left(1 + \frac{2\pi}{H\omega}\right)\left(1 - \frac{3\pi}{2H\omega}\right) \log t \leq \left(1 - \frac{3\pi}{2H\omega}\right) \log t'' \leq \log t,$$

$$\frac{2\pi}{H\omega} \leq \frac{3\pi}{2H\omega}\left(1 + \frac{2\pi}{H\omega}\right), \qquad H \leq \frac{6\pi}{\omega}.$$

But this contradicts the way to assume H. Hence, for the arbitrarily selected number $\eta > 0$, it follows from formula (3.77) that

$$T(t, f) \leq T(t', f) \leq t'^{\pi/\omega+\eta} \leq t^{(\pi/\omega+\eta)(1+2\pi/H\omega)},$$

$$\varlimsup_{\substack{t \to +\infty \\ t \notin E}} \frac{\log T(t, f)}{\log t} \leq \left(\frac{\pi}{\omega} + \eta\right)\left(1 + \frac{2\pi}{H\omega}\right).$$

Letting $H \to +\infty$ and $\eta \to 0$, we get

$$\varlimsup_{\substack{t \to +\infty \\ t \notin E}} \frac{\log T(t, f)}{\log t} \leq \frac{\pi}{\omega}.$$

Combining this with formula (3.77), we conclude that the order λ of $f(z)$ is $\leq \frac{\pi}{\omega}$, meaning that Theorem 3.6 is proved completely.

In the hypotheses of Theorem 3.6, the replacement of the distribution of zeros and poles by the distribution of the Julia directions will have the following result:

THEOREM 3.7. Let $f(z)$ be a function meromorphic on the open plane $|z| < +\infty$, having a deficient value a and q $(1 \leq q < +\infty)$ Julia directions $\Delta(\theta_k)$ $(k = 1, 2, \ldots, q; 0 \leq \theta_1 < \theta_2 < \cdots < \theta_q < 2\pi; \theta_{q+1} = \theta_1 + 2\pi)$. If the lower order μ of $f < +\infty$, then the order of f λ must be $\leq \frac{\pi}{\omega}$, where $\omega = \min_{1 \leq k \leq q}\{\theta_{k+1} - \theta_k\}$.

PROOF. According to Theorem 2.13 and the finite covering theorem, we conclude that there exist three distinct complex numbers α, β, and γ, such that for the arbitrary given small number $\varepsilon > 0$,

$$\varlimsup_{r \to +\infty} \frac{\log^+ n\{\bigcup_{k=1}^q \overline{\Omega}(\theta_k + \varepsilon, \theta_{k+1} - \varepsilon; r), f = X\}}{\log r} = 0, \qquad X = \alpha, \beta, \gamma.$$

Obviously, among the three values α, β, and γ, there are at least two of them which we may assume to be α and β that are distinct from value a. We consider the transformation

$$F(z) = \frac{f(z) - \alpha}{f(z) - \beta};$$

then $F(z)$ is a function meromorphic on the open plane $|z| < +\infty$ and has a deficient value $(a-\alpha)/(a-\beta)$. Moreover, for any arbitrarily small number $\varepsilon > 0$,

$$\varliminf_{r \to +\infty} \frac{\log^+ n\{\bigcup_{k=1}^q \overline{\Omega}(\theta_k + \varepsilon, \theta_{k+1} - \varepsilon; r), F = X\}}{\log r} = 0, \qquad X = 0, \infty.$$

Hence, we conclude by Theorem 3.6 that the order λ of $F(z)$ is $\leq \frac{\pi}{\omega}$, meaning that the order λ of $f(z)$ is $\leq \frac{\pi}{\omega}$ Theorem 3.7 is thus proved.

§3.4. The Weitsman Theorem

LEMMA 3.10. *Let $f(z)$ be a transcendental meromorphic function on the open plane $|z| < +\infty$, with the sum of its deficiencies equal to*

$$\Delta(f) = \sum_\nu \delta(a_\nu, f) = 2, \delta(a_\nu, f) > 0. \tag{3.79}$$

Then

$$\lim_{\substack{r \to +\infty \\ r \notin E_0}} \frac{N(r, 1/f')}{T(r, f')} = 0,$$

$$\lim_{\substack{r \to +\infty \\ r \notin E_0}} \frac{T(r, f')}{T(r, f)} = 2 - \delta(\infty, f),$$

where E_0 has linear measure $\operatorname{meas} E_0 \leq 2$.

PROOF. According to condition (3.79), for any arbitrarily small number $\varepsilon > 0$, provided that q is sufficiently large, then the inequality:

$$\sum_{\nu=1}^q \delta(a_\nu, f) + \delta(\infty, f) > 2 - \varepsilon \tag{3.80}$$

holds, where a_ν are the finite deficient values of $f(z)$. Now, by formula (1.45), we get

$$\sum_{\nu=1}^q m(r, a_\nu) \leq m\left(r, \sum_{\nu=1}^q \frac{1}{f - a_\nu}\right) + O(1)$$

$$\leq m\left(r, \frac{1}{f'}\right) + S(r, f) \tag{3.81}$$

$$= T\left(r, \frac{1}{f'}\right) - N\left(r, \frac{1}{f'}\right) + S(r, f),$$

and

$$T(r, f') = m(r, f') + N(r, f')$$
$$\leq m(r, f'/f) + m(r, f) + N(r, f') \qquad (3.82)$$
$$\leq T(r, f) + N(r, f) + S(r, f),$$

where $S(r, f)$ is an error term as in Theorem 1.4: namely, $S(r, f) = O\{\log[rT(r, f)]\}$ but probably excluding an exceptional set E_0 of r values, with its linear measure meas $E_0 \leq 2$. Moreover, E_0 is unrelated to a_ν ($\nu = 1, 2, \ldots, q$) and q. Furthermore, under the hypothesis that $f(z)$ is a transcendental meromorphic function, we obtain

$$S(r, f) = o\{T(r, f)\} \qquad (r \to +\infty), \ r \notin E_0. \qquad (3.83)$$

And hence, we derive from (3.82) that

$$\varlimsup_{\substack{r \to +\infty \\ r \notin E}} \frac{T(r, f')}{T(r, f)} \leq 2 - \delta(\infty, f).$$

On the other hand, from formulas (3.80) and (3.81) we get

$$2 - \delta(\infty, f) - \varepsilon \leq \varlimsup_{\substack{r \to +\infty \\ r \notin E}} \frac{T(r, f')}{T(r, f)}.$$

Since ε is arbitrarily small, we conclude that

$$\lim_{\substack{r \to +\infty \\ r \notin E_0}} \frac{T(r, f')}{T(r, f)} = 2 - \delta(a, f). \qquad (3.84)$$

Furthermore, by formula (3.81) we obtain

$$\varlimsup_{\substack{r \to +\infty \\ r \notin E_0}} \frac{N(r, 1/f')}{T(r, f')} + \varlimsup_{\substack{r \to +\infty \\ r \notin E_0}} \frac{T(r, f)}{T(r, f')} \cdot \sum_{\nu=1}^{q} \varlimsup_{r \to +\infty} \frac{m(r, a_\nu)}{T(r, f)}$$
$$\leq 1 + \varlimsup_{\substack{r \to +\infty \\ r \notin E_0}} \frac{T(r, f)}{T(r, f')} \cdot \frac{S(r, f)}{T(r, f)}.$$

Moreover, from formulas (3.80), (3.83) and (3.84) we derive

$$\varlimsup_{\substack{r \to +\infty \\ r \notin E}} \frac{N(r, 1/f')}{T(r, f')} \leq \frac{\varepsilon}{2 - \delta(\infty, f)}.$$

Since ε is arbitrarily small, we conclude that

$$\varlimsup_{\substack{r \to +\infty \\ r \notin E_0}} \frac{N(r, 1/f')}{T(r, f')} = 0.$$

Consequently, Lemma 3.10 is proved.

In the following we prove an important result of A. Weitsman [40a]:

THEOREM 3.8. *Let $f(z)$ be a meromorphic function of lower order $\mu < +\infty$ on the open plane $|z| < +\infty$. Suppose that the sum of its deficiencies is equal to*

$$\Delta(f) = \sum_\nu \delta(a_\nu, f) = 2, \qquad \delta(a_\nu, f) > 0.$$

Then the number of its deficient values p is $\leq 2\mu$.

PROOF. (1) First, according to the assumption that $\Delta(f) = 2$, p must be ≥ 2. Hence, by Theorem 3.4, we conclude that $\mu > 0$, and therefore, $f(z)$ is a transcendental meromorphic function.

We choose arbitrarily p $(p < +\infty)$ deficient values a_ν $(\nu = 1, 2, \ldots, p)$ of $f(z)$, with their corresponding deficiencies $\delta(a_\nu, f) = \delta_\nu > 0$. Without any loss of generality, we may assume that a_ν $(\nu = 1, 2, \ldots, p)$ are all finite deficient values. Otherwise, we need only to undergo a suitable fractional linear transformation. Let

$$\delta = \min_{1 \leq \nu \leq p} \{\delta_\nu\}, \qquad |a| = \max_{1 \leq \nu \leq p} \{|a_\nu|\}, \qquad |c| = \min_{1 \leq \nu \leq p} \{|c_\nu|\},$$

where c_ν is the first nonzero coefficient of the expansion of $f(z) - a_\nu$ at a neighborhood around the origin $z = 0$. We select arbitrarily numbers h $(0 < h < +\infty)$ and h_1 $(0 \leq h_1 < h)$.[7] According to the definition of lower order μ and Lemma 2.2, there exists a sequence $\{R_n\}$ such that

$$\lim_{n \to +\infty} \frac{\log T(R_n, f)}{\log R_n} = \mu,$$

and

$$T(e^h R_n, f) \leq e^{h(1 + h_1/h)\mu} T(R_n, f)(1 + o(1)) \qquad (n \to +\infty),$$
$$T(e^{h_1} R_n, f) \leq e^{h_1(1 + h_1/h)\mu} T(R_n, f)(1 + o(1)) \qquad (n \to +\infty).$$

We also select an arbitrary number σ, $0 < \sigma < \frac{1}{5}h_1$, and let $K = h(1 + \frac{h_1}{h})\mu$ and $K_1 = h_1(1 + \frac{h_1}{h})\mu$. Then there exists a positive integer n_0 such that when $n \geq n_0$,

$$R_n \geq 1, \tag{3.85}$$

$$T(R_n e^h, f) \leq e^k T(R_n, f)(1 + o(1)) \leq 2e^k T(R_n, f), \tag{3.86}$$

$$T(R_n e^{h_1}, f) \leq e^{k_1} T(R_n, f)(1 + o(1)) \leq 2e^{k_1} T(R_n, f), \tag{3.87}$$

---

[7] If we intend merely to prove Theorem 3.8, then we need only to assume $h_1 = 0$. Here the assumption is to facilitate the proof of Theorem 3.9 next.

and when $R \geq R_n$, it follows that

$$\frac{\delta}{2} T(R, f) \leq m(R, a_\nu), \qquad \nu = 1, 2, \ldots, p,$$

$$T\left(R, \frac{1}{f - a_\nu}\right) \leq 2T(R, f), \qquad \nu = 1, 2, \ldots, p,$$

$$\frac{1}{T(R, f)} \left\{ 10 \log 2 + \log^+ \log^+ |a| + \log^+ \log^+ \frac{1}{|c|} \right.$$

$$+ \frac{1}{2} h + 3K_1 + 3 \log^+ \frac{1}{e^{h/2} - e^\sigma}$$

$$+ \log^+ \frac{4(p + 1)e^\sigma}{e^\sigma - 1} + 2 \log \frac{2(2p + 1)}{h}$$

$$\left. + 2 \log R + 3 \log^+ T(R, f) \right\} < \frac{\delta}{8}.$$

Hence, according to Lemma 3.9, we conclude that there exist value t_n and a corresponding set $E_\nu(t_n)$ $(1 \leq \nu \leq p)$ of value θ $(0 \leq \theta < 2\pi)$, such that, when $\theta \in E_\nu(t_n)$,

$$\log \frac{1}{|f(t_n e^{i\theta}) - a_\nu|} \geq \frac{\delta}{4} T(t_n, f) \tag{3.88}$$

and

$$\log \frac{1}{|f'(t_n e^{i\theta})|} \geq \frac{\delta}{8} T(t_n, f), \tag{3.89}$$

as well as

$$\operatorname{meas} E_\nu(t_n) \geq M(\delta, h, \mu, \sigma). \tag{3.90}$$

Let $f'(z)$, at a neighborhood around the origin $z = 0$, have the following expansion:

$$f'(z) = c_s z^s + c_{s+1} z^{s+1} + \cdots, \qquad c_s \neq 0.$$

Let γ_k denote the zeros of $f'(z)$. Set

$$\pi(z) = \prod_{0 < |\gamma_k| < R_n e^{h - 2\sigma}} \frac{R_n e^{h - 2\sigma}(z - \gamma_k)}{(R_n e^{h - 2\sigma})^2 - \overline{\gamma}_k z},$$

$$\alpha = \frac{c_s}{\pi(0)}, \qquad G(z) = \frac{\alpha z^s \pi(z)}{f'(z)}; \tag{3.91}$$

then $G(z)$ is regular in the disk $|z| < R_n e^{h-2\sigma}$, $G(0) = 1$, and also

$$\log^+ |\alpha| \leq \log^+ |c_s| + \sum_{0 < |\gamma_k| < R_h e^{h-2\sigma}} \log \frac{R_n e^{h-2\sigma}}{|\gamma_k|}$$

$$= \log^+ |c_s| + \int_0^{R_n e^{h-2\sigma}} \frac{n(t, f' = 0) - n(0, f' = 0)}{t} \, dt \qquad (3.92)$$

$$\leq \log^+ |c_s| + N\left(R_n e^{h-2\sigma}, \frac{1}{f'}\right) - n(0, f' = 0) \log R_n e^{h-2\sigma},$$

as well as

$$\log^+ \frac{1}{|\alpha|} \leq \log^+ \frac{1}{|c_s|} + \log^+ |\pi(0)| \leq \log^+ \frac{1}{|c_s|}. \qquad (3.93)$$

(2) Now, we select a positive integer $n_1 \geq n_0$, such that when $n \geq n_1$,

$$\text{meas}[R_n e^{h-\sigma}, R_n e^h] \geq 3, \qquad (3.94)$$

and by Lemma 3.10, when $R \geq R_n$ and $R \notin E_0$,

$$\left\{ \left[\frac{1}{\sigma} \log \frac{16 e^{1+h-2\sigma}}{M(\delta, h, \mu, \sigma)} \right] \frac{N(R, 1/f')}{T(R, f')} \cdot \frac{T(R, f')}{T(R, f)} \right.$$
$$\left. + \frac{\log^+(1/|c_s|)}{T(R, f)} + \frac{n(0, f' = \infty) \log R}{T(R, f)} \right\} e^K < \frac{\delta}{32}, \qquad (3.95)$$

$$\left\{ \frac{N(R, 1/f')}{T(R, f')} \cdot \frac{T(R, f')}{T(R, f)} + \frac{\log^+ |c_s|}{T(R, f)} \right\} 2e^K < \frac{\delta}{8 \times 16}, \qquad (3.96)$$

$$\frac{T(R, f')}{T(R, f)} + \frac{\log^+(1/|c_s|)}{T(R, f)} \leq 4. \qquad (3.97)$$

In the following, we will prove the existence of values $\theta_{\nu n}$ in the set $E_\nu(t_n)$ ($n \geq n_1$, $1 \leq \nu \leq p$), such that, when $z_{\nu n} = t_n e^{i\theta_{\nu n}}$,

$$\log |G(z_{\nu n})| \geq \frac{\delta}{16} T(t_n, f), \qquad (3.98)$$

and there exists $t_n' \notin E_0$ in the interval $[R_0 e^{h-\sigma}, R_n e^h]$, such that

$$\log M(R_n e^{h-4\sigma}, G) \leq 8 \cdot \frac{e^\sigma + 1}{e^\sigma - 1} T(t_n', f). \qquad (3.99)$$

First, according to the Boutroux-Cartan Theorem, the set of points satisfying the inequality

$$\prod_{0 < |\gamma_k| < R_n e^{h-2\sigma}} |z - \gamma_k| < (H')^{n(R_n e^{h-2\sigma}, f' = 0)}$$

may be included in certain circles $(\gamma)'$, with the sum of their Euclidean radii not exceeding $2eH'$. Assume that

$$H' = \frac{1}{8e}M(\delta, h, \mu, \sigma)R_n.$$

Then, according to formula (3.90), value $\theta_{\nu n}$ exists in $E_\nu(t_n)$, such that $z_{\nu n} = t_n e^{i\theta_{\nu n}} \notin (\gamma)'$. Hence, from formulas (3.89), (3.91) and (3.93), we obtain

$$\frac{\delta}{8}T(t_n, f) \le \log\frac{1}{|f'(z_{\nu n})|}$$

$$\le \log|G(z_{\nu n})| + \log\frac{1}{|\pi(z_{\nu n})|} + \log^+\frac{1}{|\alpha|} + \log\frac{1}{|z_{\nu n}|^s}$$

$$\le \log|G(z_{\nu n})| + \log\frac{1}{|\pi(z_{\nu n})|} + \log^+\frac{1}{|c_s|} + s\log\frac{1}{t_n}.$$

Notice that

$$s = \begin{cases} n(0, f' = 0), & s \ge 0, \\ -n(0, f' = \infty), & s < 0, \end{cases} \tag{3.100}$$

and

$$\log\frac{1}{|\pi(z_{\nu n})|} \le n(R_n e^{h-2\sigma}, f' = 0)\log\frac{16e^{1+h-2\sigma}}{M(\delta, h, \mu, \sigma)}$$

$$\le \left[\frac{1}{\sigma}\log\frac{16e^{1+h-2\sigma}}{M(\delta, h, \mu, \sigma)}\right]N\left(R_n e^{h-\sigma}, \frac{1}{f'}\right).$$

We derive

$$\frac{\delta}{8}T(t_n, f) \le \log|G(z_{\nu n})|$$

$$+ \left[\frac{1}{\sigma}\log\frac{16e^{1+h-2\sigma}}{M(\delta, h, \mu, \sigma)}\right]N\left(R_n e^{h-2\sigma}, \frac{1}{f'}\right)$$

$$+ \log^+\frac{1}{|c_s|} + n(0, f' = \infty)\log t_n.$$

Moreover, by formula (3.94) and meas $E_0 \le 2$, we conclude that there exists value $t_n' \notin E_0$ in the interval $[R_n e^{h-\sigma}, R_n e^h]$, such that $t_n \le R_n e^\sigma \le R_n e^{h-\sigma} \le t_n'$, and

$$\frac{\delta}{8}T(t_n, f) \le \log|G(z_{\nu n})|$$

$$+ \left[\frac{1}{\sigma}\log\frac{16e^{1+h-2\sigma}}{M(\delta, h, \mu, \sigma)}\right]N\left(t_n', \frac{1}{f'}\right)$$

$$+ \log^+\frac{1}{|c_s|} + n(0, f' = \infty)\log t_n'.$$

Furthermore, from formulas (3.95) and (3.86), we conclude that

$$\frac{\delta}{8}T(t_n, f) \leq \log|G(z_{\nu n})| + e^{-k}\frac{\delta}{32}T(t'_n, f)$$

$$\leq \log|G(z_{\nu n})| + e^{-k}\frac{\delta}{32}T(R_n e^h, f)$$

$$\leq \log|G(z_{\nu n})| + \frac{\delta}{16}T(t_n, f),$$

$$\log|G(z_{\nu n})| \geq \frac{\delta}{16}T(t_n, f),$$

meaning that formula (3.98) holds. On the other hand, by applying the Poisson-Jensen formula, we have according to formulas (3.91) and (3.92) that

$$\log M(R_n e^{h-4\sigma}, G) \leq \frac{R_n e^{h-3\sigma} + R_n e^{h-4\sigma}}{R_n e^{h-3\sigma} - R_n e^{h-4\sigma}}m(R_n e^{h-3\sigma}, G)$$

$$\leq \frac{e^\sigma + 1}{e^\sigma - 1}\left\{m\left(R_n e^{h-3\sigma}, \frac{1}{f'}\right) + \log^+|\alpha|\right.$$

$$\left. + n(0, f' = 0)\log R_n e^{h-3\sigma}\right\}$$

$$\leq \frac{e^\sigma + 1}{e^\sigma - 1}\left\{m\left(R_n e^{h-3\sigma}, \frac{1}{f'}\right)\right.$$

$$\left. + N\left(R_n e^{h-2\sigma}, \frac{1}{f'}\right) + \log^+|c_s|\right\}$$

$$\leq \frac{2(e^\sigma + 1)}{e^\sigma - 1}\left\{T(t'_n, f') + \log^+\frac{1}{|c_s|}\right\}.$$

It then yields from formula (3.97) that

$$\log M(R_n e^{h-4\sigma}, G) \leq \frac{8(e^\sigma + 1)}{e^\sigma - 1}T(t'_n, f),$$

implying that formula (3.99) is valid.

(3) Let

$$d = \min_{1\leq\nu\neq\nu'\leq p}\{|a_\nu - a'_\nu|\} > 0.$$

We choose positive integers $n_2 \geq n_1$, such that when $n \geq n_2$,

$$e^{(\delta/16)T(t_n, f)} \geq 16,$$

$$2R_n e^{h-4\sigma}\left\{1 + \sqrt{2}\pi\sqrt{\frac{16}{\sigma}e^k T(R_n, f)}\right\}e^{-(\delta/(8\times16))T(R_n, f)} \quad (3.101)$$

$$+ e^{(\delta/2)T(R_n, f)} \leq e^{-\delta/256)T(R_n, f)} < \frac{d}{4}.$$

Applying Lemma 3.5 to $G(z)$, where we let $R = R_n e^{h-3\sigma}$, $r = R_n e^{h-4\sigma}$, $z_0 = z_{\nu n}$ and $A = e^{(\delta/16)T(t_n, f)} \geq 16$, we find that value A' exists in the interval $[\sqrt[4]{A}, \sqrt{A}]$, such that $G'(z)$ has no zeros on the level curve $|G(z)| = A'$. Moreover, for an arbitrary point z on the closure $\overline{\Omega}(z_{\nu n})$, there exists a piecewise analytic curve L with length

$$\operatorname{meas} L \leq 2R_n e^{h-4\sigma} + 2\sqrt{2}\pi R_n e^{h-4\sigma} \sqrt{\tfrac{1}{\sigma} T(R_n e^{h-3\sigma}, G)},$$

that connects points z and $z_{\nu n}$, where $\overline{\Omega}(z_{\nu n})$ is the closure of the connected branch of the set

$$\Omega(A') = \{z \mid |G(z)| > A', \ |z| < R_n e^{h-3\sigma}\} \tag{3.102}$$

that contains $z_{\nu n}$ and lies in the disk $|z| < R_n e^{h-4\sigma}$. Also when $z \in L$, then

$$|G(z)| \geq e^{(\delta/(4\cdot16))T(t_n, f)}, \qquad |z| \leq R_n e^{h-4\sigma}.$$

Hence, by formulas (3.91), (3.92), (3.97) and (3.86), we conclude that

$$\begin{aligned}
\operatorname{meas} L &\leq 2R_n e^{h-4\sigma} + 2\sqrt{2}\pi R_n e^{h-4\sigma} \sqrt{\tfrac{8}{\sigma} T(t'_n, f)} \\
&\leq 2R_n e^{h-4\sigma} \left\{1 + \sqrt{2}\pi \sqrt{\tfrac{8}{\sigma} T(R_n e^h, f)}\right\} \\
&\leq 2R_n e^{h-4\sigma} \left\{1 + \sqrt{2}\pi \sqrt{\tfrac{16}{\sigma} e^K T(R_n, f)}\right\},
\end{aligned} \tag{3.103}$$

and when $z \in L$, we obtain from formulas (3.91), (3.92), (3.96) and (3.98) that

$$\begin{aligned}
e^{(\delta/(4\cdot16))T(R_n, f)} \leq e^{(\delta/(4.16))T(t_n, f)} &\leq |G(z)| \leq \frac{|\alpha| |z|^s}{|f'(z)|} \\
&\leq \frac{1}{|f'(z)|} e^{\log|\alpha| + n(0, f'=0)\log R_n e^{h-4\sigma}} \\
&\leq \frac{1}{|f'(z)|} e^{\log^+|c_s| + N(R_n e^{h-2\sigma}, 1/f')} \\
&\leq \frac{1}{|f'(z)|} e^{\log^+|c_s| + N(t'_n, 1/f')} \\
&\leq \frac{1}{|f'(z)|} e^{(\delta/(8\cdot16))T(R_n, f)},
\end{aligned} \tag{3.104}$$

$$|f'(z)| \leq e^{-(\delta/(8\cdot16))T(R_n, f)}.$$

Therefore, when $z \in \Omega(z_{\nu n})$, according to formula (3.88) we get

$$\begin{aligned}
|f(z) - a_\nu| &\leq |f(z) - f(z_{\nu n})| + |f(z_{\nu n}) - a_\nu| \\
&\leq \int_L |f'(z)| \, |dz| + e^{-(\delta/4)T(t_n, f)} \\
&\leq 2R_n e^{h-4\sigma} \left\{1 + \sqrt{2}\pi \sqrt{\tfrac{16}{\sigma} e^K T(R_n, f)}\right\} \\
&\quad \times e^{-(\delta/8\cdot16))T(R_n, f)} + e^{-(\delta/4)T(R_n, f)}.
\end{aligned}$$

Furthermore, by formula (3.101) we conclude that

$$|f(z) - a_\nu| < e^{-(\delta/256)T(R_n, f)} < \frac{d}{4}.$$

Hence, there is no intersection among the p domains $\Omega(z_{\nu n})$ ($n \geq n_2$, $\nu = 1, 2, \ldots, p$).

(4) Since $p \geq 2$, circumference $|z| = t$ ($t \geq t_n$, $n \geq n_2$) cannot be entirely in $\Omega(z_{\nu n})$ ($1 \leq \nu \leq p$). Let $\theta_{\nu t}$ be the part of the circumference $|z| = t$ that lies in $\Omega(z_{\nu n})$ which has a linear measure $t\theta_\nu(t)$. Then, when $n \geq n_2$, an application of Theorem 3.1 and according to formulas (3.98), (3.102) and (3.99) yield.

$$\frac{\delta}{16} T(t_n, f)$$

$$\leq \log|G(z_{\nu n})| \leq \frac{1}{2} \cdot \frac{8}{16} T(t_n, f)$$

$$+ 9\sqrt{2} \exp\left\{-\pi \int_{2t_n}^{R_n e^{(h-4\sigma)/2}} \frac{dt}{t\theta_\nu(t)}\right\} \log M(R_n e^{h-4\sigma}, G),$$

$$\frac{\delta}{32} T(t_n, f) \leq 9\sqrt{2} \exp\left\{-\pi \int_{2t_n}^{R_n e^{h-4\sigma}/2} \frac{dt}{t\theta_\nu(t)}\right\} 8 \frac{e^\sigma + 1}{e^\sigma - 1} T(t'_n, f),$$

and

$$\pi \int_{2R_n e^\sigma}^{R_n e^{h-4\sigma}/2} \frac{dt}{t\theta_\nu(t)} \leq \log\left\{\frac{72 \times 32\sqrt{2}}{\delta} \cdot \frac{e^\sigma + 1}{e^\sigma - 1}\right\}$$

$$+ \log T(R_n e^h, f) - \log T(t_n, f),$$

Furthermore, by formula (3.87) the following inequalities hold:

$$\pi \int_{2R_n e^\sigma}^{R_n e^{h-4\sigma}/2} \frac{dt}{t\theta_\nu(t)} \leq \log\left\{\frac{72 \times 64\sqrt{2}}{\delta} \cdot \frac{e^\sigma + 1}{e^\sigma - 1}\right\}$$

$$+ h\left(1 + \frac{h_1}{h}\right)\mu, \qquad \nu = 1, \ldots, p. \tag{3.105}$$

Finally, we consider the following inequality:

$$p^2 = \left\{\sum_{\nu=1}^p \sqrt{\theta_\nu(\theta)} \cdot \frac{1}{\sqrt{\theta_\nu(t)}}\right\}^2$$

$$\leq \sum_{\nu=1}^p \theta_\nu(t) \cdot \sum_{\nu=1}^p \frac{1}{\theta_\nu(t)} \leq 2\pi \sum_{\nu=1}^p \frac{1}{\theta_\nu(t)},$$

$$\frac{p^2}{2} \leq \pi \sum_{\nu=1}^p \frac{1}{\theta_\nu(t)}.$$

When $n \geq n_2$, by formula (3.105) we obtain

$$\frac{p^2}{2} \int_{2R_n e^\sigma}^{R_n e^{h-4\sigma}/2} \frac{dt}{t} \leq \sum_{\nu=1}^{p} \pi \int_{2R_n e^\sigma}^{R_n e^{h-4\sigma}/2} \frac{dt}{t\theta_\nu(t)}$$

$$\leq p \log\left\{ \frac{72 \times 64\sqrt{2}}{\delta} \cdot \frac{e^\sigma + 1}{e^\sigma - 1} \right\} + ph\left(1 + \frac{h_1}{h}\right)\mu,$$

$$\frac{p}{2}(h - 5\sigma - \log 4) \leq \log\left\{ \frac{72 \times 64\sqrt{2}}{\delta} \cdot \frac{e^\sigma + 1}{e^\sigma - 1} \right\} + h\left(1 + \frac{h_1}{h}\right)\mu.$$

Since h can be chosen arbitrarily large, it follows that $p \leq 2\mu$, meaning that Theorem 3.8 is proved.

A. Weitsman proved Theorem 3.8 in 1969. However, early in 1946, A. Pfluger had already made an in-depth study on those entire functions whose deficiencies summed up to 2 [34a]. Particularly, regarding the number of finite deficient values, he proved the following result:

THEOREM 3.9. *Let $f(z)$ be an entire function of lower order $\mu < +\infty$ and the sum of its deficiencies equal to*

$$\Delta(f) = \sum_\nu \delta(a_\nu, f) = 2, \qquad \delta(a_\nu, f) > 0.$$

Then p the number of finite deficient values of $f(z)$ is $\leq \mu$.

PROOF. First, according to $\mu > 0$, $f(z)$ is a transcendental entire function. We select arbitrarily p finite deficient values a_ν $(\nu = 1, 2, \ldots, p)$, with their deficiencies being $\delta(a_\nu, f) = \delta_\nu > 0$. Let

$$\delta = \min_{1 \leq \nu \leq p}\{\delta_\nu\}, \tag{3.106}$$

$$|a| = \max_{1 \leq \nu \leq p}\{|a_\nu|\}, \tag{3.107}$$

$$d = \min_{1 \leq \nu \neq \nu' \leq p}\{|a_\nu - a'_\nu|\}. \tag{3.108}$$

Then by repeating the proof of Theorem 3.8, we have

(1) For any arbitrarily selected number h $(0 < h < +\infty)$ and number σ $(0 < \sigma < \frac{1}{5}h)$, let $h_1 = 2\sigma < h$; then there exists a sequence $\{R_n\}$, such that

$$R_n \geq 1,$$

$$T(R_n e^h, f) \leq e^K T(R_n, f)^{(1+o(1))} \leq 2e^K T(R_n, f), \tag{3.109}$$

$$K = h\left(1 + \frac{h_1}{h}\right)\mu,$$

$$T(R_n e^{h_1}, f) \leq e^{K_1} T(R_n, f)^{(1+o(1))} \leq 2e^{K_1} T(R_n, f),$$

$$K_1 = h_1\left(1 + \frac{h_1}{h}\right)\mu. \tag{3.110}$$

(2) When $n \geq n_1$, there exists points $z_{\nu n} = t_n e^{i\theta \nu n}$ ($R_n \leq t_n \leq R_n e^\sigma$), such that

$$\log |G(z_{\nu n})| \geq \frac{\delta}{16} T(t_n, f), \tag{3.111}$$

and in the interval $[R_n e^{h-\sigma}, R_n e^h]$, there exists value $t'_n \notin E_0$, such that

$$\log M(R_n e^{h-4\sigma}, G) \leq 8 \cdot \frac{e^\sigma + 1}{e^\sigma - 1} T(t'_n, f).$$

(3) When $n \geq n_2$, $n_2 \geq n_1$, there exists domain $\Omega(z_{\nu n})$ containing point $z_{\nu n}$, such that when $z \in \Omega(z_{\nu_n})$, we have

$$|f(z) - a_\nu| \leq e^{-(\delta/256)T(R_n, f)} < \frac{d}{4} \tag{3.112}$$

and when z belongs to the boundary of $\Omega(z_{\nu n})$ within the disk $|z| < R_n e^{h-4\sigma}$, the following inequality holds:

$$|G(z)| < e^{(\delta/(2 \times 16))T(t_n, f)}, \tag{3.113}$$

Moreover, there is no intersection among the p domains $\Omega(z_{\nu n})$ ($\nu = 1, 2, \ldots, p$).

We assume a sufficiently large positive integers $\tau \geq 1$, such that

$$e^{(\tau-1)\sigma} > 4,$$

$$\frac{1}{4} \cdot \frac{\delta}{18 \times 32\sqrt{2}} e^{(\tau-1)\sigma/2} > \frac{\delta}{2 \times 16}, \tag{3.114}$$

$$\frac{\delta}{4 \times 48 \times 36 \times 32\sqrt{2}} \cdot e^{(\tau-1)\sigma} \geq \frac{e^\sigma + 1}{e^\sigma - 1} 2 e^{h_1(1+h_1/h)\mu}. \tag{3.115}$$

Let $\Omega'(z_{\nu n})$ be the connected component of $\Omega(z_{\nu n})$ in the disk $|z| < Re^{\tau\sigma}$ that contains point $z_{\nu n}$, $\tilde{\theta}_{\nu t}$ be the part of the circumference $|z| = t$ in $\Omega'(z_{\nu n})$, with its linear measure being $t\tilde{\theta}_\nu(t)$. We also assume that on the set $\Omega'(z_{\nu n}) \cap (|z| = R_n e^{\tau\sigma})$ $G(z)$ attains the maximum value at point $\tilde{z}_{\nu n}$. Hence, applying Theorem 3.1 and considering formulas (3.111) and (3.113), we get

$$\frac{\delta}{16} T(t_n, f) \leq \log |G(z_{\nu n})|$$

$$\leq \frac{\delta}{2 \times 16} T(t_n, f) + 9\sqrt{2} \exp\left\{-\pi \int_{2|z_{\nu n}|}^{R_n e^{\tau\sigma}/2} \frac{dt}{t\tilde{\theta}_\nu(t)}\right\} \cdot \log |G(\tilde{z}_{\nu n})|,$$

$$\log |G(\tilde{z}_{\nu n})| \geq \frac{\delta}{18 \times 16\sqrt{2}} \exp\left\{\pi \int_{2t_n}^{Rne^{\tau\sigma}/2} \frac{dt}{t\tilde{\theta}_\nu(t)}\right\} \cdot T(t_n, f).$$

Notice that $\tilde{\theta}_\nu(t) \leq 2\pi$ and $t_n \leq R_n e^\sigma$. We conclude that

$$\log |G(\tilde{z}_{\nu n})| \geq \frac{\delta}{18 \times 32\sqrt{2}} e^{(\tau-1)\sigma/2} T(t_n, f).$$

Now, we assume positive integers $n_3 \geq n_2$, such that when $n \geq n_3$, we have

$$\frac{\delta}{e^{18 \times 32\sqrt{2}}} e^{(\tau-1)\sigma/2} T(t_n, f) \geq 16.$$

We apply Lemma 3.5 to $G(z)$, where we let $R = R_n e^{h-3\sigma}$, $r = R_n e^{h-4\sigma}$, $z_0 = \tilde{z}_{\nu n}$ and

$$A = \exp\left\{\frac{\delta}{18 \times 32\sqrt{2}} e^{(\tau-1)\sigma} T(t_n, f)\right\},$$

then there exists value A' in the interval $[\sqrt[4]{A}, \sqrt{A}]$, such that $G'(z)$ has no zeros on the level curve $|G(z)| = A'$, implying that the level curve is analytic. Moreover, for any arbitrary point z on the closure $\overline{\Omega}(\tilde{z}_{\nu n})$, there exists a piecewise analytic curve L with the length[8]

$$\text{meas } L \leq 2R_n e^{h-4\sigma} \left\{1 + \sqrt{2}\pi \sqrt{\frac{16}{\sigma} e^K T(R_n, f)}\right\},$$

that connects points z and $\tilde{z}_{\nu n}$, where $\overline{\Omega}(\tilde{z}_{\nu n})$ is the closure of the connected branch of the set

$$\Omega(A') = \{z \mid |G(z)| > A', \; |z| < R_n e^{h-3\sigma}\}$$

that contains $\tilde{z}_{\nu n}$ and lies in the disk $|z| < R_n e^{h-4\sigma}$. Also when $z \in L$, then

$$|G(z)| \geq \exp\left\{\frac{1}{4} \cdot \frac{\delta}{18 \times 32\sqrt{2}} \exp\left\{\frac{1}{2}(\tau-1)\sigma\right\} T(t_n, f)\right\}, \quad |z| < R_n e^{h-4\sigma}.$$

And hence,

$$\exp\left\{\frac{1}{4} \cdot \frac{\delta}{18 \times 32\sqrt{2}} \exp\left\{\frac{1}{2}(\tau-1)\sigma\right\} T(R_n, f)\right\} \leq |G(z)|$$

$$\leq \frac{|\alpha||z|^s}{|f'(z)|} \leq \frac{1}{|f'(z)|} \exp\left\{\frac{1}{8} \cdot \frac{\delta}{16} T(R_n, f)\right\}.[9]$$

Furthermore, by formula (3.14) we get

$$|f'(z)| \leq \exp\left\{\frac{-\delta}{8 \times 18 \times 32\sqrt{2}} \exp\left\{\frac{1}{2}(\tau-1)\sigma\right\} T(R_n, f)\right\}.$$

Therefore, when $z \in \Omega(\tilde{z}_{\nu n})$, we find that

$$|f(z) - f(z_{\nu n})| \leq \int_L |f'(z)||dz|$$

$$\leq 2R_n e^{h-4\sigma} \left\{1 + \sqrt{2}\pi \sqrt{\frac{16}{\sigma} e^K T(R_n, f)}\right\}$$

$$\times \exp\left\{-\frac{\delta}{8 \times 18 \times 32\sqrt{2}} \exp\left\{\frac{1}{2}(\tau-1)\sigma\right\} T(t_n, f)\right\}.$$

---

[8] See formula (3.103).
[9] See formula (3.104).

Taking positive integer n_4, with $n_4 \geq n_3$, so that when $n \geq n_4$ and $z \notin \Omega(\tilde{z}_{\nu n})$ $(1 \leq \nu \leq p)$, we have

$$|f(z) - f(\tilde{z}_{\nu n})| \leq \exp\left\{-\frac{\delta}{2 \times 8 \times 18 \times 32\sqrt{2}} \exp\left\{\frac{1}{2}(\tau - 1)\sigma\right\} T(R_n, f)\right\}.$$
(3.116)

Besides, by formula (3.114), we conclude that $\Omega(\tilde{z}_{\nu n}) \subset \Omega(z_{\nu n})$. Hence, point $\tilde{z}_{\nu n}$ is an interior point of $\Omega(z_{\nu n})$. When $\nu \neq \nu'$, we have according to formulas (3.108) and (3.112) that

$$\begin{aligned}
|f(\tilde{z}_{\nu n}) - f(\tilde{z}_{\nu' n})| &= |f(\tilde{z}_{\nu n}) - a_\nu + a_\nu - a_{\nu'} + a_{\nu'} - f(\tilde{z}_{\nu' n})| \\
&\geq |a_\nu - a_{\nu'}| - |f(\tilde{z}_{\nu n}) - a_\nu| - |a_{\nu'} - f(\tilde{z}_{\nu' n})| \geq \frac{d}{2}.
\end{aligned}$$
(3.117)

Moreover, from formulas (3.107) and (3.112) we get

$$|f(\tilde{z}_{\nu n})| \leq |f(\tilde{z}_{\nu n}) - a_\nu| + |a_\nu| \leq \frac{d}{4} + |a|, \qquad (1 \leq \nu \leq p). \tag{3.118}$$

We start within $\Omega(\tilde{z}_{\nu n})$ from point $\tilde{z}_{\nu n}$ along circumference $|z| = R_n e^{h-4\sigma}$ and draw a simple, continuous curve $L_{\nu n}$. Let $l_{\nu n}$ be the part of $L_{\nu n}$ between the last intersecting point of $L_{\nu n}$ with circumference $|z| = Re^{\tau\sigma}$ and the first intersecting point of $L_{\nu n}$ with circumference $|z| = R_n e^{\tau\sigma+5\pi}$. Hence, we get p curves $l_{\nu n}$ $(\nu = 1, \ldots, p)$ which divide annulus Γ_n: $R_n e^{\tau\sigma} \leq |z| \leq R_n e^{\tau\sigma+5\pi}$ into p domains Ω_{mn} $(m = 1, \ldots, p)$. Let $l_{\nu n}$ and $l_{\nu' n}$ be the boundaries of Ω_{mn} $(1 \leq m \leq p)$ in Γ_n; then according to formulas (3.117) and (3.118), an application of Theorem 3.2 yields

$$\left\{\frac{d}{4} + |a| + N\right\} \{\varepsilon_1^{1/3} + \varepsilon_2^{1/3}\} \geq \frac{d}{2},$$

where N denotes the largest value of $|f(z)|$ in $\widetilde{\Omega}_{mn}$. Moreover, by formula (3.116) we get

$$\varepsilon_1 = \varepsilon_2 = \exp\left\{-\frac{\delta}{2 \times 8 \times 18 \times 32\sqrt{2}} \exp\left\{\frac{1}{2}(\tau - 1)\sigma\right\} T(R_n, f)\right\}.$$

Hence, there exists point z'_{mn}, with $R_n e^{\tau\sigma} \leq |z'_{mn}| \leq R_n e^{\tau\sigma+5\pi}$ on $\overline{\Omega}_{mn}$, such that

$$\begin{aligned}
|f(z'_{mn})| = N &\geq \frac{d}{4} \exp\left\{\frac{\delta}{3 \times 16 \times 18 \times 32\sqrt{2}} \exp\left\{\frac{1}{2}(\tau - 1)\sigma\right\} T(R_n, f)\right\} \\
&- \frac{d}{4} - |a|.
\end{aligned}$$

We choose positive integers n_5, $n > n_4$, such that when $n \geq n_5$,

$$|f(z'_{mn})| \geq \exp\left\{\frac{\delta}{2 \times 48 \times 18 \times 32\sqrt{2}} \exp\left\{\frac{1}{2}(\tau - 1)\sigma\right\} T(R_n, f)\right\},$$

and

$$\exp\left\{\frac{\delta}{4\times 48\times 36\times 32\sqrt{2}}\exp\left\{\frac{1}{2}(\tau-1)\sigma\right\}T(R_n,f)\right\} \tag{3.119}$$
$$> \max\left\{16,\frac{d}{4}+|a|\right\}.$$

There exists point A' in the interval

$$\left[\exp\left\{\frac{\delta}{4\times 48\times 36\times 32\sqrt{2}}\exp\left\{\frac{1}{2}(\tau-1)\sigma\right\}T(R_n,f)\right\},\right.$$
$$\left.\exp\left\{\frac{\delta}{2\times 48\times 36\times 32\sqrt{2}}\exp\left\{\frac{1}{2}(\tau-1)\right\}T(R_n,f)\right\}\right],$$

such that $f'(z)$ has no zeros on the level curve $|f(z)|=A'$, meaning that the level curve is analytic. We consider the set

$$\Omega(A')\doteq\{z\,|\,|f(z)|>A'\}, \tag{3.120}$$

and let $\Omega(z'_{mn})$ be the connected branch of $\Omega(A')$ in the disk $|z|<R_ne^{h-4\sigma}$ that contains point z'_{mn}. We apply the Poisson-Jensen formula and derive

$$\log M(t_n,f)\le\frac{e^\sigma+1}{e^\sigma-1}m(R_ne^{h_1},f)\le\frac{e^\sigma+1}{e^\sigma-1}T(R_ne^{h_1},f).$$

Furthermore, from formula (3.110) we get

$$\log M(t_n,f)\le\frac{e^\sigma+1}{e^\sigma-1}2e^{h_1(1+h_1/h)\mu}T(R_n,f).$$

Combining formulas (3.115) and (3.120), we conclude that there is no intersection between $\overline{\Omega}(z'_{mn})$ and circumference $|z|=t_n$. Hence, according to the maximum modulus principle, $\overline{\Omega}(z'_{mn})$ must intersect circumference $|z|=R_n^{h-4\sigma}$, and there exist interior points in the intersecting set. On the other hand, by formulas (3.120), (3.119) and (3.112), we conclude that $\Omega(z'_{mn})$ has no intersection among $\Omega(z_{\nu n})$ and $\Omega(z_{\nu'n})$. Hence, $\Omega(z'_{mn})$ must lie between $\Omega(z_{\nu n})$ and $\Omega(z_{\nu'n})$, and there are no intersection among the $2p$ domains $\{\Omega(z_{\nu n})\}$ and $\{\Omega(z'_{mn})\}$. We assign a new, consistent notation to these $2p$ domains: $\Omega(z''_{kn})$ $(n\ge n_5,\ k=1,2,\ldots,2p)$. Let θ_{kt} be the part of the circumference $|z|=t$ that lies in $\Omega(z''_{kn})$, with $t\theta_k(t)$ being its linear measure. Then, by applying Theorem 3.1, when z''_{kn} represents a certain point $z_{\nu n}$, we have

$$\pi\int_{2R_ne^\sigma}^{R_ne^{h-4\sigma}/2}\frac{dt}{t\theta_k(t)}\le\log\left\{\frac{72\times 64\sqrt{2}}{\delta}\cdot\frac{e^\sigma+1}{e^\sigma-1}\right\}+h\left(1+\frac{h_1}{h}\right)\mu;[10]$$

---

[10] See formula (3.105).

and when z_{kn}'' represents a certain point z_{mn}',

$$\frac{\delta}{2 \times 48 \times 18 \times 32\sqrt{2}} e^{(\tau-1)\sigma/2} T(R_n, f) \leq |\log|f(z_{mn}')||$$

$$\leq \frac{\delta}{2 \times 48 \times 36 \times 32\sqrt{2}} e^{(\tau-1)\sigma/2} T(R_n, f)$$

$$+ 9\sqrt{2} \exp\left\{ -\pi \int_{2|z_{mn}'|}^{R_n e^{h-4\sigma}/2} \frac{dt}{t\theta_k(t)} \right\} \log M(R_n e^{h-4\sigma}, f).$$

Noting that $|z_{mn}'| \leq R_n e^{\tau\sigma+5\pi}$, and considering the Poisson-Jensen formula as well as formula (3.109), we find that

$$\log M(R_n e^{h-4\sigma}, f) \leq \frac{e^\sigma + 1}{e^\sigma - 1} T(R_n e^{h-3\sigma}, f) \leq \frac{e^\sigma + 1}{e^\sigma - 1} T(R_n e^h, f)$$

$$\leq \frac{e^\sigma + 1}{e^\sigma - 1} \cdot 2e^{h(1+h_1/h)\mu} T(R_n, f),$$

and accordingly we get

$$\pi \int_{2R_n e^{\tau\sigma+5\pi}}^{R_n e^{h-4\sigma}/2} \frac{dt}{t\theta_k(t)}$$

$$\leq \log\left\{ \frac{72 \times 32 \times 36 \times 48}{\delta} \cdot \frac{e^\sigma + 1}{e^\sigma - 1} \right\} + h\left(1 + \frac{h_1}{h}\right)\mu, \qquad 1 \leq k \leq 2p.$$

Finally, according to the inequalities

$$(2p)^2 = \left\{ \sum_{k=1}^{2p} \sqrt{\theta_k(t)} \cdot \frac{1}{\sqrt{\theta_k(t)}} \right\}^2 \leq \sum_{k=1}^{2p} \theta_k(t) \cdot \sum_{k=1}^{2p} \frac{1}{\theta_k(t)} \leq 2\pi \sum_{k=1}^{2p} \frac{1}{\theta_k(t)},$$

we derive

$$\frac{4p^2}{2} \int_{2R_n e^{\tau\sigma+5\pi}}^{R_n e^{h-4\sigma}/2} \frac{dt}{t} \leq \sum_{k=1}^{2p} \pi \int_{2R_n e^{\tau\sigma+5\pi}}^{R_n e^{h-4\sigma}/2} \frac{dt}{t\theta_k(t)}$$

$$\leq 2p \log\left\{ \frac{72 \times 32 \times 36 \times 48}{8} \cdot \frac{e^\sigma + 1}{e^\sigma - 1} \right\} + 2ph\left(1 + \frac{h_1}{h}\right)\mu,$$

$$p(h - \tau\sigma - 5\pi - \log 4)$$

$$\leq \log\left\{ \frac{72 \times 32 \times 36 \times 48}{8} \cdot \frac{e^\sigma + 1}{e^\sigma - 1} \right\} + h\left(1 + \frac{h_1}{h}\right)\mu.$$

Since h may be selected arbitrarily large, p must be $\leq \mu$, meaning that Theorem 3.9 is proved.

§3.5. The Edrei-Fuchs Theorem

3.5.1. Some preparations. The transformation

$$\zeta = \zeta(z) = \frac{z^{\pi/2\theta} - R^{\pi/2\theta}}{z^{\pi/2\theta} + R^{\pi/2\theta}}, \qquad 0 < \theta \leq \pi, \; 0 < R < +\infty \qquad (3.121)$$

maps domain $\Omega(-\theta, \theta)$ onto the unit circle $|\zeta| < 1$ on the ζ-plane, and the point $z = R$ into the origin $\zeta = 0$.

LEMMA 3.11. *Under the transformation of formula (3.121), the image domain of $\overline{\Omega}(-\theta+\varepsilon, \theta-\varepsilon; R_1, R_2)$ $(0 < \varepsilon < \theta, R_1 \leq R \leq R_2)$ on the ζ-plane must be in the disk $|\zeta| \leq \rho$, with*

$$\rho = 1 - \frac{\varepsilon}{2\theta}\left(\frac{R_1}{R_2}\right)^{\pi/\theta}, \tag{3.122}$$

while the pre-image domain of the disk $|\zeta| \leq \frac{1}{2}(1+\rho)$ on the z-plane must be in $\overline{\Omega}(-\theta, \theta; R'_3, R_3)$, with

$$R'_3 = \left(\frac{\varepsilon}{8\theta}\right)^{2\theta/\pi}\left(\frac{R_1}{R_2}\right)^2 R_1,$$

$$R_3 = \left(\frac{8\theta}{\varepsilon}\right)^{2\theta/\pi}\left(\frac{R_2}{R_1}\right)^2 R_2. \tag{3.123}$$

Moreover, when $|\zeta| \leq \frac{1}{2}(1+\rho)$, we have

$$\frac{2\theta}{\pi}\left(\frac{\varepsilon}{8\theta}\right)^{2\theta/\pi}\left(\frac{R_1}{R_2}\right)^2 R_1 \leq |z'(\zeta)| \leq \frac{2\theta}{\pi}\left(\frac{8\theta}{\varepsilon}\right)^{2\theta/\pi}\left(\frac{R_2}{R_1}\right)^2 R_2, \tag{3.124}$$

where $z = z(\zeta)$ represents the inverse transformation of formula (3.121).

PROOF. Let $z = re^{i\varphi} \in \overline{\Omega}(-\theta+\varepsilon, \theta-\varepsilon; R_1, R_2)$. Then

$$|\zeta| = \left|\frac{r^{\pi/2\theta}e^{i\varphi\pi/2\theta} - R^{\pi/2\theta}}{r^{\pi/2\theta}e^{i\varphi\pi/2\theta} + R^{\pi/2\theta}}\right|$$

$$= \sqrt{1 - \frac{4R^{\pi/2\theta}r^{\pi/2\theta}\cos(\varphi\pi/2\theta)}{r^{\pi/\theta} + R^{\pi/\theta} + 2R^{\pi/2\theta}r^{\pi/2\theta}\cos(\varphi\pi/2\theta)}}.$$

Notice that

$$r^{\pi/\theta} + R^{\pi/\theta} + 2R^{\pi/2\theta}r^{\pi/2\theta}\cos(\varphi\pi/2\theta) \leq 4R_2^{\pi/\theta},$$

$$4R^{\pi/2\theta}r^{\pi/2\theta}\cos(\varphi\pi/2\theta) \geq 4R_1^{\pi/\theta}\sin(\varepsilon\pi/2\theta) \geq \frac{4\varepsilon}{\theta}R_1^{\pi/\theta}.$$

Then we get

$$|\zeta| \leq \sqrt{1 - \frac{\varepsilon}{\theta}\left(\frac{R_1}{R_2}\right)^{\pi/\theta}} \leq 1 - \frac{\varepsilon}{2\theta}\left(\frac{R_1}{R_2}\right)^{\pi/\theta} = \rho,$$

meaning that formula (3.122) is proved.

On the other hand, the inverse transformation of formula (3.121) is

$$z = z(\zeta) = R\left(\frac{1+\zeta}{1-\zeta}\right)^{2\theta/\pi}.$$

Then when $|\zeta| \leq \frac{1}{2}(1+\rho)$, we have

$$|z| \leq R\left(\frac{4}{1-\rho}\right)^{2\theta/\pi} \leq \left(\frac{8\theta}{\varepsilon}\right)^{2\theta/\pi}\left(\frac{R_2}{R_1}\right)^2 R_2 = R_3,$$

and

$$|z| \geq R \left(\frac{1-\rho}{4} \right)^{2\theta/\pi} \geq \left(\frac{\varepsilon}{8\theta} \right)^{2\theta/\pi} \left(\frac{R_1}{R_2} \right)^2 R_1 = R_3',$$

meaning that formula (3.123) is proved.

Finally, according to the expression

$$z'(\zeta) = \frac{4\theta}{\pi} R \left(\frac{1+\zeta}{1-\zeta} \right)^{2\theta/\pi} \cdot \frac{1}{1-\zeta^2},$$

when $|\zeta| \leq \frac{1}{2}(1+\rho)$, we get

$$|z'(\zeta)| \leq \frac{4\theta}{\pi} R_2 \left(\frac{4}{1-\rho} \right)^{2\theta/\pi} \frac{2}{1-\rho}$$
$$= \frac{2\theta}{\pi} \left(\frac{8\theta}{\varepsilon} \right)^{1+2\theta/\pi} \left(\frac{R_2}{R_1} \right)^{2+\pi/\theta} R_2$$

and

$$|z'(\zeta)| \geq \frac{4\theta}{\pi} R_1 \left(\frac{1-\rho}{4} \right)^{2\theta/\pi} \cdot \frac{1}{2}$$
$$= \frac{2\theta}{\pi} \left(\frac{\varepsilon}{8\theta} \right)^{2\theta/\pi} \left(\frac{R_1}{R_2} \right)^2 R_1,$$

implying that formula (3.124) is proved.

LEMMA 3.12. *Let $f(z)$ be a function meromorphic in the unit disk $|z| \leq 1$, and values r, t, R and H satisfy the condition: $0 < 4eH < r < t \leq R < 1$. We also let (γ) be the Euclidean exceptional circles that correspond to these $n(1, f = 0) + \overline{n}(1, f = \infty)$ points and number H. Then, when z is in the disk $|z| \leq r$ and outside circles (γ), we have*

$$\log \left| \frac{f'(z)}{f(z)} \right| \leq \frac{R+r}{R-r} m \left(R, \frac{f'}{f} \right) - \frac{R-r}{R+r} m \left(R, \frac{f}{f'} \right)$$
$$+ \{ n(1, f = 0) + \overline{n}(1, f = \infty) \} \log \frac{2}{H}$$
$$- \frac{(R-t)^2}{4R^2} n(t, f' = 0).$$

PROOF. We denote a_i as the zeros of $f(z)$, b_j the poles of $f(z)$, and c_k the zeros of $f'(z)$. Then for an arbitrary point $z = r'e^{i\varphi}$ in the disk

$|z| < R$, we have according to the Poissen-Jensen formula:

$$\log \left| \frac{f'(z)}{f(z)} \right| = \frac{1}{2\pi} \int_0^{2\pi} \log \left| \frac{f'(Re^{i\theta})}{f(Re^{i\theta})} \right| \frac{R^2 - r'^2}{R^2 - 2Rr'\cos(\theta - \varphi) + r'^2} \, d\theta$$

$$+ \left\{ \sideset{}{'}\sum_{|a_i| < R} \log \left| \frac{R^2 - \overline{a}_i z}{R(z - a_i)} \right| + \sideset{}{'}\sum_{|b_j| < R} \log \left| \frac{R^2 - \overline{b}_j z}{R(z - b_j)} \right| \right\}$$

$$- \left\{ \sum_{|c_k| < R} \log \left| \frac{R^2 - \overline{c}_k z}{R(z - c_k)} \right| - \sideset{}{''}\sum_{|a_i| < R} \log \left| \frac{R^2 - \overline{a}_i z}{R(z - a_i)} \right| \right\},$$

where $\sum'_{|b_j| < R}$, $\sum'_{|a_i| < R}$ and $\sum''_{|a_i| < R}$ indicate that each b_j is merely counted once; each a_i is only counted once and each a_i is counted for $m_i - 1$ times (m_i is the multiplicity of a_i), respectively. Hence,

$$\sideset{}{'}\sum_{|a_i| < R} \log \left| \frac{R^2 - \overline{a}_i z}{R(z - a_i)} \right| + \sideset{}{''}\sum_{|a_i| < R} \log \left| \frac{R^2 - \overline{a}_i z}{R(z - a_i)} \right|$$

$$= \sum_{|a_i| < R} \log \left| \frac{R^2 - \overline{a}_i z}{R(z - a_i)} \right|,$$

and

$$\log \left| \frac{f'(z)}{f(z)} \right| = \frac{1}{2\pi} \int_0^{2\pi} \log \left| \frac{f'(Re^{i\theta})}{f(Re^{i\theta})} \right| \frac{R^2 - r'^2}{R^2 - 2Rr'\cos(\theta - \varphi) + r'^2} \, d\theta$$

$$+ \sum_{|a_i| < R} \log \left| \frac{R^2 - \overline{a}_i z}{R(z - a_i)} \right| + \sideset{}{'}\sum_{|b_j| < R} \log \left| \frac{R^2 - \overline{b}_j z}{R(z - b_j)} \right| \qquad (3.125)$$

$$- \sum_{|c_k| < R} \log \left| \frac{R^2 - \overline{c}_k z}{R(z - c_k)} \right|.$$

Moreover, when point z is in the disk $|z| \leq r$ and outside circles (γ), we have

$$\sum_{|a_i| < R} \log \left| \frac{R^2 - \overline{a}_i z}{R(z - a_i)} \right| + \sideset{}{'}\sum_{|b_j| < R} \log \left| \frac{R^2 - \overline{b}_j z}{R(z - b_j)} \right|$$

$$\leq \sum_{|a_i| < R} \log \frac{2}{|z - a_i|} + \sideset{}{'}\sum_{|b_j| < R} \log \frac{2}{|z - b_j|}$$

$$\leq \{ n(1, f = 0) + \overline{n}(1, f = \infty) \} \log \frac{2}{H}.$$

On the other hand, when $|z| \leq r$,

$$\sum_{|c_k|<R} \log\left|\frac{R^2 - \bar{c}_k z}{R(z - c_k)}\right| \geq \sum_{|c_k|<t} \log\left|\frac{R^2 - \bar{c}_k z}{R(z - c_k)}\right| \geq \sum_{|c_k|<t} \log\left|\frac{R^2 + |z||c_k|}{R(|z| + |c_k|)}\right|$$

$$= \sum_{|c_k|<t} \log\left\{1 + \frac{(R - |z|)(R - |c_k|)}{R(|z| + |c_k|)}\right\}$$

$$\geq n(t, f' = 0) \log\left\{1 + \frac{(R - t)^2}{2R^2}\right\}$$

$$\geq n(t, f' = 0)\frac{(R - t)^2}{4R^2}.$$

Therefore, when point $z = r'e^{i\varphi}$ is in the disk $|z| \leq r$ and outside circles (γ), we have from formula (3.125) that

$$\log\left|\frac{f'(z)}{f(z)}\right| \leq \frac{R + r'}{R - r'}m\left(R, \frac{f'}{f}\right) - \frac{R - r'}{R + r'}m\left(R, \frac{f}{f'}\right)$$

$$+ \{n(1, f = 0) + \bar{n}(1, f = \infty)\}\log\frac{2}{H}$$

$$- \frac{(R - t)^2}{4R^2}n(t, f' = 0)$$

$$\leq \frac{R + r}{R - r}m\left(R, \frac{f'}{f}\right) - \frac{R - r}{R + r}m\left(R, \frac{f}{f'}\right)$$

$$+ \{n(1, f = 0) + \bar{n}(1, f = \infty)\}\log\frac{2}{H}$$

$$- \frac{(R - t)^2}{4R^2}n(t, f' = 0),$$

implying that Lemma 3.12 is proved.

Now, we prove a lemma that plays a crucial role in the sequels:

LEMMA 3.13 [43i]. *Let $f(z)$ be a function meromorphic of order $\lambda < +\infty$ on the open plane $|z| < +\infty$, and let it satisfy the following conditions:*
(1) On $\overline{\Omega}(-\theta, \theta)$ $(0 < \theta \leq \pi)$

$$\varlimsup_{r \to +\infty} \frac{\log^+ n\{\overline{\Omega}(-\theta, \theta; r), f = X\}}{\log r} \leq \lambda' < +\infty, \qquad X = 0, \infty. \quad (3.126)$$

(2) To any selected number ε, $0 < \varepsilon < \theta$, there exists a sequence $\{R_n\}$, such that the set

$$E_n = \left\{z \,\middle|\, \log\frac{1}{|f(z) - a|} \geq N_n, |z| = R_n, -\theta + \varepsilon \leq \arg z \leq \theta - \varepsilon\right\} \quad (3.127)$$

has linear measure meas $E_n \geq \alpha R_n$ $(\alpha \geq \frac{\varepsilon}{2})$, *where a is a complex number* *that is neither zero nor* ∞, $N_n > 0$ *is a real number, and*

$$\lim_{r \to +\infty} \left\{ \left(\frac{R_{n2}}{R_{n1}} \right)^{6+2(\lambda'+\eta_0)+3\pi/\theta} R_{n2}^{\lambda'+2\eta_0} \log R_{n2} \right\} N_n^{-1} = 0, \qquad (3.128)$$

where $\eta_0 > 0$ *is a selected real number, and* $R_{n1} \leq R_n \leq R_{n2}$, $R_{n1} \to +\infty$ $(n \to +\infty)$.

Under the above assumptions, provided that n is sufficiently large, then for any point z in $\overline{\Omega}(-\theta+\varepsilon, \theta-\varepsilon; R_{n1}, R_{n2})$ and outside circles (γ),

$$\log \frac{1}{|f(z)-a|} \geq \frac{A(\alpha, \varepsilon, \theta)}{B(\theta)\log(R_{n2}/R_{n1}) + C(\alpha, \varepsilon, \theta)} \left(\frac{R_{n1}}{R_{n2}} \right)^{6+3\pi/\theta} \cdot N_n,$$

where the sum of the Euclidean radii of (γ) does not exceed $\frac{1}{8}\varepsilon R_{n1}$, the number of circles in (γ) is finite, depending on values n; $A(\alpha, \varepsilon, \theta) > 0$ and $C(\alpha, \varepsilon, \theta) < +\infty$ are constants depending on α, ε and θ but unrelated to n; and also $B(\theta) < +\infty$ is a constant depending on θ but unrelated to n.

PROOF. We consider the transformation

$$\zeta = \zeta(z) = \frac{z^{\pi/2\theta} - R_n^{\pi/2\theta}}{z^{\pi/2\theta} + R_n^{\pi/2\theta}}.$$

Then $\Omega(-\theta, \theta)$ is mapped onto a unit disk $|\zeta| < 1$ on the ζ-plane. According to Lemma 3.11, we conclude that the image domain of $\overline{\Omega}(-\infty+\varepsilon, \theta-\varepsilon; R_{n1}, R_{n2})$ on the ζ-plane must be in the disk $|\zeta| \leq \rho$, with

$$\rho = 1 - \frac{\varepsilon}{2\theta} \left(\frac{R_{n1}}{R_{n2}} \right)^{\pi/\theta}, \qquad (3.129)$$

while the pre-image domain of the disk $|\zeta| \leq \frac{1}{2}(1+\rho)$ on the z-plane must be in $\overline{\Omega}(-\theta, \theta; R_{n3})$, with

$$R_{n3} = \left(\frac{8\theta}{\varepsilon} \right)^{2\theta/\pi} \left(\frac{R_{n2}}{R_{n1}} \right)^2 R_{n2}. \qquad (3.130)$$

Moreover, when $|\zeta| \leq \frac{1}{2}(1+\rho)$ we have

$$\frac{2\theta}{\pi} \left(\frac{\varepsilon}{8\theta} \right)^{2\theta/\pi} \left(\frac{R_{n1}}{R_{n2}} \right)^2 R_{n-1} \leq |z'(\zeta)| \leq \frac{2\theta}{\pi} \left(\frac{8\theta}{\varepsilon} \right)^{1+2\theta/\pi} \left(\frac{R_{n2}}{R_{n1}} \right)^{2+\pi/\theta} R_{n2}.$$

We continue to consider the transformation $\zeta = \frac{2\zeta}{1+\rho}$, which maps the disk $|\zeta| \leq \frac{1}{2}(1+\rho)$ onto the unit disk $|\xi| \leq 1$ on the ζ-plane, and the disk $|\zeta| \leq \rho$ onto the disk $|\xi| \leq \tau$, with

$$\tau = \frac{2\rho}{1+\rho}. \qquad (3.131)$$

Moreover, when $|\xi| \leq 1$, we obtain

$$\frac{1}{2} \leq |\zeta'(\xi)| \leq 1,$$

$$\frac{\theta}{\pi}\left(\frac{\varepsilon}{8\theta}\right)^{2\theta/\pi}\left(\frac{R_{n1}}{R_{n2}}\right)^2 R_{n1} \leq |z'(\xi)| = |z'(\zeta)||\zeta'(\xi)|$$

$$\leq \frac{2\theta}{\pi}\left(\frac{8\theta}{\varepsilon}\right)^{1+2\theta/\pi}\left(\frac{R_{n2}}{R_{n1}}\right)^{2+\pi/\theta} R_{n2}. \tag{3.132}$$

Let E_ξ be the set of images of E_n on the ξ-plane. Then E_ξ is in the disk $|\xi| \leq \tau$ and is distributed on the imaginary axis, and also

$$\alpha R_n \leq \operatorname{meas} E_n = \int_{E_n} |dz| = \int_{E_\xi} |z'(\xi)||d\xi|$$

$$\leq \frac{2\theta}{\pi}\left(\frac{8\theta}{\varepsilon}\right)^{1+2\theta/\pi}\left(\frac{R_{n2}}{R_{n1}}\right)^{2+\pi/\theta} R_{n2} \operatorname{meas} E_\xi,$$

$$\operatorname{meas} E_\xi \geq \frac{\pi\alpha}{2\theta}\left(\frac{\varepsilon}{8\theta}\right)^{1+2\theta/\pi}\left(\frac{R_{n1}}{R_{n2}}\right)^{2+\pi/\theta}\frac{R_{n1}}{R_{n2}} \tag{3.133}$$

$$\geq \frac{\varepsilon\pi}{4\theta}\left(\frac{\varepsilon}{8\theta}\right)^{1+2\theta/\pi}\left(\frac{R_{n1}}{R_{n2}}\right)^{3+\pi/\theta}.$$

Let $F(\xi) = f(z(\xi))$. Then $F(\xi)$ is meromorphic on the unit disk $|\xi| \leq 1$. Moreover, according to formulas (3.129) and (3.130), for any arbitrarily selected number η, $0 < \eta \leq \eta_0$, there exists value n_0, such that, when $n \geq n_0$, then

$$n(1, F = X) \leq n\{\Omega(-\theta, \theta; R_{n3}), f = X\}$$

$$\leq A(\varepsilon, \theta)\left(\frac{R_{n2}}{R_{n1}}\right)^{2(\lambda'+\eta)} R_{n2}^{\lambda'+\eta}, \qquad X = 0, \infty, \tag{3.134}$$

where $A(\varepsilon, \theta) < +\infty$ is a constant depending on ε and θ, but unrelated to n. We construct circles using each zero and pole in the disk $|\xi| < 1$ as centers and

$$\frac{1}{n(1, F = 0) + n(1, F = \infty)} \cdot \frac{1-\tau}{4 \times 32}$$

as radius. We then let $(\gamma)_1$ be all these circles. Then the sum of their radii does not exceed $\frac{1-\tau}{4\times32}$. Hence, there exists value t_1 in the interval $[\frac{1}{4}(3+\tau), \frac{1}{8}(7+\tau)]$, and value t_2 in the interval $[\frac{1}{16}(15+\tau), \frac{1}{32}(31+\tau)]$, such that the two circumferences $|\xi| = t_1$ and $|\xi| = t_2$ do not intersect $(\gamma)_1$. We also let $(\gamma)_2$ be the Euclidean exceptional circles that correspond to these $n(1, F = 0) + \bar{n}(1, F = \infty)$ points and number H_2. We take

$$H_2 = \frac{1}{8e} \cdot \frac{\varepsilon\pi}{4\theta}\left(\frac{\varepsilon}{8\theta}\right)^{1+2\theta/\pi}\left(\frac{R_{n1}}{R_{n2}}\right)^{3+\pi/\theta}. \tag{3.135}$$

Then according to Lemma 3.12, when point ξ is the disk $|\xi| \leq \frac{1}{2}(1+\tau)$ and outside circles $(\gamma)_2$, we have

$$
\begin{aligned}
\log\left|\frac{F'(\xi)}{F(\xi)}\right| &\le \frac{R+(1+\tau)/2}{R-(1+\tau)/2}m\left(R,\frac{F'}{F}\right) - \frac{R-(1+\tau)/2}{R+(1+\tau)/2}m\left(R,\frac{F}{F'}\right) \\
&\quad + \{\bar{n}(1,F=\infty)+n(1,F=0)\}\log\frac{2}{H_2} \\
&\quad - \frac{(R-t)^2}{4R^2}n(t,F'=0) \\
&\le \frac{4}{2R-(1+\tau)}m\left(R,\frac{F'}{F}\right) - \frac{2R-(1+\tau)}{4}m\left(R,\frac{F}{F'}\right) \\
&\quad + \{n(1,F=\infty)+n(1,F=0)\}\log\frac{2}{H_2} \\
&\quad - \frac{(R-t)^2}{4R^2}n(t,F'=0),
\end{aligned}
\tag{3.136}
$$

where $\frac{1+\tau}{2} < t \le R < 1$.

In the following, for any arbitrarily selected value R, $\frac{1+\tau}{2} < t \le R \le \frac{31+\tau}{32}$, supposing there is no intersection between circumference $|\xi| = R$ and circles $(\gamma)_1$, we estimate $m(R,\frac{F'}{F})$. Let Γ be the pre-image of circumference $|\xi| = R$ on the z-plane, β_ξ be a zero or pole of $F(\xi)$ on the disk $|\xi| < 1$, β be the pre-image of β_ξ on the z-plane, l be the shortest line linking β and Γ, and l_ξ be the image of l on the ξ-plane. Notice that circumference $|\xi| = R$ is outside circles $(\gamma)_1$. We have, therefore,

$$
\begin{aligned}
\operatorname{meas} l &= \int_l |dz| = \int_{l_\xi} |z'(\xi)|\,|d\xi| \\
&\ge \min_{|\xi|<1}|z'(\xi)|\frac{1}{n(1,F=0)+n(1,F=\infty)}\cdot\frac{1-\tau}{4\times 32}.
\end{aligned}
$$

Moreover, according to formulas (3.129), (3.131), (3.132) and (3.134), we conclude that

$$
\operatorname{meas} l \ge A(\varepsilon,\theta)\left(\frac{R_{n1}}{R_{n2}}\right)^{2+(\pi/\theta)+3(\lambda'+\eta)} R_{n1}^{1-(\lambda'+\eta)},
$$

where $A(\varepsilon,\theta) > 0$ is a constant depending ε and θ but unrelated to n. Obviously, on the z-plane, the distance between any zero or pole of $f(z)$ to Γ is $\ge \operatorname{meas} l$. Now, we apply formula (1.35), where we let $r = R_{n3}$,

$\rho' = 2R_{n3}$; then when $z \in \Gamma$, we obtain

$$\log\left|\frac{f'(z)}{f(z)}\right| \leq \log^+(2R_{n1}) + 2\log^+\frac{1}{R_{n3}}$$

$$+ \log^+ T(2R_{n2}, f) + \log^+\frac{n(2R_{n3})}{R_{n1}}$$

$$+ \log\left\{\frac{1}{A(\varepsilon, \theta)}\left(\frac{R_{n2}}{R_{n1}}\right)^{2+\pi/\theta+3(\lambda'+\eta)} R_{n_1}^{\lambda'+\eta-1}\right\} + O(1),$$

where

$$n(2R_{n3}) = n(2R_{n3}, f = 0) + n(2R_{n3}, f = \infty)$$

$$\leq \frac{1}{\log 2}\left\{N\left(4R_{n3}, \frac{1}{f}\right) + N(4R_{n3}, f)\right\}$$

$$\leq \frac{2}{\log 2}T(4R_{n3}, f) + O(1).$$

Notice that the order λ of $f(z)$ is $< +\infty$. Then, when n is sufficiently large, we get

$$T(4R_{n3}, f) \leq (4R_{n3})^{\lambda+1}. \tag{3.137}$$

Hence, there exists value $n_1 \geq n_0$, such that, when $n \geq n_1$ and $z \in \Gamma$,

$$\log^+\left|\frac{f'(z)}{f(z)}\right| \leq A(\theta)\log R_{n3} \tag{3.138}$$

where $A(\theta) < +\infty$ is a constant depending on θ and unrelated to n. We conclude, therefore, that

$$\frac{1}{2\pi R}\int_{|\xi|=R}\log^+\left|\frac{f'(z)}{f(z)}\right||d\xi| \leq A(\theta)\log R_{n3}.$$

Moreover, based on the equality

$$\frac{F'(\xi)}{F(\xi)} = \frac{f'(z)}{f(z)}z'(\xi)$$

we find that

$$m\left(R, \frac{F'}{F}\right) \leq \frac{1}{2\pi R}\int_{|\xi|=R}\log^+\left|\frac{f'(z)}{f(z)}\right||d\xi|$$

$$+ \frac{1}{2\pi R}\int_{|\xi|=R}\log^+|z'(\xi)||d\xi|.$$

Furthermore, from formulas (3.130) and (3.132), we arrive at the conclusion that

$$m\left(R, \frac{F'}{F}\right) \le A(\theta) \log R_{n3} + \log\left\{\frac{2\theta}{\pi}\left(\frac{8\theta}{\varepsilon}\right)^{1+2\theta/\pi}\left(\frac{R_{n2}}{R_{n1}}\right)^{2+\pi/\theta} R_{n2}\right\}$$

$$\le A(\theta)\log\left\{\left(\frac{8\theta}{\varepsilon}\right)^{2\theta/\pi}\left(\frac{R_{n2}}{R_{n1}}\right)^2 \cdot R_{n2}\right\}$$

$$+ \log\left\{\frac{2\theta}{\pi}\left(\frac{8\theta}{\varepsilon}\right)^{1+2\theta/\pi}\left(\frac{R_{n2}}{R_{n1}}\right)^{2+\pi/\theta} R_{n2}\right\}.$$

We select sufficiently large values $n_2 \ge n_1$, such that, when $n \ge n_2$ and when there is no intersection between circumference $|\xi| = R$ $(\frac{1+\tau}{2} < t \le R \le \frac{31+\tau}{32})$ and circles $(\gamma)_1$,

$$m\left(R, \frac{F'}{F}\right) \le A(\theta)\log R_{n2}, \tag{3.139}$$

where $A(\theta) < +\infty$ is a constant depending on θ and unrelated to n.

In the following discussion, we distinguish two cases:

$$(1) \quad n(t_1, F' = 0) > \left\{\log\left[\frac{32e\theta}{\alpha\pi}\left(\frac{8\theta}{\varepsilon}\right)^{1+2\theta/\pi}\left(\frac{R_{n2}}{R_{n1}}\right)^{3+\pi/\theta}\right]\right\}^{-1}$$

$$\times \left\{\frac{\alpha}{16\pi}\left(\frac{\varepsilon}{8\theta}\right)^{1+4\theta/\pi}\left(\frac{R_{n1}}{R_{n2}}\right)^{6+\pi/\theta} N_n\right\}. \tag{3.140}$$

$$(2) \quad n(t_1, F' = 0) \le \left\{\log\left[\frac{32e\theta}{\alpha\pi}\left(\frac{8\theta}{\varepsilon}\right)^{1+2\theta/\pi}\left(\frac{R_{n2}}{R_{n1}}\right)^{3+\pi/\theta}\right]\right\}^{-1}$$

$$\times \left\{\frac{\alpha}{16\pi}\left(\frac{\varepsilon}{8\theta}\right)^{1+2\theta/\pi}\left(\frac{R_{n1}}{R_{n2}}\right)^{6+\pi/\theta} N_n\right\}. \tag{3.141}$$

First we consider case (1). According to formulas (3.134), (3.135), (3.136), (3.139) and (3.140), for the point ξ in the disk $|\xi| \le \frac{1}{2}(1 + \tau)$ and outside circles $(\gamma)_2$ we have

$$\log\left|\frac{F'(\xi)}{F(\xi)}\right| \le \frac{t_2 + (1+\tau)/2}{t_2 - (1+\tau)/2} m\left(t_2, \frac{F'}{F}\right)$$

$$+ \{n(1, F = \infty) + n(1, F = 0)\} \log\frac{2}{H} - \frac{(t_2 - t_1)^2}{4t_1^2} n(t, F' = 0)$$

$$\le \frac{2}{t_2 - (1+\tau)/2} A(\theta) \log R_{n2} + 2A(\varepsilon, \theta)\left(\frac{R_{n2}}{R_{n1}}\right)^{2(\lambda'+\eta)} R_{n2}^{\lambda'+\eta}$$

$$\times \log\left\{\frac{64e\theta}{\varepsilon\pi}\left(\frac{8\theta}{\varepsilon}\right)^{1+2\theta/\pi}\left(\frac{R_{n2}}{R_{n1}}\right)^{3+\pi/\theta}\right\}$$

$$- \frac{(t_2 - t_1)^2}{4}\left\{\log\left[\frac{32e\theta}{2\pi}\left(\frac{8\theta}{\varepsilon}\right)^{1+2\theta/\pi}\left(\frac{R_{n2}}{R_{n1}}\right)^{3+\pi/\theta}\right]\right\}^{-1}$$

$$\times \left\{\frac{\alpha}{16\pi}\left(\frac{\varepsilon}{8\theta}\right)^{1+4\theta/\pi}\left(\frac{R_{n1}}{R_{n2}}\right)^{6+\pi/\theta} N_n\right\}.$$

Notice that

$$t_2 - \frac{1+\tau}{2} \ge \frac{15+\tau}{16} - \frac{1+\tau}{2}$$

$$\ge \frac{7}{32}(1-\rho) = \frac{7}{32}\cdot\frac{\varepsilon}{2\theta}\left(\frac{R_{n1}}{R_{n2}}\right)^{\pi/\theta},$$

$$t_1 - t_2 \ge \frac{15+\tau}{16} - \frac{7+\tau}{8}$$

$$\ge \frac{1}{32}(1-\rho) = \frac{1}{32}\cdot\left(\frac{\varepsilon}{2\theta}\right)\left(\frac{R_{n1}}{R_{n2}}\right)^{\pi/\theta}.$$

Then

$$\log\left|\frac{F'(\xi)}{F(\xi)}\right| \le \frac{2\times 64\theta}{7\varepsilon}\left(\frac{R_{n2}}{R_{n1}}\right)^{\pi/\theta} A(\theta) \log R_{n2}$$

$$+ 2A(\varepsilon, \theta)\left(\frac{R_{n2}}{R_{n1}}\right)^{2(\lambda'+\eta)} R_{n2}^{\lambda'+\eta}$$

$$\times \log\left\{\frac{64e\theta}{\varepsilon\pi}\left(\frac{8\theta}{\varepsilon}\right)^{1+2\theta/\pi}\left(\frac{R_{n2}}{R_{n1}}\right)^{3+\pi/\theta}\right\}$$

$$- \left(\frac{\varepsilon}{4\times 32\theta}\right)^2\left(\frac{R_{n1}}{R_{n2}}\right)^{2\pi/\theta}$$

$$\times \left\{\log\left[\frac{32e\theta}{\alpha\pi}\left(\frac{8\theta}{\varepsilon}\right)^{1+2\theta/\pi}\frac{R_{n2}}{R_{n1}}\right]\right\}^{-1}$$

$$\times \left\{\frac{\varepsilon}{16\pi}\left(\frac{\varepsilon}{8\theta}\right)^{1+4\theta/\pi}\left(\frac{R_{n1}}{R_{n2}}\right)^{6+\pi/\theta} N_n\right\}.$$

Furthermore, according to formula (3.128), there exists $n_3 \geq n_2$, such that when $n \geq n_3$ and when point ξ is in the disk $|\xi| \leq \frac{1}{2}(1 + \tau)$ and outside circles $(\gamma)_2$,

$$\log \left| \frac{F'(\xi)}{F(\xi)} \right| \leq - \frac{A(\alpha, \theta, \varepsilon)}{B(\theta) \log(R_{n2}/R_{n1}) + C(\alpha, \theta, \varepsilon)} \left(\frac{R_{n1}}{R_{n2}} \right)^{6 + \pi/\theta} N_n, \quad (3.142)$$

where $A(\alpha, \theta, \varepsilon) > 0$ and $C(\alpha, \varepsilon, \theta) < +\infty$ are constants depending on α, θ and ε, and unrelated to n.

Next, we consider case (2). According to formulas (3.134), (3.125), (3.136) and (3.139), for point ξ in the disk $|\xi| \leq \frac{1}{2}(1 + \tau)$ and outside circles $(\gamma)_2$, we have

$$\log \left| \frac{F'(\xi)}{F(\xi)} \right| \leq \frac{t_1 + (1 + \tau)/2}{t_1 - (1 + \tau)/2} m\left(t_1, \frac{F'}{F} \right) - \frac{t_1 - (1 + \tau)/2}{t_1 + (1 + \tau)/2} m\left(t_1, \frac{F}{F'} \right)$$

$$+ \{n(1, F = \infty) + n(1, F = 0)\} \log \frac{2}{H_2}$$

$$\leq \frac{4}{2t_1 - (1 + \tau)} A(\theta) \log R_{n2} - \frac{2t_1(1 + \tau)}{4} m\left(t_1, \frac{F}{F'} \right)$$

$$+ 2A(\varepsilon, \theta) \left(\frac{R_{n2}}{R_{n1}} \right)^{2(\lambda' + \eta)} R_{n2}^{\lambda' + \eta}$$

$$\times \log \left\{ \frac{64e\theta}{\varepsilon\pi} \left(\frac{8\theta}{\varepsilon} \right)^{1 + 2\theta/\pi} \left(\frac{R_{n2}}{R_{n1}} \right)^{3 + \pi/\theta} \right\}.$$

Notice that

$$t_1 - \frac{1 + \tau}{2} \geq \frac{3 + \tau}{4} - \frac{1 + \tau}{2} \geq \frac{1}{8}(1 - \rho) = \frac{1}{8} \cdot \frac{\varepsilon}{2\theta} \left(\frac{R_{n1}}{R_{n2}} \right)^{\pi/\theta}. \quad (3.143)$$

Then

$$\log \left| \frac{F'(\xi)}{F(\xi)} \right| \leq \frac{32\theta}{\varepsilon} A(\theta) \left(\frac{R_{n2}}{R_{n1}} \right)^{\pi/\theta}$$

$$\times \log R_{n2} - \frac{\varepsilon}{32\theta} \left(\frac{R_{n1}}{R_{n2}} \right)^{\pi/\theta} m\left(t_1, \frac{F}{F'} \right)$$

$$+ 2A(\varepsilon, \theta) \left(\frac{R_{n2}}{R_{n1}} \right)^{2(\lambda' + \eta)} R_{n2}^{\lambda' + \eta}$$

$$\times \log \left\{ \frac{64e\theta}{\varepsilon\pi} \left(\frac{8\theta}{\varepsilon} \right)^{1 + 2\theta/\pi} \left(\frac{R_{n2}}{R_{n1}} \right)^{3 + \pi/\theta} \right\}.$$

We select a sufficiently large value $n_4 \geq n_3$, such that when $n \geq n_4$, and when ξ is in the disk $|\xi| \leq \frac{1}{2}(1 + \tau)$ and outside circles $(\gamma)_2$,

$$
\begin{aligned}
\log\left|\frac{F'(\xi)}{F(\xi)}\right| &\leq A(\theta, \varepsilon) \left(\frac{R_{n2}}{R_{n1}}\right)^{2(\lambda' + \eta) + \pi/\theta} R_{n2}^{\lambda' + 2\eta} \\
&\quad - \frac{\varepsilon}{32\theta} \left(\frac{R_{n1}}{R_{n2}}\right)^{\pi/\theta} m\left(t_1, \frac{F}{F'}\right).
\end{aligned}
\tag{3.144}
$$

In the following, we estimate $m(t_1, \frac{F}{F'})$. Let $(\gamma)_3$ be the Euclidean exceptional circles corresponding to these $n(t_1, F' = 0)$ points and number H_3, and assume that

$$
H_3 = \frac{1}{8e} \cdot \operatorname{meas} E_\xi \geq \frac{1}{8e} \cdot \frac{8\pi}{2\theta} \left(\frac{\varepsilon}{8\theta}\right)^{1 + 2\theta/\pi} \left(\frac{R_{n1}}{R_{n2}}\right)^{3 + \pi/\theta}
\tag{3.145}
$$

and denote E_ξ the part of $\widetilde{E}_\xi$ that lies outside circles $(\gamma)_3$. Then we have

$$
\operatorname{meas} \widetilde{E}_\xi \geq \frac{1}{2} \operatorname{meas} E_\xi \geq \frac{2\pi}{4\theta} \left(\frac{\varepsilon}{8\theta}\right)^{1 + 2\theta/\pi} \left(\frac{R_{n3}}{R_{n2}}\right)^{3 + \pi/\theta}
\tag{3.146}
$$

and when $\xi \in \widetilde{E}_\xi$, according to the Poisson-Jensen formula and formulas (3.129) and (3.131), the following inequalities hold:

$$
\begin{aligned}
\log^+\left|\frac{F(\xi)}{F'(\xi)}\right| &\leq \frac{t_1 + (1 + \tau)/2}{t_1 - (t + \tau)/2} m\left(t_1, \frac{F}{F'}\right) + n(t_1, F' = 0) \log \frac{2}{H_3} \\
&\leq \frac{32\theta}{\varepsilon} \left(\frac{R_{n2}}{R_{n1}}\right)^{\pi/\theta} m\left(t_1, \frac{F}{F'}\right) + n(t_1, F' = 0) \log \frac{2}{H_3}.
\end{aligned}
$$

And hence we get:

$$
\begin{aligned}
&\frac{1}{2\pi R_n} \int_{\widetilde{E}_\xi} \log^+\left|\frac{F(\xi)}{F'(\xi)}\right| |z'(\xi)| |d\xi| \\
&\leq \left\{ \frac{32\theta}{\varepsilon} \left(\frac{R_{n2}}{R_{n1}}\right)^{\pi/\theta} m\left(t_1, \frac{F}{F'}\right) + n(t_1, F' = 0) \log \frac{2}{H_3} \right\} \\
&\quad \times \frac{1}{2\pi R_n} \int_{\widetilde{E}_\xi} |z'(\xi)| |d\xi|.
\end{aligned}
$$

Let $\widetilde{E}_n$ denote the pre-image of $\widetilde{E}_\xi$ on the z-plane. Then by formulas (3.132) and (3.146), it follows that

$$
\begin{aligned}
2\pi R_n \geq \operatorname{meas} \widetilde{E}_n &= \int_{\widetilde{E}_n} |dz| = \int_{\widetilde{E}_\xi} |z'(\xi)| |d\xi| \\
&\geq \frac{\varepsilon}{8} \left(\frac{\varepsilon}{8\theta}\right)^{1 + 4\theta/\pi} \left(\frac{R_{n1}}{R_{n2}}\right)^{5 + \pi/\theta} R_{n1}.
\end{aligned}
\tag{3.147}
$$

And hence

$$
\frac{1}{2\pi R_n} \int_{\widetilde{E}_\xi} \log^+ \left| \frac{F(\xi)}{F'(\xi)} \right| |z'(\xi)| |d\xi|
$$

$$
\leq \frac{32\theta}{\varepsilon} \left(\frac{R_{n2}}{R_{n1}} \right)^{\pi/\theta} m\left(t, \frac{F}{F'} \right) + n(t_1, F' = 0) \log \frac{2}{H_3},
$$

$$
m\left(t_1, \frac{F}{F'} \right) \geq \frac{\varepsilon}{32\theta} \left(\frac{R_{n1}}{R_{n2}} \right)^{\pi/\theta} \frac{1}{2\pi R_n} \int_{\widetilde{E}_\xi} \log \left| \frac{F(\xi)}{F'(\xi)} \right| |z'(\xi)| |d\xi|
$$

$$
- \frac{\varepsilon}{32\theta} \left(\frac{R_{n1}}{R_{N2}} \right)^{\pi/\theta} n(t_1, F' = 0) \log \frac{2}{H_3}.
$$

(3.148)

Moreover, according to the equality

$$
\frac{f(z)}{f'(z)} = \frac{F(\xi)}{F'(\xi)} z'(\xi),
$$

we derive

$$
\frac{1}{2\pi R_n} \int_{\widetilde{E}_n} \log^+ \left| \frac{f(z)}{f'(z)} \right| |dz|
$$

$$
\leq \frac{1}{2\pi R_n} \int_{\widetilde{E}_\xi} \log^+ \left| \frac{F(\xi)}{F'(\xi)} \right| \cdot |z'(\xi)| |d\xi|
$$

$$
+ \frac{1}{2\pi R_n} \int_{\widetilde{E}_\xi} \log^+ |z'(\xi)| \cdot |z'(\xi)| |d\xi|.
$$

Noting formula (3.132), we obtain

$$
\frac{1}{2\pi R_n} \int_{\widetilde{E}_n} \log^+ \left| \frac{f(z)}{f'(z)} \right| |dz|
$$

$$
\leq \frac{1}{2\pi R_n} \int_{\widetilde{E}_\xi} \log^+ \left| \frac{F(\xi)}{F'(\xi)} \right| |z'(\xi)| |d\xi|
$$

$$
+ \log \left\{ \frac{2\theta}{\pi} \left(\frac{8\theta}{\varepsilon} \right)^{1+2\theta/\pi} \left(\frac{R_{n2}}{R_{n1}} \right)^{2+\pi/\theta} R_{n2} \right\},
$$

$$
\frac{1}{2\pi R_n} \int_{\widetilde{E}_\xi} \log \left| \frac{F(\xi)}{F'(\xi)} \right| |z'(\xi)| |d\xi|
$$

$$
\geq \frac{1}{2\pi R_n} \int_{\widetilde{E}_n} \log \left| \frac{f(z)}{f'(z)} \right| |dz|
$$

$$
- \log \left\{ \frac{2\theta}{\pi} \cdot \left(\frac{8\theta}{\varepsilon} \right)^{1+2\theta/\pi} \left(\frac{R_{n2}}{R_{n1}} \right)^{2+\pi/\theta} R_{n2} \right\}.
$$

(3.149)

Now, from the identity

$$
\frac{1}{f(z) - a} = \frac{1}{a} \cdot \frac{f(z)}{f'(z)} \left\{ \frac{f'(z)}{f(z) - a} - \frac{f'(z)}{f(z)} \right\},
$$

we derive

$$\frac{1}{2\pi R_n} \int_{\widetilde{E}_n} \log^+ \frac{1}{|f(z)-a|} |dz|$$

$$\leq \log^+ \frac{1}{|a|} + \frac{1}{2\pi R_n} \int_{\widetilde{E}_n} \log \left| \frac{f(z)}{f'(z)} \right| |dz|$$

$$+ m\left(R_n, \frac{f'}{f-a}\right) + m\left(R_n, \frac{f'}{f}\right) + \log 2,$$

$$\frac{1}{2\pi R_n} \int_{\widetilde{E}_n} \log^+ \left| \frac{f(z)}{f'(z)} \right| |dz|$$

$$\geq \frac{1}{2\pi R_n} \int_{\widetilde{E}_n} \log^+ \frac{1}{|f(z)-a|} |dz|$$

$$- \left\{ m\left(R_n, \frac{f'}{f-a}\right) + m\left(R_n, \frac{f'}{f}\right) + \log^+ \frac{1}{|a|} + \log 2 \right\}.$$

Furthermore, based on formulas (3.127) and (3.147), we get

$$\frac{1}{2\pi R_n} \int_{\widetilde{E}_n} \log^+ \left| \frac{f(z)}{f'(z)} \right| |dz|$$

$$\geq \frac{1}{2\pi R_n} \cdot \frac{\alpha}{4} \left(\frac{\varepsilon}{8\theta} \right)^{1+4\theta/\pi} \left(\frac{R_{n1}}{R_{n2}} \right)^{5+\pi/\theta} R_{n1} \cdot N_n$$

$$- \left\{ m\left(R_n, \frac{f'}{f-a}\right) + m\left(R_n, \frac{f'}{f}\right) + \log^+ \frac{1}{|a|} + \log 2 \right\} \qquad (3.150)$$

$$\geq \frac{\alpha}{8\pi} \left(\frac{\varepsilon}{8\theta} \right)^{1+4\theta/\pi} \left(\frac{R_{n1}}{R_{n2}} \right)^{6+\pi/\theta} N_n$$

$$- \left\{ m\left(R_n, \frac{f'}{f-a}\right) + m\left(R_n, \frac{f'}{f}\right) + \log^+ \frac{1}{|a|} + \log 2 \right\}.$$

We then apply Lemma 1.3, where we let $r = R_n$, $\rho = 2R_n$. It follows, therefore, that

$$m\left(R_n, \frac{f'}{f}\right) \leq 4\log^+ T(2R_n, f) + 5\log^+(2R_n) + 7\log^+ \frac{1}{R_n} + O(1),$$

$$m\left(R_n, \frac{f'}{f-a}\right) \leq 4\log^+ T(2R_n, f-a) + 5\log^+(2R_n) + 7\log^+ \frac{1}{R_n} + O(1)$$

$$\leq 4\log^+ T(2R_n, f) + 5\log^+(2R_n) + 7\log^+ \frac{1}{R_n} + O(1).$$

Since the order λ of $f(z)$ is $< +\infty$, when n is sufficiently large, we have

$$T(2R_n, f) \leq (2R_n)^{\lambda+1} \leq (2R_{n2})^{\lambda+1}. \qquad (3.151)$$

Hence, there exists value $n_5 \geq n_4$, such that when $n \geq n_5$, we have

$$m\left(R_n, \frac{f'}{f-a}\right) + m\left(R_n, \frac{f'}{f}\right) + \log^+ \frac{1}{|a|} + \log 2 \leq A \log R_{n2}, \quad (3.152)$$

where $A < +\infty$ is a constant.

To sum up our discussion for case (2), we obtain from formulas (3.141), (3.145), (3.148), (3.149), (3.150) and (3.152) that

$$m\left(t_1, \frac{F}{F'}\right) \geq \frac{\alpha}{64\pi} \left(\frac{\varepsilon}{8\theta}\right)^{2+4\theta/\pi} \left(\frac{R_{n1}}{R_{n2}}\right)^{6+2\pi/\theta} N_n$$
$$- \frac{\varepsilon}{32} \left(\frac{R_{n1}}{R_{n2}}\right)^{\pi/\theta} \left\{\log\left[\frac{2\theta}{\pi}\left(\frac{8\theta}{\varepsilon}\right)^{1+2\theta/\pi}\right.\right.$$
$$\left.\left. \times \left(\frac{R_{n2}}{R_{n1}}\right)^{2+\pi/\theta} R_{n2}\right] + A \log R_{n2}\right\}.$$

Furthermore by combining this with formula (3.144), we get

$$\log\left|\frac{F'(\xi)}{F(\xi)}\right| \leq -\frac{\alpha}{4 \times 64\pi} \left(\frac{\varepsilon}{8\theta}\right)^{3+4\theta/\pi} \left(\frac{R_{n1}}{R_{n2}}\right)^{6+3\pi/\theta} N_n$$
$$+ A(\varepsilon, \theta)\left(\frac{R_{n2}}{R_{n1}}\right)^{2(\lambda'+\eta)+\pi/\theta} R_{n2}^{\lambda'+2\eta} + \left(\frac{\varepsilon}{32\theta}\right)^2 \left(\frac{R_{n1}}{R_{n2}}\right)^{2\pi/\theta}$$
$$\times \left\{\log\left[\frac{2\theta}{\pi}\left(\frac{8\theta}{\varepsilon}\right)^{1+2\theta/\pi}\left(\frac{R_{n2}}{R_{n1}}\right)^{2+\pi/\theta} R_{n2}\right] + A \log R_{n2}\right\}.$$

Then, according to formula (3.128), there exists $n_6 \geq n_5$, such that when $n \geq n_6$, and when point ξ is in the disk $|\xi| \leq \frac{1}{2}(1 + \tau)$ and outside circles $(\gamma)_2$, it follows that

$$\log\left|\frac{F'(\xi)}{F(\xi)}\right| \leq -A(\alpha, \varepsilon, \theta)\left(\frac{R_{n1}}{R_{n2}}\right)^{6+3\pi/\theta} N_n, \quad (3.153)$$

where $A(\alpha, \varepsilon, \theta) > 0$ is a constant depending on α, θ and ε, and unrelated to n.

When case (1) is satisfied, we have (3.142), while the satisfaction of case (2) yields formula (3.153). Hence, when $n \geq n_6$ and when point ξ is in the disk $|\xi| \leq \frac{1}{2}(1 + \tau)$ and outside circles $(\gamma)_2$, it is always true that

$$\left|\frac{F'(\xi)}{F(\xi)}\right| \leq \exp\left\{-\frac{A(\alpha, \varepsilon, \theta)}{B(\theta)\log(R_{n2}/R_{n1}) + C(\alpha, \varepsilon, \theta)}\left(\frac{R_{n1}}{R_{n2}}\right)^{6+3\pi/\theta} N_n\right\},$$
$$(3.154)$$

where $A(\alpha, \theta, \varepsilon) > 0$ and $C(\alpha, \theta, \varepsilon) < +\infty$ are the constants depending on α, θ and ε, and unrelated to n, while $B(\theta) < +\infty$ is a constant depending on θ but unrelated to n.

According to formulas (3.133) and (3.135), we conclude that there exists some points $\xi_0 \notin (\gamma)_2$ on E_ξ. Let $z_0 \in E_n$ be the pre-image of ξ_0 on the z-plane and point ξ be any in the disk $|\xi| \leq \frac{1}{2}(1 + \tau)$ and outside circles $(\gamma)_2$. We use a straight line to connect them. If the connection comes across circles $(\gamma)_2$, then we use the relatively small arc to replace them. Hence, a curve L_ξ connecting point ξ and point ξ_0 is obtained, with its length measuring $\mathrm{meas}\, L_\xi \leq 2 + \pi e H_2 \leq 3$. By means of integration, we obtain from formula (3.154) that

$$
\big|\log|F(\xi)| - \log|F(\xi_0)|\big|
$$
$$
\leq |\log F(\xi) - \log F(\xi_0)| = \left| \int_{L_\xi} \frac{F'(\xi)}{F(\xi)}\, d\xi \right| \leq \int_{L_\xi} \left| \frac{F'(\xi)}{F(\xi)} \right| |d\xi|
$$
$$
\leq 3 \exp\left\{ -\frac{A(\alpha, \theta, \varepsilon)}{B(\theta)\log(R_{n2}/R_{n1}) + C(\alpha, \theta, \varepsilon)} \left(\frac{R_{n1}}{R_{n2}}\right)^{6+3\pi/\theta} N_n \right\},
$$

$$
|F(\xi)| \leq F(\xi_0)|
$$
$$
\times \exp\left\{ 3\exp\left[-\frac{A(\alpha, \theta, \varepsilon)}{B(\theta)\log(R_{n2}/R_{n1}) + C(\alpha, \theta, \varepsilon)} \right.\right.
$$
$$
\left.\left. \times \left(\frac{R_{n1}}{R_{n2}}\right)^{6+3\pi/\theta} N_n \right] \right\}.
$$

Moreover, by formula (3.127) we get

$$
|F(\xi_0)| = |f(z_0)| \leq |f(z_0) - a| + |a| \leq e^{-N_n} + |a|.
$$

Consequently, we have

$$
|F(\xi)| \leq \{ e^{-N_n} + |a| \}
$$
$$
\times \exp\left\{ 3\exp\left\{ -\frac{A(\alpha, \theta, \varepsilon)}{B(\theta)\log(R_{n2}/R_{n1}) + C(\alpha, \theta, \varepsilon)} \right.\right.
$$
$$
\left.\left. \times \left(\frac{R_{n1}}{R_{n2}}\right)^{6+3\pi/\theta} N_n \right\} \right\}.
$$

Furthermore, based on formula (3.128), there exists value $n_7 \geq n_6$, such that when $n \geq n_7$ and when point ξ is in the disk $|\xi| \leq \frac{1}{2}(1 + \tau)$ and outside circles $(\gamma)_2$, we have $|F(\xi)| \leq 2|a|$. Hence, combining this with formula (3.154), we get

$$
|F'(\xi)| \leq 2|a|
$$
$$
\times \exp\left\{ -\frac{A(\alpha, \theta, \varepsilon)}{B(\theta)\log(R_{n2}/R_{n1}) + C(\alpha, \theta, \varepsilon)} \left(\frac{R_{n1}}{R_{n2}}\right)^{6+3\pi/\theta} N_n \right\}.
$$

Again, we employ integration, and derive from formula (3.127) that:

$$|F(\xi) - F(\xi_0)| = \left| \int_{L_\xi} F'(\xi)\, d\xi \right| \leq 6|a|$$

$$\times \exp\left\{ -\frac{A(\alpha,\theta,\varepsilon)}{B(\theta)\log(R_{n2}/R_{n1}) + C(\alpha,\theta,\varepsilon)} \right.$$

$$\left. \times \left(\frac{R_{n1}}{R_{n2}}\right)^{6+3\pi/\theta} N_n \right\},$$

$$|f(z) - a| \leq |F(\xi) - F(\xi_0)| + |f(z_0) - a|$$

$$\leq 6|a| \exp\left\{ -\frac{A(\alpha,\theta,\varepsilon)}{B(\theta)\log(R_{n2}/R_{n1}) + C(\alpha,\theta,\varepsilon)} \right.$$

$$\left. \times \left(\frac{R_{n1}}{R_{n2}}\right)^{6+3\pi/\theta} N_n \right\} + e^{-N_n}$$

$$\leq 12|a| \exp\left\{ -\frac{A(\alpha,\theta,\varepsilon)}{B(\theta)\log(R_{n2}/R_{n1}) + C(\alpha,\theta,\varepsilon)} \right.$$

$$\left. \times \left(\frac{R_{n1}}{R_{n2}}\right)^{6+3\pi/\theta} N_n \right\},$$

$$\log\frac{1}{|f(z) - a|} \geq \frac{A(\alpha,\theta,\varepsilon)}{B(\theta)\log(R_{n2}/R_{n1}) + C(\alpha,\theta,\varepsilon)}$$

$$\times \left(\frac{R_{n1}}{R_{n2}}\right)^{6+3\pi/\theta} N_n - \log^+(12|a|).$$

Let $(\gamma)'$ be the set of pre-images of $(\gamma)_2$ on the z-plane. We choose $n_8 \geq n_7$, such that when $n \geq n_8$, and when point z is in $\overline{\Omega}(-\theta+\varepsilon, \theta-\varepsilon; R_{n1}, R_{n2})$ and outside $(\gamma)'$

$$\log\frac{1}{|f(z) - a|} \geq \frac{A(\alpha,\theta,\varepsilon)}{B(\theta)\log(R_{n2}/R_{n1}) + C(\alpha,\theta,\varepsilon)} \left(\frac{R_{n1}}{R_{n2}}\right)^{6+3\pi/\theta} N_n. \quad (3.155)$$

Finally, we prove that $(\gamma)'$ must be contained in some circles (γ), and the sum of radii of (γ) does not exceed $\frac{1}{8}\varepsilon R_{n1}$. Indeed, we may consider a circle $C_\xi: |\xi - \xi_1| < t$ in $(\gamma)_2$. Let C be the pre-image of C_ξ on the z-plane and z_1 be the pre-image of ξ_1. Then there exists a point z' on $\overline{C}$, such that we may construct a circle $C' \supset C$ using point z_1 as center, and the length of the straight line l' connecting point z_1 and z' as radius. We also let ξ' be the image point of z' on the ξ-plane and l'_ξ be the straight line linking ξ_1 and ξ'. Then the image of l'_ξ on the z-plane is a curve Γ' connecting z_1 and z' on the z-plane. Hence, according to formula (3.132),

we have

$$\text{meas}\, l' \le \int_{\Gamma'} |dz| = \int_{l'_\xi} |z'(\xi)|\,|d\xi| \le \frac{2\theta}{\pi} \left(\frac{8\theta}{\varepsilon}\right)^{1+2\theta/\pi} \left(\frac{R_{n2}}{R_{n1}}\right)^{2+\pi/\theta} R_{n2} t.$$

Let (γ) be all these circles (C'). Then by formula (3.135), the sum of radii of (γ) does not exceed

$$\frac{2\theta}{\pi}\left(\frac{8\theta}{\varepsilon}\right)^{1+2\theta/\pi}\left(\frac{R_{n2}}{R_{n1}}\right)^{2+\pi/\theta} R_{n2} \frac{1}{4}\frac{\varepsilon\pi}{4\theta}\left(\frac{\varepsilon}{8\theta}\right)^{1+2\theta/\pi}\left(\frac{R_{n1}}{R_{n2}}\right)^{3+\pi/\theta} \le \frac{1}{8}\varepsilon R_{n1}.$$

Accordingly, Lemma 3.13 is proved completely.

In Lemma 3.13, if we further assume that $f(z)$ is an entire function, then according to formulas (3.129) and (3.131), we conclude that

$$8eH_2 = \frac{\varepsilon\pi}{4\theta}\left(\frac{\varepsilon}{8\theta}\right)^{1+2\theta/\pi}\left(\frac{R_{n1}}{R_{n2}}\right)^{3+\pi/\theta}$$

$$\le \frac{\varepsilon\pi}{4\theta}\cdot\frac{1}{2\pi}\left(\frac{R_{n1}}{R_{n2}}\right)^{\pi/\theta} = \frac{1}{4}(1-\rho) = \frac{1+\tau}{2} - \tau.$$

Hence, there exists value t_0 is in the interval $[\tau, \frac{1}{2}(1+\tau)]$, such that circumference $|\xi| = t_0$ does not intersect circles $(\gamma)_2$. Accordingly, we conclude by formula (3.154) that

$$|n(t_0, F = 0)| = \left|\frac{1}{2\pi i}\int_{|\xi|=t_0}\frac{F'(\xi)}{F(\xi)}\,d\xi\right|$$

$$\le t_0 \exp\left\{-\frac{A(\alpha,\theta,\varepsilon)}{B(\theta)\log(R_{n2}/R_{n1}) + C(\alpha,\theta,\varepsilon)}\right.$$

$$\left.\times\left(\frac{R_{n1}}{R_{n2}}\right)^{6+3\pi/\theta} N_n\right\} \le t_0 < 1,$$

meaning that $n(t_0, F = 0) = 0$. This illustrates that $F(\xi)$ has no zeros in the disk $|\xi| \le t_0$. Therefore, formula (3.154) holds uniformly in the disk $|\xi| \le t_0$ and, in turn, we may prove that formula (3.155) holds uniformly on $\overline{\Omega}(-\theta+\varepsilon, \theta-\varepsilon; R_{n1}, R_{n2})$. Accordingly, we arrive at the following result:

LEMMA 3.14. *Let $f(z)$ be an entire function of order $\lambda < +\infty$ on the open plane $|z| < +\infty$, with order $\lambda < +\infty$, and let it satisfy the following conditions:*

(1) *On $\overline{\Omega}(-\theta, \theta)$ $(0 < \theta \le \pi)$*

$$\varlimsup_{r\to+\infty}\frac{\log^+ n\{\Omega(-\theta,\theta;r), f = 0\}}{\log r} = \lambda' < +\infty.$$

(2) *For a selected number ε, $0 < \varepsilon < \theta$, there exists a sequence $\{R_n\}$, such that the set*

$$E_n = \left\{z\,\bigg|\, \log\frac{1}{|f(z) - a|} \ge N_n, |z| = R_n, \quad -\theta+\varepsilon \le \arg z \le \theta - \varepsilon\right\}$$

has linear measure $\text{meas}\, E_n \geq \alpha R_n\ (\alpha \geq \frac{\varepsilon}{2})$, *where a is a complex number which is neither* 0 *nor* ∞, $N_n > 0$ *is a real number, and also*

$$\lim_{n \to +\infty} \left\{ \left(\frac{R_{n2}}{R_{n1}}\right)^{6+2(\lambda'+\eta_0)+3\pi/\theta} R_{n2}^{\lambda'+2\eta_0} \log R_{n2} \right\} N_n^{-1} = 0,$$

where $\eta_0 > 0$ *is a selected real number, and* $R_{n1} \leq R_n \leq R_{n2}$, $R_{n1} \to \infty$ $(n \to +\infty)$.

Under the above assumptions, provided that n *is sufficiently large, then for any point* z *in* $\overline{\Omega}(-\theta + \varepsilon,\, \theta - \varepsilon;\, R_{n1},\, R_{n2})$,

$$|f(z) - a| \leq \exp\left\{ -\frac{A(\alpha,\, \theta,\, \varepsilon)}{B(\theta)\log\frac{R_{n2}}{R_{n1}} + C(\alpha,\, \theta,\, \varepsilon)} \left(\frac{R_{n1}}{R_{n2}}\right)^{6+3\pi/\theta} N_n \right\},$$

where $A(\alpha,\, \theta,\, \varepsilon) > 0$ *and* $C(\alpha,\, \theta,\, \varepsilon)$ *are the constants depending on* α, θ *and* ε, *and is unrelated to* n, *while* $B(\theta) < +\infty$ *is a constant depending on* θ *and is unrelated to* n.

3.5.2. The Edrei-Fuchs Theorem. In the earlier 1960s, Edrei–Fuchs considered a class of meromorphic functions whose zeros and poles distribute on finitely many curves. They proved for such meromorphic functions, their number of nonzero finite deficient values are shown to be bounded by the number of the related curves. [15d, 15e]. The following two theorems are their results expressed in a simple form.

THEOREM 3.10. *Let* $f(z)$ *be a meromorphic function of order* λ $(0 < \lambda < +\infty)$ *on the open plane* $|z| < +\infty$ *and* $\Delta(\theta_k)$ $(k = 1, 2, \ldots, q;\ 0 \leq \theta_1 < \theta_2 < \cdots < \theta_q;\ \theta_{q+1} = \theta_1 + 2\pi)$ *be* q $(0 \leq q < +\infty)$ *half straight lines on the z-plane. Suppose that for any arbitrarily given small number* $\varepsilon > 0$,

$$\varlimsup_{r \to +\infty} \frac{\log^+ n\{\bigcup_{k=1}^{q} \overline{\Omega}(\theta_k + \varepsilon,\, \theta_{k+1} - \varepsilon;\, r) \cdot f = X\}}{\log r} < \lambda, (^{11}) \qquad (3.156)$$
$$X = 0,\, \infty.$$

If we let p *be the number of nonzero finite deficient values of* $f(z)$, *then* $p \leq q$.

PROOF. First we consider the case when $q = 0$. According to formula (3.156), we have

$$\varlimsup_{r \to +\infty} \frac{\log^+ n(r,\, f = X)}{\log r} = \lambda' < \lambda, \qquad X = 0,\, \infty. \qquad (3.157)$$

Suppose that $f(z)$ has a nonzero finite deficient value a with corresponding deficiency $\delta(a,\, f) = \delta > 0$. Then, according to the identity

$$\frac{1}{f(z) - a} = \frac{1}{a} \cdot \frac{f}{f'} \left(\frac{f'}{f - a} - \frac{f'}{f} \right)$$

(^{11}) When $q = 0$, $n\{\bigcup_{k=1}^{q} \overline{\Omega}(\theta_k + \varepsilon,\, \theta_{k+1} - \varepsilon;\, r),\, f = X\} = n(r,\, f = X)$.

we derive

$$m\left(r, \frac{1}{f-a}\right) \le \log^+ \frac{1}{|a|} + m\left(r, \frac{f}{f'}\right)$$
$$+ m\left(r, \frac{f'}{f-a}\right) + m\left(r, \frac{f'}{f}\right) + \log 2.$$

Moreover, according to Lemma 1.3 and $\lambda < +\infty$, we conclude that when r is sufficiently large, it follows that

$$m\left(r, \frac{f'}{f-a}\right) \le O(\log r), \tag{3.158}$$

and

$$m\left(r, \frac{f'}{f}\right) \le O(\log r). \tag{3.159}$$

And hence

$$m\left(r, \frac{f}{f'}\right) \ge m\left(r, \frac{1}{f-a}\right) - O(\log r). \tag{3.160}$$

Furthermore, based on the definition of a deficient value, when r is sufficiently large, we get

$$m\left(r, \frac{f}{f'}\right) \ge \frac{\delta}{2} T(r, f) - O(\log r). \tag{3.161}$$

We consider the transformation $\zeta = \frac{z}{2r}$ and set $F(\zeta) = f(2r\zeta)$. We also let (γ) be the Euclidean exceptional circles that correspond to these $n(1, F = 0) + \overline{n}(1, F = \infty)$ points and number H $(H = \frac{1}{16er})$ and $(\gamma)'$ be the image of (γ) on the z-plane. Hence, the radius of $(\gamma)'$ does not exceed $\frac{1}{4}$. We apply Lemma 3.12 to $F(\zeta)$, where we let $r = \frac{1}{2r}$, $R = \frac{1}{2}$. Then, when ζ is in the disk $|\zeta| \le \frac{1}{2r}$ and outside circles (γ), we have

$$\log\left|\frac{F'(\zeta)}{F(\zeta)}\right| \le \frac{(1/2)+(1/2r)}{(1/2)-(1/2r)} m\left(\frac{1}{2}, \frac{F'}{F}\right) - \frac{(1/2)-(1/2r)}{(1/2)+(1/2r)} m\left(\frac{1}{2}, \frac{F}{F'}\right)$$
$$+ \{n(1, F = 0) + \overline{n}(1, F = \infty)\} \log(32er)$$
$$\le \frac{r+1}{r-1} m\left(r, \frac{f'}{f}\right) - \frac{r-1}{r+1} m\left(r, \frac{f}{f'}\right),$$
$$+ \{n(2r, f = 0) + \overline{n}(2r, f = \infty)\} \log(32er)$$
$$+ \left\{\frac{r+1}{r-1} + \frac{r-1}{r+1}\right\} \log(2r).$$

Moreover, according to formulas (3.157), (3.159) and (3.160), when r is sufficiently large, for a point z in the disk $|z| < 1$ and outside circles $(\gamma)'$, we have

$$\log\left|\frac{f'(z)}{f(z)}\right| \le -\frac{\delta}{4} T(r, f) + O(r^{\lambda'+\eta} \log r),$$

where $0 < \eta < \frac{1}{4}(\lambda - \lambda')$. Particularly, we select a sequence r_n, $r_n < r_{n+1} \to +\infty$, $(n \to +\infty)$, such that $T(r_n, f) \geq r_n^{\lambda - \eta}$. Hence, when $n \to +\infty$, it follows that

$$\log \left| \frac{f'(z)}{f(z)} \right| \leq -\frac{\delta}{4} T(r_n, f) + O(r_n^{\lambda' + \eta} \log r_n) \to -\infty.$$

Consequently, we conclude that $f'(z)/f(z)$ is identically zero, meaning that $f(z)$ is a constant. But this contradicts with the order $\lambda > 0$. This, in turn, proves that when $q = 0$, $f(z)$ cannot possess a nonzero finite deficient value, meaning that Theorem 3.10 holds.

Now, we consider the general case: $0 < q < +\infty$. Let a_ν $(\nu = 1, 2, \ldots, p)$ be p nonzero finite deficient values of $f(z)$, with their corresponding deficiencies being $\delta(a_\nu, f) = \delta_\nu > 0$. Let

$$\omega = \min_{1 \leq k \leq q} (\theta_{k+1} - \theta_k),$$
$$\delta = \min_{1 \leq \nu \leq p} \{\delta_\nu\},$$
$$d = \min_{1 \leq \nu \neq \nu' \leq p} \{|a_\nu - a_\nu'|\},$$
$$|a| = \max_{1 \leq \nu \leq p} \{|a_\nu|\},$$
$$|c| = \min_{1 \leq \nu \leq p} \{|c_\nu|\},$$

where c_ν is the first nonzero coefficient of the expansion of $f(z) - a_\nu$ in a neighborhood of the origin $z = 0$. We arbitrarily select numbers h $(0 < h + \infty)$ and h_1 $(0 < h_1 < h)$; then according to the definition of order and Lemma 2.3, there exists a sequence $\{R_n\}$, such that

$$\lim_{n \to +\infty} \frac{\log T(R_n, f)}{\log R_n} = \lambda,$$

and

$$T(R_n e^h, f) \leq e^{h(1+h_1/h)\lambda} T(R_n, f)(1 + o(1)) \qquad (n \to +\infty),$$
$$T(R_n e^{h_1}, f) \leq e^{h_1(1+h_1/h)\lambda} T(R_n, f)(1 + o(1)) \qquad (n \to +\infty).$$

We also select arbitrarily number σ, $0 < \sigma < \frac{1}{5} h_1$, and let $k = h(1 + \frac{h_1}{h})\lambda$, $k_1 = h_1(1 + \frac{h_1}{h})\lambda$. Then there exists a positive integer n_0, such that when $n \geq n_0$, we obtain

$$R_n \geq 1,$$

$$T(R_n e^h, f) \leq e^k T(R_n, f)(1 + o(1)) \leq 2e^k T(R_n, f), \qquad (3.162)$$
$$T(R_n e^{h_1}, f) \leq e^{k_1} T(R_n, f)(1 + o(1)) \leq 2e^{k_1} T(R_n, f),$$

and when $R \geq R_n$,

$$\frac{\delta}{2} T(R, f) \leq m(R, a_\nu), \qquad \nu = 1, 2, \ldots, p,$$

$$T\left(R, \frac{1}{f - a_\nu}\right) \leq 2T(R, f), \qquad \nu = 1, 2, \ldots, p,$$

$$\frac{1}{T(R, f)} \left\{ 10 \log 2 + \log^+ \log^+ |a| + \log^+ \log^+ \frac{1}{|c|} \right.$$

$$+ \frac{1}{2} h_1 + 3k_1 + 3 \log^+ \frac{1}{e^{h_1/2} - e^\sigma}$$

$$+ \log \frac{4(p+1)e^\sigma}{e^\sigma - 1} + 2 \log \frac{2(2p+1)}{h_1}$$

$$\left. + 2 \log R + 3 \log^+ T(R, f) \right\} < \frac{\delta}{8}.$$

Hence, based on Lemma 3.9, we conclude that there exist, in the interval $[R_n, R_n e^\sigma]$, a value t_n and a corresponding set $E_\nu(t_n)$ $(1 \leq \nu \leq p)$ of values θ $(0 \leq \theta < 2\pi)$, such that when $\theta \in E_\nu(t_n)$, we have

$$\log \frac{1}{|f(t_n e^{i\theta}) - a_\nu|} \geq \frac{\delta}{4} T(t_n, f),$$

and $\operatorname{meas} E_\nu(t_n) \geq M = M(\delta, h_1, \lambda, \sigma) > 0$, where

$$M(\delta, h_1, \lambda, \sigma) = \frac{\pi \delta}{8 e^{2h_1 \lambda} \left\{ \frac{e^{h_1/2} + e^\sigma}{e^{h_1/2} - e^\sigma} + \frac{2}{h_1} \log \frac{16(p+1)e^{1+h_1/2}}{e^\sigma - 1} \right\}}.$$

We select arbitrarily number ε, $0 < \varepsilon < \min\{M/8q, \omega/2\}$. Then there exists at least one set $E_\nu(t_n) \cap [\theta_{k_\nu} + 2\varepsilon, \theta_{k_\nu+1} - 2\varepsilon]$ $(1 \leq k_\nu \leq q)$ among the q sets $E_\nu(t_n) \cap [\theta_k + 2\varepsilon, \theta_{k+1} - 2\varepsilon]$ $(k = 1, 2, \ldots, q, \theta_{q+1} = \theta_1 + 2\pi)$, with the following linear measure:

$$\operatorname{meas}\{E_\nu(t_n) \cap [\theta_{k_\nu} + 2\varepsilon, \theta_{k_\nu+1} - 2\varepsilon]\} \geq \frac{M}{2q}.$$

Moreover, according to formula (3.156), we conclude that

$$\varlimsup_{r \to +\infty} \frac{\log^+ n\{\Omega(\theta_{k_\nu} + \varepsilon, \theta_{k_\nu+1} - \varepsilon; r), f = X\}}{\log r} \leq \lambda' < +\infty, \qquad X = 0, \infty.$$

Finally, we apply Lemma 3.13, where we let $N_n = \frac{\delta}{4} T(t_n, f)$, $\alpha = \frac{M}{2q}$, $R_{n1} = R_n e^{-Q}$, $R_{n2} = R_n e^Q$, $Q > \max\{\frac{\varepsilon}{16}, \sigma\}$, and notice that

$$\lim_{n \to +\infty} \left\{ \left(\frac{R_{n2}}{R_{n1}}\right)^{6+2(\lambda'+\eta_0)+3\pi/\theta} R_{n2}^{\lambda'+2\eta_0} \log R_{n2} \right\} N_n^{-1}$$

$$\leq \lim_{n \to +\infty} \left\{ \frac{O(R_n^{\lambda'+2\eta_0} \log R_n)}{T(R_n, f)} \right\} = 0$$

where $0 < \eta_0 < \frac{1}{4}(\lambda - \lambda')$. Then when n is sufficiently large, for point z in $\overline{\Omega}(\theta_{k_\nu} + 2\varepsilon, \theta_{k_\nu + 1} - 2\varepsilon, R_n e^{-Q}, R_n e^Q)$ and outside circles (γ),

$$\log \left| \frac{1}{|f(z) - a|} \right| \geq A(\alpha, \varepsilon, Q, \sigma, \delta) T(t_n, f) > \log \frac{4}{d}, \qquad (3.163)$$

where the sum of radii of (γ) does not exceed $\frac{1}{8}\varepsilon R_n e^{-Q} < (e^Q - e^{-Q})R_n$ and $A(\alpha, \varepsilon, Q, \sigma, \delta) > 0$ is a constant unrelated to n. Hence, for the sufficiently large n, each deficient value a_ν corresponds to each set $\overline{\Omega}(\theta_{k_\nu} + 2\varepsilon, \theta_{k_\nu + 1} - 2\varepsilon; R_n e^{-Q}, R_n e^Q)$, and p deficient values correspond to p sets $\overline{\Omega}(\theta_{k_\nu} + 2\varepsilon, \theta_{k_\nu + 1} - 2\varepsilon; R_n e^{-Q}, R_n e^Q)$ $(\nu = 1, 2, \ldots, p)$. According to formula (3.163), we conclude that these p sets have no intersection among one another, and it follows that $p \leq q$. Hence, Theorem 3.10 is proved completely.

According to the hypothesis in Theorem 3.10, if we further assume that $f(z)$ is an entire function, and apply Lemma 3.14 instead of Lemma 3.13 in the proof of Theorem 3.10, then we can conclude that when n is sufficiently large, for any point z in $\overline{\Omega}(\theta_{k_\nu} + 2\varepsilon, \theta_{k_\nu + 1} - 2\varepsilon; R_n e^{-Q}, R_n e^Q)$:

$$|f(z) - a_\nu| \leq \exp\{-A(\alpha, \theta, \varepsilon, Q, \sigma, \delta) T(t_n, f)\} < \frac{d}{4} \qquad (3.164)$$

and there are no intersections among these p sets

$$\{\overline{\Omega}(\theta_{k_\nu} + 2\varepsilon, \theta_{k_\nu + 1} - 2\varepsilon; R_n e^{-Q}, R_n e^Q) | \nu = 1, 2, \ldots, p\}.$$

Again by applying Lemma 3.14, where we let $N_n = \frac{\delta}{4} T(t_n, f)$, $\alpha = \frac{M}{2q}$, $R_{n1} = R_n e^{-h+\sigma}$, $R_{n2} = R_n e^{h-\sigma}$ $(h > h_1)$. Notice that

$$\lim_{n \to +\infty} \left\{ \left(\frac{R_{n2}}{R_{n1}} \right)^{6 + 2(\lambda' + \eta_0) + 3\pi/\theta} R_{n2}^{\lambda' + 2\eta_0} \log R_{n2} \right\} N_n^{-1}$$

$$\leq \lim_{n \to +\infty} \left\{ \frac{O(R_n^{\lambda' + 2\eta_0} \log R_n)}{T(R_n, f)} \right\} = 0,$$

where $0 < \eta_0 < \frac{1}{4}(\lambda - \lambda')$; then when n is sufficiently large, for any point z in the domain $\overline{\Omega}(\theta_{k_\nu} + 2\varepsilon, \theta_{k_\nu + 1} - 2\varepsilon; R_n e^{-h+\sigma}, R_n e^{h-\sigma})$,

$$|f(z) - a_\nu| \leq \exp\{-A(\alpha, \varepsilon, h, \sigma, \delta) T(t_n, f)\}$$

where $A(\alpha, \varepsilon, \theta, h, \sigma, \delta) > 0$ is a constant unrelated to n. Furthermore, when $z \in \overline{\Omega}(\theta_{k_\nu} + 2\varepsilon, \theta_{k_\nu + 1} - 2\varepsilon; R_n e^{-h+\sigma}, R_n e^{h-\sigma})$ we get

$$|f(z)| \leq |f(z) - a_\nu| + |a_\nu| < 1 + |a| \qquad (\nu = 1, 2, \ldots, p). \qquad (3.165)$$

Without any loss of generality, we may assume that $\theta_{k_\nu} < \theta_{k_\nu + 1}$ $(\nu = 1, 2, \ldots, p)$. In the following, we prove

$$(\theta_{k_{\nu+1}} + 2\varepsilon) - (\theta_{k_\nu + 1} - 2\varepsilon) \geq \frac{\pi}{\lambda} \qquad (1 \leq \nu \leq p). \qquad (3.166)$$

In fact, otherwise, it would lead to

$$(\theta_{k_{\nu+1}} + 2\varepsilon) - (\theta_{k_{\nu}+1} - 2\varepsilon) < \frac{\pi}{\lambda}(1 - \eta), \qquad \eta > 0. \tag{3.167}$$

If in the proof of Theorem 3.10 we assume $Q \geq 3\pi$, then according to Theorem 3.2 we should obtain $\{N + |a|\}\{\varepsilon_1^{1/3} + \varepsilon_2^{1/3}\} \geq d$, where N is the largest value of $|f(z)|$ on $\overline{\Omega}(\theta_{k_{\nu}+1} - 2\varepsilon, \theta_{k_{\nu+1}} + 2\varepsilon; R_n e^{-Q}, R_n e^Q)$, and also

$$\varepsilon_1 = \varepsilon_2 = \exp\{-A(\alpha, \varepsilon, \theta, h, \sigma, \delta)T(t_n, f)\}.$$

Suppose that $|f(z_\nu)| = N$ for some z_ν on

$$\overline{\Omega}(\theta_{k_{\nu}+1} - 2\varepsilon, \theta_{k_{\nu+1}} + 2\varepsilon, R_n e^{-Q}, R_n e^Q).$$

Hence, when n is sufficiently large, we conclude that

$$|f(z_\nu)| \geq \frac{d}{2}\exp\left\{\frac{1}{3}A(\alpha, \varepsilon, \theta, Q, \sigma, \delta)T(t_n, f)\right\} - |a| \tag{3.168}$$

$$\log|f(z_\nu)| \geq \frac{1}{4}A(\alpha, \varepsilon, \theta, Q, \sigma, \delta)T(t_n, f) > |a| + 1.$$

On the other hand, according to formula (3.165), when

$$z \in \Delta(\theta_{k_{\nu}+1} - 2\varepsilon; R_n e^{-h+\sigma}, R_n e^{h-\sigma}) \cup \Delta(\theta_{k_{\nu+1}} + 2\varepsilon; R_n e^{-h+\sigma}, R_n e^{h-\sigma}),$$

we get $|f(z)| < 1 + |a|$. Hence, point

$$z_\nu \in \Omega(\theta_{k_{\nu}+1} - 2\varepsilon, \theta_{k_{\nu+1}} + 2\varepsilon; R_n e^{-h+\sigma}, R_n e^{h-\sigma}).$$

Now, an application of Lemma 2.10 yields

$$\log|f(z_\nu)| \leq \log(1 + |a|)$$
$$+ \frac{4(\frac{R_n e^{-h+\sigma}}{|z_\nu|})^{\pi/(\theta_{k_{\nu+1}} - \theta_{k_{\nu}+1} + 4\varepsilon)}}{\pi\{1 - (\frac{R_n e^{-h+\sigma}}{|z_\nu|})^{2\pi/(\theta_{k_{\nu+1}} - \theta_{k_{\nu}+1} + 4\varepsilon)}\}}\log M(R_n e^{-h+\sigma}, f)$$
$$+ \frac{4(\frac{|z_\nu|}{R_n e^{h-\sigma}})^{\pi/(\theta_{k_{\nu+1}} - \theta_{k_{\nu}+1} + 4\varepsilon)}}{\pi\{1 - (\frac{|z_\nu|}{R_n e^{h-\sigma}})^{2\pi/(\theta_{k_{\nu+1}} - \theta_{k_{\nu}+1} + 4\varepsilon)}\}}\log M(R_n e^{h-\sigma}, f). \tag{3.169}$$

Moreover, according to formulas (1.21) and (3.162), we get

$$\log M(R_n e^{-h+\sigma}, f) \leq \log M(R_n e^{h-\sigma}, f)$$
$$\leq \frac{R_n e^h + R_n e^{h-\sigma}}{R_n e^h - R_n e^{h-\sigma}}T(R_n e^h, f) \tag{3.170}$$
$$\leq \frac{e^\sigma + 1}{e^\sigma - 1}2e^{h(1 + h_1/h)\lambda}T(t_n, f).$$

And consequently, by formulas (3.167), (3.168), (3.169) and (3.170) we obtain

$$\frac{1}{4}A(\alpha, \varepsilon, \theta, Q, \sigma, \delta)T(t_n, f)$$

$$\leq \log(1 + |a|)$$

$$+ \frac{4e^{(-h+Q+\sigma)\lambda/(1-\eta)}}{\pi\{1 - e^{(-h+Q+\sigma)2\lambda/(1-\eta)}\}} 2 \cdot \frac{e^{\sigma} + 1}{e^{\sigma} - 1} e^{h(1+h_1/h)\lambda} T(t_n, f)$$

$$+ \frac{4e^{(-h+Q+\sigma)\lambda/(1-\eta)}}{\pi\{1 - e^{(-h+Q+\sigma)2\lambda/(1-\eta)}\}} 2 \frac{e^{\sigma} + 1}{e^{\sigma} - 1} e^{h(1+h_1/h)\lambda} T(t_n, f).$$

It follows by letting $n \to +\infty$,

$$\frac{1}{4}A(\alpha, \varepsilon, \theta, Q, \sigma, \delta)$$

$$\leq \frac{16(e^{\sigma} + 1)}{\pi\{1 - e^{(-h+Q+\sigma)2\lambda/(1-\eta)}\}} \exp\left\{ -\frac{\eta\lambda h}{1 - \eta} + \frac{(Q + \sigma)\lambda}{1 - \eta} + h_1\lambda \right\}.$$

But if we assume h sufficiently large, then the above formula may not be valid, meaning that formula (3.166) must hold. The validity of formula (3.166) further implies that p must be $\leq \frac{q}{2}$ and p must be $< 2\lambda$ and, in turn, we have proved the following result:

THEOREM 3.11. *Let $f(z)$ be an entire function of order λ $(0 < \lambda < +\infty)$ and $\Delta(\theta_k)$ $(k = 1, 2, \ldots, q; \ 0 \leq \theta_1 < \theta_2 < \cdots < \theta_q < 2\pi, \ \theta_{q+1} = \theta_1 + 2\pi)$ be q $(0 \leq q < +\infty)$ half straight lines on the z-plane. And suppose that for any arbitrary small number $\varepsilon > 0$,*

$$\varlimsup_{r \to +\infty} \frac{\log^+ n\{\bigcup_{k=1}^q \Omega(\theta_k + \varepsilon, \theta_{k+1} - \varepsilon; r), f = 0\}}{\log r} < \lambda.$$

If p denotes the number of nonzero finite deficient values of $f(z)$, then $p \leq \frac{q}{2}$ and $p < 2\lambda$.

Now, we may derive directly from Theorem 3.10 and Theorem 3.11 the following two results due to Yang-Le and Zhang Guanghou [42b, 42c].

THEOREM 3.12. *Let $f(z)$ be a meromorphic function of order λ $(0 < \lambda < +\infty)$ on the open plane $|z| < +\infty$, q be the number of Borel directions of order λ of $f(z)$, and p be the number of deficient values. Then $p \leq q$.*

THEOREM 3.13. *Let $f(z)$ be an entire function of order $\lambda(0 < \lambda < +\infty)$ on the open plane $|z| < \infty$, q be the number of Borel directions of order λ, and p be the number of finite deficient values. Then $p \leq \frac{q}{2}$, moreover, when $q < +\infty, p < 2\lambda$.*

Indeed, according to Theorem 2.14 and the finite covering theorem, we conclude that there exist two distinct complex numbers b and c, such that both b and c are distinct from the p deficient values a_ν $(\nu = 1, 2, \ldots, p)$, and also for any arbitrarily small number $\varepsilon > 0$, we have

$$\varlimsup_{r \to +\infty} \frac{\log^+ n\{\bigcup_{k=1}^q \Omega(\theta_k + \varepsilon, \theta_{k+1} - \varepsilon; r), f = X\}}{\log r} \leq \lambda' < \lambda, \qquad X = a, b.$$

We consider the transformation

$$F(z) = \frac{f(z) - b}{f(z) - c}.$$

Then $F(z)$ has p nonzero finite deficient values $(a_\nu - b)/(a_\nu - c)$ $(\nu = 1, 2, \ldots, p)$. Moreover, for any arbitrarily small numbers $\varepsilon > 0$,

$$\varlimsup_{r \to +\infty} \frac{\log^+ n\{\bigcup_{k=1}^q \Omega(\theta_k + \varepsilon, \theta_{k+1} - \varepsilon; r), F = X\}}{\log r} \leq \lambda' < \lambda, \qquad X = 0, \infty.$$

Hence, according to Theorem 3.10, we conclude that Theorem 3.12 is valid.

If $f(z)$ is an entire function, then we take $c = \infty$ and consider the transformation $F(z) = f(z) - b$. Consequently, according to Theorem 3.11, we conclude that Theorem 3.13 holds.

3.5.3. Improvement of the Edrei-Fuchs Theorem.

LEMMA 3.15. *Let $f(z)$ be a meromorphic function of lower order $\mu < +\infty$ on the open plane $|z| < +\infty$, and let it satisfy the following conditions:*

(1) On $\overline{\Omega}(-\theta, \theta)$ $(0 < \theta \leq \pi)$ there exists

$$\varlimsup_{r \to +\infty} \frac{\log^+ n\{\overline{\Omega}(-\theta, \theta; r), f = X\}}{\log r} \leq \mu' < +\infty, \qquad X = 0, \infty.$$

(2) For any selected number $\sigma > 0$ and sufficiently large selected number h $(0 < h < +\infty)$, there exist two sequences $\{R_n\}$ and $\{t_n\}$, such that

$$\lim_{n \to +\infty} \frac{\log T(R_n, f)}{\log R_n} = \mu, \tag{3.171}$$

$$T(R_n e^h, f) \leq e^K T(R_n, f)(1 + o(1))$$
$$\leq 2e^K T(R_n, f) \qquad (n \to +\infty), \tag{3.172}$$

$$K \leq 2h\mu,$$
$$R_n \leq t_n \leq R_n e^\sigma, \qquad n = 1, 2, \ldots. \tag{3.173}$$

Moreover, for a selected number ε, $0 < \varepsilon < \theta$, the set

$$E_n = \left\{ z \mid \log \frac{1}{|f(z) - a|} \geq N_n, \ |z| = t_n, \ -\theta + \varepsilon \leq \arg z \leq \theta - \varepsilon \right\}$$

has a linear measure $\operatorname{meas} E_n \geq \alpha R_n$ $(\alpha \geq \frac{\varepsilon}{2})$, where a is a nonzero finite complex number, $N_n > 0$ is a real number, and

$$\lim_{n \to +\infty} (R_n^{\mu' + 2\eta_0} \log R_n) N_n^{-1} = 0,$$

where $\eta_0 > 0$ is some selected real number. Then, under these hypotheses, provided that n is sufficiently large, for any point z in

$$\overline{\Omega}(-\theta + \varepsilon, \theta - \varepsilon; R_n Q^{-1}, R_n Q) \; (Q \ge e^{\sigma}) \, (^{12})$$

and outside circles (γ), we have

$$\log \frac{1}{|f(z) - a|} > A(\alpha, \varepsilon, \theta, Q) N_n,$$

where the sum of radii of (γ) does not exceed $\frac{1}{8}\varepsilon R_n$, the number of circles in (γ) is finite but depends on n, and $A(\alpha, \varepsilon, \theta, Q) > 0$ is a constant depending on α, ε, θ and Q, but is unrelated to n.

Indeed, the proof of this lemma is basically a repetition of the proof of Lemma 3.13. In the proof of Lemma 3.13, we need only to replace λ' by μ', and let $R_{n1} = R_n Q^{-1}$, $R_{n2} = R_n Q$ and $R_{n3} = (\frac{8\theta}{\varepsilon})^{2\theta/\pi} Q^5 R_n$, as well as to change R_n into t_n from formulas (3.146) to (3.139). However, the key changes are made in formulas (3.137) and (3.151), where we use order $\lambda < +\infty$. Here, we assume h sufficiently large, such that

$$h > \log \left\{ 4 \cdot \left(\frac{8\theta}{\varepsilon} \right)^{2\theta/\pi} Q^5 \right\}.$$

Then when n is sufficiently large, and according to formulas (3.170), (3.171) and (3.172), we have

$$T(2R_{n3}, f) \le T(4R_{n3}, f) \le T(R_n e^h, f)$$
$$\le 2e^K T(R_n, f) \le 2e^{2h\mu} R_n^{\mu + \eta}, \qquad \eta > 0.$$

Hence, we may replace formulas (3.137) and (3.141) by this formula. It follows that formulas (3.138) and (3.152) remain valid.

Corresponding to Lemma 3.14 we have

LEMMA 3.16. *Let $f(z)$ be an entire function of lower order $\mu < +\infty$, and let it satisfy the following conditions:*

(1) On $\overline{\Omega}(-\theta, \theta)$ $(0 < \theta \le \pi)$ there exists

$$\overline{\lim_{r \to +\infty}} \frac{\log^+ n\{\Omega(-\theta, \theta; r), f = 0\}}{\log r} \le \mu' < +\infty.$$

(2) For some selected number $\sigma > 0$ and sufficiently large h, $0 < h < +\infty$, there exist two sequences $\{R_n\}$ and $\{t_n\}$ such that

$$\lim_{n \to +\infty} \frac{\log T(R_n, f)}{\log R_n} = \mu,$$

$$T(R_n e^h, f) \le e^K T(R_n, f)(1 + o(1)) \le 2e^K T(R_n, f) \qquad (n \to +\infty),$$

$(^{12})$ $Q \ge e^{\sigma}$ guarantees that $t_n \le R_n Q$.

$$K \leq 2h\mu,$$
$$R_n \leq t_n \leq R_n e^{\sigma}, \qquad n = 1, 2, \ldots,$$

and for some selected number ε, $0 < \varepsilon < \theta$, *the set*

$$E_n = \left\{ z \mid \log \frac{2}{|f(z) - a|} \geq N_n, |z| = t_n, -\theta + \varepsilon \leq \arg z \leq \theta - \varepsilon \right\}$$

has linear measure $\operatorname{meas} E_n \geq \alpha R_n$ ($\alpha \geq \frac{\varepsilon}{2}$), *where* a *is a nonzero finite complex number*, $N_n > 0$ *is a real number, and*

$$\lim_{n \to +\infty} (R_n^{\mu' + 2\eta_0} \log R_n) N_n^{-1} = 0,$$

where $\eta_0 > 0$ *is some selected real number. Then under these hypotheses, provided that* n *is sufficiently large, for any point* z *on*

$$\overline{\Omega}(-\theta + \varepsilon, \theta - \varepsilon; R_n Q^{-1}, R_n Q) \ (Q > e^{\sigma}),$$

we have

$$|f(z) - a| \leq \exp\{-A(\alpha, \varepsilon, \theta, Q) N_n\},$$

where $A(\alpha, \varepsilon, \theta, Q) > 0$ *is a constant depending on* α, ε, θ *and* Q, *and is unrelated to* n.

Correspondingly, we may apply Lemma 3.15 and Lemma 3.16 to prove the following two results:

THEOREM 3.14. *Let* $f(z)$ *be a meromorphic function of lower order* μ ($0 < \mu < +\infty$) *on the open plane* $|z| < +\infty$, $\Delta(\theta_k)$ ($k = 1, 2, \ldots, q$; $0 \leq \theta_1 < \theta_2 < \cdots < \theta_q < 2\pi$, $\theta_{q+1} = \theta_1 + 2\pi$) *be* q ($0 \leq q < +\infty$) *half straight lines on the* z-*plane. Suppose that for any arbitrarily given small number* $\varepsilon > 0$,

$$\overline{\lim_{r \to +\infty}} \frac{\log^+ n\{\bigcup_{k=1}^q \Omega(\theta_k + \varepsilon, \theta_{k+1} - \varepsilon; r), f = X\}}{\log r} < \mu, \qquad X = 0, \infty.$$

If p *denotes the number of nonzero finite deficient values of* $f(z)$, *then* $p \leq q$.

THEOREM 3.15. *Let* $f(z)$ *be an entire function with lower order* μ ($0 < \mu < +\infty$) *on the open plane* $|z| < +\infty$, $\Delta(\theta_k)$ ($k = 1, 2, \ldots, q$; $0 \leq \theta_1 < \cdots < \theta_q$, $\theta_{q+1} = \theta_1 + 2\pi$) *be* q ($0 \leq q + \infty$) *half straight lines on the* z-*plane, and for any arbitrarily small number* $\varepsilon > 0$,

$$\overline{\lim_{r \to +\infty}} \frac{\log^+ n\{\bigcup_{k=1}^q \Omega(\theta_k + \varepsilon, \theta_{k+1} - \varepsilon; r), f = 0\}}{\log r} < \mu.$$

If p *denotes the number of nonzero finite deficient values of* $f(z)$, *then* $p \leq \frac{q}{2}$ *and* $p < 2\mu$.

Furthermore, we may deduce the following two results:

THEOREM 3.16. *Let $f(z)$ be a meromorphic function of lower order μ ($\mu <$ $+\infty$) on the open plane $|z| < +\infty$, and let it satisfy*

$$\varlimsup_{r \to +\infty} \frac{T(r, f)}{(\log r)^2} = +\infty. \tag{3.174}$$

We also assume that $\Delta(\theta_k)$ ($k = 1, 2, \ldots, q$; $0 \le \theta_1 < \theta_2 \le \cdots < \theta_q < 2\pi$, $\theta_{q+1} = \theta_1 + 2\pi$) are q ($0 \le q < +\infty$) half straight lines on the z-plane, and for any arbitrarily given small number $\varepsilon > 0$,

$$\varlimsup_{r \to +\infty} \frac{n\{\bigcup_{k=1}^q \Omega(\theta_k + \varepsilon, \theta_{k+1} - \varepsilon; r), f = X\}}{\log r} < +\infty, \qquad X = 0, \infty.$$

If p denotes the number of nonzero finite deficient values of $f(z)$, then $p \le q$.

THEOREM 3.17. *Let $f(z)$ be an entire function of lower order μ ($\mu < +\infty$) on the open plane $|z| < +\infty$, and let it satisfy*

$$\varlimsup_{r \to +\infty} \frac{T(r, f)}{(\log r)^2} = +\infty.$$

We also assume that $\Delta(\theta_k)$ ($k = 1, 2, \ldots, q$; $0 \le \theta_1 < \theta_2 < \cdots < \theta_q < 2\pi$, $\theta_{q+1} = \theta_1 + 2\pi$) are q ($0 \le q < +\infty$) half straight lines on the z-plane, and for any arbitrarily given small number $\varepsilon > 0$,

$$\varlimsup_{r \to +\infty} \frac{n\{\bigcup_{k=1}^q \Omega(\theta_k + \varepsilon, \theta_{k+1} - \varepsilon; r), f = 0\}}{\log r} < +\infty.$$

If p denotes the number of nonzero finite deficient values of $f(z)$, then $p \le \frac{q}{2}$ and $p < 2\mu$.

Indeed, we need only to prove that when $q = 0$, p must be equal to zero. This is because when $q \ge 1$, if $f(z)$ is a meromorphic function and $p \ge 2$, then according to Theorem 3.4, it follows that $\mu > 0$. Therefore, by Theorem 3.14, we conclude that $p \le q$. If $f(z)$ is an entire function and $p \ge 1$, then $\mu > 0$ holds according to Theorem 3.4. And hence, by Theorem 3.15, we conclude that $p \le \frac{q}{2}$ and $p < 2\mu$.

Now, we prove that when $q = 0$, then $p = 0$. First, according to Lemma 2.4, for any arbitrarily selected value h ($0 < h < +\infty$), there exists a sequence $\{R_n\}$, such that

$$\lim_{n \to +\infty} \frac{T(R_n, f)}{(\log R_n)^2} = +\infty, \tag{3.175}$$

$$T(R_n e^h, f) \le e^{h\mu} T(R_n, f)(1 + o(1))$$
$$\le 2e^{h\mu} T(R_n, f) \qquad (n \to +\infty). \tag{3.176}$$

On the other hand, since $q = 0$, we conclude that

$$\varlimsup_{r \to +\infty} \frac{n(r, f = X)}{\log r} < +\infty, \qquad X = 0, \infty.$$

Assume that $f(z)$ has a nonzero finite deficient value a with deficiency $\delta(a, f) = \delta > 0$; then according to the identical relation

$$\frac{1}{f(z) - a} = \frac{1}{a} \cdot \frac{f(z)}{f'(z)} \left\{ \frac{f'(z)}{f(z) - a} - \frac{f'(z)}{f(z)} \right\},$$

we derive

$$m\left(R_n, \frac{1}{f - a}\right) \leq \log^+ \frac{1}{|a|} + m\left(R_n, \frac{f}{f'}\right)$$
$$+ m\left(R_n, \frac{f'}{f - a}\right) + m\left(R_n, \frac{f'}{f}\right) + \log 2.$$

Furthermore, according to (3.176) and Lemma 1.3, where we let $r = R_n$, $\rho = R_n e^h$, it follows that

$$m\left(R_n, \frac{f'}{f - a}\right) \leq O\{\log[R_n T(R_n, f)]\},$$

$$m\left(R_n, \frac{f'}{f}\right) \leq O\{\log[R_n T(R_n, f)]\}.$$

And therefore,

$$m\left(R_n, \frac{f}{f'}\right) \geq \frac{\delta}{2} T(R_n, f) - O\{\log[R_n T(R_n, f)]\}.$$

In the following, similar to the discussion of $q = 0$ in the proof of Theorem 3.10, we may prove that when point z is in the disk $|z| \leq 1$ and outside circles $(\gamma)'$, we have

$$\log\left|\frac{f'(z)}{f(z)}\right| \leq -\frac{\delta}{4} T(R_n, f) + O\{(\log R_n)^2\},$$

where the sum of radii of $(\gamma)'$ does not exceed $\frac{1}{4}$. Hence, we conclude from formula (3.175) that $f'(z)/f(z)$ is identically zero, meaning that $f(z)$ is a constant. However, this contradicts formula (3.174). Thus, when $q = 0$, $f(z)$ cannot have any nonzero finite deficient value.

§3.6. Annotated notes

We have already made a relatively detailed discussion concerning the circumstances under which the number of deficient values should be finite. In this section, we shall, therefore, introduce briefly the important results of other aspects of the deficient value theory.[13]

3.6.1. Inverse problem. We have already proved in Chapter 1 that all deficient values $\{a\}$ of a nonconstant meromorphic function $f(z)$ on the open

---

[13] See [16b] for the important developments of the deficient value theory since Nevanlinna.

plane $|z| < +\infty$ constitute a countable set, with the sum of its deficiencies being

$$\Delta(f) = \sum_a \delta(a, f) \leq 2.$$

Then, in theory, we shall naturally raise the following inverse problem:

For any arbitrarily given set of distinct complex numbers a_i $(i = 1, 2, \ldots, q; q \leq +\infty)$ and any set of corresponding positive numbers δ_i, $0 < \delta_i \leq 1$ $(i = 1, 2, \ldots, q)$, if $\sum_i \delta_i \leq 2$ is satisfied as well, can we construct a meromorphic function $f(z)$, such that $f(z)$ only has the given set of complex numbers $\{a_i\}$ as its deficient values, with their corresponding deficiencies being $\delta(a_i, f) = \delta_i (i = 1, 2, \ldots, q)$?

For this significant and important inverse problem, ever since the efforts made by R. Nevanlinna, no striking results had been obtained, though some achievements had been made under special conditions. It was not until 1977 that D. Drasin outstandingly solved the problem by using the quasiconformal mapping theory. More concretely, he proved the following result [12b]:

Let $\{\delta_i\}$ and $\{\theta_i\}$ $(1 \leq i < N \leq +\infty)$ be two sequences of positive numbers, satisfying the following conditions:

$$0 < \delta_i + \theta_i \leq 1 \quad (1 \leq i \leq N), \qquad \sum_i (\delta_i + \theta_i) \leq 2.$$

We also assume that $\{a_i\}$ $(1 \leq i \leq N)$ is a sequence of complex numbers which satisfies the condition $a_i \neq a_j$ $(1 \leq i \neq j \leq N)$.

Under the above hypotheses, there must exist a meromorphic function $f(z)$ such that

$$\delta(a_i, f) = \delta_i, \theta(a_i, f) = \theta_i \quad (1 \leq i \leq N),$$
$$\delta(a, f) = \theta(a, f) = 0 \quad (a \notin \{a_i\})$$

Furthermore, if $\phi(r)$ is a positive, nondecreasing function, and when $r \to +\infty$, $\phi(r) \to +\infty$, then we may require that $f(z)$ satisfies the condition: When r is sufficiently large, $T(r, f) \leq r^{\phi(r)}$.

As for the entire functions, W. Fuchs and W. Hayman had already solved this inverse problem early in 1962 [17a].

In D. Drasin's result, as well as the result of W. Fuchs and W. Hayman, those meromorphic and entire functions $f(z)$ are of infinite order in general. If we require $f(z)$ to be of finite order, then, generally speaking, the above inverse problem is unsolvable. Indeed, A. Weitsman proved the following result in 1969:([14])

If $f(z)$ is a meromorphic function of finite lower order μ, and satisfies the condition $\sum_a \delta(a, f) = 2$, then $f(z)$ has at most 2μ deficient values.

In 1972 he also proved the following result:

([14]) See Theorem 3.8.

If $f(z)$ is a meromorphic function of finite lower order μ, then $\sum_a \delta^{1/3}(a, f) < +\infty$.

For the second result of A. Weitsman mentioned above, we illustrate that index $\frac{1}{3}$ is the best possible. Indeed, when $\alpha < \frac{1}{3}$, W. Hayman [21c] constructed a meromorphic function $f(z)$ of finite positive order, such that $\sum_a \delta^\alpha(a, f) = +\infty$.

Besides, we make an additional illustration: W. Hayman's result illustrates simultaneously that there exist meromorphic functions which are of finite positive order that have infinitely many deficient values. However, at earlier times. A. Gol'dberg had already constructed such examples [18a]. W. Hayman's result was, indeed, derived from A. Gol'dberg's method. For those entire functions, N. Arakelyan used the approximation theory and proved the following result [3a]:

For an arbitrarily selected number $\lambda > \frac{1}{2}$, there must exist an entire function $f(z)$ of order λ with infinitely many deficient values.

N. Arakelyan also conjectured: For an entire function $f(z)$ of finite order, it is true that

$$\sum_a \{\log(1/\delta(a, f))\}^{-1} < +\infty.$$

3.6.2. Spread relation.

In 1965, A. Edrei introduced a very useful concept, that is, the so-called concept of Pólya peak [14a]: Let $T(r)$ be a positive, nondecreasing continuous function tending to ∞ on the interval $[t_0, +\infty)$ $(t_0 \geq 0)$. The sequence $\{r_n\}$ is called a sequence of Pólya peak of $T(r)$ of order ρ $(0 \leq \rho < +\infty)$, if there exist two sequences $\{r_n'\}$ and $\{r_n''\}$, such that

$$r_n' \to +\infty, \quad \frac{r_n}{r_n'} \to +\infty, \quad \frac{r_n''}{r_n} \to +\infty \qquad (n \to +\infty),$$

and

$$\frac{T(r)}{T(r_n)} \leq \left\{\frac{r}{r_n}\right\}^\rho (1 + o(1)) \qquad (n \to +\infty, \ r_n' \leq r \leq r_n'').$$

A. Edrei proved the following result:

Suppose the lower order and order of $T(r)$ is μ $(\mu < +\infty)$ and λ $(\lambda \leq +\infty)$, respectively. Then for each finite number ρ, $\mu \leq \rho \leq \lambda$, there must exist a sequence of Pólya peak of $T(r)$ of order ρ.

Now, we introduce a famous conjecture made by A. Edrei, that is, the so-called Spread relation:

Let $f(z)$ be a meromorphic function of finite lower order μ on the open plane $|z| < +\infty$. Then there exists a sequence of Pólya peak $\{r_n\}$ of order μ. We also assume that $\Lambda(r)$ is a positive function, and satisfies the condition

$$\Lambda(r) = o\{T(r, f)\} \qquad (r \to +\infty). \tag{3.177}$$

We define

$$E_\Lambda(r, a) = \begin{cases} E\{\theta \mid |f(re^{i\theta}) - a| < e^{-\Lambda(r)}, \ -\pi < \theta \le \pi\}, & a \ne \infty, \\ E\{\theta \mid |f(re^{i\theta})| > e^{\Lambda(r)}, \ -\pi < \theta \le \pi\}, & a = \infty, \end{cases}$$

and let

$$\sigma_\Lambda(a) = \varliminf_{r \to +\infty} \text{meas}\, E_\Lambda(r, a).$$

Then the spread relation regarding complex number a is $\sigma(a) = \inf_\Lambda \sigma_\Lambda(a)$, Here the greatest lower bound is taken over all $\Lambda(r)$, which satisfy the condition of formula (3.177).

In 1967 A. Edrei conjectured that the spread relation [14a]

$$\sigma(a) \ge \min\left\{ 2\pi, \frac{4}{\mu} \sin^{-1} \sqrt{\frac{\delta(a, f)}{2}} \right\}$$

holds.

In 1973 A. Baernstein II proved the validity of this conjecture. His proof was based on the introduction of the so-called $T^*(z)$ functions, with its definition as follows:

Let $f(z)$ be a meromorphic function $\not\equiv 0$. Set

$$m^*(z) = \sup_E \frac{1}{2\pi} \int_E \log|f(re^{i\varphi})|\, d\varphi \qquad (z = re^{i\theta}, \ 0 < r < +\infty, \ 0 \le \theta \le \pi).$$

Here the least upper bound is taken over all the sets $E \subset (-\pi, \pi)$, which satisfy the condition $\text{meas}\, E = 2\theta$. Then we define $T^*(z) = m^*(z) + N(|z|, f)$.

A. Baernstein II proved that $T^*(z)$ is a subharmonic function on the upper half-plane $\text{Im}\, z > 0$, and it is a continuous function on H; $H = \{z \mid \text{Im}\, z \ge 0, \ z \ne 0\}$.

The importance of $T^*(z)$ is shown not only in the proof of Edrei's conjecture, but also in the proof of several other extreme problems [4b]. Hence, the function $T^*(z)$ is a very powerful tool, and the efforts made by A. Baernstein are of great significance.

There are several important applications of the spread relation. For example, we may derive the following famous Elliptic Theorem worked out by Edrei–Fuchs according to the spread relation [15c].

Suppose that $f(z)$ is meromorphic in the open plane $|z| < +\infty$ and of lower μ, $0 \le \mu \le 1$. Assume that the two distinct complex numbers a and b are the deficient values of $f(z)$. We also denote

$$\mu = 1 - \delta(a, f), \qquad \nu = 1 - \delta(b, f).$$

Then

$$u^2 + v^2 - 2uv \cos\mu\pi \ge \sin^2\mu\pi,$$

Moreover, when $\mu \le \cos\mu\pi$, $v = 1$, and when $v \le \cos u\pi$, $u = 1$.

Besides, according to the spread relation, we may solves the deficiency problem of meromorphic functions when their lower orders μ are ≤ 1. In

1973, A. Edrei proved the following theorem by means of the spread relation [14b]:

Let $f(z)$ be a meromorphic function of lower order μ on the open plane $|z| < +\infty$. If $0 < \mu \leq \frac{1}{2}$, then the sum of the deficiencies of $f(z)$ is

$$\sum_a \delta(a, f) < 1 - \cos \mu\pi.$$

However, there is an exceptional case; that is, when $f(z)$ has only one deficient value, its corresponding deficiency may assume any value on the interval $[0, 1]$; if $\frac{1}{2} \leq \mu \leq 1$, then the sum of deficiencies of $f(z)$ is

$$\sum_a \delta(a, f) \leq 2 - \sin \mu\pi,$$

where the "=" sign holds only when $f(z)$ has just two deficient values a_1 and a_2, with $\delta(a_1, f) = 1$ and $\delta(a_2, f) = 1 - \sin \mu\pi$.

When $\lambda > 1$, the corresponding deficiency problem remains unsolved and is a problem that is worth studying.

For any meromorphic function $f(z)$ of order $\lambda > 1$, in 1975 D. Drasin and A. Weitsman made the following conjecture about the sum of its deficiencies [13a]:

$$\sum_a \delta(a, f) \leq \Lambda = \max\{\Lambda_1, \Lambda_2\},$$

where

$$\Lambda_1 = 2 - \frac{2 \sin(\pi/2)\{2\lambda\}}{[2\lambda] + 2 \sin(\pi/2)\{2\lambda\}}, \qquad \Lambda_2 = 2 - \frac{2 \cos(\pi/2)\{2\lambda\}}{[2\lambda] + 1}.$$

Here $[x]$ indicates the integral part of x and $\{x\} = x - [x]$.

The also exhibited examples to show that the upper bound Λ is sharp.

3.6.3. F. Nevanlinna's Conjecture.

Through the examination of some examples, F. Nevanlinna made the following conjecture in 1930 [31a]:

Let $f(z)$ be a meromorphic function of order λ on the open plane $|z| < +\infty$. If the sum of the deficiencies of f is $\sum_a \delta(a, f) = 2$, then the following properties hold:

(1) The order α of $f(z)$ is ≥ 1 and is an integral multiple of $\frac{1}{2}$;

(2) The number of deficient values of $f(z)$ is $\leq 2\lambda$;

(3) The deficiency of each deficient value of $f(z)$ is an integral multiple of $\frac{1}{\lambda}$.

In 1964, A. Pfluger considered the entire functions and proved the following result [34a]:

Let $f(z)$ be an entire function of order $\lambda < +\infty$. If the sum of the deficiencies of f is $\sum_a \delta(a, f) = 2$, then the following properties hold:

(1) λ is a positive integer;

(2) The deficiency of a deficient value of $f(z)$ is an integral multiple of $\frac{1}{\lambda}$;

(3) The number of finite deficient values of $f(z)$ is $\leq \lambda$.

In 1959, A. Edrei and W. Fuchs obtained a supplementary result by proving the following property [15a]:

(4) Each deficient value of $f(z)$ is at the same time an asymptotic value.

In the sequels, when we refer to F. Nevanlinna's conjecture, we include property (4) proved by A. Edrei and W. Fuchs. In 1969 A. Weitsman outstandingly proved property (2) and, in turn, solved parts of the conjecture. For a long time, it has been difficult to solve this conjecture completely. Recently, D. Drasin in his excellent paper [12c], again by the application of the quasiconformal mapping theory, proved successfully the validity of F. Nevanlinna's conjecture.

CHAPTER 4

The Asymptotic Value Theory

The main theme of this chapter is to introduce the fundamental theory of the asymptotic value and some of its new results. The asymptotic value theory is an important component of the theory of entire and meromorphic functions. It has a close relationship with both the value distribution theory and the inverse function theory of the entire and meromorphic functions. Meanwhile, the study of the asymptotic value theory has given an impetus to the development of the geometric function in history.

§4.1. The asymptotic value and the transcendental singularity

4.1.1. The fundamental concept [32b, 39a]. Let $w = f(z)$ be a transcendental entire function or a meromorphic function on the open plane $|z| < +\infty$; $z = \varphi(w)$ be an inverse function of $f(z)$, and F be the Riemann surface covering the w-plane as defined by $\varphi(w)$, meaning that F is the Riemann surface obtained when an analytic element of $\varphi(w)$ undergoes analytic continuation of algebraic characters([1]) along all possible paths on the w-plane. Hence, $\varphi(w)$ is single-valued on F and maps F conformally onto the open plane $|z| < +\infty$. Therefore, F is a simply connected Riemann surface of parabolic-type. We denote (a) as a point on F and a the projection of (a) on the w-plane. We also denote (L_w) as a continuous curve on F and L_w the projection of (L_w) on the w-plane; L_w is a continuous curve. We call L_w or (L_w) a curve tending to a or (a), if a or (a) is an endpoint of L_w or (L_w). However, a or (a) may not belong to this curve L_w or (L_w). In the sequel, we shall prove that each boundary point of the parabolic-type Riemann surface F is an accessible boundary point (see Theorem 4.1), implying that for each boundary point (a) of F, there exists a curve (L_w) tending to (a) on F, or we can say that there is a curve L_w tending to a on the w-plane. Meanwhile, there exists an analytic element $\varphi_0(w)$ of $\varphi(w)$, such that $\varphi_0(w)$ undergoes continuation along L_w and ends at point a, where a is a nonalgebraic singularity of this continuation. We call such

---

([1]) An analytic continuation of algebraic character means that the expansion of each element of $\varphi(w)$ on the w-plane can assume the following form: $\sum_{n=n_0}^{+\infty} c_n(w - w_0)^{n/k}$ where n_0 is an integer and k is a positive integer. The continuation talked throughout this chapter has such an implication.

a singularity a a transcendental singularity defined by element $\varphi_0(w)$ and L_w. Thereafter, we use $\varphi_0(w)$ to denote not only an initial element $\varphi_0(w)$, but also any element obtained when $\varphi_0(w)$ undergoes continuation along L_w. Let $\varphi_1(w)$ be an element of $\varphi(w)$ and a_1 be a singularity defined by $\varphi_1(w)$ and L'_w. We also let $\varphi_2(w)$ be another element of $\varphi(w)$ and a_2 be a singularity defined by $\varphi_2(w)$ and L''_w. Obviously, when $a_1 \neq a_2$, the two corresponding boundary points (a_1) and (a_2) are distinctive on F. When $a_1 = a_2$, if in any neighborhood $|w - a| < \rho\,(^2)$ $(\rho > 0)$ around point a $(a = a_1 = a_2)$, there exists a curve Γ_w connecting L'_w and L''_w, such that $\varphi_1(w)$ and $\varphi_2(w)$ can undergo continuation on each other along Γ_w, then the same boundary point (a) on F is defined. Hence, (a_1) and (a_2) are nondistinctive. If there exists neighborhood of $a: |w - a| < \rho$ $(\rho > 0)$ such that undergoing continuation on each other is impossible along any curve Γ_w that connects $L_{w'}$ and $L_{w''}$, then two distinctive boundary (a_1) and (a_2) are defined on F.

Let L be a continuous curve tending to ∞ on the z-plane: $|z| < +\infty$. If point z tends to ∞ along L, $f(z)$ tends to a certain value a; then we call a the asymptotic value defined by L. Meanwhile, we call L an asymptotic path or path of definite value that corresponds to value a. Obviously, a and L determine respectively a point a and a curve L_w tending to a on the w-plane. If we note that an inverse function of $f(z)$ in a neighborhood of a certain point on L determines an element of $\varphi(w)$, then point a is a transcendental singularity determined by this element and L_w. Hence, an asymptotic value a of $f(z)$ determines a boundary point (a) on the Riemann surface F of $\varphi(w)$. Conversely, if (a) is a boundary point of F, then there exists a curve (L_w) tending to (a) on F. Therefore, there exists a curve L_w tending to a on the w-plane and an element of $\varphi(w)$, such that a is a singularity determined by this element and L_w. Obviously, L_w determines a continuous curve L stretching to ∞ on the z-plane, such that when z tends to ∞ along L, $f(z)$ tends to value a. Hence, each boundary point of F determines an asymptotic value of $f(z)$. Let a_1 and a_2 be two asymptotic values of $f(z)$, and L_1 and L_2 be two corresponding paths of definite value. If $a_1 \neq a_2$, then the two corresponding boundary points (a_1) and (a_2) on F are distinctive. If $a_1 = a_2 = a$ $(a \neq \infty)$,$(^3)$ and for a certain number $\varepsilon > 0$, there exists value R $(0 < R < +\infty)$, such that the oscillation of $f(z)$ on any curve Γ connecting L_1 and L_2 outside disk $|z| \leq R$ is greater than ε, then on the w-plane there exists a neighborhood $|w - a| < \rho$ $(0 < \rho < \frac{\varepsilon}{2})$, such that there is no continuation between an element $\varphi_1(w)$ of $\varphi(w)$ determined by L_1 and another element $\varphi_2(w)$ of $\varphi(w)$ determined by L_2 is possible, meaning that (a_1) and (a_2) are two

---

$(^2)$ When $a = \infty$, use $|w| > \frac{1}{\rho}$ to replace $|w - a| < \rho$.

$(^3)$ When $a = \infty$, by means of transformation, $\frac{1}{w}$ can be changed into $a = 0$.

distinctive boundary points of F. Hence, we say asymptotic values a_1 and a_2 satisfying one of the above-mentioned two conditions as two distinctive asymptotic values; otherwise, they are nondistinctive. Hence, we have established an identity between the asymptotic value of a transcendental entire function or meromorphic function $f(z)$ and the transcendental singularity of the inverse function $\varphi(w)$ of f or the boundary point of the Riemann surface F of $\varphi(w)$. In the sequel, without changing the asymptotic value, we may assume that the asymptotic path L is a simple, continuous curve extending from the origin $z = 0$ to ∞. Futhermore, according to the continuity of $f(z)$, we may assume that L consists of line segments, such that the endpoints of the straight line segments in L have no cluster point within the finite plane. Furthermore, we may assume that if $L_1, L_2, \ldots, L_n$ are n asymptotic paths of $f(z)$, with their asymptotic values distinctive among each other, then these n asymptotic paths, except at the origin $z = 0$, have no other intersection points in the z-plane. Hence, these n paths divide z-plane $|z| < +\infty$ into n simply-connected domains D_k $(k = 1, 2, \ldots, n)$. We may assume that L_k and L_{k+1} $(L_1 = L_{n+1})$ are adjacent and they bound the simply-connected domain D_k.

F. Iversen once classified the transcendental singularities of $\varphi(w)$ [24a]: Let a be a singularity defined by an element $\varphi_1(w)$ of $\varphi(w)$ and path L_w. If there exists a neighborhood $|w - a| < \rho$ $(\rho > 0)$ such that in this neighborhood no continuation of $\varphi_1(w)$ along any path reaching point a is possible, then we call a a nondirect transcendental singularity; otherwise we call a a nondirect transcendental singularity. Correspondingly, for a direct transcendental singularity a, there exists an unbounded connected component Ω_ρ of the open set $E\{z | |f(z) - a| < \rho\}$ on the z-plane, such that when $z \in \Omega_\rho$,

$$|f(z) - a| < \rho, \qquad f(z) \neq a,$$

and when $z \in \Gamma_\rho$,

$$|f(z) - a| = \rho,$$

where Γ_ρ denotes the finite boundary of Ω_ρ. Obviously, if the two direct transcendental singularities a_1 and a_2 are distinctive, then there must exist a certain value $\rho > 0$, such that there is no intersection between domain Ω_ρ^1 corresponding to a_1 and domain Ω_ρ^2 corresponding to a_2.

4.1.2. The Iversen Theorem. Now we prove that each boundary point of the simply parabolic-type Riemann surface is an accessible boundary point. In fact, this is a direct corollary of the following Iversen Theorem [24a]:

THEOREM 4.1. *Let F be a parabolic-type Riemann surface covering of the w-plane $|w| < +\infty$, and a be an arbitrary point on the w-plane. We also let $\rho > 0$ and (a_1) be an interior point of F with $|a_1 - a| = \rho$. Then, there* *must exists a continuous curve (L) starting from point (a_1) on F, such that L lies within the disk $|w - a| < \rho$ and tends to point a.*

PROOF. [4] Assume that $z = \varphi\{(w)\}$ [5] maps F conformally onto the whole plane $|z| < +\infty$, with its inverse transformation $w = f(z)$ being a transcendental entire or meromorphic function. In the following, we need only to prove that for any element $\varphi_1(w)$ of $\varphi(w)$ at point $w = a_1$, there must exist a continuous curve L connecting point a_1 and a in the disk $|w - a| < \rho$, such that $\varphi_1(w)$ can undergo continuation along L, but probably with the exception at point a. In fact, first we choose point $z_1 = \varphi_1(a_1)$ on the z-plane; then according to condition $|w - a| < \rho$, we identify a connected domain Ω_ρ with its boundary containing point z_1. If there exists point z_0 in Ω_ρ, such that $f(z_0) = a$, then there exists a curve L_z connecting point z_1 and z_0 in Ω_ρ. Hence, the image (L) of L_z on F is a curve originating from point (a_1). Moreover L is in the disk $|w - a| < \rho$ and L tends to point a. Hence, Theorem 4.1 holds.

On the other hand, if it is always true that $f(z) \neq a$ in Ω_ρ, then according to condition $\rho > |w - a| > \frac{\rho}{2}$, there exists a connected domain $\Omega_{\rho_1} \subset \Omega_\rho$ with its boundary containing point z_1. We prove that in Ω_ρ there exists at least a boundary point z_2 of Ω_{ρ_1} in Ω_ρ, such that $|f(z_2) - a| = \frac{\rho}{2}$. In fact, otherwise, on the finite boundary of Ω_{ρ_1} it is always true that $\frac{1}{|f(z)-a|} = \frac{1}{\rho}$. Moreover, it is always true that in Ω_{ρ_1}, $\frac{1}{|f(z)-a|} < \frac{2}{\rho}$. Hence, according to the regularity of $\frac{1}{f|z|-a}$ in Ω_{ρ_1} and the maximum modulus principle, we conclude that it is always true that $|\frac{1}{f(z)-a}| \leq \frac{1}{\rho}$ in Ω_{ρ_1}, which is, however, impossible. Hence, there exists a boundary point z_2 of Ω_{ρ_1} in Ω_ρ, such that $|f(z_2) - a| = \frac{\rho}{2}$. Analogously, according to the condition $\frac{\rho}{2} > |w - a|\frac{\rho}{4}$, there exists a connected domain $\Omega_{\rho_2} \subset \Omega_\rho$, with its boundary containing point z_2. We may also prove that there exists at least a boundary point z_3 of Ω_{ρ_2} in Ω_ρ, such that $|f(z_3) - a| = \frac{\rho}{4}$. By repeating the argument, we obtain a sequence of adjacent domains $\Omega_{\rho_1}, \Omega_{\rho_2}, \ldots,$ in Ω_ρ, such that it is always true that $\frac{1}{2^{n-1}} > |f(z) - a| > \frac{1}{2^n}$ in Ω_{ρ_n}. Moreover, Ω_{ρ_n} and $\Omega_{\rho_{n+1}}$ have a common boundary point z_{n+1}, and also $|f(z_n) - a| = \frac{\rho}{2^n}$. Now, we prove that when $n \to +\infty$, Ω_{ρ_n} must diverge to the point of infinity $z = \infty$. In fact, for any arbitrary given value $r > 0$, $|f(z) - a|$ must have a positive lower bound on $\Omega_\rho \cap (|z| \leq r)$. Hence, when n is sufficiently large, Ω_{ρ_n} can only be outside the disk $|z| \leq r$, that is when $n \to +\infty$, Ω_{ρ_n} diverges to ∞. Finally, we use successively curve L_{zn} in Ω_{ρ_n} to connect point z_n and z_{n+1}, and let $L_z = \bigcup_{n=1}^{+\infty} L_{zn}$. Then L_z is a continuous curve originated from point z_1 in Ω_ρ to ∞. Moreover, when z tends to ∞

---

[4] This proof is essentially due to G. Valiron [39a]

[5] (w) represents a point on F.

along L_z, $f(z)$ tends to value a, meaning that a is an asymptotic value of $f(z)$. Obviously, the image (L) of L_z on F is a curve originated from point (a_1). Moreover, L lies inside of the disk $|w - a| < \rho$, and L tends to point a. Hence, Theorem 4.1 is proved completely.

According to Theorem 4.1 and the Picard Theorem, we have the following corollary:

COROLLARY. *Any Picard exceptional value of a transcendental meromorphic function is also an asymptotic value. Hence, a transcendental entire function must have ∞ as its asymptotic value.*

4.1.3. The Lindelöf Theorem.
We can deduce from Theorem 3.2 directly the following classic Lindelöf Theorem [28a]:

THEOREM 4.2. *Let L_1 and L_2 be two simple, continuous curves extending from the origin $z = 0$ to ∞ on the open plane $|z| < +\infty$, and L_1 and L_2 do not intersect with each other except at the origin $z = 0$, meaning that a simply connected region D is formed. We also let $f(z)$ be regular in D and continuous on $\overline{D}$. Moreover, when z tends to ∞ along L_i $(i = 1, 2)$, $f(z)$ tends to finite limit value a_i. If $f(z)$ has an upper bound $N < +\infty$ in D, then it must be $a_1 = a_2$. Meanwhile, when z tends to ∞ on $\overline{D}$, $f(z)$ tends uniformly to value a $(a = a_1 = a_2)$.*

PROOF. Indeed, by applying a suitable conformal mapping we may assume D to be the upper half-plane: $\operatorname{Im} z > 0$, and L_1 and L_2 be the positive and negative real axis, respectively. Then we construct a sequence $R_n = e^{5n\pi}$ $(n = 1, 2, \dots)$ and use L_{in} to denote the part of L_i $(i = 1, 2)$, that is in the annulus $R_n < |z| < R_{n+1}$. Let

$$\varepsilon_{in} = \max_{z \in L_{in}} |f(z) - a_i|, \qquad i = 1, 2.$$

Then when $n \to +\infty$, $\varepsilon_{in} \to 0$. Furthermore, suppose that $M < +\infty$ is the upper bound of $|a_1|$ and $|a_2|$. Then when $a_1 \neq a_2$, and n is sufficiently large, we have

$$(M + N)(\varepsilon_{1n}^{1/3} + \varepsilon_{2n}^{1/3}) < |a_1 - a_2|.$$

By means of the transformation $\zeta = \frac{z}{R_n}$, and applying Theorem 3.2, we conclude that $a_1 = a_2$. Meanwhile, there exists a curve l_n connecting L_1 and L_2 in annulus $R_n < |z| < R_{n+1}$, such that when $z \in l_n$ it follows that

$$|f(z) - a| \leq \varepsilon_{3n}, \qquad a = a_1 = a_2,$$
$$\varepsilon_{3n} = (M + N) \max(\varepsilon_{1n}^{1/3} + \varepsilon_{2n}^{1/3}) \to 0, \qquad n \to +\infty.$$

Notice that a region D_n is formed by l_n, l_{n+1}, L_1 and L_2. Then when $z \in \overline{D}_n$,

$$|f(z) - a| \leq \max\{\varepsilon_{3n}, \varepsilon_{3n+1}, \varepsilon_{1n}, \varepsilon_{2n}, \varepsilon_{1n+1}, \varepsilon_{2n+1}\} \to 0, \qquad n \to +\infty.$$

Hence $f(z)$ tends uniformly to value a on $\overline{D}$, implying that Theorem 4.2 holds.

Let $f(z)$ be a transcendental entire function on the open plane $|z| < +\infty$, and a_1 and a_2 be two finite asymptotic values of $f(z)$, with their corresponding paths of definite value L_1 and L_2 bounding a simply connected region D. If $f(z)$ is bounded on D, then according to Theorem 4.2, we conclude that $a = a_1 = a_2$. Meanwhile, $f(z)$ tends uniformly to value a on $\overline{D}$. This shows that a_1 and a_2 are nondistinctive asymptotic values, meaning that asymptotic values a_1 and a_2 are the same. Hence, if a_1 and a_2 are two distinct finite asymptotic values of $f(z)$, then $f(z)$ must be unbounded on D. In fact, we have a stronger conclusion [43c]:

THEOREM 4.3. *Let $f(z)$ be a transcendental entire function on the open plane $|z| < +\infty$, and a_1 and a_2 be two distinct finite asymptotic values of $f(z)$, with their corresponding paths of definite value L_1 and L_2 dividing the open plane $|z| < +\infty$ into two simply connected regions D_1 and D_2. Then there exists a continuous curve l_i tending to ∞ in D_i $(i = 1, 2)$ such that*

$$\varliminf_{\substack{|z| \to +\infty \\ z \in l_i}} \frac{\log\log|f(z)|}{\log|z|} \geq \frac{1}{2}.$$

PROOF. First we choose a fixed number N such that

$$\max_{z \in L_1 \cup L_2} |f(z)| < N < +\infty,$$

and derivative $f'(z)$ has no zeros on the level curve $|f(z)| = N$, meaning that the level curve $|f(z)| = N$ is analytic. Then, since a_1 and a_2 are two distinct finite asymptotic values, and according to the Lindelöf Theorem (Theorem 4.2), we conclude that $f(z)$ is unbounded on D_i $(i = 1, 2)$. Hence, there exists a point $z_{io} \in D_i$, such that $|f(z_{io})| > eN$.

We consider the set

$$E = \{z \mid |f(z)| > N\},$$

and let Ω_i be the connected branch containing point z_{io}. Then according to the maximum modulus principle, we conclude that Ω_i is an unbounded domain, and $\Omega_i \subset D_i$. Hence, circumference $|z| = t$ $(|z_0| \leq t < +\infty)$ must intersect the boundary of Ω_i. We denote θ_{it} the part of circumference $|z| = t$ that is in Ω_i, and $t\theta_i(t)$ the linear measure of θ_{it}. We also let

$$M(t, \overline{\theta}_{it}, f) = \max_{z \in \theta_{it}} |f(z)|.$$

Then according to Theorem 3.1, when $t > 4|z_{io}|$, we have

$$\log|f(z_{io})| \leq \log N + 9\sqrt{2}\exp\left\{-\pi\int_{2|z_{io}|}^{t/2}\frac{dr}{r\theta_i(r)}\right\}\log M(t, \overline{\theta}_{it}, f),$$

$$\log eN \leq \log N + 9\sqrt{2}\exp\left\{-\pi\int_{2|z_{io}|}^{t/2}\frac{dr}{r\theta_i(r)}\right\}\log M(t, \overline{\theta}_{it}, f),$$

$$\log\log M(t, \overline{\theta}_{it}, f) \geq \pi\int_{2|z_{io}|}^{t/2}\frac{dr}{r\theta_i(r)} - \log 9\sqrt{2}.$$

Notice that $\theta_i(r) \leq 2\pi$ $(|z_{io}| < r < +\infty)$; then it follows that

$$\log\log M(t, \overline{\theta}_{it}, f) \geq \frac{1}{2}\log t - \log 36\sqrt{\frac{|z_{io}|}{2}}.$$

Therefore, whenever t is chosen sufficiently large, then, on the one hand, there exists a point z_{i1} on $\overline{\theta}_{it}$ such that

$$|z_{i1}| \geq 72^{(12\times 13)/2}, \quad |z_{i1}|^{(1/2)-(1/12)} > 4N, \quad \log|f(z_{i1})| \geq |z_{i1}|^{(1/2)-(1/12)},$$

hence $z_{i1} \in \theta_{it} \subset \Omega_i$, while on the other hand, according to the relation

$$\log M(t_1', f) = \frac{1}{4}|z_{i1}|^{(1/2)-(1/12)},$$

the associated value $t_1' > 1$.

In the following, we construct two sequences t_n $(n = 1, 2, \ldots)$ and t_n' $(n = 1, 2, \ldots)$. First, we let

$$t_1 = |z_{i1}|,$$
$$t_{n+1} = 72^{n+12}t_n^{1+\varepsilon_n}, \quad \varepsilon_n = \frac{1}{n+11}, \quad n = 1, 2, \ldots. \tag{4.1}$$

Notice that

$$t_n \geq 72^{n+11}t_{n-1} \geq 72^{(n+11)+(n+10)+\cdots+13}t_1 \geq 72^{(n+11)(n+12)/2}.$$

Then

$$72^{n+12} \leq t_n^{2\varepsilon_n}. \tag{4.2}$$

Next, we define t_n' according to the equality

$$\log M(t_n', f) = \frac{1}{4}t_n^{1/2-\varepsilon_n}, \quad n = 1, 2, \ldots. \tag{4.3}$$

Then

$$t_n' < t_{n+1}' \to +\infty \quad (n \to +\infty), \ldots, \quad n = 1, 2, \ldots.$$

Now, we prove the following fact: If a point $z_{in} \in \Omega_i$ has been chosen on circumference $|z| = t_n$ $(n \geq 1)$ such that $\log|f(z_{in})| \geq |z_{in}|^{1/2-\varepsilon_n}$, then there must exist a point $z_{in+1} \in \Omega_i$ on circumference $|z| = t_{n+1}$ such that

$$\log|f(z_{in+1})| \geq |z_{in+1}|^{1/2-\varepsilon_{n+1}}.$$

Moreover, there is a continuous curve L_{in} connecting z_{in} and z_{in+1} in Ω_i, such that when $z \in L_{in}$,

$$\log|f(z)| \geq \frac{1}{4}|z|^{1/2-3\varepsilon_n}, \qquad t'_n \leq |z| \leq t_{n+1}.$$

There exists value A_n in the interval $[\frac{1}{4}|z_{in}|^{1/2-\varepsilon_n}, \frac{1}{2}|z_{in}|^{1/2-\varepsilon_n}]$, such that the derivative $f'(z)$ has no zeros on the level curve $\log|f(z)| = A_n$, meaning that the level curve is analytic. We consider the set

$$E = \{z| \log|f(z)| > A_n\}.$$

Let Ω_{in} be the connected component containing point z_{in}. Notice that

$$\frac{1}{4}|z_{in}|^{1/2-\varepsilon_n} \geq \frac{1}{4}|z_{i1}|^{(1/2)-(1/12)} > N;$$

then $\Omega_{in} \subset \Omega_i$. Besides, according to the maximum modulus principle, we conclude that Ω_{in} is an unbounded domain. Let D_{in} be the connected component of Ω_{in} in the disk $|z| < t_{n+1}$ containing point z_{in}, and θ_{in+1} be the part of circumference $|z| = t_{n+1}$ belonging to the boundary of D_{in}. Then there must exist a point z_{in+1} on θ_{in+1}, such that

$$\log|f(z_{in+1})| \geq |z_{in+1}|^{1/2-\varepsilon_{n+1}}.$$

Indeed, otherwise, according to Theorem 3.1, we have

$$\log|f(z_{in})| \leq A_n + 9\sqrt{2}\exp\left\{-\pi\int_{2|z_{in}|}^{t_{n+1/2}+1}\frac{dr}{r\theta_i(r)}\right\}t_{n+1}^{1/2-\varepsilon_{n+1}},$$

$$|z_{in}|^{1/2-\varepsilon_n} \leq \frac{1}{2}|z_{in}|^{1/2-\varepsilon_n} + 9\sqrt{2}\exp\left\{-\pi\int_{2|z_{in}|}^{t_{n+1/2}+1}\frac{dr}{r\theta_i(r)}\right\}t_{n+1}^{1/2-\varepsilon_{n+1}},$$

$$t_n^{1/2-\varepsilon_n} \leq 18\sqrt{2}\exp\left\{-\pi\int_{2|z_{in}|}^{t_{n+1/2}+1}\frac{dr}{r\theta_i(r)}\right\}t_{n+1}^{1/2-\varepsilon_{n+1}}.$$

Notice that $\theta_i(r) \leq 2\pi$ $(0 < r < t_{n+1})$; then

$$t_n^{1/2-\varepsilon_n} \leq 36\sqrt{2}\frac{t_n^{1/2}}{t_{n+1}^{1/2}}t_{n+1}^{1/2-\varepsilon_{n+1}}, \qquad t_{n+1} < 72^{n+11}t_n^{1+\varepsilon_n},$$

which, however, contradicts Formula 4.1.

Taking the property of connectivity into account, there exists a continuous curve L_{in} connecting z_{in} and z_{in+1} in D_{in}. When $z \in L_{in}$, we have

$$\log|f(z)| \geq A_n \geq \frac{1}{4}|z_{in}|^{1/2-\varepsilon_n}, \qquad |z| \leq t_{n+1}.$$

Besides, according to Formula 4.3, we conclude that $|z| \geq t'_n$. To have a further discussion: If $|z| \leq |z_{in}|$, then

$$\log|f(z)| \geq \frac{1}{4}|z|^{1/2-\varepsilon_n} > \frac{1}{4}|z|^{1/2-3\varepsilon_n};$$

if $|z| > |z_{in}|$, it follows that

$$\log|f(z)| \geq \frac{1}{4}|z|^{1/2-\varepsilon_n} \left\{ \frac{t_n}{t_{n+1}} \right\}^{1/2-\varepsilon_n}$$

$$= \frac{1}{4}|z|^{1/2-\varepsilon_n} \left\{ \frac{1}{72^{n+12}t_n^{\varepsilon_n}} \right\}^{1/2-\varepsilon_n} .$$

Moreover, according to Formula 4.2, we get $\log|f(z)| \geq \frac{1}{4}|z|^{1/2-3\varepsilon_n}$. Hence, we prove that when $z \in L_{in}$,

$$\log|f(z)| \geq \frac{1}{4}|z|^{1/2-3\varepsilon_n}, \qquad t_n' \leq |z| \leq t_{n+1} .$$

Let $l_i = \bigcup_{n=1}^{+\infty} L_{in}$, then l_i is a continuous curve extending from point z_{i1} to ∞. Notice that when $n \to +\infty$, $\varepsilon_n \to 0$. Hence, we obtain

$$\varlimsup_{\substack{|z| \to +\infty \\ z \in l_i}} \frac{\log\log|f(z)|}{\log|z|} \geq \frac{1}{2},$$

implying that Theorem 4.3 is proved.

If a transcendental entire function $f(z)$ possesses a finite asymptotic value a, with its corresponding path of definite value being L_a, then $|f(z)|$ has an upper bound $N < +\infty$ on L_a. However, $f(z)$ is unbounded on the z-plane. Hence, there exists a point z_0, such that $|f(z_0)| > eN$. We consider the set $E = \{z | |f(z)| > N\}$ and let Ω be the connected component containing point z_0. According to the maximum modulus principle, Ω cannot be a bounded domain. Moreover, circumference $|z| = t$ $(|z_0| < t < +\infty)$ intersects definitely at the boundary of Ω. Analogously, applying the above-mentioned discussion, we may obtain the corresponding conclusion.

If $f(z)$ possesses a finite deficient value a, with its corresponding deficiency $\delta(a, f) = \delta > 0$, then when $r \geq r_0$,

$$m(r, a) = \frac{1}{2\pi} \int_0^{2\pi} \log \frac{1}{|f(re^{i\theta}) - a|} d\theta$$

$$\geq \frac{\delta}{2} T(r, f) \geq \frac{\delta}{2} T(r_0, f) .$$

Hence, there exists a point $z_r = re^{i\theta_r}$ on circumference $|z| = r$ $(r \geq r_0)$, such that

$$|f(z_r) - a| < e^{-\delta T(r_0, f)/2},$$

meaning that $f(z)$ has an upper bound on the point set $E = \{z_r | r \geq r_0\}$. Applying the analogous discussion, we may also obtain the corresponding conclusion. Hence, we have the following result [43c]:

THEOREM 4.4. *Suppose that transcendental entire function* $f(z)$ *has a finite asymptotic value or a finite deficient value; then there exists a continuous curve* L *tending to* ∞ *on the z-plane, such that*

$$\varlimsup_{\substack{|z| \to +\infty \\ z \in L}} \frac{\log \log |f(z)|}{\log |z|} \geq \frac{1}{2} \cdot (^6)$$

COROLLARY. *Under the hypothesis of Theorem 4.4, the lower order* μ *of* $f(z)$ *is* $\geq \frac{1}{2}$.

4.1.4. The Fuchs Theorem. Recently, W. Fuchs proved the following Phragmen-Lindelöf-type Theorem [16a]:

THEOREM 4.5. *Let* D *be an unbounded region on the open plane* $|z| < +\infty$, *with its finite boundary* Γ *consisting of at least one point. Assume that* $f(t)$ *is regular in* D, *and for each boundary point* $\zeta \in \Gamma$ *we have*

$$\varlimsup_{\substack{z \to \zeta \\ z \in D}} |f(z)| \leq 1. \tag{4.4}$$

Then among the following three possibilities, one and only one possibility occurs:

(1) *When* $z \in D$, $|f(z)| \leq 1$.

(2) *The point of infinity* ∞ *is a pole of* $f(z)$.

(3)
$$\varlimsup_{r \to +\infty} \frac{\log M\{D \cap (|z| = r), f\}}{\log r} = +\infty.$$

PROOF. In fact, we need only to prove if possibility (3) does not occur, then either possibility (1) or (2) will occur. Hence, we assume that possibility (3) does not occur. Then

$$\varlimsup_{r \to +\infty} \frac{\log M\{D \cap (|z| = r), f\}}{\log r} \leq p < +\infty. \tag{4.5}$$

If we further assume that possibility (1) does not occur, then

$$1 < \sup_{z \in D} |f(z)| = \alpha \leq +\infty. \tag{4.6}$$

Since derivative $f'(z)$ has at most countable zeros in D, there exists value A in the open interval $(1, \alpha)$, such that $f'(z)$ has no zeros on the level curve

$$|f(z)| = A. \tag{4.7}$$

Hence, level curve (4.7) is a simply analytic curve, and each branch of the level curve (4.7) is either a simple, closed curve or it has no endpoint in the open plane $|z| < +\infty$. Consider the set

$$D_A = \{z \,|\, |f(z)| > A, \ z \in D\}. \tag{4.8}$$

(^{6}) K. Barth, D. Brannan and W. Hayman proved a slightly more concise result. See [5a].

Then according to formula (4.6), we conclude that D_A is not an empty set. On the other hand, according to the maximum modulus principle, we conclude that each component of D_A is an unbounded domain. Taking any point z_0 from D_A, and denoting $D_A(z_0)$ the connected component containing point z_0. In the following, we distinguish two cases for discussion:

(1) The level curve (4.7) contains an unbounded component. Hence, there exists a value t_0, $t_0 \geq |z_0|$, such that when $t \geq t_0$, circumference $|z| = t$ intersects definitely the boundary of $D_A(z_0)$. We let θ_t be the part of circumference $|z| = t$ in $D_A(z_0)$, and $t\theta(t)$ be the linear measure of θ_t. Then according to formula (4.8) and Theorem 3.1, we have

$$\log|f(z_0)| \leq \log A$$
$$+ 9\sqrt{2}\exp\left\{-\pi\int_{2|z_0|}^{r/2}\frac{dt}{t\theta(t)}\right\}\log M\{D\cap(|z|=r),f\}$$
$$\leq \log A + 9\sqrt{2}\exp\left\{-\pi\int_{2t_0}^{r/2}\frac{dt}{t\theta(t)}\right\}\log M\{D\cap(|z|=r),f\}.$$

Notice that $\theta(t) \leq 2\pi$ $(t_0 \leq t < +\infty)$; then

$$\log|f(z_0)| \leq \log A + 9\sqrt{2}\left(\frac{4t_0}{r}\right)^{1/2}\log M\{D\cap(|z|=r),f\}.$$

Letting $r \to +\infty$, we conclude from formula (4.5), $|f(z_0)| \leq A$, which contradicts, however, formula (4.8). This shows that formula (4.6) does not hold, meaning that when $z \in D$, $|f(z)| \leq 1$. Hence, possibility (1) occurs.

(2) Each branch of the level curve (4.7) is bounded.

First we illustrate that circumference $|z| = r$ intersects at most finite branches of the level curve (4.7). If the intersection of circumference $|z| = r$ and D is nonempty, then the intersection set is composed of some open arcs I. Let

$$g(z) = f(z)\overline{f\left(\frac{r^2}{\bar{z}}\right)}.$$

Then $g(z)$ is regular in a neighborhood of any one of the closed arcs $J \subset I$ of I. Moreover, when $|z| = r$, $g(z) = |f(z)|^2$. Now, we assume that the level curve (4.7) possesses infinitely many branches $\{L_n\}$ that intersect circumference $|z| = r$. We take a point z_n on each branch, $|z_n| = r$. Without any loss of generality, we may assume that $z_n \to Z$, $|Z| = r$, $g(Z) = A^2 = g(z_n)$.

According to formulas (4.6) and (4.4), for each point $\zeta \in \Gamma$, there exists a neighborhood of it such that $|f(z)| < A$ holds there. Hence, Z can only be an interior point of I, and, in turn, there exists a neighborhood of point Z such that $g(z)$ is regular in this neighborhood. Moreover, when n is sufficiently large, point z_n is in such a neighborhood. Since $g(z_n) = g(Z) = A^2$, it follows that $g(z_n) = |f(z_n)|^2 = A^2$ on I. Therefore, there exist different points z_n and z'_n $(n \neq n')$ belonging to the same branch. However, this contradicts the way in which they are chosen. Hence, circumference $|z| = r$ intersects only finite branches of the level curve (4.7).

In the following, we prove that D_A is a region. Indeed, since each component of D_A is unbounded, we need only to show that for any two arbitrary points $z_1, z_2 \in D_A$, $|z_1| = |z_2| = r$, there exists a continuous curve linking point z_1 and z_2 in D_A. We use circumference $|z| = r$ to connect points z_1 and z_2. If we come across a branch L_n of the level curve (4.7), we substitute it by the part where L_n intersects circumference $|z| = r$. Since circumference $|z| = r$ intersects merely finite branches L_n $(n = 1, 2, \ldots, m)$ of the level curve (4.7), each branch is, therefore, a simple analytic curve, and the m branches have positive distances among one another. According to the above-mentioned method, we may, therefore, find a curve linking points z_1 and z_2. Then, if necessary, by moving the curve slightly, we may guarantee that this curve is within D_A. Hence, D_A is a region.

Set $D(r) = D_r \cap (c < |z| < r)$, where c is a sufficiently large fixed positive number. Then, the boundary Γ_r of $D(r)$ is composed of parts of circumference $|z| = c$ and $|z| = r$ that lie within D_A and part of the level curve (4.7) that lies within the annulus: $c < |z| < r$. Now, we apply the Gauss Theorem to harmonic function $\log |f(z)|$; then

$$\int_{\Gamma_r} \frac{\partial}{\partial n} \log |f(z)| \, |ds| = 0, \tag{4.9}$$

where $\partial/\partial n$ denotes the outer normal directional derivative of $\log |f(z)|$ with respect to Γ_r, and ds denotes the element of arc length. Notice that

$$\frac{\partial}{\partial n} \log |f(z)| \leq 0 \tag{4.10}$$

on the level curve (4.7). In fact, the equal sign cannot be held because we have at the same time

$$\frac{\partial}{\partial s} \log |f(z)| = 0$$

and

$$|f'(z)| = \left| \frac{\partial}{\partial n} \log |f(z)| - \frac{\partial}{\partial s} \log |f(z)| \right| \neq 0$$

where $\partial/\partial s$ denotes the tangential directional derivative of the level curve (4.7). Hence, on any branch L of the level curve (4.7), we have

$$\int_L \frac{\partial}{\partial n} \log |f(z)| \, |ds| < 0. \tag{4.11}$$

Furthermore, according to the Cauchy-Riemann's equation, we have

$$\int_L \frac{\partial}{\partial n} \log |f(z)| \, |ds| = \Delta_L \arg f(z),$$

where $\Delta_L \arg f(z)$ denotes the variation of $\arg f(z)$ when point z moves along L in a clockwise direction for one cycle. Notice that $f(z)$ is single-valued, therefore, $\Delta_L \arg f(z)$ is the integral multiple of 2π. Hence, according to (4.11), we have

$$\int_L \frac{\partial}{\partial n} \log |f(z)| \, |ds| < -2\pi . \tag{4.12}$$

We use $\nu(r)$ to denote the number of branches of the level curve (4.7) lying entirely in annulus: $c < |z| < r$. Then according to formulas (4.9), (4.10) and (4.12), it follows that

$$0 \le -2\pi\nu(r) + \int_{re^{i\theta} \in D_A} \frac{\partial}{\partial r} \log |f(re^{i\theta})| \cdot r \, d\theta$$

$$- \int_{ce^{i\theta} \in D_A} \left[\frac{\partial}{\partial r} \log |f(re^{i\theta})| \right]_{r=c} \cdot c \, d\theta$$

$$\le -2\pi\nu(r) + \int_{re^{i\theta} \in D_A} \frac{\partial}{\partial r} \log |f(re^{i\theta})| r \, d\theta + K \qquad (K = K(c)) .$$

Then, by means of integration, we obtain

$$2\pi \int_c^\rho \frac{\nu(r)}{r} \, dr \le K \log \frac{\rho}{c} + \iint_{re^{i\theta} \in D_A} \frac{\partial}{\partial r} \log |f(re^{i\theta})| \, dr \, d\theta . \tag{4.13}$$

In the following, we estimate the double integral. For the fixed value θ, the set $\Delta(\theta; c, \rho) \cap D_A$ is composed of some intervals $\widetilde{I}$. Moreover, when $\rho e^{i\theta} \in D_A$, $\log |f(\rho e^{i\theta})|$ is the right endpoint value of a certain interval I; when $ce^{i\theta} \in D_A$, $\log |f(ce^{i\theta})|$ is the left endpoint value of a certain interval I, and $\log A$ is the endpoint value of other intervals. Hence,

$$\int_{re^{i\theta} \in D(\rho)} \frac{\partial}{\partial r} \log |f(re^{i\theta})| \, dr \le \log^+ \left| \frac{f(\rho e^{i\theta})}{A} \right| ,$$

where we define $|f(re^{i\theta})| = A$ when $re^{i\theta} \notin D_A$. Furthermore, we obtain from formula (4.13) that

$$2\pi \int_c^\rho \frac{\nu(r)}{r} \, dr \le \int_0^{2\pi} \log^+ \left| \frac{f(\rho e^{i\theta})}{A} \right| \, d\theta + K \log \frac{\rho}{c} .$$

Then, according to formula (4.5), there exists an infinite sequence $\rho_n \colon \rho_n < \rho_{n+1} \to +\infty$ $(n \to +\infty)$, such that

$$2\pi \int_c^{\rho_n} \frac{\nu(r)}{r} \, dr \le (p + K) \log \rho_n + K_1 .$$

Since $\nu(r)$ is a nondecreasing function, we conclude, therefore, that $\nu(r) \le p + K$, meaning that the number of branches of the level curve (4.7) is finite. Moreover, notice that each branch is bounded; then there exists value ρ such that $\{\rho < |z| < +\infty\} \subset D_A$. When $|z| > \rho$, we have the Laurant's expansion

$$f(z) = \sum_{n=-\infty}^{+\infty} c_n z^n,$$

$$|c_n| = \left| \frac{R^{-n}}{2\pi} \int_{|z|=R} f(Re^{i\theta}) e^{-in\theta} \, d\theta \right| \qquad (R > \rho).$$

(4.14)

Moreover, from formulas (4.5) and (4.13) we conclude that $|c_n| \leq R^{p-n}$. Let $R \to +\infty$. Then when $n > p$, $c_n = 0$. Hence, maybe the point of infinity ∞ is a pole of $f(z)$, meaning that possibility (2) occurs; or ∞ is a removable singularity of $f(z)$, then according to the maximum modulus principle, we conclude that possibility (1) occurs. Hence, Theorem 4.5 is proved completely.

Let $w = f(z)$ be a transcendental meromorphic function, and its inverse function $z = \varphi(w)$ has ∞ as a direct transcendental singularity. Hence, there exists a number $\rho > 0$, such that the corresponding region D_ρ on the z-plane satisfies the following condition:

(1) When $z \in D_\rho$,

$$|f(z)| < \rho, \qquad f(z) \neq \infty.$$

(2) Let Γ_ρ be the finite boundary of D_ρ. Then when $z \in \Gamma_\rho$, $|f(z)| = \rho$. Therefore, according to the Fuchs Theorem,([7]) we conclude that

$$\lim_{r \to +\infty} \frac{\log M\{D_\rho \cap (|z| = r), f\}}{\log r} = +\infty.$$

(4.15)

In fact, applying formula (4.15), we may prove the following, stronger conclusion:

THEOREM 4.6. *Under the above hypothesis, there exists a curve L tending continuously to ∞ in region D_ρ, such that*

$$\lim_{\substack{|z| \to +\infty \\ z \in L}} \frac{\log |f(z)|}{\log |z|} = +\infty.$$

The proof of this theorem is completely analogous to the proof of Theorem 4.14 in the sequel.

§4.2. The Denjoy Conjecture

4.2.1. The Denjoy Conjecture. In 1907, A Denjoy made the following famous conjecture [11a]:

Let $f(z)$ be an entire function of order $\lambda < +\infty$ on the open plane $|z| < +\infty$, and let it have k distinctive finite asymptotic values. Then, $k \leq 2\lambda$.

([7]) Regarding the application of the direct transcendental singularity in the Fuchs Theorem, please see [43f].

Under special conditions, that is when k corresponding asymptotic paths are all half lines, A. Denjoy himself proved the validity of this conjecture. For general conditions, it was not until 1930 that L. Ahlfors proved this conjecture completely. In fact, he obtained a stronger conclusion [1b]:

THEOREM 4.7. *Let $f(z)$ be an entire function on the open plane $|z| < +\infty$, and let it have k distinctive finite asymptotic values. Then*

$$\lim_{r \to \infty} \frac{\log M(r, f)}{r^{k/2}} > 0.$$

PROOF. Let $a_1, a_2, \ldots, a_k$ be k distinct finite asymptotic values of $f(z)$, with its corresponding k asymptotic paths $L_1, L_2, \ldots, L_k$ dividing the open plane $|z| < +\infty$ into k simply-connected regions $D_1, D_2, \ldots, D_k$. Let

$$N = \sup_{z \in \bigcup_{i=1}^{k} L_i} |f(z)| < +\infty.$$

Then according to the Lindelöf Theorem, we conclude that there exists point z_i in D_i $(1 \leq i \leq k)$, such that $|f(z_i)| > e^2 N$, $i = 1, 2, \ldots, k$. Also there exists value N' in the open interval (N, eN), such that derivative $f'(z)$ has no zeros on the level curve $|f(z)| = N'$, meaning that the level curve is analytic. We consider the set $\Omega_i = E\{z \mid |f(z)| > N'\}$ $(1 \leq i \leq k)$, and assume Ω_i be the connected component containing point $z_i \in D_i$. Then $\Omega_i \subset D_i$. On the other hand, according to the maximum modulus principle, we conclude that Ω_i is an unbounded domain.

Let θ_{it} be the part of circumference $|z| = t$ in D_i, and $t\theta_i(t)$ be the linear measure of θ_{it}. Then we have from Theorem 3.1 that

$$\log(e^2 N) < \log |f(z_i)| \leq \log N' + 9\sqrt{2} \exp\left\{ -\pi \int_{2|z_i|}^{r/2} \frac{dt}{t\theta_i(t)} \right\} \log M(r, f),$$

$$1 \leq 9\sqrt{2} \exp\left\{ -\pi \int_{2|z_i|}^{r/2} \frac{dt}{t\theta_i(t)} \right\} \log M(r, f),$$

$$\pi \int_{2|z_i|}^{r/2} \frac{dt}{t\theta_i(t)} \leq \log\log M(r, f) + \log 9\sqrt{2}, \qquad i = 1, 2, \ldots, k.$$

Let $r_0 = \max_{1 \leq i \leq k}\{|z_i|\}$. Notice that

$$k^2 = \left(\sum_{i=1}^{k} \sqrt{\theta_i(t)} \frac{1}{\sqrt{\theta_i(t)}} \right)^2 \leq \sum_{i=1}^{k} \theta_i(t) \sum_{i=1}^{k} \frac{1}{\theta_i(t)} \leq 2\pi \sum_{i=1}^{k} \frac{1}{\theta_i(t)};$$

it follows that

$$\frac{k^2}{2}\int_{2r_0}^{r/2}\frac{dt}{t} \leq \sum_{i=1}^{k}\pi\int_{2r_0}^{r/2}\frac{dt}{t\theta_i(t)} \leq \sum_{i=1}^{k}\pi\int_{2|z_i|}^{1/2}\frac{dt}{t\theta_i(t)}$$

$$\leq k\log\log M(r,f) + k\log 9\sqrt{2},$$

$$\frac{k}{2}\log\frac{r}{4r_0} \leq \log\log M(r,f) + \log 9\sqrt{2},$$

$$\frac{\log M(r,f)}{r^{k/2}} \geq \frac{1}{(4r_0)^{k/2}9\sqrt{2}}.$$

And hence

$$\lim_{r\to+\infty}\frac{\log M(r,f)}{r^{k/2}} \geq \frac{1}{(4r_0)^{k/2}9\sqrt{2}} > 0,$$

meaning that Theorem 4.7 holds.

EXAMPLE. The order of the entire function

$$w = \int_0^z \frac{\sin t^q}{t^q}\,dt \qquad (q > 0)$$

is q, and has $2q$ distinct finite asymptotic values

$$e^{\nu\pi i/q}\int_0^\infty \frac{\sin r^q}{r^q}\,dr, \qquad \nu = 1, 2, \ldots, q,$$

with its corresponding $2q$ asymptotic paths being

$$\Delta(\theta_\nu) : \arg z = \frac{\nu\pi}{q}, \qquad \nu = 1, 2, \ldots, 2q.$$

This example shows that Denjoy's conjecture is precise.

4.2.2. Entire functions that satisfy the extreme case $k = 2\lambda$ of the Denjoy Conjecture. L. Ahlfors [1b], P. Kennedy [26b], and D. Drasin [12a] had studied these functions before, and illustrated mainly the distribution of the k paths. Here, we study mainly some other properties [43e]:

THEOREM 4.8. *Let $f(z)$ be an entire function of order $\lambda < +\infty$ on the open plane $|z| < +\infty$ and let it have k $(k \geq 1)$ distinct finite asymptotic values a_i $(i = 1, 2, \ldots, k)$, with the corresponding asymptotic paths being L_i $(i = 1, 2, \ldots, k)$. L_i and L_{i+1} $(i = 1, 2, \ldots, k, L_{k+1} = L_1)$ are adjacent, which bound a simply-connected region D_i. We also let $k = 2\lambda$. Then there exists a curve Γ_i tending to infinity in D_i $(1 \leq i \leq k)$ such that*

$$\lim_{\substack{|z|\to+\infty \\ z\in\Gamma_i}}\frac{\log\log|f(z)|}{\log|z|} = \lambda.$$

PROOF. (1) First, according to $k = 2\lambda$ and Theorem 4.7, we have

$$\log\log M(r,f) = \frac{k}{2}\log r + o(\log r). \tag{4.16}$$

Set that

$$N = \sup_{z \in L} |f(z)| < +\infty, \qquad L = \bigcup_{i=1}^{k} L_i.$$

Then according to the Lindelöf Theorem, there exist points $z_i \in D_i$ ($i = 1, 2, \ldots, k$) such that

$$|f(z_i)| \geq 2eN. \qquad (4.17)$$

Let

$$r_0 = \max\{1, |z_1|, \ldots, |z_k|\}. \qquad (4.18)$$

There exists value N' in the interval $[N, 2N]$, such that derivative $f'(z)$ has no zeros on the level curve $|f(z)| = N'$, and hence the level curve is analytic. We consider the set

$$E = \{z \mid |f(z)| > N'\}$$

and let Ω_i ($i = 1, 2, \ldots, k$) be the connected branch containing points z_i ($i = 1, 2, \ldots, k$). Then $\Omega_i \subset D_i$ ($i = 1, 2, \ldots, k$). Besides, according to the maximum modulus principle, we conclude that Ω_i ($i = 1, 2, \ldots, k$) is an unbounded domain.

(2) We let θ_{it} ($1 \leq i \leq k$; $r_0 \leq t < +\infty$) be the part of circumference $|z| = t$ in Ω_i and $t\theta_i(t)$ be the linear measure of θ_{it}.

Now, we first prove the following lemma:

LEMMA 4.1. *Given arbitrarily* ε, $0 < \varepsilon < 1$, *then there must exist value* r_ε, $r_\varepsilon \geq 4r_0$, *such that when* $r > r_\varepsilon$ *and* $2r_0 \leq R < \frac{r}{2}$,

$$\left| \int_R^{r/2} \left(\frac{\pi}{\theta_i(t)} - \frac{k}{2} \right) \frac{dt}{t} \right| \leq \varepsilon \log r, \qquad i = 1, 2, \ldots, k.$$

PROOF. According to Theorem 3.1,

$$\log |f(z_i)| \leq \log N' + 9\sqrt{2} \exp \left\{ -\pi \int_{2|z_i|}^{r/2} \frac{dt}{t\theta_i(t)} \right\} \log M(r, f),$$
$$i = 1, 2, \ldots, k.$$

Furthermore, it follows from formulas (4.16) and (4.17) that

$$\sum_{i=1}^{k} \int_{2r_0}^{r/2} \frac{\pi}{\theta_i(t)} \cdot \frac{dt}{t} \leq k \log \log M(r, f) + k \log(9\sqrt{2}),$$

$$\int_{2r_0}^{r/2} \sum_{i=1}^{k} \left\{ \frac{\pi}{\theta_i(t)} - \frac{k}{2} \right\} \frac{dt}{t} \leq k \left\{ \log \log M(r, f) - \frac{k}{2} \log r \right\}$$
$$+ \frac{k^2}{2} \log(4r_0) + k \log(9\sqrt{2}).$$

Noticing

$$\sum_{i=1}^{k} \left\{ \frac{\pi}{\theta_i(t)} - \frac{k}{2} \right\} \geq 0$$

and taking formula (4.16) into account, we conclude that

$$\int_{2r_0}^{r/2} \sum_{i=1}^{k} \left\{ \frac{\pi}{\theta_i(t)} - \frac{k}{2} \right\} \frac{dt}{t} = o(\log r).$$

Furthermore, according to the identical relation [26a]

$$\frac{\pi}{\theta_i(t)} - \frac{k}{2} = \frac{k^2}{4\pi} \left\{ \frac{2\pi}{k} - \theta_i(t) + \frac{(2\pi/k - \theta_i(t))^2}{\theta_i(t)} \right\}, \qquad (4.19)$$

we conclude that

$$\sum_{i=1}^{k} \int_{2r_0}^{r/2} \frac{((2\pi/k) - \theta_i(t))^2}{\theta_i(t)} \cdot \frac{dt}{t} = o(\log r),$$

$$\int_{2r_0}^{r/2} \frac{((2\pi/k) - \theta_i(t))^2}{\theta_i(t)} \cdot \frac{dt}{t} = o(\log r), \qquad i = 1, 2, \dots, k.$$

Hence, for any arbitrary given number ε, $0 < \varepsilon < 1$, there must exist a value r_ε, $r_\varepsilon \geq 4r_0$, such that when $r \geq r_\varepsilon$ and $2r_0 \leq R < \frac{1}{2}r$,

$$\int_R^{r/2} \frac{((2\pi/k) - \theta_i(t))^2}{\theta_i(t)} \cdot \frac{dt}{t} \leq \frac{\varepsilon^2}{k^4} \log r, \qquad i = 1, 2, \dots, k,$$

$$\int_R^{r/2} \left(\frac{2\pi}{k} - \theta_i(t) \right) \frac{dt}{t} \leq \frac{2\pi\varepsilon^2}{k^4} \log r, \qquad i = 1, 2, \dots, k. \qquad (4.20)$$

Applying the Schwarz inequality, and noting that $R \geq 2$, we obtain

$$\left| \int_R^{r/2} \left(\frac{2\pi}{k} - \theta_i(t) \right) \frac{dt}{t} \right| \leq \left\{ \int_R^{r/2} \frac{dt}{t} \right\}^{1/2},$$

$$\left\{ \int_R^{r/2} \left(\frac{2\pi}{k} - \theta_i(t) \right)^2 \frac{dt}{t} \right\}^{1/2} \leq \frac{\sqrt{2\pi}\varepsilon}{k^2} \log r, \qquad i = 1, 2, \dots, k.$$

Again, by combining formulas (4.19) and (4.20), it follows that

$$\left| \int_R^{r/2} \left(\frac{\pi}{\theta_i(t)} - \frac{k}{2} \right) \frac{dt}{t} \right| \leq \varepsilon \log r, \qquad i = 1, 2, \dots, k.$$

This completes the proof of this lemma.

(3) Let $\varepsilon_n = \frac{1}{n}$ $(n = 1, 2, \dots)$. Then there exist positive integers n_0 $(n_0 \geq 2)$, such that when $n \geq n_0$,

$$\varepsilon_n < \frac{k}{2(2k+1)}.$$

We also let

$$\varepsilon'_n = \frac{1}{1 + \varepsilon_n} \varepsilon_n^2 \qquad (n = 1, 2, \dots).$$

Then according to Lemma 4.1, there exists a sequence r_n $(n = n_0, n_0 + 1, \ldots)$ such that

$$r_{n_0} \geq \max\{K^{n_0(n_0+1)/2}, (8N)^{1/(k/2-\varepsilon_{n_0})}, (4 \log M(1, f))^{1/(k/2-\varepsilon_{n_0-1})}, r_0^{k(n_0-1)/2}\}$$

$$(K = 36 \times 2^k), \quad (4.21)$$

$$r_n < r_{n+1} \to +\infty \quad (n \to +\infty),$$

and when $r \geq \max\{4r_0, r_n\}$ and $2r_0 \leq R < \frac{r}{2}$,

$$\left| \int_R^{r/2} \left(\frac{\pi}{\theta_i(t)} - \frac{k}{2} \right) \frac{dt}{t} \right| \leq \varepsilon_n' \log r, \quad i = 1, 2, \ldots, k. \quad (4.22)$$

Now, we define sequence m_n $(n \geq n_0)$ and R_m $(m \geq m_0)$ as follows: First we choose that $R_{m_{n_0}} = r_{n_0}$, $m_{n_0} = 0$. Then we assume values n $(n \geq n_0)$; $m_n \geq 0$ and $R_{m_n} > 1$ have already been chosen. We define

$$R_{m_n+1} = K^{n+1} R_{m_n}^{1+2\varepsilon_{n-1}}, \qquad R_{m_n+l} = K^{n+1} R_{m_n+l-1}^{1+\varepsilon_n}, \qquad l = 2, 3, \ldots.$$

There must exist value l_0, $l_0 > 2$, such that when $l \geq l_0$, $R_{m_n+1} \geq r_{n+1}$. Then let $m_{n+1} = m_n + l_0$, $R_{m_{n+1}} = K^{n+1} R_{m_{n+1}-1}^{1+\varepsilon_n}$. Accordingly, we have defined sequences m_n $(n \geq n_0)$ and R_m $(m \geq m_{n_0})$. And sequences m_n $(n \geq n_0)$ and R_m $(m \geq m_{n_0})$ satisfy the following conditions:

$$\left. \begin{aligned} &R_{m_0} = r_{n_0}, \\ &R_{m+1} = K^{n+1} R_m^{1+2\varepsilon_{n-1}}, \qquad m = m_n, \\ &R_{m+1} = K^{n+1} R_m^{1+\varepsilon_n}, \qquad m = m_n + 1, m_n + 2, \ldots, m_{n+1} - 1, \\ &R_{n_0} = 0, m_{n+1} > m_n + 2, \\ &R_{m_n} \geq r_n, \qquad n = n_0, n_0 + 1, \ldots. \end{aligned} \right\} \quad (4.23)$$

Besides, when $m_n + 1 \leq m \leq m_{n+1}$, according to formulas (4.21) and (4.23), we conclude that $R_m \geq K^{(n+1)(n+2)/2}$. Finally, based on the relation

$$\log M(R_n', f) = \frac{1}{4} R_{m_n}^{k/2-\varepsilon_{n-1}}, \qquad n = n_0, n_0 + 1, \ldots, \quad (4.24)$$

a sequence R_n' $(n = n_0, n_0 + 1, \ldots)$ is defined. From $R_{m_n} \geq R_{m_0}$, $\varepsilon_{n-1} < \varepsilon_{n_0-1}$ $(n \geq n_0)$ and formula (4.21) we conclude that $R_n' \geq 1$ $(n \geq n_0)$, and also $R_n' < R_{n+1}' \to +\infty$ $(n \to +\infty)$.

(4) For this part we prove the following two lemmas. First we denote simply $\theta_{iR_m} \equiv \theta_{im}$ $(1 \leq i \leq k; m \geq 0)$.

LEMMA 4.2. *When $m = m_{n+1}$ $(n \geq n_0)$, we assume that there exists a point z_{im} on θ_{im} $(1 \leq i \leq k)$ such that*

$$\log |f(z_{im})| \geq R_m^{k/2-\varepsilon_n}.$$

Then there exists a point z_{im+1} on θ_{im+1} such that

$$\log|f(z_{im+1})| \geq R_{m+1}^{k/2-\varepsilon_{n+1}}.$$

Moreover, there is a continuous curve Γ_{im} linking z_{im} and z_{im+1} in Ω_i, such that when $z \in \Gamma_{im}$,

$$\log|f(z)| > \frac{1}{4}R_m^{k/2-\varepsilon_n}, \qquad R_{n+1}' \leq |z| \leq R_{m+1}.$$

PROOF. There exists value N'' in the interval $[\frac{1}{4}R_m^{k/2-\varepsilon_n}, \frac{1}{2}R_m^{k/2-\varepsilon_n}]$ such that derivative $f'(z)$ has no zeros on the level curve $\log|f(z)| = N''$, implying that the level curve is analytic. We consider the set

$$E = \{z \,|\, \log|f(z)| > N'', \ |z| < R_{m+1}\},$$

and let Ω_{im} be the connected branch containing point z_{im}. According to $R_m \geq r_{n_0}$, $\varepsilon_n < \varepsilon_{n+1}$ and formula (4.21), we conclude that $\Omega_{im} \subset \Omega_i$. On the other hand, based on the maximum modulus principle, $\theta_{im+1}^* \equiv \overline{\Omega}_{im} \cap (|z| = R_{m+1})$ is not an empty set. Obviously, $\theta_{im+1}^* \subset \theta_{im+1}$. In the following, we prove that there exists a point z_{im+1} on θ_{im+1}^* such that

$$\log|f(z_{im+1})| \geq R_{m+1}^{k/2-\varepsilon_{n+1}}.$$

Indeed, otherwise, according to Theorem 3.1,

$$\log|f(z_{im})| \leq N'' + 9\sqrt{2}\exp\left\{-\pi\int_{2R_m}^{R_{m+1}/2}\frac{dt}{t\theta_i(t)}\right\}R_{m+1}^{k/2-\varepsilon_{n+1}},$$

$$R_m^{k/2-\varepsilon_n} \leq \frac{1}{2}R_m^{k/2-\varepsilon_n}$$

$$+ 9\sqrt{2}\exp\left\{\int_{2R_m}^{R_{m+1}/2}\left(\frac{\pi}{\theta_i(t)} - \frac{k}{2}\right)\frac{dt}{t} - \frac{k}{2}\log\frac{R_{m+1}}{4R_m}\right\}R_{m+1}^{k/2-\varepsilon_{n+1}}.$$

Furthermore, it follows from formulas (4.22) and (4.23) that $K \leq 18\sqrt{2}\times 2^k$. But this contradicts $K = 36 \times 2^k$.

Taking connectivity into account, there exists a continuous curve Γ_{im} linking z_{im} and z_{im+1} in Ω_{im}. Moreover, when $z \in \Gamma_{im}$,

$$\log|f(z)| \geq N'' \geq \frac{1}{4}R_m^{k/2-\varepsilon_n}, \qquad |z| \leq R_{m+1}.$$

Besides, according to formula (4.24), we conclude that $|z| \geq R_{n+1}'$. Hence, Lemma 4.2 holds.

Analogously, we may prove the following lemma:

LEMMA 4.3. *When $m_n + 1 \leq m \leq m_{n+1} - 1$ $(n \geq n_0)$, suppose that there exists a point z_{im} on θ_{im} $(1 \leq i \leq k)$ such that*

$$\log|f(z_{im})| \geq R_m^{k/2-\varepsilon_n}.$$

Then there must exist a point z_{im+1} on θ_{im+1} such that

$$\log|f(z_{im+1})| \geq R_{m+1}^{k/2-\varepsilon_n}.$$

Moreover, there is a continuous curve Γ_{im} linking z_{im} and z_{im+1} on Ω_i, such that when $z \in \Gamma_{im}$,

$$\log|f(z)| \geq \frac{1}{4}R_m^{k/2-\varepsilon_n}, \qquad R_n' < |z| < R_{m+1}.$$

(5) Now, we proceed to complete the proof of Theorem 4.8. Analogous to the proof of Lemma 4.2, by applying Theorem 3.1 and according to formulas (4.17), (4.18) and (4.21)–(4.23), we may conclude that there exists a point $z_{im_{n_0}+1}$ on θ_{im} $(1 \leq i \leq k, m = m_0 + 1)$, such that

$$\log|f(z_{im_{n_0}+1})| \geq R_m^{k/2-\varepsilon_{n_0}}.$$

Next, by repeating the application of Lemmas 4.2 and 4.3, we obtain a orderly sequence of points z_{im} $(1 \leq i \leq k; m = m_{n_0} + 1, m_{n_0} + 2, \ldots)$ and sequence of curves Γ_{im} $(m = m_{n_0}+1, m_{n_0}+2, \ldots)$, such that when $z \in \Gamma_{im}$ $(m_n + 1 \leq m \leq m_{n+1}; n \geq n_0)$,

$$\log|f(z)| \geq \frac{1}{4}R_m^{k/2-\varepsilon_n}, \qquad R_n' \leq |z| \leq R_{m+1}. \tag{4.25}$$

To discuss further: when $|z| < R_m$, it follows that

$$\log|f(z)| \geq \frac{1}{4}|z|^{k/2-\varepsilon_n}; \tag{4.26}$$

and when $R_m \leq |z| \leq R_{m+1}$,

$$\log|f(z)| \geq \frac{1}{4}|z|^{k/2-\varepsilon_n}\left\{\frac{R_m}{R_{m+1}}\right\}^{k/2-\varepsilon_n}.$$

Furthermore, according to $R_m \geq K^{(n+1)(n+2)/2}$ and formula (4.23), we conclude that

$$\left\{\frac{R_m}{R_{m+1}}\right\}^{k/2-\varepsilon_n} \geq R_m^{-2k\varepsilon_n}.$$

And hence

$$\log|f(z)| \geq \frac{1}{4}|z|^{k/2-(2k+1)\varepsilon_n}.$$

Combining this with formula (4.26), when $z \in \Gamma_{im}$ $(m_n + 1 \leq m \leq m_{n+1}; n \geq n_0)$, it is always true that

$$\log|f(z)| \geq \frac{1}{4}|z|^{k/2-(2k+1)\varepsilon_n}.$$

Let

$$\Gamma_i = \bigcup_{m=m_0+1} \Gamma_{im}, \qquad 1 \leq i \leq k.$$

According to $R'_n \to +\infty$ $(n \to +\infty)$ and formula (4.25), we conclude that Γ_i is a continuous curve tending to ∞. Moreover, it is obvious that $\Gamma_i \subset \Omega_i \subset D_i$, and also

$$\lim_{\substack{|z| \to +\infty \\ z \in \Gamma_i}} \frac{\log \log |f(z)|}{\log |z|} = \lambda.$$

Hence, Theorem 4.8 is proved completely.

THEOREM 4.9. *Under the hypothesis of Theorem 4.8, the angle between any two adjacent Borel directions of $f(z)$ of order λ is $\leq \frac{\pi}{\lambda}$.*

PROOF. Suppose that Theorem 4.9 does not hold, meaning that there exist two adjacent Borel directions $\Delta(\theta_1)$ and $\Delta(\theta_2)$ $(0 \leq \theta_1 < \theta_2 < 2\pi)$ of order λ, such that $\theta_2 - \theta_1 > \frac{\pi}{\lambda}$. In the following, we shall derive a contradiction.

(1) Choose a number ε, such that

$$0 < \varepsilon < \frac{1}{2}(\theta_2 - \theta_1) \quad \text{and} \quad \frac{\pi}{\theta_2 - \theta_1 - \varepsilon} < \lambda.$$

According to Lemma 2.13, we have

$$\varlimsup_{R \to +\infty} \frac{\log^+ \log^+ M\{\overline{\Omega}(\theta_1 + \varepsilon, \theta_2 - \varepsilon; R), f\}}{\log R} \leq \lambda' < \lambda.$$

Hence, for any arbitrary small number $\eta > 0$, provided that R is sufficiently large, it follows that

$$\log M\{\overline{\Omega}(\theta_1 + \varepsilon, \theta_2 - \varepsilon; R), f\} \leq R^{\lambda' + \eta}. \tag{4.27}$$

Meanwhile, according to the definition of order, we have $\log M(R, f) \leq R^{\lambda + \eta}$. Besides, from Theorem 4.8, there exists a continuous curve Γ_i tending to ∞ as in D_i, such that when $z \in \Gamma_i$ and $|z| \geq R$, $\log |f(z)| > R^{\lambda - \eta}$.

(2) In the following we illustrate that when R is sufficiently large, there exists at most one asymptotic path L_{i_0} $(1 \leq i_0 \leq k)$ intersecting $\overline{\Gamma}(\theta_1 + \varepsilon, \theta_2 - \varepsilon; R)$. Indeed, otherwise, then there exist two asymptotic paths L_{i_1} and L_{i_2} $(1 \leq i_1 < i_2 \leq k)$ intersecting $\overline{\Gamma}(\theta_1 + \varepsilon, \theta_2 - \varepsilon; R)$. Hence Γ_{i_1} must intersect $\overline{\Gamma}(\theta_1 + \varepsilon, \theta_2 - \varepsilon; R)$. Suppose that the intersection point is z_1. Then $\log |f(z_1)| \geq R^{\lambda - \eta}$. But provided that η is suitably small, the above result contradicts formula (4.27).

Now, we consider the case when only one asymptotic path L_{i_0} $(1 \leq i_0 \leq k)$ intersects $\overline{\Gamma}(\theta_1 + \varepsilon, \theta_2 - \varepsilon; R)$. First, let z_i $(i = 1, 2, \ldots, k)$ be the intersection point of Γ_i and circumference $|z| = R$. Then

$$\log |f(z_i)| \geq R^{\lambda - \eta} \tag{4.28}$$

and also $\arg z_{i_0} > \theta_2 - \varepsilon$, $\arg z_{i_0 - 1} < \theta_1 + \varepsilon$ $(z_0 \equiv z_K)$.

Let

$$R^{\lambda - \eta} = 4R_2^{\lambda' + \eta}. \tag{4.29}$$

Then

$$\frac{R_2}{4R} = 4^{-(\lambda'+\eta+1)/(\lambda'+\eta)} R^{(\lambda-\lambda'-2\eta)/(\lambda'+\eta)} \to +\infty \qquad (R \to +\infty), \qquad (4.30)$$

and

$$\log R = \frac{\lambda'+\eta}{\lambda-\lambda'-2\eta} \log \frac{R_2}{4R} + \frac{\lambda'+\eta+1}{\lambda-\lambda'-2\eta} \log 4. \qquad (4.31)$$

There exists value N in the interval $[R_2^{\lambda'+\eta}, 2R_2^{\lambda'+\eta}]$, such that derivative $f'(z)$ has no zeros on the level curve $\log|f(z)| = N$, meaning that the level curve is analytic. We consider the set

$$E = \{z \mid \log|f(z)| > N, |z| < R_2\}$$

and let Ω_i $(i = 1, 2, \ldots, k)$ be the connected component containing point z_i $(i = 1, 2, \ldots, k)$. Obviously, provided that R is sufficiently large, we have

$$\Omega_i \subset D_i \cap (|z| < R_2), \qquad i = 1, 2, \ldots, k.$$

Moreover, Ω_{i_0} and Ω_{i_0-1} $(\Omega_0 \equiv \Omega_k)$ do not intersect $\Omega(\theta_1+\varepsilon, \theta_2-\varepsilon; R_2)$. Besides, according to the maximum modulus principle, we conclude that $\overline{\Omega}_i \cap (|z| = R_2)$ is not an empty set. We denote θ_{it} as the part of circumference $|z| = t$ in Ω_i and $t\theta_i(t)$ as the linear measure of θ_{it}. Hence, according to Theorem 3.1,

$$\log|f(z_i)| \le N + 9\sqrt{2} \exp\left\{-\pi \int_{2R}^{R_2/2} \frac{dt}{t\theta_i(t)}\right\} R_2^{\lambda+\eta}, \qquad i = 1, 2, \ldots, k.$$

Furthermore, from formulas (4.28) and (4.29) it follows that

$$\pi \int_{2R}^{R_2/2} \sum_{i=1}^{K} \frac{1}{\theta_i(t)} \cdot \frac{dt}{t} \le k(\lambda+\eta) \log R_2 - k(\lambda-\eta) \log R + k \log 18\sqrt{2}.$$

On the other hand, notice that

$$k^2 = \left\{\sum_{i=1}^{K} \sqrt{\theta_i(t)} \frac{1}{\sqrt{\theta_i(t)}}\right\}^2 \le \sum_{i=1}^{K} \theta_i(t) \cdot \sum_{i=1}^{K} \frac{1}{\theta_i(t)}$$

$$\le \{2\pi - (\theta_2 - \theta_1 - 2\varepsilon)\} \sum_{i=1}^{K} \frac{1}{\theta_i(t)},$$

$$\frac{\pi}{2\pi - (\theta_2 - \theta_1 - 2\varepsilon)} k^2 \le \pi \sum_{i=1}^{K} \frac{1}{\theta_i(t)}.$$

Then

$$\frac{\pi}{2\pi - (\theta_2 - \theta_1 - 2\varepsilon)} k^2 \log \frac{R_2}{4R} \le k(\lambda+\eta) \log R_2$$

$$- k(\lambda-\eta) \log R + k \log 18\sqrt{2}.$$

Moreover, according to formula (4.13), it follows that

$$\frac{\pi}{2\pi - (\theta_2 - \theta_1 - 2\varepsilon)}\lambda \log \frac{R_2}{4R} \leq (\lambda + \eta) \log \frac{R_2}{4R}$$

$$+ \frac{\lambda' + \eta}{\lambda - \lambda' - 2\eta} 2\eta \log \frac{R_2}{4R}$$

$$+ \log\left(18\sqrt{2} \times 4^{\lambda+\eta} \times 4^{\frac{\lambda'+\eta+1}{\lambda-\lambda'-2\eta} \cdot 2\eta}\right).$$

Let $R \to +\infty$ and $\eta \to 0$. Then according to formula (4.30) we conclude that

$$\frac{2\pi}{2\pi - (\theta_2 - \theta_1 - 2\varepsilon)}\lambda \leq \lambda.$$

Hence, a contradiction is derived.

Finally, we consider the case when no asymptotic path L_i $(1 \leq i \leq k)$ intersects $\overline{\Gamma}(\theta_1 + \varepsilon, \theta_2 - \varepsilon; R)$. $\Gamma(\theta_1 + \varepsilon, \theta_2 - \varepsilon; R)$ must lie entirely in a domain D_{i_0} $(1 \leq i_0 \leq k)$. Similarly, we have point z_i $(i = 1, 2, \ldots, k)$, and

$$\arg z_{i_0} < \theta_1 + \varepsilon \qquad \text{or} \qquad \arg z_{i_0} > \theta_2 - \varepsilon.$$

Again, followed by an analogous discussion as in the previous case, we shall derive a contradiction. Hence, Theorem 4.9 is proved completely.

COROLLARY. *Under the hypothesis of Theorem 4.8, if we let q be the number of Borel directions of $f(z)$ of order λ, then $k \leq q$.*

THEOREM 4.10. *Under the hypothesis of Theorem 4.8, if we further let $q < +\infty$, then the k asymptotic paths of $f(z)$ must asymptotically tend to the k Borel directions of $f(z)$ ([8]) of order λ, respectively. Meanwhile, among these k Borel directions, the angle between any two adjacent directions is equal to $\frac{\pi}{\lambda}$.*

PROOF. Let $\Delta(\theta_j)$ $(j = 1, 2, \ldots, q; 0 \leq \theta_1 < \theta_2 < \cdots < \theta_q < 2\pi)$ be the $q(< +\infty)$ Borel directions of $f(z)$ of order λ and

$$\omega = \min_{1 \leq j \leq q} (\theta_{j+1} - \theta_j), \qquad \theta_{q+1} = \theta_1 + 2\pi.$$

First, we prove that an asymptotic path L_i $(1 \leq i \leq k)$ must asymptotically tend to a Borel direction $\Delta(\theta_{j_i})$ $(1 \leq j_i \leq q)$ of order λ. Indeed, otherwise, then there exists at least one asymptotic path L_{i_0} $(1 \leq i_0 \leq k)$, such that L_{i_0} cannot asymptotically tend to $\Delta(\theta_j)$ $(j = 1, 2, \ldots, q)$. Hence, there exist number ε, $0 < \varepsilon < \frac{\omega}{8}$ and the corresponding infinite sequence R_n $(n = 1, 2, \ldots)$, $R_n < R_{n+1} \to +\infty$ $(n \to +\infty)$, such that L_{i_0} intersects

---

([8]) We denote a curve L continuously stretching to ∞ tends to a half line $\Delta(\theta)$. If given arbitrarily $\varepsilon > 0$, there exists value $R > 0$, such that the part of L outside disk $|z| \leq R$ lies entirely in $\Omega(\theta - \varepsilon, \theta + \varepsilon; R, +\infty)$.

at a certain $\overline{\Gamma}(\theta_{j_n} + 3\varepsilon, \theta_{j_{n+1}} - 3\varepsilon; R_n)$ $(1 \leq j_i \leq q)$. By selecting a subsequence of R_n, we may suppose that θ_{jn} is independent of n, and hence two adjacent Borel directions $\Delta(\theta_j)$ and $\Delta(\theta_{j+1})$ $(j = j_n)$ of order λ are determined.

According to Theorem 2.14, there exist two distinct finite complex numbers α and β, such that

$$\varlimsup_{R \to +\infty} \frac{\log^+ n\{\Omega(\theta_j + \varepsilon, \theta_{j+1} - \varepsilon; R), f = X\}}{\log R} \leq \lambda' < \lambda, \qquad X = \alpha, \beta.$$

Let the distances among α, β and ∞ be greater than d, $0 < d < \frac{1}{2}$. We first choose a number σ, such that σ satisfies the condition:

$$\frac{(\lambda - \lambda')(\theta_{j+1} - \theta_j - 2\varepsilon)}{\lambda(\theta_{j+1} - \theta_j - 2\varepsilon) + 2[6\pi + \lambda'(\theta_{j+1} - \theta_j - 2\varepsilon)]}$$

$$< \sigma < \frac{(\lambda - \lambda')(\theta_{j+1} - \theta_j - 2\varepsilon)}{2[6\pi + \lambda'(\theta_{j+1} - \theta_j - 2\varepsilon)]} \cdot (^9) \tag{4.32}$$

Then we apply a lemma (Lemma 5.2), which is to be proved in Chapter 5, where we let $\theta = \frac{1}{2}(\theta_{j+1} - \theta_j) - \varepsilon$, $R_1 = R_2^{1-\sigma}$, $R = R_2 = R_n$, $L = L_{i_0} \cap \overline{\Omega}(\theta_j + 3\varepsilon, \theta_{j+1} - 3\varepsilon; R_1, R_2)$, $H = \varepsilon$, $a = a_{i_0}$ and $N = 0$. Then for any arbitrary small number $\eta > 0$, provided that n is sufficiently large, for any point z on $\overline{\Omega}(\theta_j + 3\varepsilon, \theta_{j+1} - 3\varepsilon; R_1, R_2)$, we have

$$|f(z) - a_{i_0}| \leq \exp\{A(\varepsilon, \theta, d)R_n^{12\pi\sigma/(\theta_{j+1}-\theta_j-2\varepsilon)}$$
$$\times [R_n^{2(\lambda'+\eta)\sigma} \cdot R_n^{\lambda'+\eta} \cdot \sigma \log R_n + \log^+ |a_{i_0}|]\}.$$

Hence, provided that n is sufficiently large, for any point z on $\overline{\Omega}(\theta_j + 3\varepsilon, \theta_{j+1} - 3\varepsilon; R_n^{1-\sigma}, R_n)$, it follows that

$$\log |f(z)| \leq R_n^{\lambda''+\eta''},$$

$$\lambda'' = \lambda' + \frac{12\pi\sigma}{\theta_{j+1} - \theta_j - 2\varepsilon} + 2\lambda'\sigma, \qquad \eta'' = 2(\sigma + 1)\eta. \tag{4.33}$$

According to formula (4.32), we conclude that $\lambda'' < \lambda$. Besides, provided that η is sufficiently small, $\lambda'' + \eta'' < \lambda - \eta''$. We define value R, such that

$$R^{\lambda-\eta''} = 4R_n^{\lambda''+\eta''}; \tag{4.34}$$

then $R < R_n$. On the other hand, from formula (4.32) we conclude that $\lambda'' > (1 - \sigma)\lambda$. Hence, provided that η is sufficiently small, $\lambda'' + \eta'' > (1 - \sigma)(\lambda + \eta'')$. We conclude that

$$R_n^{\lambda''+\eta''} > R_n^{(1-\sigma)(\lambda+\eta'')} = R_1^{\lambda+\eta''},$$
$$R^{\lambda-\eta''} > R_1^{\lambda+\eta''}, \qquad R > R_1. \tag{4.35}$$

$(^9)$ In order to guarantee that $\sigma < 1$, we may always assume λ' sufficiently near to λ.

Now, let z_i $(1 \leq i \leq k)$ be the intersection point of Γ_i and circumference $|z| = R$. Then according to Theorem 4.8, provided that n is sufficiently large, it follows that $\log|f(z_i)| \geq R^{\lambda - \eta''}$. Obviously, according to formulas (4.33) and (4.34) we conclude that z_i $(i = 1, 2, \ldots, k)$ cannot be in $\overline{\Gamma}(\theta_j + 3\varepsilon, \theta_{j+1} - 3\varepsilon; R)$. Followed by a discussion analogous to the proof of Theorem 4.9 and notice that according to formula (4.35), here the corresponding Ω_i $(i = 1, 2, \ldots, k)$ cannot reach circumference $|z| = R_1$, then this will result in a contradiction.

Next, we prove that two distinctive asymptotic paths L_{i_1} and L_{i_2} $(1 \leq i_1 \leq i_2 \leq k)$ cannot asymptotically tend to a same Borel direction $\Delta(\theta_j)$ $(1 \leq j \leq q)$ simultaneously. Indeed, otherwise, let L_{i_1} and L_{i_2} tend asymptotically to $\Delta(\theta_j)$ simultaneously. Choose number ε, $0 < \varepsilon < \frac{\pi}{2\lambda}$; then there exists value R_0 such that the part of L_{i_1} and L_{i_2} outside disk $|z| \leq R_0$ lies entirely in $\Omega(\theta_j - \varepsilon, \theta_j + \varepsilon; R_0, +\infty)$. Let D be the simply-connected domain bounded by L_{i_1} and L_{i_2}. Then according to the Lindelöf Theorem, there exists point $z_0 \in D$ such that

$$|f(z_0)| > 2eN, \qquad N = \sup_{z \in L_{i_1} \cup L_{i_2}} |f(z)| < +\infty. \qquad (4.36)$$

There exists value N' in the interval $[N, 2N]$, such that $f'(z)$ has no zeros on the level curve $|f(z)| = N'$, meaning that the level curve is analytic. We consider the set $E = \{z \mid |f(z)| > N'\}$, and let Ω be the connected branch containing point z_0. Then $\Omega \subset D$. Moreover, according to the maximum modulus principle, we conclude that Ω is an unbounded domain. We use θ_t to denote the part of circumference $|z| = t$ in Ω and $t\theta(t)$ the linear measure of θ_t. Applying Theorem 3.1 we have

$$\log|f(z_0)| \leq \log N' + 9\sqrt{2} \exp\left\{-\pi \int_{2|z_0|}^{R/2} \frac{dt}{t\theta(t)}\right\} \log M(R, f).$$

Notice that when $\max\{2|z_0|, 2R_0\} \leq t \leq \frac{1}{2}R$, $\theta(t) \leq 2\varepsilon$, and when η is sufficiently small and R is sufficiently large, it follows that $\log M(R, f) \leq R^{\lambda + \eta}$. Hence, we obtain, based on formula (4.36), that

$$1 \leq 9\sqrt{2} \times 4^{\pi/2\varepsilon} \times \left\{\frac{\max(|z_0|, R_0)}{R}\right\}^{\pi/2\varepsilon R^{\lambda + \eta}}.$$

Furthermore, according to $\frac{\pi}{2\varepsilon} > \lambda$ and the fact that η is sufficiently small, then when R is sufficiently large, we shall derive a contradiction.

Finally, let L_i $(i = 1, 2, \ldots, k)$ tend asymptotically to $\Delta(\theta_{j_i})$ $(i = 1, 2, \ldots, k; 0 \leq \theta_{j_1} < \theta_{j_2} < \cdots < \theta_{j_k})$, respectively. We prove that

$$\theta_{j_{i+1}} - \theta_{j_i} = \frac{\pi}{\lambda}, \qquad i = 1, 2, \ldots, k, \ \theta_{j_{k+1}} = \theta_{j_1} + 2\pi.$$

Indeed, otherwise, then two typical cases occur:

(1) $\theta_{j_2} - \theta_{j_1} < \frac{\pi}{\lambda}$. First we choose a fixed number ε, such that $0 < \varepsilon < \frac{1}{2}(\frac{\pi}{\lambda} - (\theta_{j_2} - \theta_{j_1}))$. In the sequel, we may obtain analogously R_0, z_0 and the following inequality:

$$1 \le 9\sqrt{2} \times 4\pi/(\theta_{j_2} - \theta_{j_1} + 2\varepsilon) \times \left\{ \frac{\max(|z_0|, R_0)}{R} \right\}^{\pi/(\theta_{j_2} - \theta_{j_1} + 2\varepsilon)R^{\lambda + \eta}}.$$

Hence, according to

$$\frac{\pi}{\theta_{j_2} - \theta_{j_1} + 2\varepsilon} > \lambda$$

and the fact that η is sufficiently small, when R is sufficiently large, we shall derive a contradiction.

(2) $\theta_{j_2} - \theta_{j_1} > \frac{\pi}{\lambda}$. Obviously,

$$2\pi = \sum_{i=1}^{k} (\theta_{j_{i+1}} - \theta_{j_i}) = \theta_{j_2} - \theta_{j_1} + \sum_{i=2}^{k} (\theta_{j_{i+1}} - \theta_{j_i})$$
$$> \frac{\pi}{\lambda} + (k-1)\frac{\pi}{\lambda} = k\frac{\pi}{\lambda} = 2\pi.$$

Hence, we obtain a contradiction, meaning that Theorem 4.10 is proved completely.

THEOREM 4.11. *Under the hypothesis of Theorem 4.8, for any arbitrary value* θ, $0 \le \theta < 2\pi$, $\Delta(\theta)$ *is a Borel direction of* $f(z)$ *of order* λ *or there exists a number* σ, $0 < \sigma < \frac{\pi}{4}$, *such that*

$$\lim_{\substack{|z|=+\infty \\ z \in (\Omega(\theta - \sigma, \theta + \sigma) - E)}} \frac{\log\log|f(z)|}{\log|z|} = \lambda, \tag{4.37}$$

where E *is a point set in* $\Omega(\theta - \sigma, \theta + \sigma)$, *and satisfies the condition*

$$\lim_{r \to +\infty} \text{meas}\{\Lambda(\theta - \sigma, \theta + \sigma; r, +\infty) \cap E\} = 0. \tag{4.38}$$

PROOF. Without any loss of generality, we need only to prove that if $\Delta(0)$ is not a Borel direction of $f(z)$ of order λ, then there must exist a number σ, $0 < \sigma < \frac{\pi}{4}$, such that

$$\lim_{\substack{|z|=+\infty \\ z \in \{\Omega(-\sigma, \sigma) - E\}}} \frac{\log\log|f(z)|}{\log|z|} = \lambda,$$

where E is a point set in $\Omega(-\sigma, \sigma)$, and satisfies the condition

$$\lim_{r \to +\infty} \text{meas}\{\Omega(-\sigma, \sigma, r, +\infty) \cap E\} = 0.$$

If $\Delta(0)$ is not a Borel direction of $f(z)$ of order λ, then there exist numbers σ, $0 < \sigma < \frac{\pi}{4}$ and $\lambda_1 < \lambda$, and two distinct finite complex numbers α and β, such that when r is sufficiently large,

$$n\{\Omega(-2\sigma, 2\sigma, r), f = X\} \le r^{\lambda_1}, \qquad X = \alpha, \beta. \tag{4.39}$$

Let b_i $(i = 1, 2, \ldots)$ be denoted as the α-value points and β-value points in with $|b_i| \leq |b_{i+1}|$ $(i = 1, 2, \ldots)$. Then

$$\sum_{i=1}^{\infty} \frac{1}{|b_j|^{\lambda_1+1}} < +\infty.$$

We construct a disk with b_i $(i = 1, 2, \ldots)$ being the centre and $1/|b_j|^{\lambda_1+2}$ being the radius, and denote these disks as E. Obviously,

$$\text{meas}\{\Omega(-\sigma, \sigma; r, +\infty) \cap E\} \leq \sum_{|b_j|+1/|b_j|^{\lambda_1+2} \geq r} \frac{2}{|b_j|^{\lambda_1+2}}$$

$$\leq \frac{2}{r - 1/|b_1|^{\lambda_1+2}} \sum_{i=1}^{+\infty} \frac{1}{|b_j|^{\lambda_1+1}}.$$

And hence

$$\lim_{r \to +\infty} \text{meas}\{\Omega(-\sigma, \sigma; r, +\infty) \cap E\} = 0.$$

In the following, we need only to prove

$$\lim_{\substack{|z| \to +\infty \\ z \in (\Omega(-\sigma, \sigma) - E)}} \frac{\log \log |f(z)|}{\log |z|} = \lambda.$$

Indeed, otherwise, then there must exist an infinite sequence of points $z_n \in \{\Omega(-\sigma, \sigma) - E\}$ $(n = 1, 2, \ldots)$, $|z_n| < |z_{n+1}| \to +\infty$ $(n \to +\infty)$, such that

$$\log |f(z_n)| \leq |z_n|^{\lambda_2}, \qquad \lambda_2 < \lambda. \tag{4.40}$$

Let

$$\lambda_3 = \max\{\lambda_1, \lambda_2\} < \lambda,$$
$$\eta = \frac{\lambda - \lambda_3}{4\{\pi/\sigma + \lambda\}}. \tag{4.41}$$

Construct the transformation

$$\zeta = \zeta(z) = \frac{z^{\pi/4\sigma} - |z_n|^{\pi/4\sigma}}{z^{\pi/4\sigma} + |z_n|^{\pi/4\sigma}}; \tag{4.42}$$

then according to Lemma 3.1, $\Omega(-2\sigma, 2\sigma)$ is mapped onto a unit circle $|\zeta| < 1$ on the ζ-plane, and the image domain of $\overline{\Omega}(-\sigma, \sigma; |z_n|^{1-\eta}, |z_n|)$ on the ζ-plane must be included in disk $|\zeta| \leq t$, where

$$t = 1 - \frac{1}{4}|z_n|^{-\pi\eta/2\sigma}. \tag{4.43}$$

Also the image domain of disk $|\zeta| < \frac{1}{2}(1+t)$ on the z-plane must be included in $\overline{\Omega}(-2\sigma, 2\sigma; R_n', R_n)$, where

$$R_n = 16^{4\sigma/\pi}|z_n|^{1+2\eta},$$
$$R_n' = 16^{-4\sigma/\pi}|z_n|^{1-2\eta}. \tag{4.44}$$

Let ζ_n be an image point of z_n on the ζ-plane, and we construct the transformation

$$\xi = \xi(\zeta) = \frac{((1+t)/2)(\zeta - \zeta_n)}{((1+t)/2)^2 - \overline{\zeta}_n \zeta} . \tag{4.45}$$

Then disk $|\zeta| \leq \frac{1}{2}(1 + \tau)$ and disk $|\zeta| \leq \tau$ are transformed onto the unit circle $|\xi| \leq 1$ and disk $|\xi| \leq \tau$ on the ξ-plane, respectively, and

$$\tau = \frac{t(1+t)}{((1+t)/2)^2 + t^2} . \tag{4.46}$$

Let ξ_i be an image point of b_i in disk $|\xi| \leq 1$, where ζ_i is its corresponding point on the ζ-plane. In the following, we try to find a lower bound estimate of $|\xi_i|$:

$$
\begin{aligned}
|\xi_i| &= \left| \frac{((1+t)/2)(\zeta_i - \zeta_n)}{((1+t)/2)^2 - \overline{\zeta}_n \zeta_i} \right| \geq \frac{1}{2} |\zeta_i - \zeta_n| \\
&= \frac{1}{2} \left| \frac{b_i^{\pi/4\sigma} - |z_n|^{\pi/4\sigma}}{b_i^{\pi/4\sigma} + |z_n|^{\pi/4\sigma}} - \frac{z_n^{\pi/4\sigma} - |z_n|^{\pi/4\sigma}}{z_n^{\pi/4\sigma} + |z_n|^{\pi/4\sigma}} \right| \\
&= \frac{1}{2} \left| \frac{2|z_n|^{\pi/4\sigma}(z_n^{\pi/4\sigma} - b_i^{\pi/4\sigma})}{(z_n^{\pi/4\sigma} + |z_n|^{\pi/4\sigma})(b_i^{\pi/4\sigma} + |z_n|^{\pi/4\sigma})} \right| \\
&\geq \frac{1}{4 R_n^{\pi/4\sigma}} |z_n^{\pi/4\sigma} - b_i^{\pi/4\sigma}| .
\end{aligned}
\tag{4.47}
$$

Notice that points z_n and b_i are in $\Omega(-2\sigma, 2\sigma; R'_n, R_n)$, and

$$|z_n - b_i| \geq \frac{1}{|b_i|^{\lambda_1 + 2}} .$$

We consider the transformation $x = z^{\pi/4\sigma}$. Then points z_n and b_i are mapped into points $z_n^{\pi/4\sigma}$ and $b_i^{\pi/4\sigma}$ on the x-plane, respectively . We let l_x be a straight line segment connecting points $z_n^{\pi/4\sigma}$ and $b_i^{\pi/4\sigma}$, where $l \subset \Omega(-2\sigma, 2\sigma; R'_n, R_n)$ is its image on the z-plane. Hence,

$$\frac{1}{|b_i|^{\lambda_1 + 2}} \leq \int_l |dz| = \int_{l_x} |z'(x)| |dx| . \tag{4.48}$$

Notice that when n is sufficiently large, we have

$$|z'(x)| = \frac{4\sigma}{\pi} |x|^{4\sigma/\pi - 1} = \frac{4\sigma}{\pi} |z|^{1 - \pi/4\sigma} \leq \frac{4\sigma}{\pi} \cdot \frac{1}{R_n^{1\pi/4\sigma - 1}} \leq 1 .$$

Therefore, it follows from formula (4.48) that

$$\frac{1}{|b_i|^{\lambda_1 + 2}} \leq |z_n^{\pi/4\sigma} - b_i^{\pi/4\sigma}| .$$

Furthermore, according to formulas (4.48) and (4.45), when n is sufficiently large, we conclude that

$$|\xi_i| \geq \frac{1}{4R_n^{\pi/4\sigma} \cdot |b_i|^{\lambda_1+2}} \geq \frac{1}{4R_n^{\pi/4\sigma+\lambda_1+2}} \geq \frac{1}{|z_n|^{2(\lambda+3)}} = d. \tag{4.49}$$

Let $z(\zeta)$ and $\zeta(\xi)$ be the inverse transformations of formulas (4.42) and (4.45) respectively, and let

$$F(\xi) = \frac{f(z(\zeta(\xi))) - \alpha}{\beta - \alpha}. \tag{4.50}$$

Then when n is sufficiently large, we have from formula (4.39) that

$$N = n(1, F = 0) + n(1, F = 1) \leq 2R_n^{\lambda_1} \leq c \cdot |z_n|^{(1+2\eta)\lambda_3},$$

where c is a constant independent of n. Besides, according to formula (4.50), the distances between origin $\xi = 0$ and the 0-, 1-value points of $F(\xi)$ are smaller than d. Moreover, according to formula (4.41) we have

$$\log^+ |F(0)| = \log^+ |f(z_n)| \leq |z_n|^{\lambda_3}.$$

Hence, applying Theorem 2.9, we obtain

$$T\left(\frac{1+\tau}{2}, F\right) \leq \frac{c}{1-\tau}|z_n|^{(1+2^\eta)\lambda_3}\left\{\log|z_n| + \log\frac{4}{1-\tau}\right\}.$$

Furthermore, we have from formula (4.50) that

$$T\left(\frac{1+\tau}{2}, f(z(\zeta(\xi)))\right) \leq \frac{c}{1-\tau}|z_n|^{(1+2^\eta)\lambda_3}\left\{\log|z_n| + \log\frac{4}{1-\tau}\right\}.$$

And therefore

$$\log M(\tau, f(z(\zeta(\xi)))) \leq \frac{((1+\tau)/2)+\tau}{((1+\tau)/2)-\tau}T\left(\frac{1+\tau}{2}, f(z(\zeta(\xi)))\right)$$

$$\leq \frac{c}{(1-\tau)^2}|z_n|^{(1+2_\eta)\lambda_3}\left\{\log|z_n| + \log\frac{4}{1-\tau}\right\}.$$

Particularly,

$$\log M\{\overline{\Omega}(-\sigma, \sigma; |z_n|^{1-\eta}, |z_n|), f\}$$

$$\leq \frac{c}{(1-\tau)^2}|z_n|^{(1+2^\eta)\lambda_3}\left\{\log|z_n| + \log\frac{4}{1-\tau}\right\}. \tag{4.51}$$

Taking formulas (4.44) and (4.46) into account, it follows that

$$1 - \tau = \frac{((1-t)/2)^2}{((1+t)/2)^2 + t^2} \geq \frac{1}{8}(1-t)^2 = \frac{1}{128}|z_n|^{-\pi\eta/\sigma}.$$

And hence it follows from formula (4.51) that

$$\log M\{\overline{\Omega}(-\sigma, \sigma; |z_n|^{1-\eta}, |z_n|), f\} \leq c|z_n|^{2\pi\eta/\sigma+(1+2\eta)\lambda_3}\log|z_n|.$$

Furthermore, according to formula (4.42), for any arbitrary given number ε, $0 < \varepsilon < \frac{\lambda - \lambda_3}{8}$, when n is sufficiently large, we obtain

$$\log M\{\overline{\Omega}(-\sigma, \sigma; |z_n|^{1-\eta}, |z_n|), f\} \le |z_n|^{(\lambda+\lambda_3)/2+\varepsilon}. \tag{4.52}$$

Let a_i $(i = 1, 2, \ldots, k)$ be k distinctive finite asymptotic values of $f(z)$, with their corresponding asymptotic paths being L_i $(i = 1, 2, \ldots, k)$. Region D_i is bounded by L_i and L_{i+1}. According to Theorem 4.8, there exists a curve Γ_i stretching continuously to ∞ in D_i, such that

$$\lim_{\substack{|z| \to +\infty \\ z \in \Gamma_i}} \frac{\log\log|f(z)|}{\log|z|} = \lambda.$$

Let z_i' $(1 \le i \le k)$ be an intersection point of Γ_i and circumference $|z| = |z_n|^{1-\eta}$. Then when n is sufficiently large, we have

$$\log|f(z_i')| \ge |z_n|^{(1-\eta)(\lambda-\varepsilon)}, \qquad i = 1, 2, \ldots, k. \tag{4.53}$$

There exists value N' in the interval $[\frac{1}{4}|z_n|^{(1-\eta)(\lambda-\varepsilon)}, \frac{1}{2}|z_n|^{(1-\eta)(\lambda-\varepsilon)}]$, such that derivative $f'(z)$ has no zeros on the level curve $\log|f(z)| = N'$, meaning that the level curve is analytic. We consider the set

$$E = \{z \mid \log|f(z)| > N', \ |z| < |z_n|\}$$

and let Ω_i be the connected branch containing point z_i' $(1 \le i \le k)$. According to the maximum modulus principle, we conclude that $\overline{\Omega}_i \cap (|z| = |z_n|)$ is not an empty set. On the other hand, when n is sufficiently large, we have $\Omega_i \subset D_i$ $(i = 1, 2, \ldots, k)$. Moreover, based on formula (4.42) and $\varepsilon < \frac{1}{8}(\lambda - \lambda_3)$, it follows that

$$\frac{1}{4}|z_n|^{(1-\eta)(\lambda-\varepsilon)} > |z_n|^{(\lambda+\lambda_3)/2+\varepsilon}.$$

Hence, according to formula (4.53), we conclude that there is no intersection between Ω_i $(i = 1, 2, \ldots, k)$ and $\Omega(-\sigma, \sigma; |z_n|^{1-\eta}, |z_n|)$.

We denote θ_{it} $(i \le i \le k; |z_n|^{1-\eta} \le t \le |z_n|)$ as the portion of circumference $|z| = t$ in Ω_i, and $t\theta_i(t)$ as the linear measure of θ_{it}. When $t \in [|z_n|^{1-\eta}, |z_n|]$,

$$k^2 = \left\{ \sum_{i=1}^{k} \sqrt{\theta_i(t)} \cdot \frac{1}{\sqrt{\theta_i(t)}} \right\}^2 \le \sum_{i=1}^{k} \theta_i(t) \sum_{i=1}^{k} \frac{1}{\theta_i(t)}$$

$$\le (2\pi - 2\sigma) \sum_{i=1}^{k} \frac{1}{\theta_i(t)}.$$

And hence

$$\frac{\pi k^2}{2\pi - 2\sigma} \int_{2|z_n|^{1-\eta}}^{|z_n|/2} \frac{dt}{t} \le \sum_{i=1}^{k} \pi \int_{2|z_n|^{1-\eta}}^{|z_n|/2} \frac{dt}{t\theta_i(t)}. \tag{4.54}$$

Applying Theorem 3.1, we get

$$\sum_{i=1}^{k} \pi \int_{2|z_n|^{1-\eta}}^{|z_n|/2} \frac{dt}{t\theta_i(t)} \leq k \log \log M(|z_n|, f)$$

$$- k(1 - \eta)(\lambda - \varepsilon) \log|z_n| + k \log 18\sqrt{2}.$$

Combining this and formula (4.54), we have

$$\frac{\pi k}{2\pi - 2\sigma} \log \frac{|z_n|^{\eta}}{4} \leq \log \log M(|z_n|, f)$$

$$- (1 - \eta)(\lambda - \varepsilon) \log|z_n| + \log 18\sqrt{2}.$$

Therefore, when $n \to +\infty$, we conclude that

$$\frac{\pi k \eta}{2\pi - 2\sigma} \leq \lambda - (1 - \eta)(\lambda - \varepsilon).$$

Furthermore, by letting $\varepsilon \to 0$,

$$\lambda \geq \frac{\pi k}{2\pi - 2\sigma} > \frac{k}{2}.$$

But this contradicts the hypothesis, implying that Theorem 4.11 holds.

COROLLARY 1. *Under the hypothesis of Theorem 4.8, for any arbitrary value* θ, $0 \leq \theta < 2\pi$, $\Delta(\theta)$ *is a Julia direction of* $f(z)$ *or there exists a number* σ, $0 < \sigma < \frac{\pi}{4}$, *such that*

$$\lim_{\substack{|z| \to +\infty \\ z \in \Omega(\theta - \sigma, \theta + \sigma)}} \frac{\log \log |f(z)|}{\log |z|} = \lambda. \qquad (4.55)$$

It is apparent that Corollary 1 holds. When $\Delta(\theta)$ is not a Julia direction of $f(z)$, in the proof of Theorem 4.11, here the set $\{b_i\}$ of the α-, β-value points is a finite one. Hence, E is a bounded set, and therefore, formula (4.55) holds.

COROLLARY 2. *Under the hypothesis of Theorem 4.8, let* a_i *(*$i = 1, 2, \ldots,$* k) *be* k *distinct asymptotic values of* $f(z)$, *and* L_i *be an asymptotic path corresponding to* a_i *(*$1 \leq i \leq k$). *If for any arbitrary small number* $\varepsilon > 0$ *and for any arbitrary large value* r *(*$0 < r < +\infty$), *there always exist some intersections between* $\Omega(\theta - \varepsilon, \theta + \varepsilon; r, +\infty)$ *and* L *(*$L = \bigcup_{i=1}^{k} L_i$), *then* $\Delta(\theta)$ *must be a Borel direction of* $f(z)$ *of order* λ.

In fact, if $\Delta(\theta)$ is not a Borel direction of $f(z)$ of order λ, then according to Theorem 4.11, there exists number σ, $0 < \sigma < \frac{\pi}{4}$, such that formulas (4.37) and (4.38) hold. On the other hand, for any arbitrary large value r, $\Omega(\theta - \frac{\sigma}{2}, \theta + \frac{\sigma}{2}; r, +\infty)$ intersects L. Hence

$$\text{meas}\{\Omega(\theta - \sigma, \theta + \sigma; r, +\infty) \cap L\} \geq \frac{\sigma}{2} r \to +\infty \qquad (r \to +\infty),$$

meaning that there exists a point sequence in $\Omega(\theta - \sigma, \theta + \sigma; r, +\infty)$ not belonging to E that tends to ∞, such that $f(z)$ is bounded on such a point sequence. However, this contradicts formula (4.37). Hence, Corollary 2 holds.

COROLLARY 3. *Under the hypothesis of Theorem 4.8, if there exists a value* θ, $0 \le \theta < 2\pi$, *such that* $\Delta(\theta)$ *is not a Borel direction of* $f(z)$ *of order* λ, *then, when* z *tends continuously to* ∞ *along* L_i $(1 \le i \le k)$, *starting from a certain point on* L *the variation of the arguments* $\arg z$ *is smaller than* 2π.

Indeed, obviously Corollary 3 also holds. Otherwise, according to Corollary 2, we conclude that $\Delta(\theta)$ must be a Borel direction of $f(z)$ of order λ.

In Corollary 3, if merely under the hypothesis of Theorem 4.8, under general circumstances, L. Ahlfors had already proved [1b] $\arg z = o(\log|z|)$.

THEOREM 4.12. *Under the hypothesis of Theorem 4.8, for any arbitrary value* θ, $0 \le \theta < 2\pi$, *and any arbitrary small number* $\varepsilon > 0$, *it is always true that*

$$\lim_{r \to +\infty} \frac{\log \log M\{\overline{\Omega}(\theta - \varepsilon, \theta + \varepsilon; r), f\}}{\log r} = \lambda.$$

According to the proof of Theorem 4.11, we can easily prove the validity of Theorem 4.12. In fact, otherwise, then we soon come up with formula (4.52). By a similar discussion, a contradiction will result.

Before giving Theorem 4.13 and its proof, we need to prove two lemmas first.

Let $f(z)$ be a meromorphic function on the open plane $|z| < +\infty$, and $\{r_n\}$ be a monotonic sequence tending to ∞, such that

$$\lim_{r \to +\infty} \frac{\log T(r_n, f)}{\log r_n} = \nu < +\infty.$$

We also assume that $f(z)$ has a finite deficient value b, with its corresponding deficiency $\delta(b, f) = \delta > 0$. According to the definition of deficient value, there exists value r_0, such that when $r \ge r_0$, we have

$$\frac{\delta}{2} T(r, f) \le m(r, b), \tag{4.56}$$

and

$$T\left(r, \frac{1}{f-b}\right) \le 2T(r, f). \tag{4.57}$$

Applying Lemma 2.1, where we let $h = 3$, $k = 4\nu$, $h_1 = 0$, then it follows that

$$\lim_{n \to +\infty} \frac{1}{\log r_n} \int_{E[r_0, r_n]} \frac{dt}{t} \ge \frac{1}{4}, \tag{4.58}$$

where

$$E[r_0, r_n] = E \cap [r_0, r_n], \tag{4.59}$$

and

$$E = \{t | T(e^3 t, f) \le e^{4\nu} T(t, f), t \ge r_0\}. \tag{4.60}$$

LEMMA 4.4. *For any arbitrary value* $t \in E \cap (r_0, +\infty)$, *let*

$$E_t = \left\{ r \Big| \operatorname{meas} e_{r_\theta} \geq \frac{\pi\delta}{40e^{4_\nu}}, \ t \leq r \leq et \right\},$$

where

$$e_{r\theta} = \left\{ \theta \Big| \log \frac{1}{|f(re^{i\theta}) - b|} \geq \frac{\delta}{4} T(r, f), \ 0 \leq \theta < 2\pi \right\}. \tag{4.61}$$

Then

$$\int_{E_t} \frac{dt}{r} \geq \frac{1}{2}. \tag{4.62}$$

PROOF. Let b_i $(i = 1, 2, \ldots, n(te^2, b))$ be the b-value points of $f(z)$ on disk $|z| \leq e^2 t$. According to the Boutroux-Cartan Theorem, the point set satisfying the inequality

$$\prod_{i=1}^{n(e^2t, b)} |z - b_i| < H^{n(e^2t, b)}$$

can be included in at most $n(e^2t, b)$ disks (γ), with the sum of their Euclidean radii not exceeding $2eH$. Assume $H = t/4\sqrt{e}$ and let

$$\widetilde{E}_t = \{ r | (|z| = r) \cap (\gamma) = \varnothing, \quad r \in [t, et] \}.$$

Then

$$\int_{\widetilde{E}_t} \frac{dr}{r} \geq 1 - \int_t^{\sqrt{e^t}} \frac{dr}{r} = \frac{1}{2}. \tag{4.63}$$

Based on the assumption of r_0, when $r \in \widetilde{E}_t$,

$$\frac{\delta}{2} T(r, f) \leq \frac{1}{2\pi} \int_0^{2\pi} \log^+ \frac{1}{|f(re^{i\theta}) - b|} d\theta.$$

Furthermore, we derive from formula (4.61) that

$$\frac{\delta}{4} T(r, f) \leq \frac{1}{2\pi} \int_{e_{r_\theta}} \log^+ \frac{1}{|f(re^{i\theta}) - b|} d\theta. \tag{4.64}$$

Applying the Poisson-Jensen formula, we may obtain

$$\log \frac{1}{|f(re^{i\theta}) - b|} \leq \frac{e^2 + e}{e^2 - e} m \left(te^2, \frac{1}{f - b} \right)$$

$$+ \sum_{i=1}^{n(te^2, b)} \log \left| \frac{(te^2)^2 - \overline{b}_i re^{i\theta}}{te^2(re^{i\theta} - b_i)} \right|$$

$$\leq \frac{e + 1}{e - 1} m \left(te^2, \frac{1}{f - b} \right) + n(te^2, b) \log(8e^{5/2})$$

$$\leq \frac{e + 1}{e - 1} m \left(te^2, \frac{1}{f - b} \right) + N(te^3, b) \log(8e^3)$$

$$\leq \left\{ \frac{e + 1}{e - 1} + \log(8e^3) \right\} T \left(te^3, \frac{1}{f - b} \right).$$

Hence, according to formula (4.60), we conclude that

$$\log \frac{1}{|f(e^{i\theta}) - b|} \leq 2e^{4\nu} \left\{ \frac{e+1}{e-1} + \log(8e^3) \right\} T(r, f)$$

$$\leq 20e^{4\nu} T(r, f).$$

Combining this with formula (4.64) it gives

$$\frac{\delta}{4} T(r, f) \leq \frac{1}{2\pi} 20 e^{4\nu} T(r, f) \operatorname{meas} e_{r\theta},$$

it follows that

$$\operatorname{meas} e_{r\theta} \geq \frac{\pi\delta}{40 e^{4\nu}}.$$

Therefore, $\widetilde{E}_t \subset E_t$. Hence, according to formula (4.63), we conclude that formula (4.62) holds, implying that Lemma 4.4 is valid.

LEMMA 4.5. *Let*

$$\widetilde{E} = \left\{ t \mid \operatorname{meas} e_{t\theta} \geq \frac{\pi\delta}{40e^{4\nu}}, \ t \geq r_0 \right\},$$

$$\widetilde{E}[r_0 r_n] = \widetilde{E} \cap [r_0, r_n].$$

(4.65)

Then

$$\lim_{r_n \to +\infty} \frac{1}{\log r_n} \int_{\widetilde{E}[r_0, r_n]} \frac{dt}{t} \geq \frac{1}{24}.$$

PROOF. Notice the inequality

$$T(te^3, f) \leq e^{4\nu} T(r, f).$$

(4.66)

Let t_1 be the smallest value satisfying formula (4.66) in the interval $[r_0, +\infty)$. Then

$$\int_{E\cap[t_1, e^3 t_1]} \frac{dt}{t} \leq 3 = 3 \cdot 2 \cdot \frac{1}{2} \leq 6 \cdot \int_{E_{t_1}} \frac{dt}{t} \leq 6 \int_{\widetilde{E}\cap[t_1, e^3 t_1]} \frac{dt}{t}.$$

Let t_2 be the smallest value satisfying formula (4.66) in the interval $[e^3 t_1, +\infty)$. Then

$$\int_{E\cap[t_2, e^3 t_2]} \frac{dt}{t} \leq 6 \cdot \int_{\widetilde{E}\cap[t_2, e^3 t_2]} \frac{dt}{t}.$$

In general, let t_m be the smallest value satisfying formula (4.66) in interval $[e^3 t_{m-1}, +\infty)$, and also $t_{m-1} \leq r_n \leq e^3 t_m$. Then

$$\int_{E[r_0, r_n]} \frac{dt}{t} \leq \sum_{i=1}^{n-1} \int_{E\cap[t_i, e^3 t_i]} \frac{dt}{t} + 3$$

$$\leq \sum_{i=1}^{n-1} 6 \int_{\widetilde{E}\cap[t_i, e^3 t_i]} \frac{dt}{t} + 3$$

$$\leq 6 \int_{\widetilde{E}[r_0, r_n]} \frac{dt}{t} + 3.$$

Moreover, according to formula (4.58) we conclude that

$$\lim_{n \to +\infty} \frac{1}{\log r_n} \int_{\widetilde{E}[r_0, r_n]} \frac{dt}{t} \geq \frac{1}{6} \lim_{n \to +\infty} \int_{E[r_0, r_n]} \frac{dt}{t} \geq \frac{1}{24},$$

meaning that Lemma 4.5 holds.

Now, we state Theorem 4.13 and present its proof as follows:

THEOREM 4.13. *Under the hypothesis of Theorem 4.8, $f(z)$ cannot have any finite deficient value.*

PROOF. Suppose that Theorem 4.13 does not hold, meaning that $f(z)$ possesses a finite deficient value, with its corresponding deficiency $\delta(b, f) = \delta > 0$. Let a_i $(i = 1, 2, \ldots, k)$ be k distinct finite asymptotic values of $f(z)$, with their corresponding asymptotic paths being L_i $(i = 1, 2, \ldots, k)$, and simply connected region D_i be bounded by L_i and L_{i+1}. Let

$$L = \bigcup_{i=1}^{k} L_i, \qquad N = \max\left\{\sup_{z \in L} |f(z)|, |b|\right\} < +\infty.$$

Then according to the Lindelöf Theorem, there exist points $z_i \in D_i$ $(i = 1, 2, \ldots, k)$ such that

$$|f(z_i)| \geq 2eN. \tag{4.67}$$

There exists value N' in the interval $[N, 2N]$, such that derivative $f'(z)$ has no zeros on the level curve $|f(z)| = N'$, hence, the level curve is analytic. We consider the set $E = \{z \mid |f(z)| > N'\}$ and let $\Omega_i \subset D_i$ be the connected branch containing point z_i $(1 \leq i \leq k)$. According to the maximum modulus principle, we conclude that Ω_i $(i = 1, 2, \ldots, k)$ is an unbounded domain.

We denote θ_{it} $(1 \leq i \leq k; |z_i| \leq t < +\infty)$ the portion of circumference $|z| = t$ in Ω_i, and $t\theta_i(t)$ as the linear measure of θ_{it}. Applying Theorem 3.1, we have

$$\log |f(z_i)| \leq \log N' + 9\sqrt{2} \exp\left\{-\pi \int_{2|z_i|}^{r/2} \frac{dt}{t\theta_i(t)}\right\} \log M(r, f),$$

$$i = 1, 2, \ldots, k.$$

Furthermore, according to formula (4.67), we obtain

$$\pi \int_{2|z_i|}^{r/2} \frac{dt}{t\theta_i(t)} \leq \log \log M(r, f) + \log 9\sqrt{2}, \qquad i = 1, 2, \ldots, k. \tag{4.68}$$

And hence

$$\sum_{i=1}^{k} \pi \int_{2|z_i|}^{r/2} \frac{dt}{t\theta_i(t)} \leq k \log \log M(r, f) + k \log 9\sqrt{2}. \tag{4.69}$$

Choose a fixed value r_0 such that

$$r_0 \geq \max\{1, |z_1|, |z_2|, \ldots, |z_k|\},$$

and when $r \geq r_0$, formulas (4.56) and (4.57) hold. Applying Lemma 4.5, where we let $\nu = \lambda$, then, when $t \in \tilde{E}[2r_0, \frac{1}{2}r_n]$, according to formula (4.65) it follows that

$$k^2 = \left\{ \sum_{i=1}^{k} \sqrt{\theta_i(t)} \frac{1}{\sqrt{\theta_i(t)}} \right\}^2 \leq \sum_{i=1}^{k} \theta_i(t) \sum_{i=1}^{k} \frac{1}{\theta_i(t)}$$

$$\leq \left\{ 2\pi - \frac{\pi\delta}{40e^{4\lambda}} \right\} \sum_{i=1}^{k} \frac{1}{\theta_i(t)} .$$

When $t \in C\tilde{E}[2r_0, \frac{1}{2}r_n]$, where $C\tilde{E}[2r_0, r_n] = [2r_0, \frac{1}{2}r_n] - \tilde{E}(2r_0, \frac{1}{2}r_n]$, it follows that $k^2 \leq 2\pi \sum_{i=1}^{k} 1/\theta_i(t)$. Hence, we derive from formula (4.69) that

$$\frac{\pi k^2}{2\pi - \pi\delta/40e^{4\lambda}} \int_{\tilde{E}[2r_0, r_n/2]} \frac{dt}{t} + \frac{k^2}{2} \int_{C\tilde{E}[2r_0, r_n/2]} \frac{dt}{t}$$

$$\leq \sum_{i=1}^{k} \pi \int_{2r_0}^{r_n/2} \frac{dt}{t\theta_i(t)}$$

$$\leq k \log\log M(r_n, f) + k \log(9\sqrt{2})$$

$$\leq k \log T(2r_n, f) + k \log(27\sqrt{2}) ,$$

namely

$$\left\{ \frac{\pi k}{2\pi - \pi\delta/40e^{4\lambda}} - \frac{k}{2} \right\} \frac{1}{\log(r_n/2)} \int_{\tilde{E}[2r_0, r_n/2]} \frac{dt}{t} + \frac{k}{2} \left\{ 1 - \frac{\log 2r_0}{\log(r_n/2)} \right\}$$

$$\leq \frac{\log T(2r_n, f)}{\log 2r_n} \cdot \frac{\log 2r_n}{\log r_n} \left\{ 1 + \frac{\log 2}{\log(r_n/2)} \right\} + \frac{\log(27\sqrt{2})}{\log(r_n/2)} .$$

Letting $r \to +\infty$, and according to Lemma 4.5, we conclude that

$$\lambda \geq \frac{k}{2} + \frac{1}{24} \cdot \left\{ \frac{\pi k}{2\pi - \pi\delta/40e^{4\lambda}} - \frac{k}{2} \right\} > \frac{k}{2} ,$$

which is a contradiction. Hence, Theorem 4.13 is proved.

According to the proof of Theorem 4.13, we now may easily complete the proof of Theorem 3.5. In fact, if entire function $f(z)$ possesses a finite deficient value b, with its corresponding deficiency $\delta(b, f) = \delta > 0$, then when $r \geq r_0$, it follows that

$$m(r, b) = \frac{1}{2\pi} \int_0^{2\pi} \log^+ \frac{1}{|f(re^{i\theta}) - b|} d\theta$$

$$\geq \frac{\delta}{2} T(r, f) \geq \frac{\delta}{2} T(r_0, f) .$$

Hence, there exists a point $z_r = re^{i\theta_r}$ on circumference $|z| = r$ $(r \geq r_0)$, such that

$$|f(z_r) - b| \leq e^{-(\delta/2)\tau(r_0, f)} ,$$

implying that $f(z)$ has an upper bound N on the point set $\{z_r | r \geq r_0\}$. On the other hand, since $f(z)$ is unbounded on the open plane $|z| < +\infty$, we conclude that there exists a point z_1 such that $|f(z_1)| \geq 2eN$. To consider analogously, we may have corresponding region Ω_1 and corresponding formula (4.68). Besides, when applying Lemma 2.1, where we let $\nu = \mu$, μ is a lower order of $f(z)$; then we may conclude that $\mu > \frac{1}{2}$, and Theorem 3.5 holds.

§4.3. Growth of entire functions along an asymptotic path

According to the Inversen Theorem, any entire function $f(z)$ must possess ∞ as its asymptotic value, meaning that there exists a continuous curve L tending to the infinity on the open plane $|z| < +\infty$, such that when z tends to ∞ along L, $f(z)$ tends also to ∞. Then, how fast does $f(z)$ tend to ∞ along L? This is the main issue to be discussed in this section [43b].

Suppose that $\nu(r)$ is defined as a continuous monotonic function tending to ∞ on the positive real axis, and satisfies the following conditions:

(1)

$$\varlimsup_{r \to +\infty} \frac{\log \nu(r)}{\log r} = 0.$$

(2) For any arbitrarily given number $\varepsilon > 0$,

$$\varlimsup_{r \to +\infty} \frac{\nu(r^{1+\varepsilon})}{\nu(r)} = C(\varepsilon) < +\infty.$$

Here $C(\varepsilon) \geq 1$. Moreover, when $\varepsilon \to 0$, $C(\varepsilon) \to 1$. For example, we may assume $\nu(r) = \log r$, $(\log r)^2$, $\log r \cdot \log \log r$, $\ldots$.

When $\varepsilon \to 0$, $C(\varepsilon) \to 1$. Hence, there exists number $\varepsilon_0 > 0$, such that when $\varepsilon \leq \varepsilon_0$, $C(\varepsilon) < 1 + 1/25$. According to the assumption of $\nu(r)$, for any arbitrarily assumed number $\varepsilon, 0 < \varepsilon \leq \varepsilon_0$, there must exist value r_0, such that when $r \geq r_0$,

$$\nu(r) < r^{\varepsilon/3}, \qquad \nu(r^{1+\varepsilon}) < \left(C(\varepsilon) + \frac{4}{25}\right)\nu(r) < \left(1 + \frac{1}{5}\right)\nu(r).$$

Meanwhile, there exists value r_n, $r_n \geq r_0$, such that when $t \geq r_n$,

$$18\sqrt{2} \cdot t^{-\varepsilon/2} 8 \left(1 + \frac{1}{5}\right) n t^{\varepsilon/3} \leq \frac{1}{2}, \qquad (4.70)$$

where n is a positive integer.

Let

$$\tau = \max\{5^{1/\varepsilon}, r_n\}, \qquad \sigma = \tau^{1/(1+\varepsilon)}, \qquad r = \tau^{1+\varepsilon}, \qquad R = \tau^{(1+\varepsilon)^2}$$

and assume that $f(z)$ is regular on disk $|z| \leq R$, $f(0) = 1$. Moreover, there exists a point z_0 in annulus $2\sigma \leq |z| \leq \tau$ such that $|f(z_0)| \geq A$, $A \geq 16$. Then according to Lemma 3.5, we may define a region $D_r(A')$ ($\sqrt[4]{A} < A' < \sqrt{A}$) such that when $z \in D_r(A')$ it follows that $|f(z)| > A'$.

Moreover, for any two points z_2 and z_3 on $\overline{D}_r(A')$, we may find a piecewise analytic curve L linking these two points, with its length being

$$\text{meas } L \leq 2r + 2\sqrt{2}\pi r\sqrt{\left(\log\frac{R}{r}\right)^{-1} T(R, f)}. \tag{4.71}$$

Meanwhile, for the point z on L,

$$|f(z)| \geq \sqrt[4]{A}, \qquad |z| \leq r. \tag{4.72}$$

LEMMA 4.6. *There must exist a point z_1 in the intersection set between annulus $2\tau < |z| \leq r$ and $\overline{D}_r(A')$ such that*

$$|f(z_1)| \geq \min\{M(|z_1|, f), e^{8n\nu(|z_1|)}\}.$$

PROOF. If, in annulus $2\tau < |z| < r$, there exists circumference $|z| = t$ lying entirely in $D_r(A')$, then we need only to choose a point z_1 on the circumference $|z| = t$ such that $|f(z_1)| = M(|z_1|, f)$, and hence we may conclude that Lemma 4.6 holds. Therefore, we need only to consider the following case: In annulus $2\tau < |z| \leq r$, no circumference $|z| = t$ lies entirely in $D_r(A')$, meaning that each circumference $|z| = t$ intersects the boundary of $D_r(A')$. According to Theorem 3.1, where we assume $\chi = \frac{1}{2}$, notice that

$$2|z_0| \leq 2\tau < \frac{1}{2}r, \quad \theta^*(t) \leq 2\pi, \quad 2\tau < t \leq r;$$

then we may conclude that there exists at least a point z_1 on θ_r such that

$$|f(z_1)| \geq e^{8n\nu(|z_1|)}.$$

Otherwise, we would have the following estimate:

$$\log|f(z_0)| \leq \log A' + 9\sqrt{2}\exp\left\{-\pi\int_{2|z_0|}^{r/2}\frac{1}{t\theta^*(t)}\right\}8n\nu(r),$$

$$\log(A')^2 \leq \log A' + 9\sqrt{2}\exp\left\{-\pi\int_{2\tau}^{1/2}\frac{dt}{t\theta^*(t)}\right\}8n\nu(r),$$

$$1 \leq 18\sqrt{2}\cdot\tau^{-\varepsilon/2}8\left(1+\frac{1}{5}\right)n\cdot\tau^{\varepsilon/3}.$$

However, this contradicts formula (4.70). Hence, Lemma 4.6 is proved.

THEOREM 4.14. *Let $f(z)$ be an entire function on the open plane $|z| < +\infty$, and let it satisfy the condition*

$$\varlimsup_{r\to+\infty}\frac{\log M(r, f)}{\nu(r)} = +\infty.$$

Then there must exist an asymptotic path L extending from the origin and tending to ∞ such that

$$\varlimsup_{\substack{|z|\to+\infty \\ z\in L}}\frac{\log|f(z)|}{\nu(|z|)} = +\infty.$$

When $\nu(r) = \log r$, we obtain the result [21d] due to R. Boas.

PROOF. Without any loss of generality, we may assume that $|f(0)| = 1$. Since $f(z)$ satisfies the condition

$$\lim_{r \to +\infty} \frac{\log M(r, f)}{\nu(r)} = +\infty,$$

for each number n $(n = 1, 2, \ldots)$, we may define a certain value r_n, $r_0 < r_n < r_{n+1}$, such that when $r \geq r_n$,

$$M(r, f) \geq e^{8n\nu(r)},$$

$$\nu(r) \geq \max\{1, \log M(1, f)\}. \tag{4.73}$$

We choose arbitrarily a fixed number ε, $0 < \varepsilon < \min\{\varepsilon_0, \frac{1}{5}\}$. Then we choose a value R_0, $R_0 > 5^{1/\varepsilon}$, and construct a sequence $R_m = R_0^{(1+\varepsilon)^m}$, $m = 1, 2, \ldots$. Furthermore, we may define a sequence m_n $(n = 1, 2, \ldots)$, $m_n + 2 < m_{n+1}$, such that when $m \geq m_n$,

$$R_m \geq r_n, \tag{4.74}$$

and when $t \geq R_{m_n}$,

$$18\sqrt{2} \cdot t^{-\varepsilon/2} 8 \left(1 + \frac{1}{5}\right) n \cdot t^{\varepsilon/3} \leq \frac{1}{2}.$$

Finally, applying the formula

$$\log M(R'_n, f) = 2n\nu(R_{m_n}), \qquad n = 1, 2, \ldots, \tag{4.75}$$

we define a sequence R'_n $(R'_n > 1)$; then when $n \to +\infty$, $R'_n \to +\infty$.

In the following, we prove the following two facts:

(1) We may choose successively a sequence of points: z_m $(m = m_1 + 1, m_1 + 2, \ldots)$ such that

$$2R_{m-1} < |z_m| \leq R_m, \qquad m \geq m_1 + 1,$$

$$|f(z_m)| \geq e^{8n\nu(|z_m|)}, \qquad m_n < m \leq m_{n+1}. \tag{4.76}$$

(2) We may find successively the piecewise analytic curves L_m $(m = m_1 + 1, m_1 + 2, \ldots)$ linking points z_m and z_{m+1} such that when $z \in L_m$,

$$R'_n \leq |z| \leq R_{m+1}, \qquad m_n + 1 \leq m \leq m_{n+1},$$

$$|f(z)| \geq e^{n\nu(|z|)}, \qquad m_n + 1 \leq m \leq m_{n+1}.$$

In fact, we need only to consider the following few points:

1) How to choose points z_{m_1+1}?

2) After point z_{m_n} $(n \geq 2)$ has been chosen, how to choose point z_{m_n+1} and curve L_{m_n}?

3) After point z_{m_n+i} has been chosen, how to choose point z_{m_n+i+1} $(n \geq 1; 1 \leq i \leq m_{n+1} - m_n - 1)$ and curve L_{m_n+i}?

First we consider 1). There must exist point z_{m_1+1} on circumference $|z| = R_{m_1+1}$ such that $|f(z_{m_1+1})| = M(|z_{m_1+1}|, f)$. Then according to formulas (4.73) and (4.74), we obtain

$$|f(z_{m_1+1})| \geq e^{8n\nu(|z_{m_1+1}|)}.$$

Next, we consider 2). Applying Lemma 4.6, where we let $z_0 = z_{m_n}$, $r_n = R_{m_n}$, $r = R_{m_n+1}$, $\tau = R_{m_n}$, $\sigma = R_{m_n-1}$, $A = e^{8(n-1)\nu(|z_{m_n}|)}$, then there must exist a point z_{m_n+1} in the intersection set between annulus $2R_{m_n} < |z| < R_{m_n+1}$ and $\overline{D}_r(A')$ such that

$$|f(z_{m_n+1})| \geq \min\{M(|z_{m_n+1}|, f), \; e^{8n\nu(|z_{m_n+1}|)}\}.$$

We then conclude from formulas (4.73) and (4.74) that $|f(z_{m_n+1})| \geq e^{8n\nu(|z_{m_n+1}|)}$. On the other hand, according to formula (4.72), there exists a piecewise analytic curve L_{m_n} linking points z_{m_n} and z_{m_n+1}, such that when $z \in L_{m_n}$,

$$|f(z)| \geq e^{2(n-1)\nu(|z_{m_n}|)}, \qquad |z| \leq R_{m_n+1}. \tag{4.77}$$

Notice that $|z_{m_n}| > 2R_{m_n-1} > R_{m_n-1}$, and also consider formula (4.75). We conclude that $|z| \geq R'_{n-1}$. On the other hand, when $|z| \leq |z_{m_n}|$, we have $|f(z)| \geq e^{2(n-1)\nu(|z|)}$. When $|z| > |z_{m_n}|$, we have that

$$|f(z)| \geq e^{2(n-1)\nu(|z|)} \cdot e^{2(n-1)\nu(|z_{m_n}|)-2(n-1)\nu(|z|)}$$
$$\geq e^{2(n-1)\nu(|z|)} \cdot e^{2(n-1)\{\nu(R_{m_n-1})-\nu(R_{m_n+1})\}}.$$

Notice that

$$\nu(R_{m_n+1}) - \nu(R_{m_n-1})$$
$$= \nu(R_{m_n+1}) - \nu(R_{m_n}) + \nu(R_{m_n}) - \nu(R_{m_n-1})$$
$$\leq \frac{1}{5}\{\nu(R_{m_n}) + \nu(R_{m_n-1})\}$$
$$\leq \frac{1}{5}\left(2 + \frac{1}{5}\right)\nu(|z|) < \frac{1}{2}\nu(|z|).$$

It follows that $|f(z)| \geq e^{(n-1)\nu(|z|)}$.

Finally we consider 3). Again applying Lemma 4.6, where we let $z_0 = z_{m_n+i}$, $r_n = R_{m_n}$, $r = R_{m_n+i+1}$, $\tau = R_{m_n+i}$, $\sigma = R_{m_n+i-1}$, $A = e^{8n\nu(|z_{m_n+i}|)}$, then there must exist a point z_{m_n+i+1} in the intersection set between annulus $2R_{m_n+i} < |z| \leq R_{m_n+i+1}$ and $\overline{D}_r(A')$ such that

$$|f(z_{m_n+i+1})| \geq \min\{M(|z_{m_n+i+1}|, f), \; e^{8n\nu(|z_{m_n+i+1}|)}\}$$
$$= e^{8n\nu(|z_{m_n+i+1}|)}.$$

On the other hand, according to formula (4.72), there exists a piecewise analytic curve L_{m_n+i} linking points z_{m_n+i} and z_{m_n+i+1}, such that for point z on L_{m_n+i} we have

$$|f(z)| \geq e^{2n\nu(|z_{m_n+i}|)}, \qquad |z| \leq R_{m_n+i+1}. \tag{4.78}$$

Notice that $i \geq 1$, $|z_{m_n+i}| \geq R_{m_n}$, and consider formula (4.75). We conclude that $|z| \geq R'_n$. On the other hand, when $|z| \leq R_{m_n+i}$, $|f(z)| \geq e^{2n\nu(|z|)}$, and when $|z| > |z_{m_n+i}|$, $|f(z)| \geq e^{n\nu(|z|)}$.

After illustrating 1), 2), and 3), facts (1) and (2) are obvious. Now, we use a straight line L_{m_1} to connect the origin $z = 0$ and point z_{m_1+1}. Then we let

$$L = \bigcup_{m=m_1}^{+\infty} L_m. \tag{4.79}$$

Hence, L is a piecewise analytic curve extending from the origin and tending to infinity. Moreover,

$$\lim_{\substack{|z|\to+\infty \\ z\in L}} \frac{\log|f(z)|}{\nu(|z|)} = +\infty.$$

Hence, Theorem 4.14 is proved.

If we let $\nu(r) = \log r$ and replace $M(r, f)$ by $M\{D \cap (|z| = r), f\}$, then by repeating the proof of Theorem 4.14, we may complete the proof of Theorem 4.6.

Now, we consider another kind of theorem relating to the property of growth. A. Huber once proved the following result [23a]:

If $f(z)$ is a transcendental entire function, then for each $\rho > 0$, there exists a path L_ρ tending to ∞ such that

$$\int_{L_\rho} |f(z)|^{-\rho}|dz| < +\infty.$$

W. Hayman once asked whether or not a path not depending on ρ exists [21d]. For this question, when order $\lambda < +\infty$,([10]) we may obtain an affirmative answer.

THEOREM 4.15. *Suppose that $f(z)$ is a transcendental entire function of order $\lambda < +\infty$. Then there must exist a path L tending to ∞ such that for each value $\rho > 0$,*

$$\int_L |f(z)|^{-\rho}|dz| < +\infty.$$

___

([10]) Recently, J. Lewis, J. Rossi and A. Weitsman had successfully eliminated the hypothetical condition that order $\lambda < +\infty$.

PROOF. According to the assumption that $f(z)$ is a transcendental entire function, we have

$$\varlimsup_{r \to +\infty} \frac{\log M(r, f)}{\log r} = +\infty.$$

On the other hand, according to the assumption that the order of $f(z)$ is $\lambda < +\infty$, for any arbitrarily given number ε, $0 < \varepsilon < \frac{1}{5}$, there exists value t_0, such that when $t \geq t_0$, $T(t, f) \leq t^{\lambda + \varepsilon}$. We repeat the proof of Theorem 4.14, where we assume particularly $\nu(r) = \log r$ and $R_0 \geq \max\{5^{1/\varepsilon}, t_0\}$. According to formulas (4.77) and (4.78), we conclude that when $z \in L_m$ $(m_n \leq m \leq m_{n+1})$,

$$|f(z)| \geq |z_m|^{2n}. \tag{4.80}$$

Meanwhile, it follows from formula (4.71) that

$$\operatorname{meas} L_m \leq 2R_{m+1} + 2\sqrt{2}\pi R_{m+1}$$
$$\times \sqrt{[\log(R_{m+2}/R_{m+1})]^{-1} T(R_{m+2}, f)} \tag{4.81}$$
$$\leq 2(1 + \sqrt{2}\pi) R_{m+1}^{1+(\lambda/2)+((\lambda+\varepsilon+1)/2) \cdot \varepsilon}.$$

Let $L = \bigcup_{m=m_1+1}^{\infty} L_m$. Then L is the required path tending to ∞. In fact, for each value $\rho > 0$, there exists a certain positive integers n_0, $n_0 = n_0(\rho)$, such that when $n \geq n_0$,

$$2\rho n - (1 + \varepsilon)^2 \left\{ 1 + \frac{\lambda}{2} + \frac{\lambda + 1 + \varepsilon}{2} \cdot \varepsilon \right\} \leq 2.$$

We also let

$$L' = \bigcup_{m=m_1+1}^{m_{n_0}} L_m, \qquad L'' = \bigcup_{m=m_{n_0}+1}^{+\infty} L_m.$$

Then according to formulas (4.77), (4.80) and (4.81), we conclude that

$$\int_L \frac{1}{|f(z)|^\rho} d|z| = \int_{L'} \frac{1}{|f(z)|^\rho} |dz| + \int_{L''} \frac{1}{|f(z)|^\rho} |dz| \leq \int_{L'} \frac{|dz|}{|f(z)|^\rho}$$
$$+ 2(1 + \sqrt{2}\pi) \sum_{m=m_{n_0}+1} R_{m+1}^{1+(\lambda/2)+((\lambda+1+\varepsilon)/2)\varepsilon} \frac{1}{2^{2n\rho} R_{m-1}^{2n\rho}}$$
$$\leq \int_{L'} \frac{|dz|}{|f(z)|^\rho} + 2(1 + \sqrt{2}\pi) \sum_{m=m_{n_0}+1}^{+\infty} \frac{1}{R_{m-1}^2} < +\infty,$$

meaning that Theorem 4.15 follows.

Now, we consider the case of meromorphic functions. W. Hayman once raised the following question [21d]:

Let $f(z)$ be a meromorphic function of order μ on the open plane $|z| < +\infty$. Then when $n(r, \infty) = O(r^k)$, and $k < \frac{1}{2} < \mu$, does the corresponding Boas Theorem still hold?

In fact, we have a stronger conclusion:

THEOREM 4.16. *Let* $f(z)$ *be a meromorphic function on the open plane* $|z| < +\infty$, *and let it satisfy the following conditions:*([11])

$$\lim_{r \to +\infty} \frac{\log T(r, f)}{\log r} = \mu, \qquad \overline{\lim_{r \to +\infty}} \frac{\log^+ n(r, \infty)}{\log r} = k,$$

$$k < \rho, \qquad \rho = \min\left\{\mu, \frac{1}{2}\right\}.$$

Then there must exist a path of definite value L *tending to* ∞ *such that*

$$\lim_{\substack{|z| \to +\infty \\ z \in L}} \frac{\log \log |f(z)|}{\log |z|} \geq \rho.(^{12})$$

PROOF. Without any loss of generality, we may assume that $|f(0)| = 1$. Let $\varepsilon_n = \frac{1}{n}$ $(n = 1, 2, \ldots)$. Then there exists value n_0, $n_0 \geq 3$, such that when $n \geq n_0$, it follows that

$$(1 + 4\varepsilon_n)^3 (k + \varepsilon_n) + \varepsilon_n - \rho < 0, \qquad \varepsilon_n < \frac{\rho}{5\rho + 1}.$$

Take

$$K = 72^2 \sigma, \qquad \sigma = 2^{1/(\rho - \varepsilon_{n_0})}. \tag{4.82}$$

For each ε_n $(n \geq n_0)$, there coresponds an increasing sequence r_n $(n = n_0, n_0 + 1, \ldots)$ such that

$$r_{n_0} \geq \max\{K^{n_0(n_0 - 1)/2}, [48T(2, f)]^{1/(\rho - \varepsilon_{n_0})}\}.$$

Meanwhile, when $r \geq r_n$, $T(r, f) \geq r^{\rho - \varepsilon_n}$,

$$N(1, f) + n(r, \infty) \log r \leq r^{k + \varepsilon_n}, \tag{4.83}$$

$$1 - 2K^{2(k+1)} \cdot r^{(1 + 4\varepsilon_n)^3(k + \varepsilon_n) + \varepsilon_n - \rho} \geq \frac{1}{2}.$$

We define sequences m_n $(n \geq n_0)$ and R_m $(m \geq m_{n_0})$ such that([13])

$$\left.\begin{array}{l} R_{m_{n_0}} = r_{n_0}, \qquad R_{m+1} = K^n R_m^{1 + \varepsilon_{n-1}}, \qquad m = m_n, \\[2mm] R_{m+1} = R_m^{1 + \varepsilon_n}, \qquad m = m_n + 1, m_n + 2, \ldots, m_{n+1} - 1, \\[2mm] R_{n_0} = 0, \qquad R_{m_n} \geq r_n, \qquad n = n_0, n_0 + 1, \ldots; \qquad m_{n+1} > m_n + 3. \end{array}\right\} \tag{4.84}$$

([11]) Under more precise conditions, W. Hayman proved that ∞ must be an asymptotic value. See [21g].

([12]) When $f(z)$ is an entire function and has no zeros, this estimate is precise. See [5a].

([13]) See formula (4.23).

Notice that

$$R_{m_n+1} \geq K^n R_{m_n} \geq K^n R_{m_{n-1}+1} \geq K^{n+(n-1)} R_{m_{n-2}+1} \geq \cdots$$

$$\geq K^{n+(n-1)+(n-2)+\cdots+(n_0+1)} R_{m_{n_0}+1} \geq K^{n+(n-1)+\cdots+n_0} R_{m_{n_0}}$$

$$\geq K^{n+(n-1)+\cdots+n_0} r_{n_0} \geq K^{n+(n-1)+\cdots+n_0} K^{n_0(n_0-1)/2} = K^{(n+1)n/2}.$$

Then

$$K^n \leq R_{m_n+1}^{2/n+1}, \qquad n \geq n_0. \tag{4.85}$$

Moreover, according to the relations

$$T(2R'_n, f) = \frac{1}{48} R_{m_n}^{\rho-\varepsilon_n}, \qquad n = n_0, n_0 + 1, \ldots \tag{4.86}$$

we then obtain a sequence R'_n; it follows that $R'_n < R'_{n+1} \to +\infty$, $n \to +\infty$. Notice that

$$T(2R'_n f) = \frac{1}{48} R_{m_n}^{\rho-\varepsilon_n} \geq \frac{1}{48} R_{m_{n_0}}^{\rho-\varepsilon_n} \geq \frac{1}{48} r_{n_0}^{\rho-\varepsilon_n} \geq T(2, f).$$

We conclude that $R'_n \geq 1$ $(n \geq n_0)$. On the other hand, we also have

$$2R'_n < R_{m_n}, \qquad n \geq n_0. \tag{4.87}$$

In fact, if $2R'_n \geq r_n$, it follows that

$$T(2R'_n, f) \geq (2R'_n)^{\rho-\varepsilon_n}, \qquad \frac{1}{48} R_{m_n}^{\rho-\varepsilon_n} \geq (2R'_n)^{\rho-\varepsilon_n},$$

and

$$R_{m_n} \geq 48^{1/(\rho-\varepsilon_n)} 2R'_n > 2R'_n.$$

If $2R'_n < r_n$, we have $2R'_n < R_{m_n}$.

Let a_i denote a pole of $f(z)$ and

$$\Pi_m(z) = \prod_{|a_i|<R_{m+2}} \frac{R_{m+2}(z-a_i)}{R_{m+2}^2 - \bar{a}_i z}, \qquad F_m(z) = f(z)\Pi_m(z).$$

Then $F_m(z)$ is regular in disk $|z| < R_{m+2}$.

Now we prove the next three lemmas.

LEMMA 4.7. *When* $r \in [R_m, R_{m+1}]$ $(m_n \leq m \leq m_{n+1} - 1, n \geq n_0)$, *we have*

$$\log M(r, F_m) \geq \frac{1}{2} r^{\rho-\varepsilon_n}.$$

PROOF. First, according to $r \geq R_{m_n} \geq r_n$, $|f(0)| = 1$ and

$$T(r, f) = m(r, f) + N(r, f)$$

$$\leq m(r, F_m) + m\left(r, \frac{1}{\Pi_m}\right) + N(r, f)$$

$$\leq m(r, F_m) + m(r, \Pi_m) + \log \frac{1}{|\Pi_m(0)|}$$

$$= m(r, F_m) + N(R_{m+2}, f),$$

we obtain

$$\log M(r, F_m) \geq r^{\rho - \varepsilon_n} - \{N(1, f) + n(R_{m+2}, f) \log R_{m+2}\}$$
$$\geq r^{\rho - \varepsilon_n} - R_{m+2}^{k+\varepsilon_n}.$$

When $m = m_n$, if $n \geq n_0 + 1$, then

$$R_{m+2} = R_{m+1}^{1+\varepsilon_n} = \{K^n R_{m_n}^{1+\varepsilon_{n-1}}\}^{1+\varepsilon_n} = K^{n+1} R_{m_n}^{(n+1)/(n-1)}$$
$$\leq K^2 R_{m_{n-1}+1}^{2\varepsilon_n} \cdot R_{m_n}^{1+2\varepsilon_{n-1}}$$
$$\leq K^2 R_{m_n}^{1+2\varepsilon_n + 2\varepsilon_{n-1}} \leq K^2 R_{m_n}^{(1+4\varepsilon_n)^2}.$$

If $n = n_0$, then

$$R_{m+2} = R_{m+1}^{1+\varepsilon_{n_0}} = \{K^{n_0} R_{m_{n_0}}^{1+\varepsilon_{n_0-1}}\}^{1+\varepsilon_{n_0}}$$
$$= K^{n_0+1} R_{m_{n_0}}^{(n_0+1)/(n_0-1)} \leq K^2 r_{n_0}^{2\varepsilon_{n_0}} R_{m_{n_0}}^{1+2\varepsilon_{n_0-1}}$$
$$\leq K^2 R_{m_{n_0}}^{1+2\varepsilon_{n_0}+2\varepsilon_{n_0-1}} \leq K^2 R_{m_{n_0}}^{(1+4\varepsilon_{n_0})^2},$$

implying that

$$R_{m+2} \leq K^2 R_{m_n}^{(1+4\varepsilon_n)^2}, \qquad n \geq n_0.$$

When $m_n + 1 \leq m \leq m_{n+1} - 2$, we have

$$R_{m+2} = R_m^{(1+\varepsilon_n)^2} \leq K^2 R_m^{(1+4\varepsilon_n)^2};$$

and when $m = m_{n+1} - 1$,

$$R_{m+2} = K^{n+1} R_{m_{n+1}}^{1+\varepsilon_n} = K^{n+1} R_{m_{n+1}-1}^{(1+\varepsilon_n)^2}$$
$$\leq K R_{m_{n+1}}^{2\varepsilon_{n+1}} R_m^{(1+\varepsilon_n)^2} \leq K^2 R_m^{(1+4\varepsilon_n)^2}.$$

Hence, we conclude that

$$\log M(r, F_m) \geq r^{\rho - \varepsilon_n} - K^{2(n+\varepsilon_n)} R_m^{(1+4\varepsilon_n)^2(k+\varepsilon_n)}$$
$$\geq r^{\rho - \varepsilon_n} - K^{2(n+1)} r^{(1+4\varepsilon_n)^2(k+\varepsilon_n)}$$
$$= r^{\rho - \varepsilon_n}\{1 - K^{2(k+1)} r^{(1+4\varepsilon_n)^2(k+\varepsilon_n)+\varepsilon_{n-\lambda}}\}.$$

Furthermore, we notice formula (4.83); then Lemma 4.7 follows.

LEMMA 4.8. *Suppose that there exists a point* z_{m+1} *in annulus* $R_m \leq |z| \leq R_{m+1}$ *(*$m_n \leq m \leq m_{n+1} - 1$; $n \geq n_0$*) such that*

$$\log |F_m(z_{m+1})| \geq \frac{1}{2} |z_{m+1}|^{\lambda - \varepsilon_n}. \tag{4.88}$$

Then $\log |F_{m+1}(z_{m+1})| \geq \frac{1}{4} |z_{m+1}|^{\lambda - \varepsilon_n}$.

PROOF. First, according to the identity

$$\frac{\Pi_m(z)}{\Pi_{m+1}(z)} = \prod_{|a_i|<R_{m+2}} \frac{R_{m+2}(z-a_i)}{R_{m+2}^2 - \overline{a}_i z} \cdot \frac{R_{m+3}^2 - \overline{a}_i z}{R_{m+3}(z-a_i)}$$

$$\times \prod_{R_{m+2}<|\alpha_i|<R_{m+3}} \frac{R_{m+3}^2 - \overline{a}_i z}{R_{m+3}(z-a_i)},$$

we derive

$$\log\left|\frac{\Pi_m(z)}{\Pi_{m+1}(z)}\right| \le n(R_{m+3}, \infty) \log \frac{2R_{m+3}}{R_{m+2} - R_{m+1}}, \qquad |z| < R_{m+1}.$$

When $m_n \le m \le m_{n+1} - 2$ $(n \ge n_0)$, we have from formula (4.85)

$$R_{m+2} - R_{m+1} = R_{m+1}(R_{m+1}^{\varepsilon_n} - 1) \ge R_{m_n}(R_{m_n+1}^{\varepsilon_n} - 1)$$

$$\ge R_{m_n}(K^{\frac{1}{2}(n+1)} - 1) > 2;$$

and when $m = m_{n+1} - 1$, it follows from formula (4.85) that

$$R_{m+2} - R_{m+1} = R_{m+1}(K^{n+1} - 1) > 2.$$

Hence, we conclude that

$$\log\left|\frac{\Pi_m(z)}{\Pi_{m+1}(z)}\right| \le n(R_{m+3}, \infty) \log R_{m+3} \le R_{m+3}^{k+\varepsilon_n}, \qquad |z| \le R_{m+1}. \tag{4.89}$$

Notice that

$$F_{m+1}(z) = f(z)\Pi_{m+1}(z) = F_m(z) \cdot \frac{\Pi_{m+1}(z)}{\Pi_m(z)}.$$

Then according to formulas (4.88) and (4.89), we obtain

$$\frac{1}{2}|z_{m+1}|^{\rho-\varepsilon_n} \le \log|F_m(z_{m+1})|$$

$$= \log|F_{m+1}(z_{m+1})| + \log\left|\frac{\Pi_m(z)}{\Pi_{m+1}(z)}\right|$$

$$\le \log|F_{m+1}(z_{m+1})| + R_{m+3}^{k+\varepsilon_n},$$

it follows that

$$\log|F_{m+1}(z_{m+1})| \ge \frac{1}{2}|z_{m+1}|^{\rho-\varepsilon_n} - R_{m+3}^{k+\varepsilon_n}$$

$$\ge \frac{1}{2}|z_{m+1}|^{\rho-\varepsilon_n}\left\{1 - \frac{2R_{m+3}^{k+\varepsilon_n}}{R_m^{\rho-\varepsilon_n}}\right\}.$$

When $m = m_n$, if $n \geq n_0 + 1$, then according to formulas (4.84) and (4.85), we have

$$R_{m+3} = \{K^n R_m^{1+\varepsilon_{n-1}}\}^{(1+\varepsilon_n)^2} \leq K^2 R_m^{(1+\varepsilon_{n-1}+2\varepsilon_n)(1+\varepsilon_n)^2}$$
$$\leq K^2 R_m^{(1+4\varepsilon_n)^3}.$$

If $n = n_0$, then

$$R_{m+3} = \{K^{n_0} R_m^{1+\varepsilon_{n_0}-1}\}^{(1+\varepsilon_{n_0})^2}$$
$$\leq \{K \cdot r_{n_0}^{2\varepsilon_{n_0}} R_m^{1+\varepsilon_{n_0}-1}\}^{(1+\varepsilon_{n_0})^2} \leq K^2 R_m^{(1+4\varepsilon_{n_0})^3},$$

implying that $R_{m+3} \leq K^2 R_m^{(1+4\varepsilon_n)^3}$, $n \geq n_0$. When $m_n + 1 \leq m \leq m_{n+1} - 3$, $R_{m+3} = R_m^{(1+\varepsilon_n)^3} \leq R_m^{(1+4\varepsilon_n)^3}$, and when $m = m_{n+1} - 2$, we have from formulas (4.84) and (4.85) that

$$R_{m+3} = k^{n+1} R_m^{(1+\varepsilon_n)^3} \leq k R_{m_{n+1}}^{2\varepsilon_{n+1}} \cdot R_m^{(1+\varepsilon_n)^3} \leq K R_m^{(1+4\varepsilon_n)^3}.$$

When $m = m_{n+1} - 1$, according to formulas (4.84) and (4.85), we obtain

$$R_{m+3} = \{K^{n+1} R_m^{(1+\varepsilon_n)^2}\}^{1+\varepsilon_{n+1}}$$
$$= K^{n+2} R_m^{(1+\varepsilon_n)^2(1+\varepsilon_{n+1})}$$
$$\leq K^2 R_{m_n+1}^{2\varepsilon_{n+1}} R_m^{(1+\varepsilon_n)^2(1+\varepsilon_n+1)} \leq K^2 R_m^{(1+4\varepsilon_n)^3}.$$

Hence, we conclude that

$$\log|F_{m+1}(z_{m+1})| \geq \frac{1}{2}|z_{m+1}|^{\rho-\varepsilon_n}\{1 - 2K^{2(k+1)} R_m^{(1+4\varepsilon_n)^3(k+\varepsilon_n)+\varepsilon_n-\rho}\}.$$

Furthermore, we notice formula (4.83); then Lemma 4.8 follows.

Suppose that there exists a point z_m in annulus $R_{m-1} < |z| \leq R_m$ ($m_n + 1 \leq m \leq m_{n+1}$; $n \geq n_0$) such that $\log|F_m(z_m)| \geq \frac{1}{4}|z_m|^{\rho-\varepsilon_n}$. Since derivative $F_m'(z)$ of $F_m(z)$ has at most finite zeros in disk $|z| \leq R_{m+2}$, there exists value A_m in the interval $[\frac{1}{16}|z_m|^{\rho-\varepsilon_n}, \frac{1}{8}|z_m|^{\rho-\varepsilon_n}]$, such that $F_m'(z)$ has no zeros on the level curve $\log|F_m(z)| = A_m$ in the disk $|z| \leq R_{m+2}$, meaning that the level curve is analytic. We consider the set

$$D(A_m) = \{z | \log|F_m(z)| > A_m, \ |z| < R_{m+2}\}.$$

Let $D_m(A_m)$ be the connected branch of $D(A_m)$ in the disk $|z| < R_{m+1}$ that contains point z_m. Then the boundary of $D_m(A_m)$ is a piecewise analytic curve. There are finite components of complementary sets of the closure $\overline{D}_m(A_m)$ with respect to the closed plane $|z| \leq \infty$. Let θ_m be the part of circumference $|z| = R_{m+1}$ in $D(A_m)$ which also is a part of the boundary of $D_m(A_m)$. We also let θ_t be the part of circumference $|z| = t$ $(t < R_{m+1})$

in $D_m(A_m)$, and $t\theta(t)$ be the linear measure of θ_t $(t \leq R_{m+1})$. According to formulas (4.82) and (4.85), we have

$$R_{m+1} \geq R_m^{1+\varepsilon_n} = R_m R_m^{\varepsilon_n} \geq R_m R_{m_n+1}^{\varepsilon_n} \geq K^{n/2} \cdot R_m$$
$$> K R_m > 2\sigma R_m.$$

Hence, $\sigma R_m < \frac{1}{2} R_{m+1}$.

LEMMA 4.9. *When* $m_n + 1 \leq m \leq m_{n+1} - 1$ $(n \geq n_0)$, *there exists point* z_{m+1} *in annulus* $\max\{\sigma|z_m|, R_m\} < |z| \leq R_{m+1}$ *(when* $\sigma|z_m| < R_m$, *we make the supplementary assumption that there does not exist value* t *in the interval* $[\sigma|z_m|, R_m]$, *so that circumference* $|z| = t$ *lies entirely in* $D_m(A_m))$ *such that*

$$\log|F_{m+1}(z_{m+1})| \geq \frac{1}{4}|z_{m+1}|^{\rho - \varepsilon_n}.$$

Meanwhile, there exists a continuous curve L_m *in* $D_m(A_m)$ *linking point* z_m *and* z_{m+1},[14] *such that when* $z \in L_m$, *we have* $\log|F_m(z)| \geq \frac{1}{16}|z_m|^{\rho - \varepsilon_n}$. *Besides, when* $|z_{m+1}| < R_{m+1}$, *in annulus* $|z_{m+1}| \leq |z| \leq R_{m+1}$, *there does not exist circumference* $|z| = t$ *lying entirely in* $D_{m+1}(A_{m+1})$.

PROOF. The following two cases are discussed:

(1) In annulus $\sigma|z_m| \leq |z| \leq R_{m+1}$, there exists circumference $|z| = t$ lying entirely in $D_m(A_m)$. Let

$$T = \sup\{t \mid |z| = t \subset D_m(A_m), \ t < R_{m+1}\}.$$

Then $\max\{\sigma|z_m|, R_m\} < T \leq R_{m+1}$. Choose point z_{m+1}, $|z_{m+1}| = T$, such that $\log|F_m(z_{m+1})| = \log M(T, F_m)$. Notice that $|z_{m+1}| > R_m > R_{m_n} \leq r_n$; then according to Lemma 4.7, we have

$$\log|F_m(z_{m+1})| \geq \frac{1}{2}|z_{m+1}|^{\rho - \varepsilon_n}. \tag{4.90}$$

Moreover, from Lemma 4.8, we conclude that $\log|F_{m+1}(z_{m+1})| \geq \frac{1}{4}|z_{m+1}|^{\rho - \varepsilon_n}$. Notice that

$$\frac{1}{16}|z_{m+1}|^{\rho - \varepsilon_n} \geq \frac{1}{16}(\sigma|z_m|)^{\rho - \varepsilon_n}$$
$$\geq \frac{1}{16}\sigma^{\rho - \varepsilon_n}|z_m|^{\rho - \varepsilon_n} = \frac{1}{8}|z_m|^{\rho - \varepsilon_n},$$
$$|F_m(z)| \geq |F_{m+1}(z)|.$$

Then when $|z_{m+1}| < R_{m+1}$, it follows that

$$D_{m+1}(A_{m+1}) \cap (|z| < R_{m+1}) \subset D_m(A_m),$$

meaning that value t cannot be in the interval $[|z_{m+1}|, R_{m+1}]$, such that circumference $|z| = t$ lies entirely in $D_{m+1}(A_{m+1})$.

---

[14] Point z_{m+1} may lie on the boundary of $D_m(A_m)$.

On the other hand, according to formula (4.90), we have

$$\log|F_m(z_{m+1})| \geq \frac{1}{2}|z_{m+1}|^{\rho-\varepsilon_n} \geq \frac{1}{2}|z_m|^{\rho-\varepsilon_n} \geq A_m.$$

Furthermore, taking the property of continuity of $\log|F_m(z)|$ into account, we may obtain a small circle C with z_{m+1} being its centre, such that $\log|F_m(z)| > A_m$ in C. According to the definition of T, in the disk $|z| < R_{m+1}$, there must exist circumference $|z| = t$ lying entirely in $D_m(A_m)$ and intersecting, in the meantime, the interior part of circle C. We assume an intersection point z'. According to the property of connectivity, there exists a continuous curve L'_m in $D_m(A_m)$ linking z_m and z'. Then, we use a straight line segment L''_m to connect points z' and z_{m+1}. Hence, $L_m = L'_m \cup L''_m$ is a continuous curve linking z_m and z_{m+1}. Moreover, when $z \in L_m$, we have

$$\log|F_m(z)| > A_m > \frac{1}{16}|z_m|^{\rho-\varepsilon_n},$$

and Lemma 4.9 follows.

(2) In annulus $\sigma|z_m| \leq |z| < R_{m+1}$, there does not exist circumference $|z| = t$ lying entirely in $D_m(A_m)$. Applying Theorem 3.1, we may conclude that there exists a point z_{m+1} in θ_m such that

$$\log|F_m(z_{m+1})| \geq \frac{1}{2}|z_{m+1}|^{\rho-\varepsilon_n}.$$

Otherwise, we would have

$$\log|F_m(z_m)| \leq A_m + 9\sqrt{2}\exp\left\{-\pi\int_{\sigma|z_m|}^{R_{m+1}/2}\frac{dt}{t\theta(t)}\right\}\frac{1}{2}R_{m+1}^{\rho-\varepsilon_n},$$

$$\frac{1}{4}|z_m|^{\rho-\varepsilon_n} \leq \frac{1}{8}|z_m|^{\rho-\varepsilon_n} + 9\sqrt{2}\exp\left\{-\pi\int_{\sigma|z_m|}^{R_{m+1}/2}\frac{dt}{t\theta(t)}\right\}\frac{1}{2}R_{m+1}^{\rho-\varepsilon_n}.$$

Notice that $\theta(t) \leq 2\pi$; then

$$\frac{1}{8}|z_m|^{\rho-\varepsilon_n} \leq 9\sigma^{1/2}\left\{\frac{|z_m|}{R_{m+1}}\right\}^{1/2-(\rho+\varepsilon_n)} \cdot R_{m+1}$$

$$\frac{1}{8} \leq 9\sigma^{1/2}\left\{\frac{|z_m|}{R_{m+1}}\right\}^{1/2-\rho+\varepsilon_n} \leq 9\sigma^{1/2}\left(\frac{R_m}{R_{m+1}}\right)^{\varepsilon_n}$$

$$= 9\sigma^{1/2}R_m^{-\varepsilon_n^2} \leq 9\sigma^{1/2}R_{m_n+1}^{-\varepsilon_n^2}.$$

Moreover, according to formula (4.85), we conclude that

$$\frac{1}{8} < 9\sigma^{1/2}K^{-(n+1)/2n} < 9\sigma^{1/2}K^{-1/2}.$$

Furthermore, noticing formula (4.82), we may immediately derive a contradiction.

According to Lemma 4.8, we have $\log|F_{m+1}(z_{m+1})| \geq \frac{1}{4}|z_{m+1}|^{\rho-\varepsilon_n}$. Meanwhile, based on the property of connectivity, there exists a continuous curve L_m in $D_m(A_m)$ linking points z_m and z_{m+1}. Moreover, when $z \in L_m$, we have $\log|F_m(z)| \geq A_m \geq \frac{1}{16}|z_m|^{\rho-\varepsilon_n}$, meaning that Lemma 4.9 holds.

LEMMA 4.10. *When* $m = m_{n+1}$ $(n \geq n_0)$, *there exists point* z_{m+1} *in annulus* $\max\{\sigma|z_m|, R_m\} < |z| \leq R_{m+1}$ *(when* $\sigma|z_m| < R_m$, *we make the supplementary assumption that there does not exist value* t *in the interval* $[\sigma|z_m|, R_m]$ *such that circumference* $|z| = t$ *lies entirely in* $D_m(A_m)$*) such that*

$$\log|F_{m+1}(z_{m+1})| \geq \frac{1}{4}|z_{m+1}|^{\rho-\varepsilon_n}.$$

Meanwhile, there exists a continuous curve L_m *in* $D_m(A_m)$ *linking points* z_m *and* z_{m+1} *such that when* $z \in L_m$,

$$\log|F_m(z)| \geq \frac{1}{16}|z_m|^{\rho-\varepsilon_n}.$$

Besides, when $|z_{m+1}| < R_{m+1}$, *in annulus* $|z_{m+1}| \leq |z| \leq R_{m+1}$, *there does not exist circumference* $|z| = t$ *entirely lying in* $D_{m+1}(A_{m+1})$.

PROOF. We discuss the following two cases separately:

(1) In annulus $\sigma|z_m| \leq |z| < R_{m+1}$, there exists circumference $|z| = t$ lying entirely in $D_m(A_m)$. We construct a proof analogous to the proof of condition (1) of Lemma 4.9. We need only to note that the point z_{m+1} satisfies $|z_{m+1}| > R_{m_{n+1}} \geq r_{n+1}$. Hence, by replacing formula (4.90), we obtain

$$\log|F_m(z_{m+1})| \geq \frac{1}{2}|z_{m+1}|^{\rho-\varepsilon_n}.$$

(2) In annulus $\sigma|z_m| \leq |z| < R_{m+1}$, there does not exist circumference $|z| = t$ lying entirely in $D_m(A_m)$. Applying Theorem 3.1, we may conclude that there exists a point z_{m+1} on θ_m such that

$$\log|F_m(z_{m+1})| \geq \frac{1}{2}|z_{m+1}|^{\rho-\varepsilon_{n+1}}.$$

Otherwise, we would have

$$\frac{1}{4}|z_m|^{\rho-\varepsilon_n} \leq \frac{1}{8}|z_m|^{\rho-\varepsilon_n} + 9\sigma^{1/2}\left(\frac{|z_m|}{R_{m+1}}\right)^{1/2} R_{m+1}^{\rho-\varepsilon_{n+1}},$$

$$\frac{1}{8} \leq 9\sigma^{1/2}|z_m|^{1/2-\rho+\varepsilon_n} R_{m+1}^{\rho-1/2-\varepsilon_{n+1}}$$

$$\leq 9\sigma^{1/2} R_m^{1/2-\rho+\varepsilon_n} R_{m+1}^{\rho-1/2-\varepsilon_{n+1}}$$

$$\leq 9\sigma^{1/2} R_m^{\varepsilon_n} R_{m+1}^{-\varepsilon_{n+1}}.$$

Moreover, according to formulas (4.84) and (4.82), we conclude that

$$\frac{1}{8} \leq 9\sigma^{1/2}K^{-1} = 9\sigma^{1/2}\frac{1}{72\sigma} < \frac{1}{8},$$

thus a contradiction is derived. Then followed by arguments analogous to the proof of condition (2) of Lemma 4.9, we may conclude that Lemma 4.10 holds.

Now, we proceed to complete the proof of Theorem 4.16. First we assume point z_1, $|z_1| = R_1$, such that $\log|F_1(z_1)| = \log M(R_1, F_1)$. Then, by repeating the application of Lemmas 4.9 and 4.10, we obtain successively a sequence of points z_m $(m = 1, 2, \ldots)$ and sequence of curves L_m $(m = 1, 2, \ldots)$. Notice that when $m_n + 1 \leq m \leq m_{n+1}$ $(n \geq n_0 + 1)$ and $z \in L_m$, we have

$$\log|f(z)| \geq \log|F_m(z)| \geq \frac{1}{16}|z_m|^{\rho-\varepsilon_n},$$

$$|z_m| > R_{m-1}, \qquad |z| \leq R_{m+1}.$$

On the other hand, according to formulas (4.86) and (4.87), we have

$$\log M(R'_n, F_m) \leq 3m(2R'_n, F_m) \leq 3T(2R'_n, f)$$

$$= \frac{1}{16}R_{m_n}^{\rho-\varepsilon_n} \leq \frac{1}{16}|z_m|^{\rho-\varepsilon_n} \leq \log|F_m(z)|.$$

Hence, we conclude that $|z| \geq R'_n$.

Notice that when $|z| \leq |z_m|$,

$$\log|f(z)| \geq \frac{1}{16}|z|^{\rho-\varepsilon_n} \geq \frac{1}{16K^{2\rho}}|z|^{\rho-(5\rho+1)\varepsilon_n},$$

and when $|z| < |z_m|$,

$$\log|f(z)| \geq \frac{1}{16}|z|^{\rho-\varepsilon_n}\left(\frac{R_{m-1}}{R_{m+1}}\right)^{\rho-\varepsilon_n}.$$

Furthermore, when $m = m_n + 1$, according to formulas (4.84) and (4.85), we have

$$R_{m+1} = \{K^n R_{m_n}^{1+\varepsilon_{n-1}}\}^{1+\varepsilon_n} = K^{n+1} R_{m_n}^{(1+\varepsilon_{n-1})(1+\varepsilon_n)}$$

$$\leq K^2 R_{m_{n-1}+1}^{2\varepsilon_n} R_{m_n}^{(1+\varepsilon_{n-1})(1+\varepsilon_n)} \leq K^2 R_{m-1}^{1+5\varepsilon_n},$$

meaning that

$$\left(\frac{R_{m-1}}{R_{m+1}}\right)^{\rho-\varepsilon_n} \geq \left(\frac{1}{K^2 R_{m-1}^{5\varepsilon_n}}\right)^{\rho-\varepsilon_n}$$

$$> \frac{1}{K^{2\rho}R_{m-1}^{5\rho\varepsilon_n}} > \frac{1}{K^{2\rho}|z_m|^{5\rho\varepsilon_n}} > \frac{1}{K^{2\rho}|z|^{5\rho\varepsilon_n}}.$$

When $m_n + 2 \leq m \leq m_{n+1} - 1$,

$$\left(\frac{R_{m-1}}{R_{m+1}}\right)^{\rho-\varepsilon_n} \geq \frac{1}{|z|^{3\rho\varepsilon_n}}.$$

When $m = m_{n+1}$, we have according to formulas (4.84) and (4.85),

$$R_{m+1} = K^{n+1} R_{m_{n+1}}^{1+\varepsilon_n} = K^{n+1} (R_{m_{n+1}-1})^{(1+\varepsilon_n)^2}$$
$$\leq K R_{m_n+1}^{2\varepsilon_{n+1}} R_{m_{n+1}-1}^{(1+\varepsilon_n)^2} \leq K R_{m_{n+1}-1}^{1+5\varepsilon_n},$$

implying that

$$\left(\frac{R_{m-1}}{R_{m+1}}\right)^{\rho-\varepsilon_n} > \left(\frac{1}{K R_{m-1}^{5\varepsilon_n}}\right)^{\rho-\varepsilon_n} > \frac{1}{K^\rho |z|^{5\rho\varepsilon_n}}.$$

Hence, we conclude that

$$\log |f(z)| > \frac{1}{16 K^{2\rho}} |z|^{\rho-(5\rho+1)\varepsilon_n}.$$

Let

$$L = \bigcup_{m=m_{n_0}+1}^{+\infty} L_m.$$

Then L is a continuous curve tending to ∞. Moreover, when $z \in L_m$ and $m_n + 1 \leq m \leq m_{n+1}$ $(n \geq n_0 + 1)$,

$$\log |f(z)| \geq \frac{1}{K^{2\rho}} \cdot |z|^{\rho-(5\rho+1)\varepsilon_n}.$$

By letting $\varepsilon_n \to 0$, we have

$$\lim_{\substack{|z| \to +\infty \\ z \in L}} \frac{\log \log |f(z)|}{\log |z|} \geq \rho.$$

This also completes the proof of Theorem 4.16.

§4.4. An estimate on the length of the asymptotic path of an entire function

P. Erdös once raised the following question [21f]:

Let $f(z)$ be an entire function of finite order on the open plane $|z| < +\infty$ and let a be a finite asymptotic value of $f(z)$. Then can we find a path of definite value L corresponding to a, such that the part of L_r of L in the disk $|z| \leq r$ has its length meas $L_r = O(r)$?

At an earlier time, W. Hayman raised a similar question for the case when $a = \infty$ [21d]. When $a = \infty$, if $T(r, f) = O\{(\log r)^2\}$, then L may be a half line originating from the origin $z = 0$. However, A. Gol'dberg and A. Eremenko proved that if $T(r, f)/(\log r)^2$ tends to ∞, then in general the estimate meas $L_r = O(r)$ does not hold [19a].

In this chapter, we shall give a general estimate on meas L_r. Concretely speaking, we have the following results [43d]:

THEOREM 4.17. *Let $f(z)$ be an entire function of order $\lambda < +\infty$ on the open plane $|z| < +\infty$, and a be a finite asymptotic value of $f(z)$. Then there must exist a rectifiable curve L continuously tending to ∞ such that*

$$\lim_{\substack{|z| \to +\infty \\ z \in L}} f(z) = a.$$

Meanwhile, the length of the part L_r of L in the disk $|z| \le r$ satisfies

$$\varlimsup_{r \to +\infty} \frac{\operatorname{meas} L_r}{r^{1+\lambda/2+\varepsilon}} = 0,$$

where $\varepsilon > 0$.

PROOF. Based on the assumption that a is a finite asymptotic value of $f(z)$, hence, there exists a path Γ continuously tending to ∞ on the z-plane such that

$$\lim_{\substack{|z| \to +\infty \\ z \in \Gamma}} f(z) = a.$$

According to the property of continuity of a function, without changing the asymptotic value, we may adjust Γ suitably, such that Γ satisfies the following properties: Γ is a simple polygon line extending from the origin and tending to ∞, and the endpoints of these line segments that compose Γ do not have a cluster point in the open plane $|z| < +\infty$.

We assume arbitrarily a value k, $k > 1$, and construct a sequence R_m:

$$R_m = K^m, \qquad m = 1, 2, \ldots. \tag{4.91}$$

Therefore, Γ must intersect each circumference $|z| = R_m$ $(m = 1, 2, \ldots)$. Moreover, there are at most finite intersection points on circumference $|z| = R_m$ $(m \ge 1)$. Hence, all the intersection points constitute a countable set E. We follow the method below to denote each of the points in E: When extending from the origin and tending to ∞ along Γ, according to the order of intersection, we denote the intersection points in E as $z_1, z_2, \ldots$. Besides, we let Γ_m be the part of Γ in annulus $R_m \le |z| \le R_{m+1}$. Obviously, Γ_m is composed of finite zigzag line segments Γ_{mj} $(j = 1, 2, \ldots, k_m; \ k_m < +\infty)$.

Let

$$\varepsilon_m = \max_{z \in \Gamma_m} |f(z) - a|.$$

Then when $m \to +\infty$, $\varepsilon_m \to 0$. On the other hand, we assume that $f(z) = a + b_n z^n + b_{n+1} z^{n+1} + \cdots$, $b_n \ne 0$, $n \ge 0$. Construct the transformation

$$g(z) = \frac{f(z) - a}{b_n z^n}. \tag{4.92}$$

Then $g(0) = 1$, and also

$$T(r, f) - \log^+ |b_n r^n| - \log^+ |a| - \log 2$$
$$\le T(r, g) \le T(r, f) + \log^+ \frac{1}{|b_n r^n|} + \log^+ |a| + \log 2.$$

Hence

$$\overline{\lim_{r \to +\infty}} \frac{\log T(r, g)}{\log r} = \lambda. \tag{4.93}$$

We also let $\varepsilon'_m = \max_{z \in \Gamma_m} |g(z)|$. Then

$$\varepsilon'_m \leq \frac{\varepsilon_m}{|b_n| R_m^n}. \tag{4.94}$$

Accordingly, when $m \to +\infty$, $\varepsilon'_m \to 0$.

According to formulas (4.93) and (4.94), there must exist positive integer m, such that when $m \geq m_0$,

$$\left. \begin{array}{l} T(R_m, g) \leq R_m^{\lambda+\varepsilon}, \qquad \varepsilon > 0, \\ \varepsilon'_m < \dfrac{1}{3}. \end{array} \right\} \tag{4.95}$$

We let $I_m = [2\varepsilon'_m, 3\varepsilon'_m]$ $(m \geq m_0)$. When $t \in I_m$ and $0 \leq \phi < 2\pi$, we have $|g(0) - te^{j\Phi}| \geq |g(0)| - t \geq 1 - 3\varepsilon'_m > 0$. And therefore

$$n\left(R_{m+1}, \frac{1}{g(z) - te^{j\Phi}}\right) \leq \left(\log \frac{R_{m+2}}{R_{m+1}}\right)^{-1},$$

$$N\left(R_{m+2}, \frac{1}{g(z) - te^{j\Phi}}\right) = (\log K)^{-1} N\left(R_{m+1}, \frac{1}{g(z)/t - e^{j\Phi}}\right).$$

Then, again according to the Cartan's identity, we derive

$$\frac{1}{2\pi} \int_0^{2\pi} n\left(R_{m+1}, \frac{1}{g(z) - te^{i\Phi}}\right) d\Phi$$

$$\leq (\log K)^{-1} \left\{ T\left(R_{m+2}, \frac{g(t)}{t}\right) - \log^+ \frac{1}{t} \right\}$$

$$\leq (\log K)^{-1} T(R_{m+2}, g).$$

Moreover, from the Length-Area principle, we conclude that

$$\int_{2\varepsilon'_m}^{3\varepsilon'_m} \frac{l_m^2(t)}{t} dt \leq 2\pi^2 R_{m+1}^2 (\log K)^{-1} T(R_{m+2}, g) = K_0,$$

where $l_m(t)$ is the total length of the part of the level curve $|g(z)| = t$ in disk $|z| < R_{m+1}$.

Let J_m be the set values t satisfying the condition

$$\frac{l_m^2(t)}{t} \geq \frac{2K_0}{3\varepsilon'_m - 2\varepsilon'_m}$$

in I_m. Then

$$K_0 \geq \int_{J_m} \frac{l_m^2(t)}{t} dt \geq \frac{2K_0}{\varepsilon'_m} \text{ meas } J_m.$$

And hence meas $J_m \leq \varepsilon'_m/2$. We let $I^*_m = I_m - J_m$. Then meas $I^*_m \geq \varepsilon'_m/2$, and when $t \in I^*_m$, it follows that

$$\frac{l^2_m(t)}{t} < \frac{2}{\varepsilon'_m} K_0,$$

$$l^2_m(t) < 12\pi^2 R^2_{m+1} (\log K)^{-1} T(R_{m+1}, g),$$

$$l_m(t) < 2\sqrt{3}\pi R_{m+1} \sqrt{(\log K)^{-1} T(R_{m+1}, g)}.$$

Moreover, according to formula (4.95), we obtain

$$l_m(t) < 2\sqrt{3}\pi R_{m+1} \sqrt{(\log K)^{-1} T(R_{m+1}, g)}$$

$$\leq 2\sqrt{3}\pi R_{m+1} \sqrt{(\log K)^{-1} R^{\lambda+\varepsilon}_{m+1}} \qquad (4.96)$$

$$= 2\sqrt{3}\pi R_{m+1} \sqrt{(\log K)^{-1} K^{\lambda+3} R^{\lambda+\varepsilon}_m}.$$

Derivative $g'(z)$ has at most finite zeros in disk $|z| \leq R_{m+1}$. Hence, there exists value ε''_m in I^*_m such that there are no zeros of $g'(z)$ on the part of level curve $|g(z)| = \varepsilon''_m$ in disk $|z| \leq R_{m+1}$, meaning that the level curve is analytic and rectifiable. We consider the set

$$D(\varepsilon''_m) = \{z \mid |g(z)| < \varepsilon''_m, \ R_m < |z| < R_{m+1}\}$$

and let D_{mj} $(j = 1, 2, \ldots, k_m)$ be the connected component containing Γ_{mj} $(j = 1, 2, \ldots, k_m)$. Apparently, any two of these k_m connected components may have no intersection, or may coincide. Hence, there exist k'_m $(k'_m \leq k_m)$ distinct connected components D'_{mj} $(j = 1, 2, \ldots, k'_m)$, the boundary of D'_{mj} consists of either arcs of two circles $|z| = R_m$ and $|z| = R_{m+1}$ or part of the level curve: $|g(z)| = \varepsilon''_m$. If $j \neq j'$, then the boundaries of D'_{mj} and $D'_{mj'}$ cannot intersect. In fact, otherwise, there exists a point z_0 on the level curve $|g(z)| = \varepsilon''_m$ such that point z_0 is on both boundaries of D'_{mj} and $D'_{mj'}$. Since $g'(z_0) \neq 0$, $g(z)$ is univalent in some neighborhood of point z_0. Hence, this neighborhood is divided into two parts by the level curve $|g(z)| = \varepsilon''_m$, with one part satisfying $|g(z)| < \varepsilon''_m$ on it, and the other $|g(z)| > \varepsilon''_m$ on it. Therefore, we conclude that D'_{mj} and $D'_{mj'}$ intersect each other, and hence, D'_{mj} and $D'_{mj'}$ coincide. However, D'_{mj} and $D'_{mj'}$ are distinct. We derive, therefore, a contradiction. Besides, we denote E'_{mj} as the set of intersection points between E and the boundary of D'_{mj}; then each of the points in E'_{mj} must belong to the same boundary branch of D'_{mj}. In fact, otherwise, suppose that there exist two points z' and z'' in $E'_{mj'}$ belonging to two different boundary branches C' and C'' of D'_{mj} respectively. We consider arbitrarily a branch, for example C'. Since D'_{mj} is a bounded domain, C' is a closed curve, and hence, C'

divides z-plane into two parts. We denote D_1' as the bounded part and D_2' as the unbounded one; then the origin $z = 0$ cannot be in D_1'. Otherwise, according to the maximum modulus principle, we would have

$$1 = |g(0)| \leq \varepsilon_m'' \leq 3\varepsilon_m' < 1.$$

And hence, we derive a contradiction. Therefore, D_1' must lie entirely in annulus $R_m < |z| < R_{m+1}$. On the other hand, point z' is on circumferences $|z| = R_m$ or $|z| = R_{m+1}$. Moreover, a part of the circular arc containing point z' in itself belongs to C'. Therefore, D_{mj}' must lie entirely in D_1'. In fact, otherwise, then z' cannot be a boundary point of D_{mj}'. Obviously, C'' lies in D_1'. Hence, z'' is in D_1', and in turn, it is in annulus $R_m < |z| < R_{m+1}$. However, z'' can only be on circumferences $|z| = R_m$ or $|z| = R_{m+1}$ which, in turn, leads to a contradiction.

Now, we consider the set $D = \{D_{mj}' | m = m_0 + 1, m_0 + 2, \ldots; j = 1, 2, \ldots, k_m'\}$. First, we assume a point z_{i_0} from E, such that z_{i_0} is the last intersection point between Γ and circumference $|z| = R_{m_0}$. Obviously, the polygon line part of Γ between points z_{i_0} and z_{i_0+1} must belong to a domain D_{i_0} of D. Let E_{i_0} be the set of intersection points between E and the boundary of D_{i_0}, and the largest suffix of points in E_{i_0} be i_1, $i_1 \geq i_0+1$. Notice that since z_{i_0} and z_{i_1} belong to the same boundary branch of D_{i_0}, there exists boundary curve L_1 connecting points z_{i_0} and z_{i_1}. Meanwhile, L_1 is a piecewise analytic rectifiable curve. Then, we consider the polygon line part of Γ between points z_{i_1} and z_{i_1+1}. Analogously, we may define a domain D_{i_1} of D and a subset E_{i_1} of E. Let i_2, $i_2 \geq i_1+1$, be the largest suffix of points in E_{i_1} and L_2 be a piece of boundary curve connecting z_{i_1} and z_{i_2}. Repeating the same process, we obtain a sequence of curves $L_1, L_2, \ldots$.

Let

$$L = \bigcup_{j=1}^{+\infty} L_j.$$

Then L is a rectifiable curve continuously tending to ∞. Now, we estimate meas L_r when $r > R_{m_0}$. First, there exists a positive integer m_1 such that $R_{m_1} < r \leq R_{m_1+1}$. And hence

$$\text{meas } L_r \leq \text{meas } L_{R_{m_1+1}} \leq \sum_{m=m_0}^{m_1} \{4\pi(R_m + R_{m+1}) + l_m(\varepsilon_m'')\}.$$

Moreover, according to formula (4.96),

$$\text{meas } L_r \leq \sum_{m=m_0}^{m_1} \left\{ 4\pi(R_m + R_{m+1}) + 2\sqrt{3}\pi R_{m+1} \sqrt{(\log K)^{-1} K^{\lambda+\varepsilon} R_m^{\lambda+\varepsilon}} \right\}.$$

Furthermore, from formulas (4.91) and (4.95), we derive

$$\text{meas}\, L_r \leq \sum_{m=m_0}^{m_1} \left\{ 4\pi(L+K) + 2\sqrt{3}\pi K^{1+(\lambda+\varepsilon)/2} \cdot \sqrt{(\log K)^{-1}} \right\} R_m^{1+(\lambda+\varepsilon)/2}.$$

Notice that

$$\frac{R_m^{1+(\lambda+\varepsilon)/2}}{R_{m_1}^{1+(\lambda+\varepsilon)/2}} = \left(\frac{1}{K^{1+(\lambda+\varepsilon)/2}} \right)^{m_1-m}, \qquad K > 1;$$

then

$$\text{meas}\, L_r \leq 4\pi \left\{ 1 + K + K^{1+(\lambda+\varepsilon)/2}\sqrt{(\log K)^{-1}} \right\}$$

$$\times R^{1+(\lambda+\varepsilon)/2} mes_{m_1} \sum_{m=0}^{+\infty} \left(\frac{1}{K^{1+(\lambda+\varepsilon)/2}} \right)^m$$

$$\leq 4\pi \left\{ 1 + K + K^{1+(\lambda+\varepsilon)/2}\sqrt{(\log K)^{-1}} \right\}$$

$$\times R_{m_1}^{1+(\lambda+\varepsilon)/2} \cdot \frac{K^{1+(\lambda+\varepsilon)/2}}{K^{1+(\lambda+\varepsilon)/2} - 1}$$

$$\leq 4\pi \left\{ 1 + K + K^{1+(\lambda+\varepsilon)/2}\sqrt{(\log K)^{-1}} \right\}$$

$$\times \frac{K^{1+(\lambda+\varepsilon)/2}}{K^{1+(\lambda+\varepsilon)/2} - 1} r^{1+(\lambda+\varepsilon)/2}$$

and hence

$$\varlimsup_{r \to +\infty} \frac{\text{meas}\, L_r}{r^{1+\lambda/2+\varepsilon}} = 0.$$

Finally, we prove that

$$\lim_{\substack{|z| \to +\infty \\ z \in L}} f(z) = a. \tag{4.97}$$

In fact, if we denote L_m as the part of L in annulus $R_m \leq |z| \leq R_{m+1}$, then $\max|g(z)| \leq \varepsilon_m'' \leq 3\varepsilon_m'$. Furthermore, from formulas (4.92) and (4.94), we conclude that

$$\max_{z \in L_m}|f(z) - a| \leq |b_n|R_{m+1}^n \max_{z \in L_m}|g(z)|$$

$$\leq |b_n|R_{m+1}^n \cdot 3 \cdot \frac{\varepsilon_n}{|b_n|R_m^n} = 3K^n\varepsilon_m.$$

Notice that $m \to +\infty$, $\varepsilon_m \to 0$. Hence, formula (4.97) holds. Thus Theorem 4.17 is proved.

THEOREM 4.18. *Let $f(z)$ be a transcendental entire function of order $\lambda < +\infty$ on the open plane $|z| < +\infty$. Then there must exist a rectifiable curve L continuously tending to ∞, such that*

$$\varlimsup_{\substack{|z| \to +\infty \\ z \in L}} \frac{\log|f(z)|}{\log|z|} = +\infty.$$

Meanwhile, for the length of L_r in disk $|z| \leq r$, we have

$$\varlimsup_{r \to +\infty} \frac{\operatorname{meas} L_r}{r^{1+\lambda/2+\varepsilon}} = 0,$$

where $\varepsilon > 0$.

PROOF. First, according to Theorem 4.14, where we let $\nu(r) = \log r$, then there exists a curve Γ continuously tending to ∞ such that

$$\varlimsup_{\substack{|z| \to +\infty \\ z \in \Gamma}} \frac{\log |f(z)|}{\log |z|} = +\infty.$$

Taking into account the property of a continuous function, we may assume that Γ is a simple polygon line originating from the origin $z = 0$ and tending to ∞. Moreover, the endpoints of these line segments which compose Γ have no cluster points in the open plane $|z| < +\infty$.

We select arbitrarily a value K, $K > 1$, and construct the sequence R_m:

$$R_m = K^m, \qquad m = 1, 2, \dots . \tag{4.98}$$

Hence, there are at most finite intersection points between Γ and each circumference $|z| = R_m$ $(m \geq 1)$. We let all the intersection points be E. Completely similar to the proof of Theorem 4.17, we denote the points in the point set E as $z_1, z_2, \dots$, and use similar methods to define Γ_m and Γ_{mj} $(j = 1, 2, \dots, k_m; k_m < +\infty)$. Let

$$A_m = \min_{z \in \Gamma_m} \left\{ \frac{\log |f(z)|}{\log |z|} \right\}.$$

Then when $m \to +\infty$, $A_m \to +\infty$. On the other hand, we assume that

$$f(z) = b_n z^n + b_{n+1} z^{n+1} + \cdots, \qquad b_n \neq 0, \; n \geq 0.$$

Consider the formation

$$g(z) = \frac{f(z)}{b_n z^n}. \tag{4.99}$$

Then $g(0) = 1$, and also

$$\varlimsup_{r \to +\infty} \frac{\log T(r, g)}{\log r} = \lambda. \tag{4.100}$$

We also let

$$A'_m = \min_{z \in \Gamma_m} \left\{ \frac{\log |g(z)|}{\log |z|} \right\}.$$

Then

$$A'_m \geq \begin{cases} A_m - n - \dfrac{\log |b_n|}{\log R_m}, & |b_n| \geq 1, \\[2ex] A_m - n - \dfrac{\log |b_n|}{\log R_{m+1}}, & |b_n| < 1. \end{cases} \tag{4.101}$$

Hence, when $m \to +\infty$, $A'_m \to +\infty$. According to formulas (4.100) and (4.101), for any arbitrarily given number $\varepsilon > 0$, there must exist positive integer m_0 such that when $m \geq m_0$,

$$T(R_m, g) \leq R_m^{\lambda+\varepsilon}, \qquad R_m^{A'_m} > 4.$$

We let $I_m = [\frac{1}{4}R_m^{A'_m}, \frac{1}{2}R_m^{A'_m}]$ $(m \geq m_0)$. Analogous to the discussion of the proof of Theorem 4.17, we may conclude that there exists value A''_m in I_m such that derivative $g'(z)$ in disk $|z| \leq R_{m+1}$ has no zeros on the level curve $|g(z)| = A''_m$, and also the total length of the level curve in disk $|z| < R_{m+1}$ is

$$l_m(A''_m) \leq 2\sqrt{2}\pi R_{m+1}\sqrt{(\log K)^{-1}k^{\lambda+\varepsilon}R_{m+1}^{\lambda+\varepsilon}}. \tag{4.102}$$

We consider the set

$$D(A''_m) = \{z \mid |g(z)| > A''_m, \ R_m < |z| < R_{m+1}\}$$

and let D_{mj} $(j = 1, 2, \ldots, k_m)$ be the connected components containing Γ_{mj} $(j = 1, 2, \ldots, k_m)$. In these k_m connected components, there exist k'_m $(k'_m \leq k_m)$ distinct connected components D'_{mj} $(j = 1, 2, \ldots, k'_m)$, with its boundary composed of either the circular arc on the two circumferences $|z| = R_m$ and $|z| = R_{m+1}$ or part of the level curve $|g(z)| = A''_m$. Furthermore, the boundaries of the two distinct connected components D'_{mj} and $D'_{mj'}$ $(j \neq j')$ have no intersections. Besides, we let E'_{mj} be the set of intersection points between E and the boundary of D'_{mj}. Then each point in E'_{mj} $(j = 1, 2, \ldots, k'_m)$ must belong to the same boundary branch of D'_{mj}, but with the possible exception of at most one such set. In fact, if there exists a point set E'_{mj} $(1 \leq j \leq k'_m)$ such that the two points z' and z'' of E'_{mj} belong to two different boundary branches C' and C'' of D'_{mj}, respectively, then we may conclude that an annular domain is bounded by C' and C'' in annulus $R_m < |z| < R_{m+1}$. Moreover, the origin $z = 0$ and the point of infinity belong to two different components of complementary set of these annular domains, respectively. We first consider C'. And C' is a simple closed curve, dividing the z-plane into a bounded part D'_1 and unbounded part D'_2. D'_1 must contain the origin $z = 0$. Otherwise, analogous to the discussion of the proof of Theorem 4.17, we may conclude that z'' must be in annulus $R_m < |z| < R_{m+1}$ which, in turn, leads to a contradiction. Similarly, the bounded part D''_1 bounded by C'' must also contain the origin $z = 0$. Hence, the annular domain bounded by C' and C'' assumes the required property. Next, we notice that z' and z'' are on two circumferences $|z| = R_m$ and $|z| = R_{m+1}$, respectively. Moreover, there exists a curve C connecting these two points on D'_{mj}. Now, we may prove that no other set $E'_{mj'}$ $(j \neq j')$ can exist such that the points in $E'_{mj'}$ belong to different boundary components of $D'_{mj'}$, respectively. Otherwise, the annular domain corresponding to $E'_{mj'}$ must intersect C, and hence, D'_{mj} and $D'_{mj'}$ have intersections and thus coincide with each other. However, since D'_{mj} and $D'_{mj'}$ are distinctive, we derive, therefore, a contradiction.

We consider the set $D = \{D'_{mj} | m = 1, 2, \ldots; j = 1, 2, \ldots, k'_m\}$. First, we assume a point z_{i_0} from E such that z_{i_0} is the last intersection point between Γ and circumference $|z| = R_{m_0}$. Obviously, the polygon line part of Γ between points z_{i_0} and z_{i_0+1} must belong to a domain D_{i_0} of D. Let E_{i_0} be the set of intersection points between the boundaries of E and D_{i_0}, and also i_1, $i_1 \geq i_0 + 1$, be the largest suffix of points in E_{i_0}. If z_{i_0} and z_{i_1} belong to the same boundary branch of D_{i_0}, then there exists a piece of boundary curve L_1 connecting z_{i_0} and z_{i_1}. Meanwhile, L_1 is piecewise, analytic and rectifiable. If z_{i_0} and z_{i_1} belong to different boundary branches C'_{i_0} and C''_{i_0} of D_{i_0}, respectively, then according to the above-mentioned discussion, there exists a point z' on C'_{i_0} such that $\arg z' = \arg z_{i_1}$. We use a straight line C to connect points z' and z_{i_1}. If C intersects the complementary set of D_{i_0}, then the part of C in the complementary set is replaced by a boundary curve of D_{i_0}. Hence, we may obtain a rectifiable curve L_1 connecting z_{i_0} and z_{i_1}. Then, we consider the polygon line part of Γ between z_{i_1} and z_{i_1+1}. Applying similar discussion, we may obtain a rectifiable curve L_2 connecting z_{i_1} and z_{i_2} ($i_2 \geq i_1 + 1$). By repeating the process, we may obtain a sequence of curves: $L_1, L_2, \ldots$. Let

$$L = \bigcup_{i=1}^{+\infty} L_i.$$

Then L is a rectifiable curve tending to ∞. Now we estimate meas L_1 when $r \geq R_{m_0}$. First, there exists a positive integer m_1 such that $R_{m_1} < r \leq R_{m_1+1}$, and hence

$$\operatorname{meas} L_r \leq \operatorname{meas} L_{R_{m_1+1}} \leq \sum_{m=m_0}^{m_1} \{4\pi(R_1 + R_{m+1}) + R_{m+1} + l_m(A''_m)\}.$$

Moreover, according to formulas (4.98) and (4.102), we have

$$\operatorname{meas} L_r \leq \sum_{m=m_0}^{m_1} \left\{ 4\pi + (4\pi + 1)K + 2\sqrt{2}\pi K^{\lambda+1+\varepsilon} \sqrt{(\log K)^{-1}} \right\} R_m^{1+(\lambda+\varepsilon)/2}.$$

Notice that

$$\frac{R_m^{1+(\lambda+\varepsilon)/2}}{R_{m_1}^{1+(\lambda+\varepsilon)/2}} = \left\{ \frac{1}{K^{1+(\lambda+\varepsilon)/2}} \right\}^{m_1-m_0}, \qquad K > 1.$$

Then

$$\text{meas } L_r \leq \left\{ 4\pi + (4\pi + 1)K + 2\sqrt{2}\pi K^{\lambda+1+\varepsilon} \sqrt{(\log K)^{-1}} \right\}$$

$$\times R_{m_1}^{1+(\lambda+\varepsilon)/2} \sum_{m=0}^{\infty} \left(\frac{1}{K^{1+(\lambda+\varepsilon)/2}} \right)^m$$

$$\leq \left\{ 4\pi + (4\pi + 1)K + 2\sqrt{2}\pi K^{\lambda+1+\varepsilon} \sqrt{(\log K)^{-1}} \right\}$$

$$\times \frac{K^{1+(\lambda+\varepsilon)/2}}{K^{1+(\lambda+\varepsilon)/2-1}} r^{1+(\lambda+\varepsilon)/2}.$$

And therefore,

$$\lim_{r \to +\infty} \frac{\text{meas } L_r}{r^{1+\lambda/2+\varepsilon}} = 0.$$

Finally, we prove

$$\lim_{\substack{|z| \to +\infty \\ z \in L}} \frac{\log |f(z)|}{\log |z|} = +\infty. \tag{4.103}$$

In fact, if we denote L_m as the part of L in annulus $R_m \leq |z| \leq R_{m+1}$, then

$$\min_{z \in L_m} |g(z)| \geq A_m'' \geq \frac{1}{4} R_m^{A_m'}. \tag{4.104}$$

Furthermore, according to formulas (4.99) and (4.104), we conclude that

$$\min_{z \in L_m} |f(z)| \geq \frac{1}{4} |b_n| R_m^{A_m'+n}.$$

Hence, when $z \in L_m$,

$$|f(z)| \geq \frac{1}{4} |b_n| |z|^{A_m'+n} \left\{ \frac{R_m}{|z|} \right\}^{A_m'+n}$$

$$\geq \frac{1}{4} |b_n| |z|^{A_m'+n} \left(\frac{1}{k} \right)^{A_m'+n},$$

$$\log |f(z)| \geq \log \frac{|b_n|}{4} + (A_m' + n) \left(1 - \frac{\log K}{\log |z|} \right) \log |z|.$$

Notice that when $m \to +\infty$, $A_m' \to +\infty$. Hence, formula (4.103) holds. Thus Theorem 4.18 is proved.

From the proofs of Theorem 4.17 and Theorem 4.18, we may find that if we perceive how fast the entire function $f(z)$ tends to an asymptotic value a (no matter whether or not it is finite) along an asymptotic path, then generally speaking, we may obtain a rectifiable asymptotic path L such that L and Γ define the same asymptotic value a. Moreover, for the length meas L_r of L_r, we may now have its estimation. Meanwhile, the speed for $f(z)$ tending to a along L will not be "changed". Hence, we may supplement some results regarding the property of growth as illustrated in §4.3 such that there is an estimate on the length of the corresponding asymptotic path. In fact, we only cite Theorem 4.18 as an example to illustrate the above-mentioned fact.

§4.5. Direct transcendental singularities

4.5.1. The Ahlfors Theorem.
In 1932 L. Ahlfors proved the following result [1c]:

THEOREM 4.19. *Let* $w = f(z)$ *be a meromorphic function of order* λ *on the open plane* $|z| < +\infty$, *and* $z = \varphi(w)$ *be the inverse function of* $f(z)$. *We also let* l *be the number of distinct direct transcendental singularities of* $\varphi(w)$. *Then when* $\lambda < 1$, $0 \le l \le 1$ *and when* $\lambda \ge 1$, $l \le 2\lambda$.

PROOF. When $\lambda = +\infty$, it is obvious that Theorem 4.19 holds. Therefore, we may assume that $\lambda < +\infty$. Then, we let a_i $(i = 1, 2, \ldots, l; 2 \le l < +\infty)$ be l distinct direct transcendental singularities of $\varphi(w)$. Hence, there exists a value $\rho > 0$ such that there is no intersection among the l regions Ω_ρ^i $(i = 1, 2, \ldots, l)$ corresponding to a_i $(i = 1, 2, \ldots, l)$ on the z-plane. Moreover, when $z \in \Omega_\rho^i$ $(1 \le i \le l)$, we have([15])

$$\frac{1}{|f(z) - a_i|} > \frac{1}{\rho}, \qquad f(z) \ne a_i,$$

and when $z \in \Gamma_\rho^i$, it follows that

$$\frac{1}{|f(z) - a_i|} = \frac{1}{\rho},$$

where Γ_ρ^i is the finite boundary of Ω_ρ^i, and is also an analytic curve.

We assume arbitrarily a point z_i from domain Ω_ρ^i $(1 \le i \le l)$. Then, we let

$$r_0 = \max\{1, |z_1|, |z_2|, \ldots, |z_l|\}$$

and denote θ_{it} as the part of circumference $|z| = t$ $(t \ge r_0)$ in Ω_ρ^i and $t\theta_i(t)$ the linear measure of θ_{it}. According to Theorem 3.1, when r is sufficiently large, we have

$$\log \frac{1}{|f(z_i) - a_i|} \le \log \frac{1}{\rho} + 9\sqrt{2} \exp\left\{-\pi \int_{2|z_i|}^{r/2} \frac{dt}{t\theta_i(t)}\right\}$$
$$\times \log M\left\{\overline{\Omega}_\rho^i \cap (|z| = r), \frac{1}{f - a_i}\right\},$$

$$\pi \int_{2r_0}^{r/2} \frac{dt}{t\theta_i(t)} \le \log\log M\left\{\overline{\Omega}_\rho^i \cap (|z| = r), \frac{1}{f - a_i}\right\} \qquad (4.105)$$
$$+ \log\left\{9\sqrt{2}\left[\log \frac{\rho}{|f(z_i) - a_i|}\right]^{-1}\right\}$$
$$i = 1, 2, \ldots, l.$$

([15]) When $a_i = \infty$, we use $f(z)$ to replace $1/(f(z) - a_i)$.

Moreover, applying Lemma 3.8, where we let $f(z)$ be $1/(f(z) - a_i)$, r be $2r$, $R' = 3r$, $R = 4r$, and $H = \frac{r}{8e}$, we conclude that when $|z| \le 2r$ and $z \notin (\gamma)$,

$$\log \frac{1}{|f(z) - a_i|} \le \left\{ 5 + \frac{\log(48e)}{\log(4/3)} \right\} T\left(4r, \frac{1}{f - a_i} \right)$$
$$\le \left\{ 5 + \frac{\log(48e)}{\log(4/3)} \right\} \{T(4r, f) + O(1)\}.$$

On the one hand, we assume value t from the interval $[r, 2r]$ such that there is no intersection between circumference $|z| = t$ and (γ). On the other hand, according to the maximum modulus principle, we conclude that $\log M\{\overline{\Omega}_\rho^i \cap (|z| = r), 1/(f - a_i)\}$ is a monotonically increasing function of r. Hence,

$$\log M \left\{ \overline{\Omega}_\rho^i \cap (|z| = r), \frac{1}{f - a_i} \right\}$$
$$\le \log M \left\{ \overline{\Omega}_\rho^i \cap (|z| = t), \frac{1}{f - a_i} \right\}$$
$$\le \left\{ 5 + \frac{\log(48e)}{\log(4/3)} \right\} \{T(4r, f) + O(1)\}.$$

Combining this with formula (4.105), we get

$$\pi \int_{2r_0}^{r/2} \frac{dt}{t\theta_i(t)} \le \log T(4r, f) + O(1), \qquad i = 1, 2, \ldots, l,$$
$$\sum_{i=1}^{l} \pi \int_{2r_0}^{r/2} \frac{dt}{t\theta_i(t)} \le l \log T(4r, f) + O(1).$$

Notice that

$$l^2 = \left\{ \sum_{i=1}^{l} \sqrt{\theta_i(t)} \cdot \frac{1}{\sqrt{\theta_i(t)}} \right\}^2$$
$$\le \sum_{i=1}^{l} \theta_i(t) \sum_{i=1}^{l} \frac{1}{\theta_i(t)} \le 2\pi \sum_{i=1}^{l} \frac{1}{\theta_i(t)}.$$

Hence

$$\frac{l^2}{2} \le \sum_{i=1}^{l} \frac{\pi}{\theta_i(t)},$$
$$\frac{l^2}{2} \log \frac{r}{4r_0} = \frac{l^2}{2} \int_{2r_0}^{r/2} \frac{dt}{t} \le \sum_{i=1}^{l} \pi \int_{2r_0}^{r/2} \frac{dt}{t\theta_i(t)}.$$

Thus

$$\frac{l^2}{2} \log \frac{r}{4r_0} \leq l \log T(4r, f) + O(1),$$

$$\frac{l}{2} \leq \frac{\log T(4r, f)}{\log 4r} \cdot \frac{\log 4r}{\log(r/4r_0)} + \frac{O(1)}{\log(r/4r_0)}.$$

By letting $r \to +\infty$ we conclude that $l \leq 2\lambda$, meaning that $\lambda \geq 1$. Therefore, when $\lambda < 1$, we can only have $0 \leq l \leq 1$, and hence, Theorem 4.19 is proved completely.

THEOREM 4.20. *Let* $w = f(z)$ *be a meromorphic function on the open plane* $|z| < +\infty$, *and let it satisfy the following conditions:*

(1) *The inverse function* $z = \varphi(w)$ *contains a direct transcendental singularity* a.

(2) *When* $a \neq \infty$,

$$\min_{0 \leq \theta < 2\pi} \frac{1}{|f(re^{i\theta}) - a|} = O(1);$$

and when $\alpha = \infty$,

$$\min_{0 \leq \theta < 2\pi} |f(re^{i\theta})| = O(1).$$

Then there exists a continuous curve L *tending to* ∞ *on the* z-plane, *such that when* $a \neq \infty$,

$$\lim_{\substack{|z| \to +\infty \\ z \in L}} \frac{\log \log(1/|f(z) - a|)}{\log |z|} \geq \frac{1}{2};$$

and when $a = \infty$,

$$\lim_{\substack{|z| \to +\infty \\ z \in L}} \frac{\log \log |f(z)|}{\log |z|} \geq \frac{1}{2}.$$

PROOF. Without any loss of generality, we need only to consider the case when $\alpha = \infty$. According to the definition of direct transcendental singularities and condition (2), and based on the fact that the number of zeros and poles of derivative $f'(z)$ can at most be countable, we may find a value ρ, $\rho > 0$, such that the corresponding region Ω_ρ on the z-plane has the following properties:

(1) When $z \in \Omega_\rho$, we have $|f(z)| > \frac{1}{\rho}$ and $f(z) \neq \infty$.

(2) The finite boundary part Γ_ρ of Ω_ρ is an analytic curve, and when $z \in \Gamma_\rho$, $|f(z)| = \frac{1}{\rho}$.

(3) When t is sufficiently large, each circumference $|z| = t$ intersects Γ_ρ.

First we assume a point $z_0 \in \Omega_\rho$ such that when $t \geq |z_0|$, circumference $|z| = t$ intersect Γ_ρ. We let θ_t be the arc of circumference $|z| = t$ in Ω_ρ and $t\theta(t)$ be the linear measure of θ_t. According to the maximum modulus principle, we conclude that $M\{\overline{\Omega}_\rho \cap (|z| = t), f\}$ is a monotonically increasing function of t. Applying Theorem 3.1, when $t > 4|z_0|$, we have

$$\log |f(z_0)| \le \log \frac{1}{\rho} + 9\sqrt{2} \exp \left\{ -\pi \int_{2|z_0|}^{r/2} \frac{dr}{r\theta_i(r)} \right\}$$
$$\times \log M\{\overline{\Omega}_\rho \cap (|z| = t), f\}.$$

Notice that $|f(z_0)| > \frac{1}{\rho}$ and $\theta(r) \le 2\pi$ $(r \ge |z_0|)$. Then we derive

$$\log \log M\{\overline{\Omega}_\rho \cap (|z| = t), f\} \ge \frac{1}{2} \log t - \text{const}.$$

Hence, when t is sufficiently large, there exists, on the other hand, a point $z_1 \in \Omega_\rho$ in θ_t such that

$$|z_1| \ge 72^{(12 \times 13)/2}, \quad |z_1|^{(1/2 - 1/12)} > \frac{4}{\rho}, \quad \log |f(z_1)| \ge |z_1|^{(1/2 - 1/12)}.$$

On the other hand, according to the formula

$$\log M\{\overline{\Omega}_\rho \cap (|z| = t_1'), f\} = \frac{1}{4} |z_1|^{(1/2 - 1/12)}$$

the associated value $t_1' > 1$.

Now, we construct two sequences t_n $(n = 1, 2, \ldots)$ and t_n' $(n = 1, 2, \ldots, c)$. Concretely speaking, we let

$$\begin{cases} t_1 = |z_1|, \\ t_{n+1} = 72^{n+12} t_n^{1+\varepsilon_n}, \quad \varepsilon_n = \frac{1}{11 + n}, \quad n = 1, 2, \ldots. \end{cases}$$

And according to the formula

$$\log M\{\overline{\Omega}_\rho \cap (|z| = t_n'), f\} = \frac{1}{4} t_n^{1/2 - \varepsilon_n}, \quad n = 1, 2, \ldots$$

we define sequence t_n'. Obviously, we have

$$t_n' < t_{n+1}' \to +\infty \quad (n \to +\infty), \quad t_n' > 1, \quad n = 1, 2, \ldots.$$

Followed by a discussion analogous to the proof of Theorem 4.4 and we conclude immediately that Theorem 4.20 holds.

COROLLARY 1. *Under the assumption of Theorem 4.20, the lower order μ of $f(z)$ is $\ge \frac{1}{2}$.*

COROLLARY 2. *Let $w = f(z)$ be a meromorphic function of lower order $\mu < \frac{1}{2}$ on the open plane $|z| < +\infty$, and the inverse function $z = \varphi(w)$ of $f(z)$ possesses a direct transcendental singularity a. Then there exists a monotonically increasing sequence $\{r_n\}$ tending to ∞ such that*

$$\begin{cases} \lim\limits_{n \to +\infty} \min\limits_{0 \le \theta < 2\pi} |f(r_n e^{i\theta})| = +\infty, & a = \infty, \\ \lim\limits_{n \to +\infty} \min\limits_{0 \le \theta < 2\pi} \dfrac{1}{|f(r_n e^{i\theta}) - a|} = +\infty, & a \ne \infty. \end{cases}$$

Corollary 2 is an extension of the classical Wiman Theorem [41a].

4.5.2. Two lemmas.

LEMMA 4.11. *Let* $w = f(z)$ *be a meromorphic function of lower order* $\mu <$ $+\infty$ *on the open plane* $|z| < +\infty$ *and let it satisfy the following conditions:*

(1) *The inverse function* $z = \varphi(w)$ *possesses a direct transcendental singularity* a.

(2) *When* $a \neq \infty$,

$$\min_{0 \leq \theta < 2\pi} \frac{1}{|f(re^{i\theta}) - a|} = O(1);$$

and when $a = \infty$,

$$\min_{0 \leq \theta < 2\pi} |f(re^{i\theta})| = O(1).$$

Then for any arbitrarily assumed η, $0 < \eta < \frac{1}{2}$, *there exists a region* Ω *with its boundary containing point* ∞ *on the* z-*plane, such that* Ω *possesses the following properties:*

(1) *When* $z \in \Omega$,

$$\begin{cases} \log \dfrac{1}{|f(z) - a|} \geq |z|^{1/2 - \eta}, & f(z) \neq a \ (a \neq \infty), \\[2mm] \log |f(z)| \geq |z|^{1/2 - \eta}, & f(z) \neq \infty \ (a = \infty). \end{cases}$$

(2) *If sequence* r_n $(n = 1, 2, \ldots)$ *satisfies the condition*

$$\lim_{n \to +\infty} \frac{\log T(r_n, f)}{\log r_n} = \mu,$$

then for sufficiently large n, *there exists value* t_n *in the interval* $[r_n^{1-\eta}, r_n]$ *such that*

$$\theta(t_n) \geq K(\eta, \mu) = \frac{(1 - \eta)\eta\pi}{2(\mu + 1)}.$$

Here $\theta(t)$ *is defined implicitly as follows: If we let* θ_t *be the arc segment of circumference* $|z| = t$ *in* Ω, *then* $t\theta(t)$ *is the linear measure of* θ_t.

PROOF. Without any loss of generality, we need only to consider the case when $a = \infty$. Analogous to the proof of Theorem 4.20, we may find a value $\rho > 0$ such that the corresponding region Ω_ρ on the z-plane possesses the following properties:

(1) When $z \in \Omega_\rho$, we have $|f(z)| > \frac{1}{\rho}$ and $f(z) \neq \infty$.

(2) The finite boundary part Γ_ρ of Ω_ρ is an analytic curve. Moreover, when $z \in \Gamma_\rho$, $|f(z)| = \frac{1}{\rho}$.

(3) When t is sufficiently large, each circumference $|z| = t$ intersects Γ_ρ.

We take a point $z_0 \in \Omega_\rho$ such that when $t \geq |z_0|$, circumference $|z| = t$ intersects Γ_ρ. We let θ_{ρ_t} be the part of circumference $|z| = t$ in Ω_ρ, and $t\theta_\rho(t)$ be the linear measure of θ_{ρ_t}. Then analogous to the proof of Theorem 4.20, we may conclude that when $t > 4|z_0|$,

$$\log \log M\{\overline{\Omega}_\rho \cap (|z| = t), f\} \geq \frac{1}{2} \log t - \text{const}.$$

Hence, when t is sufficiently large, there exists at least one point z_1 on θ_{ρ_t} such that

$$|z_1| \geq \max \left\{ 72^{4/\eta^2}, \, 32^{4(\mu+1)/\eta(1-\eta)^2} \left[18\sqrt{2} \left(3 + \frac{\log 64e}{\log 2} \right)^{2/(1-\eta)} \right] \right\}$$

$$|z_1|^{(1/2)-(\eta/2)} > \frac{4}{\rho}, \qquad \log|f(z_1)| > |z_1|^{(1/2)-(\eta/2)}.$$

(4.106)

Then, we define sequence R_m:

$$\begin{cases} R_1 = |z_1|, \\ R_{m+1} = R_{m+1}^{1+\eta/2}, \qquad m = 1, 2, \ldots. \end{cases}$$

(4.107)

Suppose that there exists a point $z_m \in \Omega_\rho$ on circumference $|z| = R_m$ ($m \geq 1$) such that

$$\log|f(z_m)| \geq |z_m|^{(1/2)-(\eta/2)}.$$

(4.108)

Obviously, there exists value A_m in the interval $[\frac{1}{4}|z_m|^{(1/2)-(\eta/2)}, \frac{1}{2}|z_m|^{(1/2)-(\eta/2)}]$ such that derivative $f'(z)$ has no zeros on the level curve $\log|f(z)| = A_m$, meaning that the level curve is analytic. We consider the set $D = \{z \mid \log|f(z)| > A_m\}$ and let D_m be the connected branch containing point z_m. According to the inequality

$$A_m \geq \frac{1}{4}|z_m|^{(1/2)-(\eta/2)} \geq \frac{1}{4}|z_1|^{(1/2)-(\eta/2)} > \frac{1}{\rho},$$

we conclude that $D_m \subset \Omega_\rho$. Hence, $f(z)$ is regular on D_m. On the other hand, according to the maximum modulus principle, we conclude that D_m is an unbounded domain. Let Ω_m be the connected component of D_m in the disk $|z| < R_{m+1}$ containing point z_m, θ_{m+1} be the intersection point between circumference $|z| = R_{m+1}$ and Ω_m, θ_{mt} be the part of circumference $|z| = t$ ($t < R_{m+1}$) in Ω_m, and $t\theta_m(t)$ be the linear measure of θ_{mt}. In the following, we prove that there exists at least one point z_{m+1} on θ_{m+1} such that

$$\log|f(z_{m+1})| \geq |z_{m+1}|^{(1/2)-(\eta/2)}.$$

Otherwise, applying Theorem 3.1, we have

$$\log|f(z_m)| \leq A_m + 9\sqrt{2} \exp\left\{ -\pi \int_{2|z_m|}^{R_{m+1}/2} \frac{dt}{t\theta_m(t)} \right\} R_{m+1}^{(1/2)-(\eta/2)}.$$

Moreover, noting

$$\theta_m(t) \leq \theta_\rho(t) \leq 2\pi, \qquad A_m < \frac{1}{2}|z_m|^{(1/2)-(\eta/2)},$$

and formula (4.108), we derive

$$|z_m|^{(1/2)-(\eta/2)} \geq \frac{1}{2}|z_m|^{(1/2)-(\eta/2)} + 9\sqrt{2} \left\{ 4\frac{|z_m|}{R_{m+1}} \right\}^{1/2} R_{m+1}^{(1/2)-(\eta/2)},$$

$$R_{m+1}^{\eta/2} \leq 72|z_m|^{\eta/2}, \qquad R_{m+1} < 72^{2/\eta} R_m, \qquad R_m < 72^{4/\eta^2}.$$

And hence $|z_1| = R_1 \leq R_m < 72^{4/\eta^2}$. But this contradicts formula (4.106).
When $z \in \Omega_m$,

$$\log|f(z)| > A_m \geq \frac{1}{4}|z_m|^{(1/2)-(\eta/2)}, \qquad |z| < R_{m+1}.$$

We make a further discussion: When $|z| \leq |z_m|$,

$$\log|f(z)| \geq |z|^{1/2-\eta}\frac{1}{4}|z_m|^{\eta/2} \geq |z|^{1/2-\eta} \cdot \frac{1}{4}|z_1|^{\eta/2} > |z|^{1/2-\eta};$$

and when $|z| > |z_m|$,

$$\log|f(z)| \geq \frac{1}{4}|z|^{(1/2)-(\eta/2)}\left\{\frac{|z_m|}{R_{m+1}}\right\}^{(1/2)-(\eta/2)}$$

$$= \frac{1}{4}|z|^{(1/2)-(\eta/2)}|z_m|^{-(\eta/4)+(\eta^2/4)}$$

$$\geq \frac{1}{4}|z|^{(1/2)-(3\eta/4)} \geq |z|^{1/2-\eta} \cdot \frac{1}{4}|z_m|^{\eta/4}$$

$$\geq |z|^{1/2-\eta}\frac{1}{4}|z_1|^{\eta/4} \geq |z|^{1/2-\eta},$$

it follows that when $z \in \Omega_m$, we conclude that $\log|f(z)| \geq |z|^{1/2-\eta}$.

Let $\Omega = \bigcup_{m=1}^{\infty} \Omega_m$. Then $\Omega \subset \Omega_\rho$ is an unbounded domain. Meanwhile, ∞ is its boundary point, and when $z \in \Omega$, $\log|f(z)| > |z|^{1/2-\eta}$.

In the following, we prove the second part of Lemma 4.11. According to the assumption that sequence r_n $(n = 1, 2, \ldots)$ satisfies the condition

$$\lim_{n \to +\infty} \frac{\log T(r_n, f)}{\log r_n} = \mu,$$

it follows that there exists a positive integer n_0 such that when $n \geq n_0$,

$$T(r_n, f) < r_n^{\mu+1}, \qquad r_n \geq R_1. \tag{4.109}$$

Now, we prove that for value n, $n \geq n_0$, there must exist value m_n such that $r_n^{1-\eta} \leq R_{m_n} < R_{m_n+1} < r_n$. In fact, if R_m satisfies the inequality $R_m < r_n^{1-\eta} \leq R_{m+1}$, then

$$R_{m+2} = R_m^{(1+\eta/2)^2} \leq r^{(1-\eta)(1+\eta/2)^2} = r_n^{1-(3\eta^2/4)-(\eta^3/4)} < r_n.$$

Hence, we need only to define $m_n = m + 1$.

Now, applying Lemma 3.8, where we let $r = \frac{1}{4}R_{m_n+1}$, $R' = \frac{1}{2}R_{m_n+1}$, $R = R_{m_n+1}$ and $H = \frac{1}{64e}R_{m_n+1}$, then when $|z| \leq \frac{1}{4}R_{m_n+1}$ and $z \notin (\gamma)$,

$$\log|f(z)| \leq \left\{3 + \frac{\log(64e)}{\log 2}\right\} T(R_{m_n+1}, f)$$

$$\leq \left\{3 + \frac{\log(64e)}{\log 2}\right\} T(r_n, f).$$

Particularly, we assume value R in the interval $[\frac{1}{8}R_{m_n+1}, \frac{1}{4}R_{m_n+1}]$, such that there is no intersection between circumference $|z| = R$ and (γ). Then, we let $z = Re^{i\theta}$ and apply formula (4.109), we conclude that

$$\log|f(Re^{i\theta})| \leq \left\{3 + \frac{\log(64e)}{\log 2}\right\} T(r_n, f). \tag{4.110}$$

Let Ω'_{m_n} be the connected branch of Ω_{m_n} in the disk $|z| < R$ containing point z_{m_n}, θ_{Rt} be the part of circumference $|z| = t$ $(t < R)$ in Ω'_{m_n}, and $t\theta_R(t)$ be the linear measure of θ_{Rt}. Now, we prove that there exists at least one value t_n in the interval $[2R_{m_n}, \frac{1}{2}R]$ such that

$$\theta_R(t_n) \geq \frac{(1-\eta)\eta\pi}{2(\mu+1)}. \tag{4.111}$$

In fact, otherwise, when $t \in [2R_{m_n}, \frac{1}{2}R]$ it is always true that

$$\theta_R(t) < \frac{(1-\eta)\eta\pi}{2(\mu+1)}. \tag{4.112}$$

In the following, we shall derive a contradiction. First, according to Theorem 3.1 and formula (4.110), we have

$$\log|f(z_{m_n})| \leq A_{m_n} + 9\sqrt{2}\exp\left\{-\pi\int_{2|z_{m_n}|}^{R/2} \frac{dt}{r\theta_R(r)}\right\}\left\{3 + \frac{\log(64e)}{\log 2}\right\} r_n^{\mu+1}.$$

Then, noting

$$\log|f(z_{m_n})| \geq |z_{m_n}|^{(1/2)-(\eta/2)}, \qquad A_{m_n} \leq \frac{1}{2}|z_{m_n}|^{(1/2)-(\eta/2)}$$

and formula (4.108), we derive

$$|z_{m_n}|^{(1/2)-(\eta/2)} \leq \frac{1}{2}|z_{m_n}|^{(1/2)-(\eta/2)} + 9\sqrt{2}\left\{4 \cdot \frac{|z_{m_n}|}{R}\right\}^{2(\mu+1)/\eta(1-\eta)}$$
$$\times \left\{3 + \frac{\log(64e)}{\log 2}\right\} r_n^{\mu+1}.$$

Moreover, according to $R > \frac{1}{8}R_{m_n+1}$ and $|z_1| < |z_{m_n}|$, we conclude that

$$|z_{m_n}|^{(1/2)-(\eta/2)} \leq 18\sqrt{2} \cdot 32^{2(\mu+1)/(1-\eta)^\eta} R_{m_n}^{-(\mu+1)(1-\eta)}\left\{3 + \frac{\log(64e)}{\log 2}\right\} r_n^{\mu+1}$$
$$\leq 18\sqrt{2} \cdot \left\{3 + \frac{\log(64e)}{\log 2}\right\} 32^{2(\mu+1)/\eta(1-\eta)},$$
$$|z_1| \leq \left\{18\sqrt{2}\left(3 + \frac{\log(64e)}{\log 2}\right)\right\}^{2/(1-\eta)} 32^{4(\mu+1)/\eta(1-\eta)^2}.$$

But this contradicts formula (4.106).

Now according to the inequality

$$r_n^{1-\eta} \leq 2R_m < \frac{1}{2}R \leq r_n, \qquad \theta(t_n) \geq \theta_R(t_n)$$

and formula (4.111), we conclude that the second part of Lemma 4.11 holds. Hence, Lemma 4.11 is proved completely.

Analogously, we may prove the following result:

LEMMA 4.12. *Let* $w = f(z)$ *be a meromorphic function of order* $\lambda < +\infty$ *on the open plane* $|z| < +\infty$, *and let it satisfy the following conditions:*

(1) *The inverse function* $z = \varphi(w)$ *has a direct transcendental singularity* a.

(2) *When* $a \neq \infty$,

$$\min_{0 \leq \theta < 2\pi} \frac{1}{|f(re^{i\theta}) - a|} = O(1);$$

and when $a = \infty$,

$$\min_{0 \leq \theta < 2\pi} |f(re^{i\theta})| = O(1).$$

Then for arbitrarily assumed number η, $0 < \eta < \frac{1}{2}$, *there exists a region* Ω *on the z-plane, with its boundary containing point* ∞, *such that* Ω *possesses the following properties:*

(1) *When* $z \in \Omega$,

$$\log \frac{1}{|f(z) - a|} \geq |z|^{(1/2)-(\eta/2)}, \qquad f(z) \neq a \quad (a \neq \infty);$$

$$\log |f(z)| \geq |z|^{(1/2)-(\eta/2)}, \qquad f(z) \neq \infty \quad (a = \infty).$$

(2) *There exists a value* r_0 *such that when* $r \geq r_0$, *there exists a value* t *in the interval* $[r^{1-\eta}, r]$ *such that*

$$\theta(t) \geq K(\eta, \lambda) = \frac{(1 - \eta)\eta\pi}{2(\lambda + 1)}.$$

Here $\theta(t)$ *contains the following implication: If we let* θ_t *be the arc segment of circumference* $|z| = t$ *in* Ω, *then* $t\theta(t)$ *represents the linear measure of* θ_t.

Finally, we point out that the hypothetical condition (2) of Theorem 4.20, Lemma 4.11 and Lemma 4.12 can be replaced by one of the following conditions:

(1) $f(z)$ has a deficient value b, and $b \neq a$.

(2) $f(z)$ has an asymptotic value b, and $b \neq a$.

(3) $\varphi(w)$ has a direct transcendental singularity b that is distinct from a.

In fact, among the above-mentioned three conditions, if any one of these conditions holds, then we must have

$$\begin{cases} \min_{0 \leq \theta < 2\pi} \dfrac{1}{|f(re^{i\theta}) - a|} = O(1) & (a \neq \infty); \\[2mm] \min_{0 \leq \theta < 2\pi} |f(re^{i\theta})| = O(1) & (a = \infty). \end{cases}$$

4.5.3. The distribution of zeros and poles of a meromorphic function and the direct transcendental singularities of its inverse function. The Ahlfors Theorem illustrates that the order of a meromorphic function can bound the number of direct transcendental singularities of its inverse function. Now, we prove a result of another kind.

THEOREM 4.21. *Let* $w = f(z)$ *be a meromorphic function on the open plane* $|z| < +\infty$ *and* $z = \varphi(w)$ *be the inverse function of* $f(z)$. *We also let the order of* $f(z)$ *be* $\lambda < +\infty$, *and there exist* q $(1 \leq q \leq +\infty)$ *half lines* $\Delta(\theta_k)$ $(0 \leq \theta_1 < \theta_2 < \cdots < \theta_q < 2\pi \,; \; \theta_{q+1} = \theta_1 + 2\pi)$ *such that, for any arbitrarily given number* $\varepsilon > 0$,

$$\varlimsup_{r \to +\infty} \frac{\log^+ n\{\bigcup_{k=1}^q \Omega(\theta_k + \varepsilon, \theta_{k+1} - \varepsilon \,; r), f = z\}}{\log r} \leq \nu < \frac{1}{2}, \qquad z = 0, \infty.$$

Then the number of finite distinct nonzero direct transcendental singularities of $\varphi(w)$ *is* $l \leq q$.

PROOF. In fact, since $q \geq 1$, we need only to consider the case when $l \geq 2$. Hence, according to the Ahlfors Theorem, we conclude that $\lambda \geq 1$. Let

$$\omega = \min_{1 \leq k < q} \{\theta_{k+1} - \theta_k\},$$

and for any arbitrarily assumed number η,

$$0 < \eta < \frac{(1 - 2\nu)\omega}{60\pi}.$$

We construct sequence $r_n = 2^{(1+\eta)^n}$ $(n = 1, 2, \ldots)$. Let a_i $(i = 1, 2, \ldots, l;$ $2 \leq l < +\infty)$ be any l distinct finite nonzero direct transcendental singularities of $\varphi(w)$. Then there exists a number $\rho > 0$ such that the region Ω_ρ^i corresponding to a_i $(1 \leq i \leq l)$ on the z-plane satisfies the following conditions:

(1) When $z \in \Omega_\rho^i$, $1/|f(z) - a_i| > 1/\rho$ and $f(z) \neq a_i$.

(2) The finite boundary part Γ_ρ^i of Ω_ρ^i is an analytic curve. Moreover, when $z \in \Gamma_\rho^i$,

$$\frac{1}{|f(z) - a_i|} = \frac{1}{\rho}.$$

(3) When t is sufficiently large, circumference $|z| = t$ intersects every Γ_ρ^i.

(4) There is no intersection among the l regions Ω_ρ^i.

Furthermore, according to $l \geq 2$ and Lemma 4.12, we may conclude that there exists a region $\Omega_i \subset \Omega_\rho^i$ corresponding to a_i $(1 \leq i \leq l)$ and containing point ∞ in its boundary on the z-plane, such that Ω_i possesses the following properties:

(1) When $z \in \Omega_i$,

$$\log \frac{1}{|f(z) - a_i|} \geq |z|^{1/2 - \eta}.$$

(2) When n is sufficiently large, there exists value t_{in} in the interval $[r_n^{1-\eta}, r_n]$ such that

$$\theta_i(t_{in}) \geq k = \frac{(1 - \eta)\eta\pi}{2(\lambda + 1)};$$

here $\theta_i(t)$ contains the following implications: If we let θ_{it} be the arc segment of circumference $|z| = t$ in Ω_i, then $t\theta_i(t)$ represents the linear measure of θ_{it}. Hence, when $\theta \in \theta_i(t_{in})$,

$$\log \frac{1}{|f(t_{in} e^{i\theta}) - a_i|} \geq t_{in}^{1/2 - \eta}.$$

We select arbitrarily a fixed number ε,

$$0 < \varepsilon < \min \left\{ \frac{\omega}{8}, \frac{k}{8q} \right\}.$$

Then, according to the assumption, we have

$$\varlimsup_{r \to +\infty} \frac{\log^+ n \{ \bigcup_{k=1}^{q} \Omega(\theta_k + \varepsilon, \theta_{k+1} - \varepsilon; r), f = X \}}{\log r} \leq \gamma < \frac{1}{2}, \qquad X = 0, \infty.$$

On the other hand, when n is sufficiently large, there exists at least one region $\Omega(\theta_{k_i} + 2\varepsilon, \theta_{k_i+1} - 2\varepsilon; r_n^{i-\eta}, r_n)$ in the q regions $\Omega(\theta_k + 2\varepsilon, \theta_{k+1} - 2\varepsilon; r_n^{1-\eta}, r_n)$ $(k = 1, 2, \ldots, q)$, such that

$$\text{meas } E_{in} \geq \frac{K}{2q} t_{in},$$

where

$$E_{in} = \{ z \, | \, \arg z \in \theta_i(t_{in}) \cap [\theta_{k_i} + 2\varepsilon, \theta_{k_i+1} - 2\varepsilon], \, |z| = t_{in} \}.$$

According to the selection of η and $r_n^{1-\eta} \leq t_{in} \leq r_n$, we conclude that

$$\lim_{n \to +\infty} \left\{ \left(\frac{r_n}{r_n^{1-\eta}} \right)^{6 + 2(\nu + \eta) + 3\pi/\omega} \cdot r_n^{\nu + 2\eta} \log r_n \right\} \cdot t_{in}^{-1/2 + \eta} = 0.$$

Hence, when n is sufficiently large, applying Lemma 3.13 we get

$$\log \frac{1}{|f(z) - a_i|} \geq \frac{A}{B\eta \log r_n + C} \left\{ \frac{r_n^{1+\eta}}{r_n} \right\}^{6 + 3\pi/\omega}$$
$$\times t_{in}^{1/2 - \eta} \to +\infty \qquad (n \to +\infty),$$

where point z is in $\overline{\Omega}(\theta_{k_i} + 2\varepsilon, \theta_{k_i+1} - 2\varepsilon; r_n^{1-\eta}, r_n)$ and outside some circles $(\gamma)_i$, $(\gamma)_i$ has finite numbers of circles, with the sum of their radii not exceeding $\frac{1}{8}\varepsilon r_n^{1-\eta}$. Besides, A, B, C and n are constants independent of n.

When n is sufficiently large, each a_i corresponds to a region $\Omega(\theta_{\varepsilon_i} + 2\varepsilon, \theta_{k_i+1} - 2\varepsilon; r_n^{1-\eta}, r_n)$. Hence, l singularities a_i correspond to l regions $\Omega(\theta_{k_i} + 2\varepsilon, \theta_{k_i+1} - 2\varepsilon; r_n^{1-\eta}, r_n)$. In the following, we prove that these l regions do not coincide among each other. In fact, otherwise, let $k_i = k_{i'} = k$ $(i \neq i')$; then according to the sum of the radii of $(\gamma)_i$ and $(\gamma)_{i'}$ not exceeding $\frac{1}{4}\varepsilon r_n^{1-\eta}$, we conclude that there exists point z_n in $\overline{\Omega}(\theta_k + 2\varepsilon, \theta_{k+1} - 2\varepsilon; r_n^{1-\eta}, r_n)$ such that

$$\log \frac{1}{|f(z_n) - a_i|} \geq \frac{A}{B\eta \log r_n + C}.$$

Meanwhile, we have

$$\log \frac{1}{|f(z_n) - a_i|} \geq \frac{A}{B\eta \log r_n + C}$$
$$\times r_n^{-(6+3\pi/\omega)^\eta + (1-\eta)(1/2-\eta)} \to +\infty, \qquad (n \to +\infty),$$

and hence, we must have $a_i = a_{i'} = a$. Moreover, according to the selection of ε, and the sum of radii of $(\gamma)_i \cup (\gamma)_{i'}$ not exceeding $\frac{1}{4}\varepsilon r_n^{1-\eta}$, we may select a point z_n' on E_{in}. In the meantime, we select a point z_n'' on E_{in} such that z_n' and z_n'' are outside $(\gamma)_i \cup (\gamma)_{i'}$. Then, we use a straight line to connect z_n' and z_n''; if we come across $(\gamma)_i \cup (\gamma)_{i'}$, then we use a circular arc to replace it. Hence, we obtain a curve L_n connecting point z_n' and z_n''. Obviously, L_n is in $\overline{\Omega}(\theta_k + 2\varepsilon, \theta_{k+1} - 2\varepsilon; r_n^{1-\eta}, r_n)$ and outside circles $(\gamma)_i \cup (\gamma)_{i'}$. Therefore, when $z \in L_n$, we have

$$\log \frac{1}{|f(z) - a|} \geq \frac{A}{B\eta \log r_n + C}$$
$$\times r_n^{-(6+3\pi/\omega)\eta + (1-\eta(1/2-\eta)} \to +\infty \qquad (n \to +\infty),$$

meaning that

$$\lim_{\substack{n \to +\infty \\ z \in L_n}} f(z) = a.$$

On the other hand, L_n must intersect Γ_ρ^i and $\Gamma_\rho^{i'}$ at the same time. Moreover, when $z \in \Gamma_\rho^i \cup \Gamma_\rho^{i'}$, it follows that $|f(z) - a| = \rho$. Hence, when n is sufficiently large, we derive a contradiction, and consequently, Theorem 4.21 is proved.

THEOREM 4.22. *Let $w = f(z)$ be a meromorphic function on the open plane $|z| < +\infty$, and $z = \varphi(w)$ be the inverse function of $f(z)$. We also assume that the order λ of $f(z)$ is $< +\infty$, and $f(z)$ has q Julia directions.*

Then the number of distinctive direct transcendental singularities of $\varphi(w)$ is $l \leq q$.

PROOF. When $q = +\infty$, it is obvious that Theorem 4.22 holds. When $l \geq 1$, according to Theorem 2.15, we have $q \geq 1$. Hence, we need only to consider the case when $l \geq 2$ and $q < +\infty$. Let $\Delta(\theta_k)$ $(k - 1, 2, \ldots, q)$ be q Julia directions of $f(z)$. Then according to Theorem 2.13 and the finite covering theorem, we may conclude that there exist two distinct finite complex numbers b and c, such that b and c are not the direct transcendental singularities of $\varphi(w)$, and meanwhile, for any arbitrarily assumed number $\varepsilon > 0$, we have

$$\varlimsup_{r \to +\infty} \frac{\log^+ n\{\bigcup_{k=1}^{q} \overline{\Omega}(\theta_k + \varepsilon, \theta_{k+1} - \varepsilon; r), f = X\}}{\log r} = 0, \qquad X = b, c.$$

Consider the transformation

$$F(z) = \frac{f(z) - b}{f(z) - c}.$$

Then the inverse function of $F(z)$ contains l distinct finite nonzero direct transcendental singularities; moreover,

$$\varlimsup_{r \to +\infty} \frac{\log^+ n\{\bigcup_{k=1}^{q} \Omega(\theta_k + \varepsilon, \theta_{k+1} - \varepsilon; r), F = X\}}{\log r} = 0, \qquad X = 0, \infty.$$

Hence, according to Theorem 4.21, we conclude that $l \leq q$. This also proves Theorem 4.22.

CHAPTER 5

The Relationship between Deficient Values
and Asymptotic Values of an Entire Function

In 1929, by examining some examples, R. Nevanlinna recognized that there is an intrinsic link between the problem of exceptional values and the asymptotic values theory. Moreover, he conjectured that a deficient value is, in the meantime, an asymptotic value [32a]. However, early in 1939, O. Teichmüller negated this conjecture [37a]. In fact, to date many results illustrate that R. Nevanlinna's conjecture is incorrect [3a, 18a].

Rejecting R. Nevanlinna's concrete conjecture does not mean that the idea conceived by R. Nevanlinna regarding the existence of an intrinsic link between exceptional values and asymptotic values is wrong. In fact, in this chapter, we shall establish for some important kinds of functions, some general formulas regarding the relationship between the number of deficient values and the number of asymptotic values. These formulas illustrate that there is a close relationship between deficient values and asymptotic values, meaning that R. Nevanlinna's concept regarding the existence of an intrinsic link between these two values is still correct.

§5.1. The theorem of the bound
and its application regarding functions meromorphic in the unit disk

5.1.1. The theorem of the bound.

THEOREM 5.1. *Let $f(z)$ be a meromorphic function on the unit disk $|z| \leq 1$, and a, b, and c be three complex numbers, with the spherical distances among them greater than d, $0 < d < \frac{1}{2}$. Moreover,*

$$n(1, f = a) \leq n, \quad n(1, f = b) \leq n, \quad n(1, f = c) \leq n.$$

Besides, let α be a finite complex number and $N \geq 0$ be a real number. Assume the point set E_α of $f(z)$ in the disk $|z| \leq \tau$ $(0 < \tau < 1)$ satisfying the inequality $|f(z) - \alpha| \leq e^{-N}$ cannot be included in some circles, with the sum of their radii not exceeding $2eh$ $(h \leq 0.01)$. Then for point z in the *disk* $|z| \leq r$ $(0 < r < 1)$ *and outside* (γ), *we have*

$$\log^+ \frac{1}{|f(z) - \alpha|} \geq A \left\{ \log \frac{1}{(1 - \tau)(1 - r)h} \right\}^{-1}$$

$$\times (1 - \tau)(1 - r)N - \frac{1}{(1 - \tau)^2 (1 - r)^2}$$

$$\times \left\{ Bn \log \frac{1}{h} + C \log \frac{2}{(1 - \tau)(1 - r)} \right.$$

$$\left. + D \log \frac{1}{d} + E \log^+ |\alpha| \right\},$$

where (γ) *represents some circles, with the sum of their Euclidean radii not exceeding* $2eh$ *and the number of circles not more than* $3n$. *Besides,* A $(A > 0)$, B, C, D *and* E *are constants.*

PROOF. Let (γ) be the pseudo-non-Euclidean exceptional circles corresponding to these

$$n(1, f = a) + n(1, f = b) + n(1, f = c)$$

points and number h $(h \leq 0.01)$. We assume arbitrarily a point z_0 in the disk $|z| \leq r$ and outside circles (γ). Consider the transformation

$$\zeta = \frac{z - z_0}{1 - \bar{z}_0 z},$$

and its inverse transformation

$$z = \frac{\zeta + z_0}{1 + \bar{z}_0 \zeta}.$$

Then function $F(\zeta) = f((\zeta + z_0)/(1 + \bar{z}_0 \zeta))$ is meromorphic on the disk $|\zeta| \leq 1$. Moreover,

$$n(1, F = a) \leq n, \quad n(1, F = b) \leq n, \quad n(1, F = c) \leq n.$$

Circles (γ) are transformed into corresponding exceptional circles $(\gamma)_\zeta$ on the ζ-plane. Notice that point $\zeta = 0 \notin (\gamma)_\zeta$ and $F(0) = f(z_0)$. Besides, according to the inequality

$$|\zeta| \leq \frac{|z| + |z_0|}{1 + |z_0||z|},$$

we conclude that the image domain of disk $|z| \leq \tau$ on the ζ-plane must be included in disk $|\zeta| \leq t$, with

$$t = \frac{\tau + r}{1 + \tau r}.$$

Let $(\gamma)'_\zeta$ be the pseudo-non-Euclidean exceptional circles corresponding to these $n(\frac{1+t}{2}, F = \alpha)$ points and number h, and $(\gamma)'$ be the corresponding exceptional circles of $(\gamma)'_\zeta$ on the z-plane. Notice that the Euclidean radius of a circle does not exceed its pseudo-non-Euclidean radius. Therefore, according to the assumption, there exists a point $z_1 \in E_\alpha$ in the disk $|z| \le \tau$ and outside circles $(\gamma)'$, such that $|f(z_1) - \alpha| \le e^{-N}$. Let ζ_1 be the image of point z_1 on the ζ-plane. Then ζ_1 is in the disk $|\zeta| \le t$ and outside circles $(\gamma)'_\zeta$. Besides, we have $F(\zeta_1) = f(z_1)$.

Let α_i $(i = 1, 2, \ldots, n(\frac{1+t}{2}, F = \alpha))$ be the α-value points of $F(\zeta)$ in the disk. Then according to the Poisson-Jensen formula, we derive

$$
\begin{aligned}
\log \frac{1}{|F(\zeta_1) - \alpha|} &\le \frac{(1+t)/2 + t}{(1+t)/2 - t} m\left(\frac{1+t}{2}, \frac{1}{F-\alpha}\right) \\
&\quad + \sum_i \log\left|\frac{((1+t)/2)^2 - \bar{\alpha}_i \zeta_1}{((1+t)/2)(\zeta_1 - \alpha_i)}\right| \\
&\le \frac{4}{1-t} m\left(\frac{1+t}{2}, \frac{1}{F-\alpha}\right) + \sum_i \log\left|\frac{1 - \bar{\alpha}_i \zeta_1}{\zeta_1 - \alpha_i}\right| \\
&\quad + \sum_i \log\left|\frac{((1+t)/2)^2 - \bar{\alpha}_i \zeta_1}{((1+t)/2)(1 - \bar{\alpha}_i \zeta_1)}\right|.
\end{aligned}
$$

Moreover, notice that point $\zeta_1 \notin (\gamma)'_\zeta$, $|F(\zeta_1) - \alpha| = |f(z_1) - \alpha| \le e^{-N}$, and

$$
\left|\frac{((1+t)/2)^2 - \bar{\alpha}_i \zeta_1}{((1+t)/2)(1 - \bar{\alpha}_i \zeta_1)}\right| \le \frac{4}{1-t};
$$

then

$$
\begin{aligned}
N &\le \frac{4}{1-t} m\left(\frac{1+t}{2}, \frac{1}{F-\alpha}\right) + n\left(\frac{1+t}{2}, F = \alpha\right) \log \frac{2}{h} \\
&\quad + n\left(\frac{1+t}{2}, F = \alpha\right) \log \frac{4}{1-t}.
\end{aligned}
$$

We may assume that $F(0) \ne \alpha$. Otherwise, it is obvious that Theorem 5.1 holds. Hence

$$
\begin{aligned}
n\left(\frac{1+t}{2}, F = \alpha\right) &\le \left\{\log \frac{(3+t)/4}{(1+t)/2}\right\}^{-1} \int_{(1+t)/2}^{(3+t)/4} \frac{n(R, F = \alpha)}{R} dR \\
&\le \frac{4}{1-t} N\left(\frac{3+t}{4}, \frac{1}{F-\alpha}\right).
\end{aligned}
$$

We obtain, therefore,

$$
\begin{aligned}
N &\le \frac{4}{1-t}\left(1+\log\frac{8}{(1-t)h}\right) T\left(\frac{3+t}{4},\frac{1}{F-\alpha}\right) \\
&= \frac{4}{1-t}\left(1+\log\frac{8}{(1-t)h}\right) \\
&\quad \times \left\{ T\left(\frac{3+t}{4}, F-\alpha\right) + \log\frac{1}{|F(0)-\alpha|}\right\} \\
&\le \frac{4}{1-t}\left(1+\log\frac{8}{(1-t)h}\right) \\
&\quad \times \left\{ T\left(\frac{3+t}{4}, F\right) + \log\frac{1}{|F(0)-\alpha|} + \log^+|\alpha| + \log 2\right\}.
\end{aligned}
\tag{5.1}
$$

Now, we estimate $T(\frac{3+t}{4}, F)$. First, since the spherical distances among a, b, and c are greater than d, we may assume that $|a| \le \frac{3}{d}$ and $|b| \le \frac{3}{d}$. Consider the transformation

$$
G(\zeta) = \frac{F(\zeta)-a}{F(\zeta)-b} \cdot \frac{c-b}{c-a}.
$$

Then function $G(\zeta)$ is meromorphic on the disk $|\zeta| \le 1$. Moreover,

$$
n(1, G=0) \le n, \quad n(1, G=1) \le n, \quad n(1, G=\infty) \le n.
$$

Notice that point $\zeta = 0 \notin (\gamma)'_\zeta$. Besides, we may assume that $|G(0)| \le 1$. Otherwise, we need only to exchange the position of a and b. Hence, applying Theorem 2.8, we get

$$
\begin{aligned}
T\left(\frac{3+t}{4}, G\right) &\le \frac{1}{(1-(3+t)/4)^2} \\
&\quad \times \left\{ An\log\frac{1}{h} + B\log\frac{2}{1-(3+t)/4}\right\} \\
&\le \frac{1}{(1-t)^2}\left\{ An\log\frac{1}{h} + B\log\frac{2}{1-t}\right\},
\end{aligned}
$$

where A and B are constants.

On the other hand, according to the identity

$$
\frac{1}{F(\zeta)-b} = \frac{1}{b-a}\left\{ G(\zeta)\left(1+\frac{b}{c-b}-\frac{a}{c-b}\right) - 1\right\},
$$

we derive

$$T\left(\frac{3+t}{4}, \frac{1}{F-b}\right) \le T\left(\frac{3+t}{4}, G\right) + A\log\frac{1}{d},$$

$$T\left(\frac{3+t}{4}, F-b\right) \le \frac{1}{(1-t)^2}\left\{An\log\frac{1}{h} + B\log\frac{2}{1-t}\right\}$$
$$+ C\log\frac{1}{d} + \log|F(0)-b|,$$

$$T\left(\frac{3+t}{4}, F\right) \le \frac{1}{(1-t)^2}\left\{An\log\frac{1}{h} + B\log\frac{2}{1-t}\right\}$$
$$+ C\log\frac{1}{d} + \log^+|F(0)|.$$

Combining this with formula (5.1) we obtain

$$N \le \frac{4}{1-t}\left(1 + \log\frac{8}{(1-t)h}\right)$$
$$\times\left\{\frac{1}{(1-t)^2}\left(An\log\frac{1}{h} + B\log\frac{2}{1-t}\right) + C\log\frac{1}{d} + D\log^+|\alpha|\right.$$
$$\left. + \log^+|F(0)| + \log\frac{1}{|F(0)-\alpha|}\right\},$$

$$N \le \frac{A}{1-t}\log\frac{2}{(1-t)h}\left\{\frac{1}{(1-t)^2}\left(Bn\log\frac{1}{h} + C\log\frac{2}{1-t}\right)\right.$$
$$\left. + D\log\frac{1}{d} + E\log^+|\alpha| + \log^+\frac{1}{|F(0)-\alpha|}\right\}.$$

If we notice that

$$1 - t = 1 - \frac{\tau+r}{1+\tau r} \ge \frac{1}{2}(1-\tau)(1-r),$$

$F(0) = f(z_0)$ and z_0 is any point in the disk $|z| \le r$ and outside circles (γ), and also (γ) contains at most $3n$ circles, with the sum of their Euclidean radii not exceeding $2eh$, then Theorem 5.1 follows.

By examining the proof of Theorem 5.1, we may easily find that if $f(z)$ has an upper bound $M < +\infty$ in the unit disk $|z| \le 1$, then for $T(\frac{3+t}{4}, F)$ in formula (5.1), we may have a direct estimation:

$$T\left(\frac{3+t}{4}, F\right) \le \log^+ M.$$

Besides, notice that z_0 may be any points on $|z| \le r$. Hence, we have the following result:

THEOREM 5.2. *Let $f(z)$ be a regular function on the unit disk $|z| \le 1$ with $f(z)$ (or $|f(z)|$) being bounded by $M < +\infty$. Besides, we let α be a* *finite complex number and $N \geq 0$ be a real number. Suppose that the point set E_α of $f(z)$ in the disk $|z| \leq \tau$ $(0 < \tau < 1)$ satisfying the inequality*

$$|f(z) - \alpha| \leq e^{-N}$$

cannot be included in some circles with the sum of their radii not exceeding $2eh$ $(h \leq 0.01)$. Then for point z in the disk $|z| \leq r$ $(0 < r < 1)$, we have

$$|f(z) - \alpha| \leq \exp\left\{ (\log^+ M + \log^+ |\alpha| + \log 2) \right.$$

$$\left. - A \left(\log \frac{1}{(1-\tau)(1-r)h} \right)^{-1} (1-\tau)(1-r)N \right\},$$

where $A > 0$ is a constant.

THEOREM 5.3. *Let $f(z)$ be a regular function on the unit disk $|z| \leq 1$, and a and b be two finite complex numbers, with the spherical distances among a, b and ∞ greater than d, $0 < d < \frac{1}{2}$. Furthermore*

$$n(1, f = a) \leq n, \qquad n(1, f = b) \leq n.$$

Besides, suppose that α is a finite complex number, $N \geq 0$ is a real number, and the point set E_α of $f(z)$ in disk $|z| \leq \tau$ $(0 < \tau < 1)$ satisfying the inequality

$$|f(z) - \alpha| \leq e^{-N}$$

cannot be included in some circles, with the sum of their radii not exceeding $2eh$ $(h \leq 0.01)$. Then for point z in the disk $|z| \leq r$ $(\tau \leq r < 1)$, we have

$$|f(z)-\alpha| \leq \exp\left\{ \frac{1}{(1-r)^6} \left(An \log \frac{1}{h} + B \log \frac{2}{1-r} + C \log \frac{1}{d} + D \log^+ |\alpha| \right) \right.$$

$$\left. - E \left(\log \frac{1}{(1-\tau)(1-r)h} \right)^{-1} (1-\tau)(1-r)N \right\},$$

where A, B, C, D and E $(E > 0)$ are constants.

PROOF. Let (γ) be the pseudo-non-Euclidean exceptional circles corresponding to these $n(1, f = a) + n(1, f = b)$ points and number h $(h \leq 0.01)$. According to the assumption, there exists a point $z_0 \notin (\gamma)$ on the set E_α. Consider the transformation

$$\zeta = \frac{z - z_0}{1 - \bar{z}_0 z},$$

with its inverse transformation

$$z = \frac{\zeta + z_0}{1 + \bar{z}_0 \zeta}.$$

Then function

$$F(\zeta) = f\left(\frac{\zeta + z_0}{1 + \bar{z}_0 \zeta} \right)$$

is regular on the disk $|\zeta| \leq 1$. Moreover,

$$n(1, F = a) \leq n, \qquad n(1, F = b) \leq n.$$

Circles (γ) are transformed into corresponding pseudo-non-Euclidean exceptional circles $(\gamma)_\zeta$ on the ζ-plane. Notice that point $\zeta = 0 \notin (\gamma)_\zeta$, $F(0) = f(z_0)$ and $z_0 \in E_\alpha$, then

$$F(0) = |f(z_0)| \leq |f(z_0) - \alpha| + |\alpha| \leq |\alpha| + 1. \tag{5.2}$$

On the other hand, according to the inequality

$$|\zeta| < \frac{|z| + |z_0|}{1 + |z_0||z|},$$

we conclude that the image domain of disk $|z| \leq \frac{1}{2}(1 + r)$ on the ζ-plane must be included in disk $|\zeta| \leq t$, with

$$t = \frac{1 + r}{1 + ((1 + r)/2)^2}.$$

Consider the transformation

$$G(\zeta) = \frac{F(\zeta) - a}{F(\zeta) - b}.$$

We may assume that $|G(0)| \leq 1$. Otherwise, we need only to exchange the position of a and b, and it follows from Theorem 2.8 that

$$T\left(\frac{1+t}{2}, G\right) \leq \frac{1}{(1 - ((1+t)/2))^2}\left\{An \log \frac{1}{h} + B \log \frac{2}{1 - (1+t)/2}\right\}$$

$$\leq \frac{1}{(1 - t)^2}\left\{An \log \frac{1}{h} + B \log \frac{2}{1 - t}\right\}.$$

On the other hand, according to the equality

$$\frac{1}{F(\zeta) - b} = \frac{1}{b - a}\{G(\zeta) - 1\},$$

we derive

$$T\left(\frac{1+t}{2}, \frac{1}{F - b}\right) \leq T\left(\frac{1+t}{2}, G\right) + A \log \frac{1}{d},$$

$$T\left(\frac{1+t}{2}, F\right) \leq \frac{1}{(1 - t)^2}\left\{An \log \frac{1}{h} + B \log \frac{2}{1 - t}\right\}$$

$$+ C \log \frac{1}{d} + \log^+ |F(0)|.$$

Combining this with formula (5.2), we obtain

$$T\left(\frac{1+t}{2}, F\right) \leq \frac{1}{(1 - t)^2}\left\{An \log \frac{1}{h} + B \log \frac{2}{1 - t}\right\}$$

$$+ C \log \frac{1}{d} + \log^+ |\alpha|.$$

Moreover, notice that

$$\log M\left(\frac{1+r}{2}, f\right) \le \log M(t, F)$$

$$\le \frac{(1+t)/2+t}{(1+t)/2-t} T\left(\frac{1+t}{2}, F\right).$$

We conclude that

$$\log M\left(\frac{1+r}{2}, f\right) \le \frac{1}{(1-r)^6}\left\{An\log\frac{1}{h} + B\log\frac{2}{1-r} + C\log\frac{1}{d} + \log^+|\alpha|\right\}.$$

By constructing the transformation $\xi = \frac{2z}{1+r}$, disk $|z| \le \frac{1+r}{z}$ is transformed onto the unit disk $|\xi| \le 1$ on the ξ-plane. Meanwhile, function $\varphi(\xi) = f(\frac{1+r}{2}\xi)$ is regular on disk $|\xi| \le 1$ and has an upper bound M, with

$$M = \exp\left\{\frac{1}{(1-r)^6}\left(An\log\frac{1}{h} + B\log\frac{2}{1-r} + C\log\frac{1}{d} + \log^+|\alpha|\right)\right\}.$$

Besides, disk $|z| \le \tau$ is transformed onto disk $|\xi| \le R$, $R = \frac{2\tau}{1+r} < 1$ on the ξ-plane, and the point set E_α is transformed into point set E_ξ on the ξ-plane. Meanwhile, E_ξ cannot be included in some circles, with the sum of their Euclidean radii not exceeding $2eh$. Hence, applying Theorem 5.2, for the point ξ in disk $|\xi| \le \frac{2r}{1+r}$, we have

$$|\varphi(\xi) - \alpha| \le \exp\left\{\frac{1}{(1-r)^6}\left\{An\log\frac{1}{h} + B\log\frac{2}{1-r} + C\log\frac{1}{d} + D\log^+|\alpha|\right\}.\right.$$

$$- E\left(\log\frac{1}{(1 - 2\tau/(1+r))(1 - 2r/(1+r))h}\right)^{-1}$$

$$\left. \times \left(1 - \frac{2\tau}{1+r}\right)\left(1 - \frac{2r}{1+r}\right) N\right\}.$$

Moreover, noticing that $\tau \le r$, we conclude that for point z in the disk $|z| \le r$ $(\tau \le r < 1)$, it follows that

$$|f(z) - \alpha|$$

$$\le \exp\left\{\frac{1}{(1-r)^6}\left(An\log\frac{1}{h} + B\log\frac{2}{1-r} + C\log\frac{1}{d} + D\log^+|\alpha|\right)\right.$$

$$\left. - E\left(\log\frac{1}{(1-\tau)(1-r)h}\right)^{-1}(1-\tau)(1-r)N\right\}.$$

Hence, Theorem 5.3 is proved.

5.1.2. Applications. As an application of Theorem 5.1, we prove the following result [43c]:

LEMMA 5.1. *Let* $f(z)$ *be a meromorphic function on* $\overline{\Omega}(-\theta,\theta)$ $(0<\theta\leq\pi)$ *and let it satisfy the following conditions:*

(1) *There exist three distinct complex numbers* a, b, *and* c, *such that the spherical distances among* a, b *and* c *are greater than* d $(0<d<\frac{1}{2})$. *Meanwhile,*

$$\varlimsup_{r\to+\infty}\frac{\log^+ n\{\overline{\Omega}(-\theta_1,\theta;r),f=X\}}{\log r}\leq\nu<+\infty,\qquad X=a,b,c.$$

(2) *There exists a point set* E_α *in* $\Gamma(-\theta+\varepsilon,\theta-\varepsilon;R)$ $(0<\varepsilon<\theta;$ $0<R<+\infty)$, *with its linear measure* $\mathrm{meas}\,E_\alpha\geq HR$ $(H\geq\frac{\varepsilon}{2})$, *or there exists a continuum* L *in* $\overline{\Omega}(-\theta+\varepsilon,\theta-\varepsilon;R_1,R_2)$, *with its diameter* $\geq HR_1$ $(0<\varepsilon<\theta;$ $R_1\leq R\leq R_2)$. *Moreover, when* $z\in E_\alpha$ *or* $z\in L$,

$$|f(z)-\alpha|\leq e^{-N},$$

where α *is a finite complex number and* $N\geq 0$ *is a real number.*

Then, for any arbitrarily assumed number $\eta>0$, *provided that* R_1 *is sufficiently large, for any point* z *in* $\overline{\Omega}(-\theta+\varepsilon,\theta-\varepsilon;R_1,R_2)$ *and outside* (γ), *we have*

$$\log^+\frac{1}{|f(z)-\alpha|}\geq\frac{A(\varepsilon,\theta)}{B(\theta)\log\frac{R_2}{R_1}+C(\varepsilon,\theta)}$$

$$\times\left(\frac{R_1}{R_2}\right)^{\frac{2\pi}{\theta}}\cdot N-D(\varepsilon,\theta)\left(\frac{R_2}{R_1}\right)^{\frac{4\pi}{\theta}}\tag{5.3}$$

$$\times\left\{\left(\frac{R_2}{R_1}\right)^{2(\nu+\eta)}R_2^{(\nu+\eta)}\left(1+\log\frac{R_2}{R_1}\right)+\log^+|\alpha|\right\},$$

where (γ) *represents some circles, with the sum of their Euclidean radii not more than* $\frac{\varepsilon}{4}R_1$, *and the number of circles does not exceed* $3n$,

$$n=\left(\frac{8\theta}{\varepsilon}\right)^{\frac{2\theta}{\pi}(\nu+\eta)}\cdot\left(\frac{R_2}{R_1}\right)^{2(\nu+\eta)}R_2^{\nu+\eta}.$$

Besides, $A(\varepsilon,\theta)>0$, $C(\varepsilon,\theta)<+\infty$ *and* $D(\varepsilon,\theta)$ *are constants depending on* ε *and* θ, *and* $B(\theta)<+\infty$ *is a constant depending on* θ *only.*

PROOF. Consider the transformation

$$\zeta=\frac{z^{\pi/2\theta}-R^{\pi/2\theta}}{z^{\pi/2\theta}+R^{\pi/2\theta}}.$$

Then $\Omega(-\theta,\theta)$ is transformed onto the unit disk $|\zeta|<1$ on the ζ-plane. According to Lemma 3.11, we conclude that the image domain of $\overline{\Omega}(-\theta+\varepsilon,\theta-\varepsilon;R_1,R_2)$ on the ζ-plane must be included in disk $|\zeta|\leq\rho$, with

$$\rho=1-\frac{\varepsilon}{2\theta}\left(\frac{R_1}{R_2}\right)^{\frac{\pi}{\theta}}.\tag{5.4}$$

Moreover, the image domain of disk $|\zeta| \leq \frac{1}{2}(1 + \rho)$ on the z-plane must be included in $\overline{\Omega}(-\theta, \theta; R_3)$, with

$$R_3 = \left(\frac{8\theta}{\varepsilon}\right)^{2\theta/\pi} \left(\frac{R_2}{R_1}\right)^2 R_2,$$

and when $|\zeta| \leq \rho$,

$$|z'(\zeta)| \leq \frac{2\theta}{\pi} \left(\frac{4\theta}{\varepsilon}\right)^{1+2\theta/\pi} \left(\frac{R_2}{R_1}\right)^{2+\pi/\theta} R_2.$$

Consider the transformation $\xi = 2\zeta/(1 + \rho)$. Then disk $|\zeta| \leq \frac{1}{2}(1 + \rho)$ is transformed onto the unit disk $|\xi| \leq 1$ on the ξ-plane, disk $|\zeta| \leq \rho$ is transformed into disk $|\xi| \leq \tau$, with

$$\tau = \frac{2\rho}{1 + \rho}, \tag{5.5}$$

and when $|\xi| \leq \tau$,

$$|\zeta'(\xi)| \leq \frac{1 + \rho}{2} < 1,$$

$$|z'(\xi)| = |z'(\zeta)| |\zeta'(\xi)| \leq \frac{2\theta}{\pi} \left(\frac{4\theta}{\varepsilon}\right)^{1+2\theta/\pi} \left(\frac{R_2}{R_1}\right)^{2+\pi/\theta} R_2.$$

On the other hand, the point set E_α may be transformed into point set E_ξ on the ξ-plane; then E_ξ is in disk $|\xi| \leq \tau$ and distributed on the imaginary axis. Moreover,

$$HR \leq \text{meas } E_\alpha = \int_{E_\alpha} |dz| = \int_{E_\xi} |z'(\xi)| |d\xi|$$

$$\leq \frac{2\theta}{\pi} \cdot \left(\frac{4\theta}{\varepsilon}\right)^{1+2\theta/\pi} \left(\frac{R_2}{R_1}\right)^{2+\pi/\theta} R_2 \text{ meas } E_\xi,$$

$$\text{meas } E_\xi \geq \frac{H\pi}{2\theta} \left(\frac{\varepsilon}{4\theta}\right)^{1+2\theta/\pi} \left(\frac{R_1}{R_2}\right)^{3+\pi/\theta},$$

or a continuum L is transformed into a continuum L_ξ on the ξ-plane, then L_ξ lies in disk $|\xi| \leq \tau$. Furthermore, we suppose that line segment l' connects points z_1 and z_2 on L, with the length meas l' of l' equal to the diameter of L. We also let ξ_1 and ξ_2 be the image points of z_1 and z_2 on the ξ-plane. Obviously, the image l of the line segment l_ξ connecting points ξ_1 and ξ_2 on the z-plane is a curve linking points z_1 and z_2. Hence, we have

$$HR_1 \leq \text{meas } l' \leq \int_l |dz| = \int_{l_\xi} |z'(\xi)| |d\xi|$$

$$\leq \frac{2\theta}{\pi} \left(\frac{4\theta}{\varepsilon}\right)^{1+2\theta/\pi} \left(\frac{R_2}{R_1}\right)^{2+\pi/\theta} R_2 \text{ meas } l_\xi,$$

$$\text{meas}\, l_\xi \geq \frac{H\pi}{2\theta} \left(\frac{\varepsilon}{4\theta}\right)^{1+2\theta/\pi} \left(\frac{R_1}{R_2}\right)^{3+\pi/\theta}.$$

If we notice that the length of the diameter of L_ξ is $\geq$ meas l_ξ and we assume

$$h = \frac{1}{8e} \frac{\pi\varepsilon}{4\theta} \left(\frac{\varepsilon}{4\theta}\right)^{1+2\theta/\pi} \left(\frac{R_1}{R_2}\right)^{3+\pi/\theta}, \tag{5.6}$$

then we conclude that E_ξ or L_ξ cannot be included in some circles, with the sum of their Euclidean radii not more than $2eh$.

Let $F(\xi) = f(z(\zeta(\xi)))$. Then $F(\xi)$ is meromorphic on the unit disk $|\xi| \leq 1$. Moreover, according to condition (1), for any arbitrarily assumed number $\eta > 0$, provided that R_1 is sufficiently large, we have

$$n(1, F = X) \leq n, \quad X = a, b, c,$$

$$n = R_3^{\nu+\eta} = \left(\frac{8\theta}{\varepsilon}\right)^{\frac{2\theta}{\pi}(\nu+\eta)} \left(\frac{R_2}{R_1}\right)^{2(\nu+\eta)} R_2^{\nu+\eta}. \tag{5.7}$$

Applying Theorem 5.1, we conclude that for the point ξ in the disk $|\xi| \leq \tau$ and outside circles $(\gamma)_\zeta$, we have

$$\log^+ \frac{1}{|F(\xi) - \alpha|} \geq A \left(\log \frac{1}{(1-\tau)^2 h}\right)^{-1} (1-\tau)^2 N$$
$$- \frac{1}{(1-\tau)^4} \left\{ Bn\log \frac{1}{h} + C\log \frac{1}{(1-\tau)^2} + D\log \frac{1}{d} + E\log^+ |\alpha| \right\}, \tag{5.8}$$

where $(\gamma)_\zeta$ represents some circles, with the sum of their Euclidean radii not exceeding $2eh$, and the number of circles not more than $3n$.

In the following, we prove that the image set of circles $(\gamma)_\xi$ on the z-plane must be included in some circles (γ), with the sum of the Euclidean radii of (γ) not exceeding $\frac{1}{4}\varepsilon R_1$, and the number of circles included in (γ) not more than $3n$. In fact, we consider arbitrarily a circle $\Gamma_\xi: |\xi - \xi_0| < \sigma$ in $(\gamma)_\xi$. Let Γ be the image of Γ_ξ on the z-plane and point z_0 be the image of ξ_0; then there exists a point z' in $\overline{\Gamma}$, such that the circle Γ', with point z_0 being the center and the line segment l' linking points z_0 and z' being the radius, contains Γ in itself. Let ξ' be the image point of z' on the ξ-plane. Obviously, the line segment l_ξ linking points ξ_0 and ξ' on the z-plane is a curve l linking points z_0 and z'. Hence, we have

$$\text{meas}\, l' \leq \int_l |dz| = \int_{l_\xi} |z'(\xi)|\,|d\xi|$$
$$\leq \frac{4\theta}{\pi} \left(\frac{4\theta}{\varepsilon}\right)^{1+2\theta/\pi} \left(\frac{R_2}{R_1}\right)^{2+\pi/\theta} R_2 \sigma.$$

Therefore, if we let (γ) be the whole $\{\Gamma'\}$, then the sum of the Euclidean radii of (γ) does not exceed $\frac{1}{4}\varepsilon R_1$, and it contains less than $3n$ circles.

Finally, according to formulas (5.4)–(5.8), we conclude that when point z is in $\overline{\Omega}(-\theta + \varepsilon, \theta - \varepsilon; R_1, R_2)$ and outside circles (γ), formula (5.3) holds. Hence, Lemma 5.1 is proved.

Analogously, applying Theorem 5.3, we may prove the following result:

LEMMA 5.2. *Let* $f(z)$ *be a regular function on* $\overline{\Omega}(-\theta, \theta)$ $(0 < \theta \le \pi)$, *and let it satisfy the following conditions:*

(1) *There exist two distinct finite complex numbers* a *and* b, *such that the spherical distances among* a, b *and* ∞ *are greater than* d $(0 < d < \frac{1}{2})$. *Meanwhile,*

$$\varlimsup_{r \to +\infty} \frac{\log^+ n\{\overline{\Omega}(-\theta, \theta, r), f = X\}}{\log r} \le \nu < +\infty, \qquad X = a, b.$$

(2) *There exists a point set* E_α *in* $\Gamma(-\theta + \varepsilon, \theta - \varepsilon; R)$ $(0 < \varepsilon < \theta;$ $R_1 \le R \le R_2)$, *with its linear measure* $\operatorname{meas} E_\alpha \ge HR$ $(H \ge \frac{\varepsilon}{2})$, *or there exists a continuum* L *in* $\overline{\Omega}(-\theta + \varepsilon, \theta - \varepsilon; R_1, R_2)$, *with its diameter* $\ge HR_1$. *Moreover, when* $z \in E_\alpha$ *or* $z \in L$,

$$|f(z) - \alpha| \le e^{-N},$$

where α *is a finite complex number and* $N \ge 0$ *is a real number.*

Then for any arbitrarily assumed number $\eta > 0$, *provided that* R_1 *is sufficiently large, for any point* z *on* $\overline{\Omega}(-\theta + \varepsilon, \theta - \varepsilon; R_1, R_2)$

$$|f(z) - \alpha| \le \exp\left\{ A(\varepsilon, \theta)\left(\frac{R_2}{R_1}\right)^{6\pi/\theta}\left[\left(\frac{R_1}{R_1}\right)^{2(\nu+\eta)} R_2^{\nu+\eta}\left(1 + \log\frac{R_2}{R_1}\right) + \log^+|\alpha|\right] \right.$$
$$\left. - \frac{B(\varepsilon, \theta)}{C(\theta)\log(R_2/R_1) + D(\varepsilon, \theta)}\left(\frac{R_1}{R_2}\right)^{2\pi/\theta} N \right\},$$

where $A(\varepsilon, \theta) < +\infty$, $B(\varepsilon, \theta) > 0$ *and* $D(\varepsilon, \theta) < +\infty$ *are constants depending on* ε *and* θ, *and* $C(\theta)$ *is a constant depending on* θ *only.*

§5.2. Entire functions of finite lower order [43c]

THEOREM 5.4. *Let* $f(z)$ *be an entire function of lower order* $\mu < +\infty$ *on the open plane* $|z| < +\infty$, q *be the number of Julia directions of* $f(z)$, l *be the number of distinct finite asymptotic values, and* p *be the number of finite deficient values, where* l' *deficient values are, in the meantime, asymptotic values. Then we have the following formula:*

$$2p - l' + l \le q.$$

PROOF. (1) First we consider the following case: $q < +\infty$, $p < +\infty$, $l < +\infty$, and p and l cannot be zero simultaneously.

Since p and l cannot be zero at the same time, then according to the corollary of Theorem 4.4, we conclude that $\mu \ge \frac{1}{2}$. Besides, according to Theorem 2.15, we have $q \ge 1$. In the following, we let a_i $(i = 1, 2, \ldots, p)$

be p finite deficient values of $f(z)$, with its corresponding deficiencies $\delta(a_i, f) = \delta_i > 0$, and b_j $(j = 1, 2, \ldots, l)$ be the l distinct finite asymptotic values of $f(z)$, with its corresponding paths of fixed values being L_j. According to Lemma 2.2 (where we let $h_1 = 0$) and Lemma 3.9 (where we let $k = 3\mu$, $\sigma = \log 2$), we conclude that there must exist two sequences r_n $(n = 1, 2, \ldots)$ and t_n $(n = 1, 2, \ldots)$ such that([1])

$$\lim_{n \to +\infty} \frac{\log T(r_n, f)}{\log r_n} = \mu, \tag{5.9}$$

and

$$r_n \leq t_n \leq 2r_n. \tag{5.10}$$

Moreover,

$$\operatorname{meas} E_i^n \geq K, \qquad K = K(\delta, \mu, p) > 0, \quad i = 1, 2, \ldots, p,$$

where

$$E_i^n = \left\{ \theta \,\bigg|\, \log \frac{1}{|f(t_n e^{i\theta}) - a_i|} \geq \frac{\delta}{4} T(t_n, f), \ 0 \leq \theta < 2\pi \right\}, \qquad \delta = \min\{\delta_i\}.$$

Let $\Delta(\theta_k)$ $(k = 1, 2, \ldots, q; \ 0 \leq \theta_1 < \theta_2 < \cdots < \theta_q < 2\pi)$ be q Julia directions of $f(z)$, and assume that

$$\omega = \min_{1 \leq k \leq q} \{\theta_{k+1} - \theta_k\}, \ \theta_{q+1} = \theta_1 + 2\pi,$$
$$M = \max\{1, |a_1|, \ldots, |a_p|, |b_1|, \ldots, |b_l|\}.$$

Then we select a number η,

$$0 < \eta < \frac{\min\{\omega, \mu\omega\}}{360(\mu + 1)\pi}.$$

Then according to formulas (5.9) and (5.10), when n is sufficiently large,

$$T(t_n, f) \geq T(r_n, f) \geq r_n^{\mu - \eta} \tag{5.11}$$

and

$$T(r_n, f) \leq r_n^{\mu + \eta}. \tag{5.12}$$

Finally, we assume a fixed number ε,

$$0 < \varepsilon < \min\left\{ \frac{\pi\eta}{6(\mu + 1)}, \frac{K}{12q}, \frac{\omega}{8} \right\}$$

and let

$$\overline{\Omega} = \bigcup_{k=1}^{q} \overline{\Omega}(\theta_k + \varepsilon, \theta_{k+1} - \varepsilon).$$

Then according to Theorem 2.13 and the finite covering theorem, we conclude that there must exist two distinct finite complex numbers α and β, such that

$$\varlimsup_{r \to +\infty} \frac{\log^+ n(r, \overline{\Omega}, f = X)}{\log r} = 0, \qquad X = \alpha, \beta.$$

([1]) When $p = 0$, we consider only the sequence $\{r_n\}$ which satisfies formula (5.9).

Since α and β are finite numbers, there exists a number d, $0 < d < \frac{1}{2}$, such that the spherical distances among α, β and ∞ are greater than d.

(2) We need also the following lemma.

LEMMA 5.3. *Suppose that two simply continuous curves L_{nm} $(m = 1, 2)$ connect point A_{nm} $(m = 1, 2)$ on disk $|z| = r_n^{1-3\eta}n$ and point B_{nm} $(m = 1, 2)$ on disk $|z| = \frac{1}{2}r_n$, respectively. Meanwhile, these two curves have no intersection and are all in $\overline{\Omega}(-3\varepsilon, 3\varepsilon; r_n^{1-3\eta}, \frac{1}{2}r_n)$. Hence, a region $\Omega_n \subset \overline{\Omega}(-3\varepsilon, 3\varepsilon; r_n^{1-3\eta}, \frac{1}{2}r_n)$ is bounded by L_{n1}, L_{n2} and the circular arc of disk $|z| = r_n^{1-3\eta}$ between points A_{n1} and A_{n2}, and the circular arc of disk $|z| = \frac{1}{2}r_n$ between points B_{n_1} and B_{n_2}. Furthermore, we assume that*

$$\max_{z \in L_{nm}} \log|f(z)| \leq M_{nm} < +\infty, \qquad m = 1, 2,$$

$$\Omega_n' = \Omega_n \cap \Gamma_n',$$

where Γ_n' represents annulus $r_n^{1-2n} \leq |z| \leq r_n^{1-\eta}$.

Then when n is sufficiently large, for the point z in Ω_n', we have

$$\log|f(z)| \leq M_{n1} + M_{n2} + 1.$$

PROOF. First, applying the Poisson-Jensen formula and according to formula (5.12), we obtain

$$\log M(r_n^{1-3\eta}, f) \leq \log M(\tfrac{1}{2}r_n, f) \leq 3T(r_n, f) \leq 3r_n^{\mu+\eta}. \tag{5.13}$$

Next, applying Lemma 2.10 and according to formula (5.13), when $z \in \Omega_n'$,

$$\log|f(z)| \leq M_{n1} + M_{n2} + \frac{4r_n^{-\pi\eta/6\varepsilon}}{\pi[1 - r_n^{-\pi\eta/3\varepsilon}]}3r_n^{\mu+\eta}$$
$$+ \frac{4 \cdot 2^{\pi/6\varepsilon} \cdot r_n^{-\pi\eta/6\varepsilon}}{\pi[1 - 2^{\pi/3\varepsilon}r_n^{-\pi\eta/3\varepsilon}]}3r_n^{\mu+\eta}.$$

Furthermore, according to the selection of ε, we have

$$-\frac{\pi\eta}{6\varepsilon} + \mu + \eta < -(\mu + 1) + (\mu + \eta) = -1 + \eta < 0.$$

Hence, when n is sufficiently large, we conclude that $\log|f(z)| \leq M_{n1} + M_{n2} + 1$. Thus Lemma 5.3 holds.

(3) We consider sequence of annuli: Γ_n: $r_n^{1-3\eta} \leq |z| \leq 2r_n$ $(n = 1, 2, \ldots)$. Since the paths of fixed value L_j $(j = 1, 2, \ldots, l)$ is a simply continuous curve originating from the origin and tending to ∞, L_j intersects both circumference $|z| = r_n^{1-3\eta}$ and circumference $|z| = 2r_n$. We assume the last intersection point on circumference $|z| = r_n^{1-3\eta}$ as well as the first intersection point on circumference $|z| = 2r_n$. Moreover, we let L_{jn} be the part of L_j between these two intersection points. Obviously, each curve

L_{jn} $(1 \leq j \leq l)$ intersects at least one $\overline{\Omega}(\theta_{k_j} + 3\varepsilon, \ \theta_{k_j+2} - 3\varepsilon; r_n^{1-3\eta}, 2r_n)$ $(1 \leq k_j \leq q)$ among the q closed domains $\overline{\Omega}(\theta_k + 3\varepsilon, \ \theta_{k+2} - 3\varepsilon; r_n^{1-3\eta}, 2r_n)$ $(k = 1, 2, \ldots, q; \ \theta_{q+1} = \theta_1 + 2\pi, \ \theta_{q+2} = \theta_2 + 2\pi)$, and hence, merely one Julia direction $\Delta \ (\theta_{k_j+1})$ will be determined. Accordingly, following this way, each curve L_{jn} corresponds to one Julia direction $\Delta b(\theta_{k_j+1}) = \Delta(\theta_{k_j+1})$, l curves L_{jn} $(j = 1, 2, \ldots, l)$ correspond to l Julia directions $\Delta b(\theta_{k_j+1})$ $(j = 1, 2, \ldots, l)$. Now, we prove that any two Julia directions of the set $\{\Delta b(\theta_{k_j+1}) \,|\, j = 1, 2, \ldots, l\}$ do not coincide with each other.

In fact, otherwise, suppose that the two Julia directions $\Delta b(\theta_{k_j+1})$ and $\Delta b(\theta_{k_{j'}+1})$ coincide with each other. Hence, L_{jn} and $L_{j'n}$ $(j \neq j')$ intersect simultaneously $\overline{\Omega}(\theta_k + 3\varepsilon, \ \theta_{k+2} - 3\varepsilon; r_n^{1-3\eta}, 2r_n)$ $(k = k_j = k_{j'})$. In the following, we prove that this is impossible, meaning that a contradiction is derived. Hence, we need only to consider the following types of typical cases:

1) L_{jn} and $L_{j'n}$ intersect at $\overline{\Omega}(\theta_k + 3\varepsilon, \ \theta_{k+1} - 3\varepsilon; r_n^{1-3\eta}, 2r_n)$ simultaneously. Under this case, $L_{jn} \cap \overline{\Omega}(\theta_k + 2\varepsilon, \ \theta_{k+1} - 2\varepsilon, r_n^{1-3\eta}, 2r_n)$ must contain a part of the continuous curve L'_{jn}, such that the diameter of L'_{jn} is $\geq \frac{\varepsilon}{2} r_n^{1-3\eta}$. Moreover, when n is sufficiently large,

$$|f(z) - b_j| < 1, \qquad z \in L'_{jn}.$$

Hence, applying Lemma 5.2, where we let $\theta = \frac{1}{2}(\theta_{k+1} - \theta_k) - \varepsilon$, $R_1 = r_n^{1-3\eta}$, $R_2 = 2r_n$, $L = L'_{jn}$, $H = \frac{\varepsilon}{2}$, $\alpha = b_j$, $N = 0$ and $\nu = 0$, then when $z \in \overline{\Omega}(\theta_k + 2\varepsilon, \ \theta_{k+1} - 2\varepsilon; r_n^{1-3\eta}, 2r_n)$, we obtain

$$|f(z) - b_j| \leq \exp\left\{ A(\varepsilon, \theta) 2^{6\pi/((\theta_{k+1}-\theta_k)/2-\varepsilon)} r_n^{18\pi\eta/((\theta_{k+1}-\theta_k)/2-\varepsilon)} \right.$$
$$\left. \times [2^{2\eta} r_n^{6\eta^2} 2^\eta r_n^\eta \cdot \log(2r_n^{3\eta}) + \log^+ |b_j|] \right\}$$
$$\leq \exp\left\{ A(\varepsilon, \theta, \eta) r_n^{18\pi\eta/\omega/2-\varepsilon} [r_n^{6\eta^2+\eta} \log r_n + \log M] \right\}.$$

Furthermore, if we notice that $\varepsilon < \frac{\omega}{8} < \frac{\omega}{4}$, $\eta < 1$, and

$$|f(z)| \leq |f(z) - b_j| + |b_j| \leq |f(z) - b_j| + M,$$

then we may conclude that

$$\log |f(z)| \leq A(\varepsilon, \theta, \eta) r_n^{\frac{72\pi\eta}{\omega}+8\eta}. \tag{5.14}$$

According to the selection of η, we have

$$\frac{1}{1-3\eta} \left\{ \frac{72\pi\eta}{\omega} + 8\eta \right\} < \frac{1}{4}.$$

And hence
$$\log|f(z)| \leq \{r_n^{1-3\eta}\}^{1/4} \leq |z|^{1/4}.$$

Besides, there must exist a curve L_n linking L_{jn} and $L_{j'n}$ on $\overline{\Omega}(\theta_k + 3\varepsilon, \theta_{k+1} - 3\varepsilon; r_n^{1-3\eta}, 2r_n)$. Hence, when $z \in L_n$ in particular,

$$\log|f(z)| \leq |z|^{\frac{1}{4}}. \tag{5.15}$$

On the other hand, according to Theorem 4.3, we conclude that there exists a continuous curve L tending to ∞ between L_j and $L_{j'}$, such that

$$\lim_{\substack{|z| \to +\infty \\ z \in L}} \frac{\log\log|f(z)|}{\log|z|} \geq \frac{1}{2}.$$

Since L must intersect some point z_n on L_n, when n is sufficiently large, we have, on the one hand, $\log|f(z_n)| \geq |z_n|^{1/3}$. On the other hand, according to formula (5.15), we have $\log|f(z_n)| \leq |z_n|^{1/4}$. Hence, we derive a contradiction.

2) L_{jn} intersects within $\overline{\Omega}(\theta_k + 3\varepsilon, \theta_{k+1} - 3\varepsilon; r_n^{1-3\eta}, 2r_n)$. Meanwhile, $L_{j'n}$ intersects at $\overline{\Omega}(\theta_{k+1} + 3\varepsilon, \theta_{k+2} - 3\varepsilon; r_n^{1-3\eta}, 2r_n)$. Similar to the discussion of case (1), we can conclude particularly that when

$$z \in \Delta(\theta_{k+1} - 3\varepsilon; r_n^{1-3\eta}, \tfrac{1}{2}r_n), (^2)$$

$$\log|f(z)| \leq \{r_n^{1-3\eta}\}^{1/4} = M_{n1},$$

and when $z \in \Delta(\theta_{k+1} + 3\varepsilon; r_n^{1-3\eta}, \tfrac{1}{2}r_n)$,

$$\log|f(z)| \leq \{r_n^{1-3\eta}\}^{1/4} = M_{n2}.$$

Furthermore, applying Lemma 5.3, we conclude that when

$$z \in \overline{\Omega}(\theta_{k+1} - 3\varepsilon, \theta_{k+1} + 3\varepsilon; r_n^{1-2\eta}, r_n^{1-\eta}),$$
$$\log|f(z)| \leq 2\{r_n^{1-3\eta}\}^{1/4} + 1.$$

Obviously, there exists a curve L_n' linking point $r_n^{1-\eta}e^{i(\theta_{k+1}-3\varepsilon)}$ and L_{jn} in $\overline{\Omega}(\theta_k + 3\varepsilon, \theta_{k+1} - 3\varepsilon; r_n^{1-3\eta}, 2r_n)$, and also a curve L_n'' linking point $r_n^{1-\eta}e^{i(\theta_{k+1}+3\varepsilon)}$ and $L_{j'n}$ in $\overline{\Omega}(\theta_{k+1} + 3\varepsilon, \theta_{k+2} - 3\varepsilon; r_n^{1-3\eta}, 2r_n)$. Let

$$L_n = L_n' \cup \Gamma(\theta_{k+1} - 3\varepsilon, \theta_{k+1} + 3\varepsilon, r_n^{1-\eta})\,(^3) \cup L_n''.$$

Then L_n is a curve linking L_{jn} and $L_{j'n}$. Moreover, when $z \in L_n$, provided that n is sufficiently large, then it follows that $\log|f(z)| \leq 3|z|^{1/4}$. In the

---

(^{2}) This notation means that straight line segment $\arg z = \theta_{k+1} - 3\varepsilon$, $r_n^{1-3\eta} < |z| < \tfrac{1}{2}r_n$.

(^{3}) This notation means that circular arc, $|z| = r_n^{1-\eta}$, $\theta_k - 3\varepsilon < \arg z < \theta_{k+1} + 3\varepsilon$.

following, we make a discussion similar to case 1), and a contradiction will be derived.

3) L_{jn} intersects at $\overline{\Omega}(\theta_k + 3\varepsilon, \theta_{k+1} - 3\varepsilon; r_n^{1-3\eta}, 2r_n)$. Meanwhile, $L_{j'n}$ lies entirely in $\overline{\Omega}(\theta_{k+1} - 3\varepsilon, \theta_{k+1} + 3\varepsilon; r_n^{1-3\eta}, 2r_n)$.

4) L_{jn} and $L_{j'n}$ lie entirely in $\overline{\Omega}(\theta_{k+1} - 3\varepsilon, \theta_{k+1} + 3\varepsilon; r_n^{1-3\eta}, 2r_n)$ simultaneously.

Under cases 3) and 4), if we notice that when n is sufficiently large, we have

$$|f(z) - b_j| < 1, \quad z \in L_{jn}, \quad |f(z) - b_{j'}| < 1, \quad z \in L_{j'n},$$

or

$$|f(z)| \leq M + 1, \quad z \in L_{jn} \cup L_{j'n}.$$

Then applying Lemma 5.3, similar to the discussion of cases 1) and 2), we may derive a contradiction.

(4) When n is sufficiently large, each deficient value a_i $(1 \leq i \leq p)$ corresponds to each set E_i^n, meas $E_i^n \geq K$. Moreover, according to the selection of ε, we conclude that there exists at least one set $E_i^n \cap [\theta_{k_i} + 3\varepsilon, \theta_{k_i+1} - 3\varepsilon]$ $(1 \leq k_i \leq q)$ among the q sets $E_i^n \cap [\theta_k + 3\varepsilon, \theta_{k+1} - 3\varepsilon]$ $(k = 1, 2, \ldots, q; \ \theta_{q+1} = \theta_1 + 2\pi)$, such that its measure is

$$\text{meas}\{E_i^n \cap [\theta_{k_i} + 3\varepsilon, \theta_{k_i+1} - 3\varepsilon]\} \geq \frac{K}{2q},$$

and, further, we can uniquely determine two adjacent Julia directions $\Delta(\theta_{k_i})$ and $\Delta(\theta_{k_i+1})$.[4] Hence, following this way, each deficient value a_i corresponds to two adjacent Julia directions $\Delta_a(\theta_{k_i}) = \Delta(\theta_{k_i})$ and $\Delta_a(\theta_{k_i+1}) = \Delta(\theta_{k_i+1})$, p deficient values correspond to pairs of Julia directions $\{\Delta_a(\theta_{k_i}), \Delta_a(\theta_{k_i+1}) \mid i = 1, 2, \ldots, p\}$. Now, we prove that any two Julia directions from the set $\{\Delta_a(\theta_{k_i}), \Delta_a(\theta_{k_i+1}) \mid i = 1, 2, \ldots, p\}$ do not coincide with each other.

In fact, otherwise, suppose that two Julia directions coincide with each other, we shall derive a contradiction. Hence, we need only to consider the following two typical cases:

1) meas$\{E_i^n \cap [\theta_{k_i} + 3\varepsilon, \theta_{k_i+1} - 3\varepsilon]\} \geq K/2q$ and

meas$\{E_i^n \cap [\theta_{k'_i} + 3\varepsilon, \theta_{k'_i+1} - 3\varepsilon]\} \geq K/2q$ $(k_i = k_{i'} = k, \ 1 \leq k \leq q)$.

We apply Lemma 5.2, where we let $\theta = \frac{1}{2}(\theta_{k+1} - \theta_k) - \varepsilon$, $R = t_n$, $R_1 = r_n^{1-3\eta}$, $R_2 = 2r_n$, $E_\alpha = \{t_n e^{i\varphi} \mid \varphi \in E_i^n \cap [\theta_{k_i} + 3\varepsilon, \theta_{k_i+1} - 3\varepsilon]\}$, $H = \frac{K}{2q} \geq \frac{\varepsilon}{2}$, $\alpha = a_i$, $N = \frac{\delta}{4}T(t_n, f)$ and $\nu = 0$, then when n is sufficiently large, for

_____

[4] When $p \geq 1$, q must be ≥ 2. This fact is a direct corollary of Theorem 6.2. Hence, there must exist two adjacent Julia directions.

point z in $\overline{\Omega}(\theta_k + 2\varepsilon, \ \theta_{k+1} - 2\varepsilon; \ r_n^{1-3\eta}, \ 2r_n)$,

$$|f(z) - a_i| \leq \exp\left\{A(\varepsilon, \theta)2^{12\pi/(\theta_{k+1}-\theta_k-2\varepsilon)}r_n^{36\pi\eta/(\theta_{k+1}-\theta_k-2\varepsilon)}\right.$$

$$\times [2^{2\eta}r_n^{6\eta^2}2^{\eta}r_n^{\eta}\log(2r_n^{3\eta}) + \log^+|a_i|]$$

$$- \frac{B(\varepsilon, \theta)}{C(\theta)\log(2r_n^{3\eta}) + D(\varepsilon, \theta)}r_n^{-12\pi\eta/(\theta_{k+1}-\theta_k-2\varepsilon)}$$

$$\left.\times 2^{-4\pi/(\theta_{k+1}-\theta_k-2\varepsilon)}\frac{\delta}{4}T(t_n f)\right\}$$

$$\leq \exp\{A(\varepsilon, \theta, \eta)r_n^{72\pi\eta/\omega+8\eta} - B(\varepsilon, \theta, \eta, \delta)r_n^{-24\pi\eta/\omega-\eta}T(t_n, f)\}.$$

Moreover, according to the selection of η,

$$\frac{72\pi\eta}{\omega} + 8\eta - \left\{-\frac{24\pi\eta}{\omega} - \eta + (\mu-\eta)\right\} < -\frac{2}{3} < 0,$$

$$\mu - \eta - \frac{24\pi\eta}{\omega} - 2\eta > \frac{\mu}{2}.$$

Hence, from formula (5.11), we conclude that

$$\log|f(z) - a_i| \leq -r_n^{\mu/2}. \tag{5.16}$$

On the other hand, there exists a point $z_n = t_n e^{i\varphi}$, $\varphi \in E_i^n \cap [\theta_{k_i'} + 3\varepsilon, \ \theta_{k_{i'}+1} - 3\varepsilon]$ on $\overline{\Omega}(\theta_k + 2\varepsilon, \ \theta_{k+1} - 2\varepsilon; \ r_n^{1-3\eta}, \ 2r_n)$ such that

$$\log|f(z_n) - a_{i'}| \leq -\frac{\delta}{4}T(t_n, f) \leq -\frac{\delta}{4}r_n^{(\mu-\eta)} \leq -r_n^{\mu/2}.$$

If we notice that $a_i \neq a_{i'}$, and

$$|a_i - a_{i'}| \leq |f(z_n) - a_i| + |f(z_n) - a_{i'}| \leq 2e^{-r_n\mu/2},$$

then when n is sufficiently large, we derive a contradiction.

2) $\text{meas}\{E_i^n \cap [\theta_{k_i} + 3\varepsilon, \ \theta_{k_i+1} - 3\varepsilon]\} \geq \frac{K}{2q}$, and

$$\text{meas}\{E_i^n \cap [\theta_{k_{i'}} + 3\varepsilon, \ \theta_{k_{i'}+1} - 3\varepsilon]\} \geq \frac{K}{2q} \quad (k_i + 1 = k_{i'} = k, \ 2 \leq k \leq q+1).$$

Similar to the discussion of case 1), we may obtain, in particular, when $z \in \Delta(\theta_k - 3\varepsilon, r_n^{1-3\eta}, \frac{1}{2}r_n)$,

$$\log|f(z) - a_i| \leq -r_n^{\mu/2}, \qquad \log|f(z)| \leq \log M + 1 = M_{n1},$$

and when $z \in \Delta(\theta_k + 3\varepsilon, r_n^{1-3\eta}, \frac{1}{2}r_n)$

$$\log|f(z) - a_{i'}| \leq -r_n^{\mu/2}, \qquad \log|f(z)| \leq \log M + 1 = M_{n2}.$$

Moreover, applying Lemma 5.3, we conclude that when

$$z \in \overline{\Omega}(\theta_k - 3\varepsilon, \theta_k + 3\varepsilon; r_n^{1-2\eta}, r_n^{1-\eta}),$$

$$\log|f(z)| \leq 2\log M + 3, \ |f(z)| \leq e^3 M^2 = N.$$

Finally, if we notice that $p < +\infty$, then the distances among the p deficient values a_i $(i = 1, 2, \ldots, p)$ have a positive lower bound $d' > 0$. Hence, applying the transformation $\zeta = z/r_n^{1-2\eta}$ and Theorem 3.2, we conclude that $a_i = a_{i'}$. However, according to the assumption, $a_i \neq a_{i'}$, and hence, we derive a contradiction.

(5) We prove the following lemma.

LEMMA 5.4. *Suppose there exists a set* $E_i \cap [\theta_k + 3\varepsilon, \theta_{k+1} - 3\varepsilon]$ *with*

$$\text{meas}\{E_i^n \cap [\theta_k + 3\varepsilon, \theta_{k+1} - 3\varepsilon]\} \geq \frac{K}{2q} \quad (1 \leq k \leq q, \ \theta_{q+1} = \theta_1 + 2\pi),$$

and a certain curve L_{jn} *intersects at* $\overline{\Omega}(\theta_k + 3\varepsilon, \theta_{k+1} - 3\varepsilon; r_n^{1-3\eta}, 2r_n)$ $(\theta_{q+2} = \theta_2 + 2\pi)$ *or at*

$$\overline{\Omega}(\theta_{k-1} + 3\varepsilon, \theta_{k+1} - 3\varepsilon; r_n^{1-3\eta}, 2r_n) \quad (\theta_0 = \theta_q - 2\pi).$$

Then when n *is sufficiently large,* a_i *must be equal to* b_j. *Meanwhile, there exists a continuous curve* L_n *linking* L_{jn} *and an arbitrary point* $z_n \in \{t_n e^{i\varphi} \mid \varphi \in E_i^n \cap [\theta_k + 3\varepsilon, \theta_{k+1} - 3\varepsilon]\}$, *such that when* $z \in L_n$,

$$|f(z) - a| \leq \varepsilon_n, \quad a = a_i = b_j, \ \varepsilon_n \to 0, \ n \to +\infty.$$

PROOF. We need only to consider the following few typical cases:
1) $\text{meas}\{E_i^n \cap [\theta_k + 3\varepsilon, \theta_{k+1} - 2\varepsilon]\} \geq \frac{K}{2q}$, and L_{jn} intersects at

$$\overline{\Omega}(\theta_k + 3\varepsilon, \theta_{k+1} - 3\varepsilon; r_n^{1-3\eta}, 2r_n).$$

First, according to the way we derive formula (5.16), we conclude that when n is sufficiently large, for the point z on

$$\overline{\Omega}(\theta_k + 2\varepsilon, \theta_{k+1} - 2\varepsilon; r_n^{1-3\eta}, 2r_n),$$

we have

$$\log|f(z) - a_i| \leq -r_n^{\mu/2}. \tag{5.17}$$

Next, we let

$$\varepsilon_{jn} = \max_{z \in L_{jn}} |f(z) - b_j|. \tag{5.18}$$

Then when $n \to +\infty$, $\varepsilon_{jn} \to 0$. We assume arbitrarily a point $z_0 \in L_{jn} \cap \overline{\Omega}(\theta_k + 3\varepsilon, \theta_{k+1} - 3\varepsilon; r_n^{1-3\eta}, 2r_n)$, then

$$|a_i - b_j| \leq |f(z_0) - a_i| + |f(z_0) - b_j| \leq e^{-r_n^{\mu/2}} + \varepsilon_{jn} \to 0 \quad (n \to +\infty).$$

Hence, we conclude that $a_i = b_j$. We assume arbitrarily a point z_n on the set $\{t_n e^{i\varphi} \mid \varphi \in E_i^n \cap [\theta_k + 3\varepsilon, \theta_{k+1} - 3\varepsilon]\}$. Then, we use a straight line L_n linking points z_n and z_0. It follows when $z \in L_n$,

$$|f(z) - a| \le e^{-r_n^{\mu/2}}, \qquad a = a_i = b_j.$$

2) $\text{meas}\{E_i^n \cap [\theta_k + 3\varepsilon, \theta_{k+1} - 3\varepsilon]\} \ge \frac{K}{2q}$, and also L_{jn} intersects at $\overline{\Omega}(\theta_{k+1}, 3\varepsilon, \theta_{k+2} - 3\varepsilon; r_n^{1-3\eta}, 2r_n)$.

First, following the way we derive formula (5.14), we may conclude that when n is sufficiently large, for the point z on

$$\overline{\Omega}(\theta_{k+1} + 2\varepsilon, \theta_{k+2} - 2\varepsilon; r_n^{1-3\eta}, 2r_n),$$

$$\log|f(z)| \le A(\varepsilon, \theta, \eta) r_n^{72\pi\eta/\omega + 8\eta}. \tag{5.19}$$

Hence, in particular, when $z \in \Delta(\theta_{k+1} + 3\varepsilon; r_n^{1-3\eta}, \frac{1}{2}r_n)$,

$$\log|f(z)| \le A(\varepsilon, \theta, \eta) r_n^{72\pi\eta/\omega + 8\eta} = M_{n1}.$$

Next, according to formula (5.17), when $z \in \overline{\Omega}(\theta_k + 2\varepsilon, \theta_{k+1} - 2\varepsilon; r_n^{1-3\eta}, 2r_n)$,

$$\log|f(z)| \le \log^+ |f(z) - a_i| + \log^+ |a_i| + \log 2 \\ \le \log M + \log 2 = \log 2M. \tag{5.20}$$

Hence, in particular, when $z \in \Delta(\theta_{k+1} - 3\varepsilon; r_n^{1-3\eta}, \frac{1}{2}r_n)$

$$\log|f(z)| \le \log 2M = M_{n2}.$$

Moreover, applying Lemma 5.3, when $z \in \overline{\Omega}(\theta_{k+1} - 3\varepsilon, \theta_{k+1} + 3\varepsilon, r_n^{1-2\eta}, r_n^{1-\eta})$, we conclude that

$$\log|f(z)| \le A(\varepsilon, \theta, \eta) r_n^{72\pi\eta/\omega + 8\eta} + \log 2M + 1 < r_n^{72\pi\eta/\omega + 9\eta}. \tag{5.21}$$

Hence, according to formulas (5.19)–(5.21), when point z is on disk $|z - \frac{1}{2}(r_n^{1-2\eta} + r_n^{1-\eta}) \cdot e^{i\theta_{k+1}}| < 4\varepsilon \cdot \frac{1}{2}(r_n^{1-2\eta} + r_n^{1-\eta})$, we have

$$\log|f(z)| \le r_n^{72\pi\eta/\omega + 9\eta}.$$

Besides, according to formula (5.17), when

$$z \in \Gamma(\theta_{k+1} - 3\varepsilon, \theta_{k+1} - 2\varepsilon, \frac{1}{2}(r_n^{1-2\eta} + r_n^{1-\eta})),$$

we have that $|f(z) - a_i| \le e^{-r_n^{\mu/2}}$. Consider the transformation

$$\zeta = \frac{z - (1/2)(r_n^{1-2\eta} + r_n^{1-\eta}) e^{i\theta_{k+1}}}{2\varepsilon(r_n^{1-\eta} + r_n^{1-2\eta})}.$$

Then disk $|z - \frac{1}{2}(r_n^{1-2\eta} + r_n^{1-\eta}) e^{i\theta_{k+1}}| \le 2\varepsilon(r_n^{1-2\eta} + r_n^{1-\eta})$ is transformed onto the unit disk $|\zeta| \le 1$ on the ζ-plane, and $\Gamma(\theta_{k+1} - 3\varepsilon, \theta_{k+1} - 2\varepsilon; \frac{1}{2}(r_n^{1-2\eta} + r_n^{1-\eta}))$

is transformed into a set E_a on the ζ-plane. Obviously, E_a lies in disk $|\zeta| \leq \frac{3}{4}$.

Moreover, if we assume $h = \frac{1}{16e}$, then E_a cannot be included in some circles, with the sum of their Euclidean radii not exceeding $2eh$. Let $g(\zeta) = f(2\varepsilon(r_n^{1-2\eta} + r_n^{1-\eta})\zeta + \frac{1}{2}(r_n^{1-2\eta} + r_n^{1-\eta})e^{i\theta_{k+1}})$. Then $g(\zeta)$ is regular on disk $|\zeta| \leq 1$, and has an upper bound $r_n^{72\pi\eta/\omega+9\eta}$. Meanwhile, when $z \in E_a$,

$$|g(\zeta) - a_i| \leq e^{-r_n^{\mu/2}}.$$

If we further apply Theorem 5.2, where we let $M = \exp\{r_n^{72\pi\eta/\omega+9\eta}\}$, $N = r_n^{\mu/2}$, $a = a_i$, $\tau = r = \frac{3}{4}$, then we may conclude that for point ζ on disk $|\zeta| \leq \frac{3}{4}$,

$$|g(\zeta) - a_i| \leq \exp\left\{ (r_n^{72\pi\eta/\omega+9\eta} + \log^+ |a| + \log 2) \right.$$
$$\left. - A\left(\log \frac{1}{(1/4)^2(1/16e)} \right)^{-1} \left(\frac{1}{4}\right)^2 r_n^{\mu/2} \right\}$$
$$\leq \exp\{(r_n^{72\pi\eta/\omega+9\eta} + \log M + \log 2 - Ar_n^{\mu/2}\}.$$

Then, according to the selection of η, we have

$$\frac{72\pi\eta}{\omega} + 9\eta - \frac{\mu}{2} < -\frac{\mu}{4}.$$

Hence, when n is sufficiently large, for point z on disk

$$|z - \frac{1}{2}(r_n^{1-2\eta} + r_n^{1-\eta})e^{i\theta_{k+1}}| \leq 2\varepsilon(r_n^{1-2\eta} + r_n^{1-\eta}),$$

$$|f(z) - a_i| \leq \exp\{-Ar_n^{\mu/2}\},$$

where $A > 0$ is a constant. Particularly, when

$$z \in \Gamma(\theta_{k+1} + 2\varepsilon, \theta_{k+1} + 3\varepsilon; \frac{1}{2}(r_n^{1-2\eta} + r_n^{1-\eta})),$$

$$|f(z) - a_i| \leq \exp\{-Ar_n^{\mu/2}\}.$$

Applying Lemma 5.2, where we let $\theta = \frac{1}{2}(\theta_{k+2} - \theta_{k+1}) - \varepsilon$, $R_1 = r_n^{1-3\eta}$, $R_2 = 2r_n$, $R = \frac{1}{2}(r_n^{1-2\eta} + r_n^{1-\eta})$, $E_a = \Gamma(\theta_{k+1} + 2\varepsilon, \theta_{k+1} + 3\varepsilon; \frac{1}{2}(r_n^{1-2\eta} + r_n^{1-\eta}))$, $H = \varepsilon$, $\alpha = a_i$, $N = Ar_n^{\mu/2}$ and $\nu = 0$, then when

$$z \in \overline{\Omega}(\theta_{k+1} + 2\varepsilon, \theta_{k+2} - 2\varepsilon; r_n^{1-3\eta}, 2r_n),$$

$$|f(z) - a_i| \le \exp\left\{ A(\varepsilon, \theta) 2^{12\pi/(\theta_{k+2} - \theta_{k+1} - 2\varepsilon)} r_n^{18\pi\eta/(\theta_{k+2} - \theta_{k+1} - 2\varepsilon)} \right.$$

$$\times [2^{2\eta} r_n^{6\eta^2} 2^\eta r_n^\eta \log(2r_n^{3\eta}) + \log^+ |a_i|]$$

$$- \frac{B(\varepsilon, \theta)}{C(\theta) \log(2r_n^{3\eta}) + D(\varepsilon, \theta)} (2^{4\pi/(\theta_{k+2} - \theta_{k+1} - 2\varepsilon)})^{-1}$$

$$\left. \times r_n^{-12\pi\eta/(\theta_{k+2} - \theta_{k+1} - 2\varepsilon)} A r_n^{\mu/2} \right\}$$

$$\le \exp\{ A(\varepsilon, \theta, \eta) r_n^{72\pi\eta/\omega + 8\eta} - B(\theta, \varepsilon) r_n^{-(24\pi\eta/\omega) - \eta + (\mu/2)} \}.$$

Moreover, according to the selection of η, we have

$$\frac{72\pi\eta}{\omega} + 8\eta - \left(\frac{\mu}{2} - \frac{24\pi\eta}{\omega} - \eta \right) < -\frac{\mu}{6},$$

$$\frac{\mu}{2} - \frac{24\pi\eta}{\omega} - \eta > \frac{\mu}{3}.$$

Hence, $|f(z) - a_i| \le e^{-r_n^{\mu/3}}$. Now, we assume point

$$z_0 \in L_{jn} \cap \overline{\Omega}(\theta_{k+1} + 3\varepsilon, \theta_{k+2} - 3\varepsilon, r_n^{1-3\eta}, 2r_n);$$

then according to formula (5.18), $|f(z_0) - b_j| \le \varepsilon_{jn}$. Since $p < +\infty$ and $l < +\infty$, there exist only finite distinct values among these p deficient values a_i $(i = 1, 2, \ldots, p)$ and l asymptotic values b_j $(j = 1, 2, \ldots, l)$. Hence, the distances among these finite distinct values must have a positive lower bound $d' > 0$. On the other hand, we have

$$|a_i - b_j| \le |f(z_0) - a_i| + |f(z_0) - b_j| \le e^{-r_n^{\mu/3}} + \varepsilon_{jn}.$$

Hence, when n is sufficiently large, $|a_i - b_j| < d'$. Therefore, we conclude that $a_i = b_j$. We select arbitrarily a point $z_n = t_n e^{i\varphi}$, $\varphi \in E_i^n \cap [\theta_k + 3\varepsilon, \theta_{k+1} - 3\varepsilon]$, and then use a line L_n' to connect points z_n and $\frac{1}{2}(r_n^{1-2\eta} + r_n^{1-\eta}) e^{i(\theta_{k+1} - 3\varepsilon)}$. Meanwhile, we use another line L_n'' to connect points z_0 and $\frac{1}{2}(r_n^{1-2\eta} + r_n^{1-\eta}) \cdot e^{i(\theta_{k+1} + 3\varepsilon)}$. Let

$$L_n = L_n' \cup \Gamma(\theta_{k+1} - 3\varepsilon, \theta_{k+1} + 3\varepsilon; \tfrac{1}{2}(r_n^{1-2\eta} + r_n^{1-\eta})) \cup L_n''.$$

Then when $z \in L_n$,

$$|f(z) - a| \le e^{-r_n^{\mu/3}}, \qquad a = a_i = b_j.$$

3) $\mathrm{meas}\{E_i^n \cap [\theta_k + 3\varepsilon, \theta_{k+1} - 3\varepsilon]\} \ge \frac{K}{2q}$, and L_{jn} lies entirely in

$$\overline{\Omega}(\theta_{k+1} - 3\varepsilon, \theta_{k+1} + 3\varepsilon; r_n^{1-3\eta}, 2r_n).$$

First, when n is sufficiently large, for point z on

$$\overline{\Omega}(\theta_k + 2\varepsilon, \theta_{k+1} - 2\varepsilon, r_n^{1-3\eta}, 2r_n),$$

we have $\log|f(z) - a_i| \le -r_n^{\mu/2}$, and hence, $\log|f(z)| \le \log M + \log 2 = M_{n1}$.
Next, suppose that A_{n2} is the intersection point of L_{jn} and circumference
$|z| = r_n^{1-3\eta}$. Starting from A_{n2} and tending to ∞ along L_j, we let B_{n2} be
the first intersection point between L_{jn} and circumference $|z| = \frac{1}{2}r_n$, and
L'_{jn} be the part of L_{jn} between A_{n2} and B_{n2}. Then according to formula
(5.18), when $z \in L'_{jn}$,

$$\log|f(z) - b_j| \le \log \varepsilon_{jn},$$

and hence,

$$\begin{aligned}\log|f(z)| &\le \log^+|f(z) - b_j| + \log^+|b_j| + \log 2\\ &\le \log M + \log 2 = M_{n2}.\end{aligned}$$

A simply connected domain

$$\Omega'_n \subset \Omega(\theta_{k+1} - 3\varepsilon;\, \theta_{k+1} + 3\varepsilon;\, r_n^{1-3\eta},\, \tfrac{1}{2}r_n)$$

is formed, which is bounded by parts of the arc on L'_{jn} and

$$\Delta(\theta_{k+1} - 3\varepsilon;\, r_n^{1-3\eta},\, \tfrac{1}{2}r_n)$$

as well as $\Gamma(\theta_{k+1} - 3\varepsilon,\, \theta_{k+1} + 3\varepsilon;\, \tfrac{1}{2}r_n)$ and $\Gamma(\theta_{k+1} - 3\varepsilon,\, \theta_{k+1} + 3\varepsilon;\, \tfrac{1}{2}r_n)$.
Suppose that the conformal mapping $\zeta = \varphi_n(z)$ transforms Ω'_n onto

$$\Omega_\zeta(\theta_{k+1} - 3\varepsilon,\, \theta_{k+1} + 3\varepsilon;\, r_n^{1-3\eta},\, R_n)$$

on the ζ-plane; point A_{n2} is transformed into point $r_n^{1-3\eta}e^{i(\theta_{k+1}+3\varepsilon)}$, point
$r_n^{1-3\eta}e^{i(\theta_{k+1}-3\varepsilon)} = A_{n1}$ into $r_n^{1-3\eta}e^{i(\theta_{k+1}-3\varepsilon)}$; point B_{n2} into point $R_n e^{i(\theta_{k+1}+3\varepsilon)}$
and $\frac{1}{2}r_n e^{i(\theta_{k+1}-3\varepsilon)} = B_{n1}$ into point $R_n e^{i(\theta_{k+1}-3\varepsilon)}$. Then according to Lemma
3.4, we conclude that

$$R_n \ge \tfrac{1}{2}r_n. \tag{5.22}$$

Let $z = \varphi_n^{-1}(\zeta)$ be the inverse transformation of $\zeta = \varphi_n(z)$ and $F_n(\zeta) = f(\varphi_n^{-1}(\zeta))$. Then when $\zeta \in \Delta_\zeta(\theta_{k+1} - 3\varepsilon;\, r_n^{1-3\eta},\, R_n)$,

$$\log|F_n(\zeta) - a_i| \le -r_n^{\mu/2},$$
$$\log|F_n(\zeta)| \le \log 2M = M_{n1}.$$

And when $z \in \Delta_\zeta(\theta_{k+1} + 3\varepsilon;\, r_n^{1-3\eta},\, R_n)$,

$$\log|F_n(\zeta) - b_j| \le \log \varepsilon_{jn},$$
$$\log|F_n(\zeta)| \le \log 2M = M_{n2}.$$

Moreover, when $\zeta \in \Gamma_\zeta(\theta_{k+1} - 3\varepsilon,\, \theta_{k+1} + 3\varepsilon;\, r_n^{1-3\eta})$, according to formula
(5.13) we have $\log|F_n(\zeta)| \le 3r_n^{\mu+\eta}$, and when

$$\zeta \in \Gamma_\zeta(\theta_{k+1} - 3\varepsilon,\, \theta_{k+1} + 3\varepsilon;\, R_n),$$

$\log|F_n(\zeta)| \leq 3r_n^{\mu+\eta}$. Hence, applying Lemma 2.10, when $\zeta \in \Omega_\zeta(\theta_{k+1} - 3\varepsilon, \theta_{k+1} + 3\varepsilon; r_n^{1-2\eta}, r_n^{1-\eta})$, we obtain

$$\log|F_n(\zeta)| \leq M_{n1} + M_{n2} + \left\{ \frac{4(r_n^{1-3\eta}/|\zeta|)^{\pi/6\varepsilon}}{\pi[1 - (r_n^{1-3\eta}/|\zeta|)^{\pi/3\varepsilon}]} \right.$$
$$\left. + \frac{4(|\zeta|/R_n)^{\pi/6\varepsilon}}{\pi[1 - (|\zeta|/R_n)^{\pi/3\varepsilon}]} \right\} 3r_n^{\mu+\eta}.$$

Moreover, from formula (5.22) we have

$$\log|F_n(\zeta)| \leq M_{n1} + M_{n2} + \left\{ \frac{4r_n^{-\pi\eta/6\varepsilon}}{\pi[1 - r_n^{-\pi\eta/3\varepsilon}]} + \frac{4 \cdot 2^{\pi/6\varepsilon} r_n^{-\pi\eta/6\varepsilon}}{\pi[1 - 2^{\pi/3\varepsilon} r_n^{-\pi\eta/3\varepsilon}]} \right\} 3r_n^{\mu+\eta}.$$

Furthermore, based on the selection of ε, we get

$$-\frac{\pi\eta}{6\varepsilon} + \mu + \eta < -1 + \eta < 0.$$

And hence, when n is sufficiently large,

$$\log|F_n(\zeta)| \leq 2\log 2M + 1 = \log(4eM^2) = \log N.$$

Now, applying the transformation $\xi = \zeta/r_n^{1-2\eta}$ and Theorem 3.2, we conclude that $a_i = b_j$. Meanwhile, there exists a continuous curve l_n' linking $\Delta_\zeta(\theta_{k+1} - 3\varepsilon, r_n^{1-2\eta}, r_n^{1-\eta})$ and $\Delta_\zeta(\theta_{k+1} + 3\varepsilon; r_n^{1-2\eta}, r_n^{1-\eta})$ on $\overline{\Omega}_\zeta(\theta_{k+1} - 3\varepsilon, \theta_{k+1} + 3\varepsilon; r_n^{1-2\eta}, r_n^{1-\eta})$, such that when $\zeta \in l_n'$,

$$|F_n(\zeta) - a| \leq \varepsilon_{jn3}, \qquad a = a_i = b_j,$$

$$\varepsilon_{n3} = (M + N)\max\{e^{-(1/3)r_n^{\mu/2}}, \varepsilon_{jn}^{1/3}\} \to 0, \qquad n \to +\infty.$$

Let L_n' be the image of l_n' on the z-plane. Then L_n' is a continuous curve linking $\Delta(\theta_{k+1} - 3\varepsilon, r_n^{1-3\eta}, 2r_n)$ and L_{jn}. Moreover, when $z \in L_n'$, $|f(z) - a| \leq \varepsilon_{n3}$. Let z_0 be an intersection point of L_n' and $\Delta(\theta_{k+1} - 3\varepsilon, r_n^{1-3\eta}, 2r_n)$. Then, we use a line L_n'' to connect point z_0 and an arbitrary point $z_n = t_n e^{i\varphi}$, $\varphi \in E_\eta^\eta \cap [\theta_k + 3\varepsilon, \theta_{k+1} - 3\varepsilon]$. Let $L_n = L_n' \cup L_n''$. Then when $z \in L_n$, we conclude that $|f(z) - a| \leq \varepsilon_{n3}$. According to the above discussion, if we assume $\varepsilon_n = \varepsilon_{n3}$, then Lemma 5.4 follows.

(6) Now, we prove that there are at most l' Julia directions that coincide among one another between the two groups of Julia direction sets $\{\Delta_a(\theta_{k_i}), \Delta_a(\theta_{k_i+1}) \mid i = 1, 2, \ldots, p\}$ and $\{\Delta_b(\theta_{k_j+1}) \mid j = 1, 2, \ldots, l\}$. In fact, if $a_i \neq b_j$, then according to Lemma 5.4, we conclude that $\Delta_b(\theta_{k_j+1})$ can neither coincide with $\Delta_a(\theta_{k_i})$ nor $\Delta_a(\theta_{k_i+1})$. Hence, between the above-mentioned two groups of Julia direction sets, if the number of coincided Julia directions is $> l'$, then there must exist two Julia directions $\Delta_b(\theta_{k_j+1})$ and $\Delta_b(\theta_{k_{j'}+1})$ $(j \neq j')$ coinciding with $\Delta_a(\theta_{k_i})$ and $\Delta_a(\theta_{k_i+1})$ $(a_i = b_j = b_{j'})$,

respectively. Applying Lemma 5.4, we may find a continuous curve L_n linking L_{jn} and $L_{j'n}$, such that when $z \in L_n$,

$$|f(z) - b| \leq \varepsilon_n, \qquad b = b_j = b_{j'}, \quad \varepsilon_n \to 0, \quad n \to +\infty.$$

However, this contradicts the assumption that b_j and $b_{j'}$ are distinct asymptotic values. Hence, there are at most l' Julia directions that coincide among one another between these two groups of sets $\{\Delta_a(\theta_{k_i}), \Delta_a(\theta_{k_i+1}) \mid i = 1, 2, \ldots, p\}$ and $\{\Delta_b(\theta_{k_i+1}) \mid i = 1, 2, \ldots, l\}$. Therefore, we conclude that $2p - l' + l \leq q$.

(7) Finally, we consider other cases:

When $q = +\infty$ or p and l are simultaneously equal to zero, it is obvious that the theorem holds.

When $q < +\infty$, then it is impossible that $p = +\infty$ or $l = +\infty$. In fact, when $q < +\infty$, $p = +\infty$ and $l < +\infty$, we assume $p' < +\infty$, such that $2p' - l' + l > q$. On the other hand, according to our initial consideration of the proof of case (1), it must be that $2p' - l' + l \leq q$. Hence, we obtain a contradiction. When $p < +\infty$, $l = +\infty$ or $p = +\infty$, $l = +\infty$ we obtain analogously a contradiction. Hence, Theorem 5.4 is proved completely.

Theorem 5.4 has the following corollaries:

COROLLARY 1. $p + l \leq q$ and $l \leq q$.

COROLLARY 2. $2p \leq q$.

COROLLARY 3. If $2p = q$, then $l = l'$.

COROLLARY 4. If $l = q$, then $p = 0$.

Can we construct an entire function $f(z)$ whose lower order μ is finite, such that $f(z)$ has q Julia directions, l distinct finite asymptotic values, p finite deficient values, where l' deficient values are also asymptotic values, and satisfy the equality $2p - l' + l = q$? Under certain special cases, such an example exists. For instance, function $\int_0^z e^{-z^q} dz$ has q finite deficient values, $2q$ Julia directions. Moreover, $l = l'$. Hence, the equality $2p - l' + l = q$ holds. Besides, function

$$\int_0^z \frac{\sin z^q}{z^q} dz$$

has $2q$ distinct finite asymptotic values, $2q$ Julia directions. Moreover, $p = 0$. Hence, the equality $2p - l' + l = 2q$ holds.

THEOREM 5.5. *Suppose that $f(z)$ is an entire function of lower order $\mu < +\infty$. Let q be the number of Julia directions, l be the number of distinct finite asymptotic values, and p be the number of finite deficient values, where l' deficient values are simultaneously asymptotic values. We also assume that $q < +\infty$ and $2p - l' + l = q$. Then $p = l'$ and $\lambda = \mu$, where λ is the order of $f(z)$.*

PROOF. (1) First, according to the condition that $2p - l' + l = q < +\infty$, we conclude that $p < +\infty$ and $l < +\infty$. Next, since $f(z)$ is an entire function, we have $q \geq 1$. Hence, this equality $2p - l' + l = q \geq 1$ illustrates that p and l cannot be zero at the same time and, in turn, according to the corollary of Theorem 4.4, we conclude that $\mu \geq \frac{1}{2}$. On the other hand, since $q < +\infty$ and $\mu < +\infty$, according to the Corollary 1 of Theorem 2.17, we conclude the order $\lambda < +\infty$.

In the following, we first assume that $p \geq 1$ and let a_i $(i = 1, 2, \ldots, p)$ be p finite deficient values of $f(z)$, with $\delta(a_i, f) = \delta_i > 0$ being their corresponding deficiencies, b_j $(j = 1, 2, \ldots, l)$ be l distinct finite asymptotic values of $f(z)$, with L_j being their corresponding paths of fixed values, and $\Delta(\theta_k)$ $(k = 1, 2, \ldots, q; \ 0 \leq \theta_1 < \theta_2 < \cdots \theta_q < 2\pi)$ be q Julia directions of $f(z)$. Let

$$\omega = \min_{1 \leq k \leq q} \{\theta_{k+1} - \theta_k\}, \quad \theta_{q+1} = \theta_1 + 2\pi, \quad \delta = \min_{1 \leq i \leq p} \{\delta_i\},$$

$$M = \max\{1, |a_1|, \ldots, |a_p|, |b_1|, \ldots, |b_l|\}$$

and assume arbitrarily a number η,

$$0 < \eta < \frac{\min(\omega, \mu\omega)}{360(\lambda + 1)\pi}.$$

Then, we construct a sequence $r_n = 2^{(1+\eta)^n}$ $(n = 1, 2, \ldots)$. According to Lemma 2.5 and Lemma 3.9 (where we let $K = 2h(1 + \lambda)(1 + \eta)\eta^{-1}$, $h = 3$, $\sigma = \log 2$), provided that n is sufficiently large, then when $n \geq n_0$, there must exist a value t_n in the interval $[r_{n-1}, 2r_n]$ such that

$$\text{meas } E_i^n \geq K, \quad K = K(\delta, \lambda, p, \eta) > 0, \quad i = 1, 2, \ldots, p,$$

where

$$E_i^n = \left\{\theta \left| \log \frac{1}{|f(t_n e^{i\theta}) - a_i|} \geq \frac{\delta}{4} T(t_n, f), \ 0 \leq \theta < 2\pi \right.\right\},$$

$$i = 1, 2, \ldots, p.$$

Besides, according to the definition of order λ and lower order μ, provided that n_0 is sufficiently large, when $r \geq r_{n_0}$, $r^{\mu-\eta} \leq T(r, f) \leq r^{\lambda+\eta}$. Finally, we assume a number ε,

$$0 < \varepsilon < \min\left\{\frac{\pi\eta}{6(\lambda + 1)}, \frac{K}{12q}, \frac{\omega}{8}\right\}.$$

Let

$$\overline{\Omega} = \bigcup_{k=1}^{q} \overline{\Omega}(\theta_k + \varepsilon, \theta_{k+1} - \varepsilon).$$

Then according to Theorem 2.13 and the finite covering theorem, we conclude that there exist two distinct finite values α and β such that

$$\varlimsup_{r \to +\infty} \frac{\log^+ n\{r, \overline{\Omega}, f = X\}}{\log r} = 0, \quad X = \alpha, \beta.$$

Since α and β are finite values, there exists a number d, $0 < d < \frac{1}{2}$, such that the spherical distances among α, β and ∞ are greater than d.

(2) We consider the sequence of annuli: $\Gamma_n: r_n^{1-3\eta} \leq |z| \leq 2r_n$ ($n = 1, 2, \ldots$). Starting from the origin and tending to ∞ along L_j ($j = 1, 2, \ldots, l$), we assume the last intersection point of L_j and circumference $|z| = r_n^{1-3\eta}$, and also the first intersection point of L_j and circumference $|z| = 2r_n$. Then, we let L_{jn} be the part of L_j between these two intersection points. Analogous to the proof of Theorem 5.4, provided that n_0 is sufficiently large, for each value n ($n = n_0, n_0 + 1, \ldots$), we may let each curve L_{jn} ($j = 1, 2, \ldots, l$) correspond to one Julia direction $\Delta(\theta_{k_{jn}+1}) = \Delta_b(\theta_{k_{jn}+1})$ ($1 \leq k_{jn} \leq q$; $j = 1, 2, \ldots l$), L_{jn} corresponds to $\Delta_b(\theta_{k_{jn}+1})$ means that L_{jn} intersects $\Omega(\theta_{k_{jn}} + 3\varepsilon, \theta_{k_{jn}+2} - 3\varepsilon; r_n^{1-3\eta}, 2r_n)$ ($\theta_{q+1} = \theta_1 + 2\pi$). Simultaneously, we may prove that any two Julia directions in the set $\{\Delta_b(\theta_{k_{jn}+1}) | j = 1, 2, \ldots, l\}$ do not coincide with each other.

Analogous to the proof of Theorem 5.4, provided that n_0 is sufficiently large, for each value n ($n = n_0, n_0 + 1, \ldots$), we may let each deficient value a_i ($i = 1, 2, \ldots, p$) correspond to one pair of Julia directions $\{\Delta(\theta_{k_{in}}) = \Delta_a(\theta_{k_{in}}), \Delta(\theta_{k_{in}+1}) = \Delta_a(\theta_{k_{in}+1})\}$ ($i = 1, 2, \ldots, p$). a_i corresponds to $\{\Delta_a(\theta_{k_{in}}), \Delta_a(\theta_{k_{in}+1})\}$ means that

$$\text{meas}\{E_i^n \cap [\theta_{k_{in}} + 3\varepsilon, \theta_{k_{in}+1} - 3\varepsilon]\} \geq \frac{K}{2q}.$$

Similarly, we may prove that any two Julia directions in the set $\{\Delta_a(\theta_{k_{in}}), \Delta_a(\theta_{k_{in}+1}) | i = 1, 2, \ldots, p\}$ do not coincide with each other.

Finally, analogous to the proof of Theorem 5.4, we may conclude that at most l' Julia directions coincide with each other between the two sets $\{\Delta_a(\theta_{k_{in}}), \Delta_a(\theta_{k_{in}+1}) | i = 1, 2, \ldots, p\}$ and $\{\Delta_b(\theta_{k_{jn}+1}) | j = 1, 2, \ldots, l\}$. Hence,

$$2p - l' + l \leq q.$$

On the other hand, according to the assumption that $2p - l' + l = q$, we conclude that each Julia direction $\Delta(\theta_k)$ ($k = 1, 2, \ldots, q$) must belong to $\{\Delta_b(\theta_{k_{in}+1}) | j = 1, 2, \ldots, l\}$ or $\{\Delta_a(\theta_{k_{in}}), \Delta_a(\theta_{k_{in}+1}) | i = 1, 2, \ldots, p\}$. Otherwise, we would have $2p - l' + l \leq q - 1$, and hence, a contradiction is derived.

(3) We prove that $p = l'$. When $p = 0$, it must be that $p = l' = 0$. Hence, we need only to consider the case when $p \geq 1$. In fact, if $p \neq l'$, then $p > l'$. We may assume that deficient value a_1 is not an asymptotic value of $f(z)$. For each value n ($n = n_0, n_0 + 1, \ldots$), a_1 corresponds to one pair of Julia directions $\{\Delta_a(\theta_{k_{1n}}), \Delta_a(\theta_{k_{1n}+1})\}$ and, in turn, we have

$$\text{meas}\{E_1^n \cap [\theta_{k_{1n}} + 3\varepsilon, \theta_{k_{1n}+1} - 3\varepsilon]\} \geq \frac{K}{2q}.$$

In the following, we prove that $k_{1n} = k_{1n+1}$ $(n = n_0, n_0 + 1, \dots)$. In fact, otherwise, since each Julia direction $\Delta(\theta_k)$ $(k = 1, 2, \dots, q)$ must belong to $\{\Delta_a(\theta_{k_{in}}), \Delta_a(\theta_{k_{in}+1}) \mid i = 1, 2, \dots, p\}$ or $\{\Delta_b(\theta_{k_j+1}) \mid j = 1, 2, \dots, l\}$ $(n = n_0, n_0 + 1, \dots)$, we need only to consider the following few typical cases.

1) $q > 3$ and $k_{1n} = k_{in+1}$, $i \neq 1$.

Applying Lemma 5.2, where we let $\theta = \frac{1}{2}(\theta_{k_{1n}+1} - \theta_{k_{1n}}) - \varepsilon$, $R = t_n$, $R_1 = r_{n-1}$, $R_2 = 2r_{n+1}$, $E_\alpha = \{t_n e^{i\varphi} \mid \varphi \in E_1^n \cap [\theta_{k_{1n}} + 3\varepsilon, \theta_{k_{1n}+1} - 3\varepsilon]\}$, $H = \frac{K}{2q}$, $\alpha = a_1$, $N = \frac{\delta}{4} T(t_n, f)$ and $\nu = 0$, then provided that n_0 is sufficiently large, when $n \geq n_0$, for the point z on $\overline{\Omega}(\theta_{k_{1n}} + 2\varepsilon, \theta_{k_{1n}+1} - 2\varepsilon; r_{n-1}, 2r_{n+1})$,

$$|f(z) - a_1| \leq \exp\left\{ A(\varepsilon, \theta) \left(\frac{2r_{n+1}}{r_{n-1}}\right)^{6\pi/\theta} \right.$$
$$\times \left[\left(\frac{2r_{n+1}}{r_{n-1}}\right)^{2\eta} (2r_{n+1})^\eta \log \frac{2r_{n+1}}{r_{n-1}} + \log^+ |a_1| \right]$$
$$- \frac{B(\varepsilon, \theta)}{C(\theta) \log(2r_{n+1}/r_{n-1}) + D(\varepsilon, \theta)}$$
$$\left. \times \left(\frac{r_{n-1}}{2r_{n+1}}\right)^{\pi/\theta} \frac{\delta}{4} T(t_n, f) \right\}.$$

Moreover, notice that

$$\theta = \frac{1}{2}(\theta_{k_{1n}+1} - \theta_{k_{1n}}) - \varepsilon \geq \frac{\omega}{2} - \varepsilon \geq \frac{\omega}{4},$$
$$\frac{r_{n+1}}{r_{n-1}} = r_{n+1}^{1-(1+\eta)^{-2}} = r_{n+1}^{(2+\eta)(1+\eta)^{-2} \cdot \eta} \leq r_{n+1}^{3\eta},$$
$$T(t_n, f) \geq r_n^{\mu-\eta} \geq r_{n-1}^{\mu-\eta} \geq r_{n+1}^{(1-2\eta)(\mu-\eta)}.$$

Then

$$|f(z) - a_1| \leq \exp\left\{ A(\varepsilon, \theta, \eta) r_{n+1}^{72\pi\eta/\omega + 8\eta} \right.$$
$$\left. - B(\varepsilon, \theta, \omega, \eta, \delta) r_{n+1}^{(1-2\eta)(\mu-\eta)-(24\pi\eta/\omega)-\eta} \right\}.$$

Furthermore, according to the selection of η, we have

$$\frac{72\pi\eta}{\omega} + 8\eta - \left\{ (1 - 2\eta)(\mu - \eta) - \frac{24\pi\eta}{\omega} - \eta \right\} < -\frac{2}{3}\mu < 0,$$
$$(1 - 2\eta)(\mu - \eta) - \frac{24\pi\eta}{\omega} - \eta > \frac{\mu}{2}.$$

And hence,

$$|f(z) - a_1| \leq e^{-r_{n+1}^{\mu/2}}. \tag{5.23}$$

On the other hand, since

$$\text{meas}\{E_i^{n+1} \cap [\theta_{k_{in+1}} + 3\varepsilon, \theta_{k_{in+1}+1} - 3\varepsilon]\} \geq \frac{K}{2q},$$

there exists a point $z_0 = t_{n+1}e^{i\varphi}$, $\varphi \in E_i^{n+1} \cap [\theta_{k_{in+1}} + 3\varepsilon, \theta_{k_{in+1}+1} - 3\varepsilon]$, such that

$$\log \frac{1}{|f(z_0) - a_i|} \geq \frac{\delta}{4} T(t_{n+1}, f)$$

or

$$|f(z_0) - a_i| < e^{-\delta/4 T(t_{n+1}, f)}.$$

Hence, combining this with formula (5.23), we conclude that

$$|a_i - a_1| \leq |f(z_0) - a_1| + |f(z_0) - a_i|$$
$$\leq e^{-r_{n+1}^{\mu/2}} + e^{-(\delta/4)T(t_{n+1}, f)}.$$

Accordingly, provided that n_0 is sufficiently large, when $n \geq n_0$, it must be that $a_1 = a_i$, and hence, a contradiction is derived.

2) $q > 3$ and $k_{1n} = k_{in+1} + 1$, $i \neq 1$.

Analogous to the discussion of the proof of case 2) of Theorem 5.4 at paragraph (4), provided that n_0 is sufficiently large, when $n \geq n_0$, it must be that $a_1 = a_i$, and hence, we derive a contradiction

3) $q \geq 2$ and L_{jn+1} intersects $\overline{\Omega}(\theta_{k_{1n}-1} + 3\varepsilon, \theta_{k_{1n}+1} - 3\varepsilon; r_{n+1}^{1-3\eta}, 2r_{n+1})$.

Analogous to the proof of Lemma 5.4, provided that n_0 is sufficiently large, when $n \geq n_0$, it must be that $a_1 = b_j$, and hence, a contradiction is derived.

4) $q = 2$ and $k_{1n} = k_{1n+1} + 1$.

First, analogous to the discussion of case 1), for the point z on $\overline{\Omega}(\theta_{k_{1n}} + 2\varepsilon, \theta_{k_{1n}+1} - 2\varepsilon, r_{n-1}, 2r_{n+1})$, we have

$$|f(z) - a_1| < e^{-r_{n+1}^{\mu/2}}. \tag{5.24}$$

Particularly, when $z \in \Delta(\theta_{k_{1n}} + 2\varepsilon, r_{n-1}, r_{n+1})$,

$$|f(z) - a_1| \leq e^{-r_{n+1}^{\mu/2}},$$

and when $z \in \Delta(\theta_{k_{1n}+1} - 2\varepsilon, r_{n-1}, r_{n+1})$.

$$|f(z) - a_1| \leq e^{-r_{n+1}^{\mu/2}}.$$

Next, we have

$$\text{meas}\{E_i^{n+1} \cap [\theta_{k_{1n+1}} + 3\varepsilon, \theta_{k_{1n+1}+1} - 3\varepsilon]\} \geq \frac{K}{2q}.$$

Hence, analogous to the discussion of the proof of case 1) of Theorem 5.4 at paragraph (4), when $z \in \overline{\Omega}(\theta_{k_{1n+1}} + 2\varepsilon, \theta_{k_{1n+1}+1} - 2\varepsilon; r_{n-1}, 2r_{n+1})$, we obtain

$$|f(z) - a_1| \leq e^{-r_{n+1}^{\mu/2}}. \tag{5.25}$$

Particularly, when $z \in \Delta(\theta_{k_{1n+1}} - 2\varepsilon; r_{n-1}, r_{n+1})$,

$$|f(z) - a_1| \le e^{-r_{n+1}^{\mu/2}},$$

and when $z \in \Delta(\theta_{k_{1n+1}} + 2\varepsilon; r_{n-1}, r_{n+1})$,

$$|f(z) - a_1| \le e^{-r_{n+1}^{\mu/2}}.$$

Besides, provided that n_0 is sufficiently large, applying the Poisson-Jensen formula, we conclude that

$$\log M(r_{n-1}, f - a_1) \le 3T(2r_{n-1}, f - a_1)$$
$$\le 3\{T(2r_{n-1}, f) + \log M + \log 2\} \le r_{n-1}^{\lambda+\eta},$$

$$\log M(r_{n+1}, f - a_1) \le r_{n+1}^{\lambda+\eta}.$$

Now, applying Lemma 2.10, when $z \in \Gamma(\theta_{k_{1n}} - 2\varepsilon, \theta_{k_{1n}} + 2\varepsilon; r_n)$, we get

$$\log|f(z) - a_1| \le \left\{1 - \left[\frac{4(r_{n-1}/r_n)^{\pi/4\varepsilon}}{\pi(1 - (r_{n-1}/r_n)^{\pi/2\varepsilon})} + \frac{4(r_n/r_{n+1})^{\pi/4\varepsilon}}{\pi(1 - (r_n/r_{n+1})^{\pi/2\varepsilon})}\right]\right\}(-r_{n+1}^{\mu/2})$$
$$+ \frac{4(r_{n-1}/r_n)^{\pi/4\varepsilon}}{\pi(1 - (r_{n-1}/r_n)^{\pi/2\varepsilon})}r_n^{\lambda+\eta} + \frac{4(r_n/r_{n+1})^{\pi/4\varepsilon}}{\pi(1 - (r_n/r_{n+1})^{\pi/2\varepsilon})}r_{n+1}^{\lambda+\eta},$$

$$\log|f(z) - a_1| \le \left\{1 - \frac{4r_{n-1}^{-\pi\eta/4\varepsilon}}{\pi(1 - r_{n-1}^{-\pi\eta/2\varepsilon})} - \frac{4r_n^{-\pi\eta/4\varepsilon}}{\pi(1 - r_n^{-\pi\eta/2\varepsilon})}\right\}(-r_{n+1}^{\mu/2})$$
$$+ \frac{4r_{n-1}^{-\pi\eta/4\varepsilon+\lambda+\eta}}{\pi(1 - r_{n-1}^{-\pi\eta/2\varepsilon})} + \frac{4}{\pi(1 - r_n^{\pi\eta/2\varepsilon})}r_n^{-\pi\eta/4\varepsilon+(\lambda+\eta)(1+\eta)}.$$

Notice that provided that n_0 is sufficiently large, when $n \ge n_0$, we have

$$1 - \frac{4r_{n-1}^{-\pi\eta/4\varepsilon}}{\pi(1 - r_{n-1}^{-\pi\eta/2\varepsilon})} - \frac{4r_n^{-\pi\eta/4\varepsilon}}{\pi(1 - r_n^{-\pi\eta/2\varepsilon})} \ge \frac{1}{2}.$$

Moreover, according to the selections of ε and η, we have

$$-\frac{\pi\eta}{4\varepsilon} + \lambda + \eta \le -\frac{3}{2}\lambda - \frac{3}{2} + \lambda + \eta \le -\frac{1}{2}\lambda - 1,$$

$$-\frac{\pi\eta}{4\varepsilon} + (\lambda + \eta)(1 + \eta) \le -\frac{3}{2}\lambda - \frac{3}{2} + \lambda + \eta \le -\frac{1}{2}\lambda - 1.$$

Hence, we conclude that

$$\log|f(z) - a_1| \le r_{n+1}^{\mu/2} + 1. \tag{5.26}$$

Analogously, when $z \in \Gamma(\theta_{k_{1n}+1} - 2\varepsilon; \theta_{k_{1n}+1} + 2\varepsilon; r_n)$, we conclude that

$$\log|f(z) - a_1| \le -r_{n+1}^{\mu/2} + 1. \tag{5.27}$$

Hence, according to formulas (5.24)–(5.27), we have

$$\log |M(r_n, f - a_1) \le -\tfrac{1}{2} r_{n+1}^{\mu/2} + 1.$$

If there exists an integral sequence n_m $(m = 1, 2, \dots)$, such that $k_{1n_m} = k_{1m_m} + 1$ $(m = 1, 2, \dots)$, then

$$\log M(r_{n_m}, f - a_1) \le -\tfrac{1}{2} r_{n_m+1}^{\mu/2} + 1 \qquad (m = 1, 2, \dots).$$

Hence, we conclude that $f(z) \equiv a_1$ which is a contradiction.

Now, we let $k_1 = k_{1n}$ $(n = n_0, n_0 + 1, \dots)$. Hence, for each value n $(n \ge n_0)$, $\Delta_a(\theta_{k_{1n}})$ must coincide with $\Delta_a(\theta_{k_1})$ and $\Delta_a(\theta_{k_{1n}} + 1)$ must coincide with $\Delta_a(\theta_{k_1+1})$. Therefore, $f(z)$ converges uniformly to value a_1 on $\overline{\Omega}(\theta_{k_1} + 2\varepsilon, \theta_{k_1+1} - 2\varepsilon)$, meaning that a_1 is an asymptotic value of $f(z)$. However, this contradicts the assumption that a_1 is not an asymptotic value, and hence, we prove that $p = l'$.

(4) We prove that $\lambda = \mu$.

1) According to the assumption that $2p - l' + l = q$, for the corresponding two Julia direction sets $\{\Delta_a(\theta_{k_{in}}), \Delta_a(\theta_{k_{in}+1}) \,|\, i = 1, 2, \dots, p\}$ and $\{\Delta_b(\theta_{k_{jn}+1}) \,|\, j = 1, 2, \dots, l\}$ of each value n $(n = n_0, n_0 + 1, \dots)$, there must only be l' Julia directions coinciding among one another between these two sets. Moreover, notice that $p = l'$. Analogous to the discussion of Theorem 5.4 at paragraph (6), we further conclude that there must exist one and only one Julia direction coinciding with each other between each pair of Julia directions $\{\Delta_a(\theta_{k_{in}}), \Delta_a(\theta_{k_{in}+1})\}$ $(i = 1, 2, \dots, p)$ and sets $\{\Delta_b(\theta_{k_{jn}+1}) \,|\, j = 1, 2, \dots, l\}$.

In the following, we prove that $k_{in} = k_{in+1}$ $(i = 1, 2, \dots, p, n = n_0, n_0 + 1, \dots)$. In fact, otherwise, we need only to consider the following few typical cases:

(i) $q = 2$.

Under this case, it must be that $p = 1$. Hence, analogous to the discussion of case 1) in (3), we may prove that there exists an integral sequence n_m $(m = 1, 2, \dots)$ such that $k_{1n_m} = k_{1n_m+1} + 1$ $(m = 1, 2, \dots)$. Meanwhile, we have

$$\log M(r_{n_m}, f - a_1) \le -\tfrac{1}{2} r_{n_m}^{\mu/2} + 1 \qquad (m = 1, 2, \dots).$$

Hence, we conclude that $f(z) \equiv a_1$, which is a contradiction.

(ii) $q > 3$ and $k_{in} = k_{i'n+1}$ or $k_{in} = k_{i'n+1} + 1$ $(i \ne i')$.

Analogous to the discussion of cases 1) and 2) of (3), we conclude that $a_i = a_{i'}$, and hence, a contradiction is derived.

(iii) $q > 3$ and $k_{in} = k_{jn+1} + 1$ and $k_{in} + 1 = k_{j'n+1} + 1$ $(j \ne j')$.

Analogous to the discussion of case 1) of (3) and the proof of Theorem 5.4 at paragraph (6), we conclude that b_j and $b_{j'}$ are nondistinct asymptotic values, and hence, we derive a contradiction.

(iv) $q > 3$ and $k_{in} = k_{in+1} + 1$, and $k_{in} + 1 = k_{jn+1} + 1$.

First, there must exist a Julia direction $\Delta_b(\theta_{k_{j'n+1}+1})$ in the set $\{\Delta_b(\theta_{k_{jn+1}-1})|$ $j = 1, 2, \ldots, l\}$, such that $\Delta_b(\theta_{k_{j'n+1}+1})$ coincides with $\Delta_a(\theta_{k_{in+1}})$ or $\Delta_a(\theta_{k_{in+1}+1})$. Analogous to the discussion of cases 1) and 4) of (3) and the proof of Lemma 5.4, we may conclude that b_j and $b_{j'}$ are nondistinct asymptotic values, and hence, a contradiction is derived.

Now, let $k_i = k_{in}$ $(1 \le i \le p; \; n = n_0, n_0 + 1, \ldots)$; then $f(z)$ tends uniformly to a_i on $\overline{\Omega}(\theta_{k_i} + 2\varepsilon, \theta_{k_i+1} - 2\varepsilon)$ $(1 \le i \le p)$. Particularly, there exists a path of fixed value L'_i on $\overline{\Omega}(\theta_{k_i} + 2\varepsilon, \theta_{k_i+1} - 2\varepsilon)$ $(1 \le i \le p)$ such that

$$\lim_{\substack{z \in L'_i \\ |z| \to +\infty}} f(z) = a_i.$$

We may assume that L'_i $(1 \le i \le p)$ and the path of fixed value L_i defines the same asymptotic value, meaning that $a_i = b_i$ $(i = 1, 2, \ldots, p)$.

2) Now, we prove that provided that n_0 is sufficiently large, the part of L_j $(p + 1 \le i \le l)$ outside disk $|z| \le r_{n_0}$ must entirely lie within a certain $\Omega(\theta_{k_j} - 3\varepsilon, \theta_{k_j} + 3\varepsilon, r_{n_0}, +\infty)$ $(1 \le k_j \le q, \; k_j \ne k_i, \; k_j \ne k_i + 1,$ $i = 1, 2, \ldots, p)$. In fact, otherwise, then there must exist value n, $n \ge n_0$, such that L_j must intersect a certain closed domain $\overline{\Omega}(\theta_k + 3\varepsilon, \theta_{k+1} - 3\varepsilon; r_n^{1-3\eta}, 2r_n)$ $(1 \le k \le q)$. Hence, $k \ne k_{jn}$ or $k + 1 \ne k_{jn}$. We let $k \ne k_{jn}$. Besides, $\Delta(\theta_k)$ belongs to $\{\Delta_a(\theta_{k_{in}}), \Delta_a(\theta_{k_{in}+1}) | i = 1, 2, \ldots, p\}$ or $\{\Delta_b(\theta_{k_{jn}+1}) | j = 1, 2, \ldots, l\}$. Hence, analogous to the discussion of the proof of Theorem 5.4 at paragraph (3), we may conclude that b_j and a certain a_i $(1 \le i \le p)$ are nondistinct asymptotic values or b_j and a certain $b_{j'}$ $(p + 1 \le j' \le l, \; j' \ne j)$ are nondistinct asymptotic values, and hence, a contradiction is derived. Finally, $l - p$ domains $\Omega(\theta_{k_j} - 3\varepsilon, \theta_{k_j} + 3\varepsilon; r_{n_0}, +\infty)$ $(j = p + 1, p + 2, \ldots, l)$ are distinct among one another.

3) Now, we prove that $\lambda = \mu$.

Since $f(z)$ possesses ∞ as its deficient value, with its corresponding deficiency $\delta(\infty, f) = 1$, then according to Lemma 2.3 (where we let $h_1 = 0$) and Lemma 3.9 (where we let $k = 2\lambda$, $\sigma = \log 2$), we conclude that there exist two sequences $\{R'_m\}$ and $\{R_m\}$, such that

$$\lim_{m \to +\infty} \frac{\log T(R'_m, f)}{\log R'_m} = \lambda, \qquad R'_m \le R_m \le 2R_m.$$

Meanwhile, R_m satisfies the following property: If

$$E_m = \{\theta | \log|f(R_m e^{i\theta})| > \tfrac{1}{4}T(R_m, f), 0 \le \theta < 2\pi\} \tag{5.28}$$

then

$$\text{meas } E_m \ge \widetilde{K}, \qquad \widetilde{K} = \widetilde{K}(\lambda) > 0,$$

where $\widetilde{K}$ is a constant depending on order λ but unrelated to m.

As for the selection of ε, we make a further demand: $\varepsilon < \frac{\widetilde{K}}{12q}$. Hence, there exists at least one interval $[\theta_{km} + 3\varepsilon, \theta_{k_m+1} - 3\varepsilon]$ in the q intervals $[\theta_k + 3\varepsilon, \theta_{k+1} - 3\varepsilon]$ $(k = 1, 2, \ldots, q)$ such that

$$\operatorname{meas}\{E_m \cap [\theta_{km} + 3\varepsilon, \theta_{k_m+1} - 3\varepsilon]\} \geq \frac{\widetilde{K}}{2q}.$$

We may assume that $k_m = k_{m+1}$ $(m = 1, 2, \ldots)$. Otherwise, we need only to select a suitable subsequence.

In the following, we distinguish three cases for discussion:

(i) $q = 1$.

First, according to Corollary 2 of Theorem 2.17, we conclude that $\lambda \leq \frac{1}{2}$. Next, according to the assumption that $2p - l' + l = q$, it must be that $l = 1$. Hence, according to the corollary of Theorem 4.4, we conclude that $\mu \geq \frac{1}{2}$. Hence, we have $\lambda = \mu$.

(ii) $q = 2$.

According to the condition that $2p - l' + l = q$, we conclude that only two cases may happen:

i) $p = l = 1$.

We may assume that

$$\lim_{\substack{|z| \to +\infty \\ z \in \overline{\Omega}(\theta_1 + 2\varepsilon, \theta_2 - 2\varepsilon)}} f(z) = a_1. \tag{5.29}$$

Hence, we can only have

$$\operatorname{meas}\{E_m \cap [\theta_2 + 3\varepsilon, \theta_3 - 3\varepsilon]\} \geq \frac{\widetilde{K}}{4}, \qquad m = 1, 2, \ldots,$$

where $\theta_3 = \theta_1 + 2\pi$. In the following, we prove that $\lambda \leq \pi/(\theta_3 - \theta_2 - \varepsilon)$. In fact, otherwise, then $\lambda > \pi/(\theta_3 - \theta_2 - \varepsilon)$. Since $f(z)$ has no Julia direction on $\Omega(\theta_2, \theta_3)$, therefore, according to Lemma 2.10, we have

$$\varlimsup_{R \to +\infty} \frac{\log^+ \log^+ M\{\overline{\Omega}(-(\theta_3 - \theta_2)/2 + \varepsilon, (\theta_3 - \theta_2)/2 - \varepsilon; R), f\}}{\log R}$$
$$\leq \frac{\pi}{\theta_3 - \theta_2 - \varepsilon}.$$

Moreover, according to formula (5.28), we conclude that $\lambda \leq \pi/(\theta_3 - \theta_2 - \varepsilon)$, and hence, we derive a contradiction.

On the other hand, according to formula (5.29), we have, in particular,

$$\lim_{\substack{|z| \to +\infty \\ z \in \Delta(\theta_1 + 2\varepsilon)}} f(z) = a_1, \qquad \lim_{\substack{|z| \to +\infty \\ z \in \Delta(\theta_2 - 2\varepsilon)}} f(z) = a_1.$$

Let

$$\widetilde{M} = \sup_{z \in \Delta(\theta_1 + 2\varepsilon) \cup \Delta(\theta_2 - 2\varepsilon)} |f(z)|$$

and assume a point $z_0 = R_m e^{i\varphi}$, $\varphi \in E_m \cap [\theta_2 + 2\varepsilon, \theta_3 - 2\varepsilon]$, such that

$$\log|f(z_0)| > \frac{1}{4} T(R_m, f) > 1 + 2\log\widetilde{M}. \tag{5.30}$$

Applying Lemma 2.9, we have

$$\log|f(z_0)| \le 2\log\widetilde{M} + \frac{4(|z_0|/R)^{\pi/(\theta_3-\theta_2+4\varepsilon)}}{\pi\{1 - (|z_0|/R)^{2\pi/(\theta_3-\theta_2+4\varepsilon)}\}}$$
$$\times \log M(R, f).$$

Moreover, from formula (5.30), we conclude that

$$\frac{\pi}{\theta_3 - \theta_2 + 4\varepsilon} \le \varliminf_{R\to+\infty} \frac{\log\log M(R, f)}{\log R} = \mu$$

or $\frac{\pi}{\mu} \le \theta_3 - \theta_2 + 4\varepsilon$. On the other hand, we have $\frac{\pi}{\lambda} \ge \theta_3 - \theta_2 - \varepsilon$. Hence, $\frac{\pi}{\mu} \le \frac{\pi}{\lambda} + 5\varepsilon$. Since we may select ε sufficiently small, it must be that $\lambda \le \mu$. It follows that $\lambda = \mu$.

ii) $l = 2$.

Under this case, the two paths L_1 and L_2 outside disk $|z| \le r_{n_0}$ must be in $\Omega(\theta_1 - 3\varepsilon, \theta_1 + 3\varepsilon; r_{n_0}, +\infty)$ and $\Omega(\theta_2 - 3\varepsilon, \theta_2 + 3\varepsilon; r_{n_0}, +\infty)$, respectively. Hence, analogous to the discussion of (1), we may conclude that $\lambda = \mu$.

iii) $q \ge 3$.

Since there must exist paths of fixed value in $\Omega(\theta_{k_m} - 3\varepsilon, \theta_{k_n} + 3\varepsilon; r_{n_0}, +\infty)$ and $\Omega(\theta_{k_m+1} - 3\varepsilon, \theta_{k_m+1} + 3\varepsilon; r_{n_0}, +\infty)$, respectively, we may analogously conclude that $\lambda = \mu$. Hence, this proves completely Theorem 5.5.

§5.3. On entire functions having a finite number of Julia directions [43h]

THEOREM 5.6. *Let $f(z)$ be an entire function (on the open plane $|z| < +\infty$). We also let q be the number of Julia directions of $f(z)$, l be the number of distinct finite asymptotic values and p be the number of finite deficient values, where l' deficient values are simultaneously the asymptotic values. If $q < +\infty$, then*

$$p - l' + l \le 2\mu,$$

where μ is the lower order of $f(z)$.

PROOF. When $\mu = +\infty$, it is obvious that Theorem 5.6 holds. Hence, we need only to consider the case when $\mu < +\infty$.

When $\mu < +\infty$, according to $q < +\infty$ and Corollary 1 of Theorem 2.17, we conclude that the order λ of $f(z)$ is $< +\infty$. On the other hand, when $p - l' = 0$, according to Theorem 4.7, we conclude that Theorem 5.6 holds. Hence, we need only to consider the case when $p - l' \ge 1$. When $p - l' \ge 1$, according to the corollary of Theorem 4.4, we conclude that the lower order $\mu \ge \frac{1}{2}$.

In the following, we need only to prove that when $\frac{1}{2} \leq \mu \leq \lambda < +\infty$ and $p - l' \geq 1$, Theorem 5.6 holds. In fact, otherwise, then there exist integer p_1, $1 \leq p_1 \leq p - l'$, $p_1 < +\infty$ and integer l_1, $0 \leq l_1 \leq l$, $l_1 < +\infty$, such that

$$p_1 + l_1 \geq [2\mu] + 1.$$

(1) We select p_1 deficient but nonasymptotic values a_i $(i = 1, 2, \ldots, p_1)$ of $f(z)$, with its corresponding deficiencies $\delta(a_i, f) = \delta_i > 0$, and l_1 distinct finite asymptotic values b_j $(j = 1, 2, \ldots, l_1)$, with its corresponding paths of fixed value L_j. Without any loss of generality, we may assume that L_j $(1 \leq j \leq l_1)$ is a simply continuous curve originating from the origin and tending to ∞. Meanwhile, it is a line segment in the part of disk $|z| \leq 2$. Moreover, l_1 curves L_j $(j = 1, 2, \ldots, l_1)$ have no intersection among one another except at the origin. L_j and L_{j+1} $(1 \leq j \leq l_1; L_{l_1+1} \equiv L_1)$ are adjacent, bounding a simply connected domain D_j. Let

$$M_0 = \sup_{z^\varepsilon \bigcup_{j=1}^{l_1} L_j} |f(z)| < +\infty, \tag{5.32}$$

$$M_1 = \max\{1, |a_1|, \ldots, |a_{p_1}|, |b_1|, \ldots, |b_{l_1}|\}, \tag{5.33}$$

$$M_2 = \min\{1, |a_i - a_{i'}|, |b_j - b_{j'}|, |a_i - b_j| \,|\, 1 \leq i \\ \neq i' \leq p_1, \ 1 \leq j \neq j' \leq l_1, b_j \neq b_{j'}\}. \tag{5.34}$$

According to Theorem 4.3, we conclude that there exists a continuous curve Γ_j tending to ∞ in D_j $(1 \leq j \leq l_1)$ such that

$$\lim_{\substack{|z| \to +\infty \\ z \in \Gamma_j}} \frac{\log \log |f(z)|}{\log |z|} \geq \frac{1}{2}. \tag{5.35}$$

Hence, we may select a point z'_j on Γ_j $(1 \leq j \leq l_1)$ such that $|f(z'_j)| \geq M_0$. Let

$$r'_0 = \max\{|z'_1|, |z'_2|, \ldots, |z'_{l_1}|\}.$$

When $r > r'_0$, let $\Omega_j(r)$ be a connected component of D_j $(1 \leq j \leq l_1)$ in disk $|z| < r$ and containing point z'_j. Notice that

$$M\{\overline{\Omega}_j(r) \cap (|z| = r), f\} = \max_{z \in \overline{\Omega}_j(r) \cap (|z|=r)} |f(z)|, \qquad j = 1, 2, \ldots, l_1;$$

then $M\{\overline{\Omega}_j(r) \cap (|z| = r), f\}$ $(1 \leq j \leq l_1)$ is a monotonic increasing function of r $(r \geq r'_0)$. Moreover, according to formula (5.35), we have

$$\varliminf_{r \to +\infty} \frac{\log \log M\{\overline{\Omega}_j(r) \cap (|z| = r), f\}}{\log r} \geq \frac{1}{2}. \tag{5.36}$$

On the other hand, we choose a point z''_j on $D_j \cap (|z| = 1)$ $(1 \leq j \leq l_1)$. Then, according to connectivity, there exists a curve L'_j linking points z'_j

and z_j'' in D_j $(1 \leq j \leq l_1)$. We then assume value r_0'', $r_0'' > r_0'$, such that l_1 curves L_j' $(j = 1, 2, \ldots, l_1)$ are in disk $|z| < r$ and $\Omega_j(r)$ is the connected component containing point z_j''. Let $L_j(r)$ be the part of L_j $(1 \leq j \leq l_1)$ between origin $z = 0$ and the first intersection point of itself and circumference $|z| = r$. Then a simply connected region $D_j(r)$ is bounded by $L_j(r)$ and $L_{j+1}(r)$ in disk $|z| < r$. Obviously, when $r \geq r_0''$, $\Omega_j(r) \subset D_j(r)$ $(j = 1, 2, \ldots, l_1)$.

(2) Let $\Delta(\theta_k)$ $(k = 1, 2, \ldots, q; \ 0 \leq \theta_1 < \theta_2 < \cdots < \theta_q < 2\pi)$ be q Julia directions of $f(z)$, and

$$\omega = \min_{1 \leq k \leq q} (\theta_{k+1} - \theta_k), \ \theta_{q+1} = \theta_1 + 2\pi,$$

$$\delta = \min_{1 \leq i \leq p_1} \{\delta_i\}.$$

We select arbitrarily a number η,

$$0 < \eta < \frac{\omega}{2(31 \times 32 + 84)\pi}$$

and number r_0, $r_0 > r_0''$. Then, we construct a sequence $r_m = r_0^{(1+\eta)^{2m}}$, $m = 1, 2, \ldots$. According to Lemma 2.4 (where we let $h_1 = 0$) and Lemma 3.9 (where we let $\sigma = \log 2$), there exists integer m_0, $m_0 \geq 2$, such that when $r \geq r_{m_0}$,

$$T(r, f) \leq r^{\lambda+\eta}, \ \log M(r, f) \leq r^{\lambda+\eta}, \tag{5.37}$$

and

$$T(r, f) \geq r^{\mu-\eta}. \tag{5.38}$$

Moreover, when $m \geq m_0$, there exists value t_m in the interval $[r_m, 2r_m^{1+\eta}]$ such that the set

$$E_i(t_m) = \left\{ \theta \, \middle| \, \log \frac{1}{|f(t_m e^{i\theta}) - a_i|} \geq \frac{\delta}{4} T(t_m, f), \ 0 \leq \theta < 2\pi \right\},$$

$$(i = 1, 2, \ldots, p_1)$$

has its measure satisfying:

$$\text{meas} \, E_i(t_m) \geq K = K(\delta, \lambda, p_1, \eta) > 0.$$

Finally, we choose a fixed number ε,

$$0 < \varepsilon < \min \left\{ \frac{\pi\eta}{24(\lambda + 1)}, \frac{K}{12q}, \frac{\omega}{40} \right\}$$

and let $\overline{\Omega} = \bigcup_{k=1}^{q} \overline{\Omega}(\theta_k + \varepsilon, \theta_{k+1} - \varepsilon)$. Then according to Theorem 2.13 and the finite covering theorem, there exist two distinct finite values α and β, such that

$$\varlimsup_{r \to +\infty} \frac{\log^+ n(r, \overline{\Omega}, f = X)}{\log r} = 0, \qquad X = \alpha, \beta.$$

Obviously, there exists some number d, $0 < d < \frac{1}{2}$, such that the distances among α, β and ∞ are greater than d.

(3) When $m \geq m_0$, according to the selection of ε, for each value i ($1 \leq i \leq p_1$), there exists at least one set $E_i(t_m) \cap [\theta_{k_i} + 3\varepsilon, \theta_{k_i+1} - 3\varepsilon]$ ($1 \leq k_i \leq q$) among the q sets $E_i(t_m) \cap [\theta_k + 3\varepsilon, \theta_{k+1} - 3\varepsilon]$ ($k = 1, 2, \ldots, q$) such that

$$\text{meas}\{E_i(t_m) \cap [\theta_{k_i} + 3\varepsilon, \theta_{k_i+1} - 3\varepsilon]\} \geq \frac{K}{2q}.$$

Moreover, according to formula (5.38) and the selection of η, applying Lemma 5.2, where we let $\theta = \frac{1}{2}(\theta_{k_i+1} - \theta_{k_i}) - \varepsilon$, $R_1 = t_{m-1}$, $R = t_m$, $R_2 = t_{m+1}$, $E_\alpha = \{t_m e^{i\varphi} \mid \varphi \in E_i(t_m) \cap [\theta_{k_i} + 3\varepsilon, \theta_{k_i+1} - 3\varepsilon]\}$, $H = \frac{K}{2q} \geq \frac{\varepsilon}{2}$, $\alpha = a_i$, $N = \frac{\delta}{4}T(t_m, f)$ and $\nu = 0$, we conclude that there exists value m_1, $m_1 \geq m_0$, such that when $m \geq m_1$ and

$$z \in \overline{\Omega}(\theta_{k_i} + 2\varepsilon, \theta_{k_i+1} - 2\varepsilon; t_{m-1}, t_{m+1}),$$

$$\log|f(z) - a_i| \leq -At_m^{-(31+8\pi/\omega+1)\eta} T(t_m, f), \tag{5.39}$$

where $A > 0$ is a constant independent of m. Let

$$\varepsilon_m' = e^{-At_m^{-((31 \times 8\pi)/\omega+1)\eta T(t_m, f)}}.$$

Then there exists value m_2, $m_2 \geq m_1$, such that when $m \geq m_2$,

$$\varepsilon_m' < \frac{1}{4}M_2 < 1. \tag{5.40}$$

Hence, the p_1 sets $\overline{\Omega}(\theta_{k_i} + 2\varepsilon, \theta_{k_i+1} - 2\varepsilon; t_{m-1}, t_{m+1})$ ($i = 1, 2, \ldots, p_1$) are distinct among one another.

Let L_{jm} be the part of L_j ($1 \leq j \leq l_1$) between the last intersection point of itself and circumference $|z| = t_{m-1}$, and the first intersection point of itself and the circumference $|z| = t_{m+1}$. The l_1 curves L_{jm} ($j = 1, 2, \ldots, l_1$) divide annulus $C_m : t_{m-1} < |z| < t_{m+1}$ into l_1 regions D_{jm} ($j = 1, 2, \ldots, l_1$). Moreover, L_{jm} and L_{j+1m} are the boundary part of D_{jm}. Let

$$\varepsilon_m'' = \max_{1 \leq j \leq l_1} \left\{ \sup_{z \in L_j \cap C_m} |f(z) - b_j| \right\}. \tag{5.41}$$

Then when $m \to +\infty$, $\varepsilon_m'' \to 0$. Hence, there exists value m_3, $m_3 \geq m_2$, such that when $m \geq m_3$,

$$\varepsilon_m'' < \frac{1}{4}M_2 < 1. \tag{5.42}$$

Therefore, line segment $\Delta((\theta_{k_i} + \theta_{k_i+1})/2; t_{m-1}, t_{m+1})$ may lie entirely in D_{jm} ($1 \leq j \leq l_1$), or it may have no intersection with D_{jm}. Let s_j ($0 \leq s_j \leq p_1$) line segments $\Delta((\theta_{k_i} + \theta_{k_i+1})/2; t_{m-1}, t_m)$ ($i = 1, 2, \ldots, s_j$) be in D_{jm}. Then D_{jm} is divided into $s_j + 1$ domains $D_{jm\nu}$ ($\nu = 1, 2, \ldots, s_j + 1$). We assume that L_{jm} is the boundary part of D_{jm1} and L_{j+1m} is the boundary part of D_{jms_j+1}. Obviously, we have $\sum_{j=1}^{l_1} s_j = p_1$.

(4) We prove if $s_j \neq 0$ $(1 \leq j \leq l_1)$, then when m is sufficiently large, there exists a point $z_{jm\nu}$ in $D_{jm\nu} \cap (e^{-30^\varepsilon} t_m \leq |z| \leq e^{30^\varepsilon} t_m)$ $(1 \leq \nu \leq s_j+1)$ and a constant $A > 0$ independent of m, such that

$$\log |f(z_{jm\nu})| \geq AT(t_m, f).$$

In fact, we need only to consider a typical case, that is to prove the result for D_{jms_j+1}. Notice that L_{j+1m} is the boundary part of D_{jms_j+1}. Besides, suppose that $\Delta(\frac{\theta_k+\theta_{k+1}}{2}; t_{m-1}, t_{m+1})$ $(1 \leq k \leq q)$ is another boundary part of D_{jms_j+1}. Moreover, when $m \geq m_3$ and

$$z \in \overline{\Omega}(\theta_k + 2\varepsilon, \theta_{k+1} - 2\varepsilon; t_{m-1}, t_{m+1}),$$

$$\log |f(z) - a_i| \leq \log \varepsilon'_m = -At_m^{-(31 \times 8\pi/\omega+1)\eta} T(t_m, f) \quad (1 \leq i \leq p_1), \quad (5.43)$$

where $A > 0$ is a constant independent of m. According to formulas (5.40)–(5.43), we conclude that L_{j+1m} cannot intersect at

$$\overline{\Omega}(\theta_k + 2\varepsilon, \theta_{k+1} - 2\varepsilon; t_{m-1}, t_{m+1}).$$

1) We further prove that L_{j+1m} cannot be entirely in

$$\Omega(\theta_{k+1} - 3\varepsilon, \theta_{k+1} + 3\varepsilon; t_{m-1}, t_{m+1}).$$

In fact, otherwise, suppose that a region

$$\Omega' \subset \Omega(\theta_{k+1} - 3\varepsilon, \theta_{k+1} + 3\varepsilon; t_{m-1}, t_{m+1})$$

is bounded by L_{j+1m} and $\Delta(\theta_{k+1} - 3\varepsilon; t_{m-1}, t_{m+1})$ in annulus C_m. Moreover, we let A_1 be the intersection point between $\overline{\Delta}(\theta_{k+1} - 3\varepsilon; t_{m-1}, t_{m+1})$ and circumference $|z| = t_{m-1}$, and B_1 be the intersection point between $\overline{\Delta}(\theta_{k+1} - 3\varepsilon; t_{m-1}, t_{m+1})$, and also A_2 be the intersection point between L_{j+1m} and circumference $|z| = t_{m-1}$, and B_2 be the intersection point between L_{j+1m} and circumference $|z| = t_{m+1}$. We also assume that $\zeta = \varphi(z)$ maps Ω' conformally onto a region

$$\Omega_\zeta(\theta_{k+1} - 3\varepsilon, \theta_{k+1} + 3\varepsilon; t_{m-1}, t'_{m+1})$$

on the ζ-plane, and points A_1, A_2, B_1 and B_2 are transformed into points $t_{m-1}e^{i(\theta_{k+1}-3\varepsilon)}$, $t_{m-1}e^{i(\theta_{k+1}+3\varepsilon)}$, $t'_{m+1}e^{i(\theta_{k+1}-3\varepsilon)}$ and $t'_{m+1}e^{i(\theta_{k+1}+3\varepsilon)}$. Then according to Lemma 2.10, we conclude that $t'_{m+1} \geq t_{m+1}$ respectively. Let $g(\zeta) = f(\varphi^{-1}(\zeta))$. Then according to formulas (5.33) and (5.43), when

$$\zeta \in \Delta_\zeta(\theta_{k+1} - 3\varepsilon; t_{m-1}, t'_{m+1}),$$

we have

$$\log |g(\zeta)| = \log |f(z)| \leq \log^+ |f(z) - a_i| + \log^+ |a_i| + \log 2$$
$$\leq \log^+ \varepsilon'_m + \log M_1 + \log 2 = \log 2M_1.$$

On the other hand, from formulas (5.33) and (5.41), when

$$\zeta \in \Delta_\zeta(\theta_{k+1} + 3\varepsilon; t_{m-1}, t'_{m+1}),$$

$$\log|g(\zeta)| = \log|f(z)| \leq \log^+|f(z) - b_{j+1}| + \log^+|b_{j+1}| + \log 2$$

$$\leq \log^+ \varepsilon''_m + \log M_1 + \log 2 = \log 2M_1.$$

Besides, there exists value m_4, $m_4 \geq m_3$, such that when $m \geq m_4$, $t_{m-1} < e^{-3\pi} t_m < e^{3\pi} t_m < \frac{1}{4} t_{m+1}$. Hence, applying Lemma 2.10, we conclude that when $\zeta \in \Omega_\zeta(\theta_{k+1} - 3\varepsilon, \theta_{k+1} + 3\varepsilon; e^{-3\pi} t_m, e^{3\pi} t_m)$,

$$\log|g(\zeta)| \leq 2\log 2M_1 + \frac{4(t_{m-1}/|\zeta|)^{\pi/6\varepsilon}}{\pi(1 - (t_{m-1}/|\zeta|)^{\pi/3\varepsilon})}$$

$$\times \log M(t_{m-1}, f) + \frac{4(|\zeta|/t_{m+1})^{\pi/6\varepsilon}}{\pi(1 - (|\zeta|/t_{m+1})^{\pi/3\varepsilon})} \log M(t_{m+1}, f).$$

Furthermore, according to formula (5.37) and the selection of ε, there exists value m_5, $m_5 \geq m_4$, such that when $m \geq m_5$ and

$$\zeta \in \Omega_\zeta(\theta_{k+1} - 3\varepsilon, \theta_{k+1} + 3\varepsilon; e^{-3\pi} t_m, e^{3\pi} t_m),$$

$$\log|g(\zeta)| \leq 2\log 2M_1 + 1, \qquad |g(\zeta)| \leq 4e M_1^2.$$

Obviously, there exists value m_6, $m_6 \geq m_5$, such that when $m \geq m_6$,

$$(M_1 + 4e M_1^2)(\varepsilon'^{1/3}_m + \varepsilon''^{1/3}_m) < M_2.$$

Hence, by applying the transformation $\xi = e^{3\pi} t_m^{-1} \zeta$ and Lemma 3.2, we conclude that $a_i = b_{j+1}$, and hence, a contradiction is derived.

2) We prove that when m is sufficiently large, $L_{j+1\,m}$ cannot intersect at $\overline{\Omega}(\theta_{k+1} + 3\varepsilon, \theta_{k+2} - 3\varepsilon; t_{m-1}, t_{m+1})$ either. In fact, otherwise, then according to formula (5.33), applying Lemma 5.2, where we let $\theta = \frac{1}{2}(\theta_{k+2} - \theta_{k+1}) - \varepsilon$, $R_1 = t_{m-1}$, $R = t_m$, $R_2 = t_{m+1}$, $L = L_{j+1\,m}$, $H = \varepsilon > \frac{\varepsilon}{2}$, $\alpha = b_{j+1}$, $N = 0$ and $\gamma = 0$, we conclude that there exists value m_7, $m_7 \geq m_6$, such that when $m \geq m_7$ and $z \in \overline{\Omega}(\theta_{k+1} + 2\varepsilon, \theta_{k+1} - 2\varepsilon; t_{m-1}, t_{m+1})$,

$$\log|f(z)| \leq \log^+|f(z) - b_{j+1}| + \log^+|b_j| + \log 2$$

$$\leq A t_m^{((31\times24\pi)/\omega+40)\eta},$$

where $A > 0$ is a constant independent of m. Particularly, when $z \in \Delta(\theta_{k+1} + 3\varepsilon, t_{m-1}, t_{m+1})$,

$$\log|f(z)| \leq A t_m^{((31\times21\pi)/\omega+40)\eta}.$$

On the other hand, according to formulas (5.33) and (5.43), when $z \in \Delta(\theta_{k+1} - 3\varepsilon, t_{m-1}, t_{m+1})$, $\log|f(z)| \leq \log 2M_1$. Analogously, applying Lemma 2.10, when $m \geq m_7$ and $z \in \Omega(\theta_{k+1} - 3\varepsilon, \theta_{k+1} + 3\varepsilon; e^{-3\pi} t_m, e^{3\pi} t_m)$,

$$\log|f(z)| \leq A t_m^{((31\times24\pi)/\omega+40)\eta} + \log 2M_1 + 1.$$

Furthermore, when $z \in \Omega(\theta_{k+1} - 4\varepsilon, \theta_{k+1} + 4\varepsilon; e^{-3\pi}t_m, e^{3\pi}t_m)$,

$$|f(z)| \leq \exp\{At_m^{((31\times24\pi)/\omega+40)\eta} + \log 2M_1 + 1\}.$$

We construct the transformation

$$\zeta = \frac{1}{4\varepsilon}(\log z - \log t_m - i\theta_{k+1})$$

and let $g(\zeta) = f(z)$. Then $g(\zeta)$ is regular on disk $|\zeta| \leq 1$, and the image set E of $\Gamma(\theta_{k+1} - 3\varepsilon, \theta_{k+1} - 2\varepsilon; t_m)$ on the ζ-plane is included in disk $|\zeta| \leq \frac{3}{4}$. If we assume $h = \frac{1}{16e}$, then E cannot be included in some circles with the sum of their Euclidean radii not exceeding $2eh$. From formula (5.43), when $\zeta \in E$,

$$|g(\zeta) - a_i| \leq \exp\{-At_m^{-((31\times8\pi)/\omega+1)\eta}T(t_m, f)\}.$$

According to formula (5.33) and the selection of η, applying Theorem 5.2, where we let $M = \exp\{At_m^{((31\times24\pi)/\omega+40)\eta} + \log 2M_1 + 1\}$, $\alpha = a_i$, $N = At_m^{-((31\times8\pi)/\omega+1)\eta} \times T(t_m, f)$, $\tau = r = \frac{3}{4}$, we conclude that there exists value m_8, $m_8 \geq m_7$, such that when $m \geq m_8$ and $|\zeta| \leq \frac{3}{4}$,

$$\log|g(\zeta) - a_i| \leq -At_m^{-((31\times8\pi)\omega+1)\eta}T(t_m, f),$$

where $A > 0$ is a constant independent of m. Particularly, when $m \geq m_8$ and $z \in \Gamma(\theta_{k+1} + 2\varepsilon, \theta_{k+1} + 3\varepsilon, t_m)$,

$$\log|f(z) - a_i| \leq -At_m^{-((31\times8\pi)/\omega+1)\eta}T(t_m, f), \quad A > 0.$$

Analogously, applying Lemma 5.2, we conclude that there exists value m_9, $m_9 \geq m_8$, such that when $m \geq m_9$ and

$$z \in \overline{\Omega}(\theta_{k+1} + 2\varepsilon, \theta_{k+2} - 2\varepsilon; t_{m-1}, t_{m+1}),$$

$$|f(z) - a_i| \leq \exp\{-At_m^{-2((31\times8\pi)\omega+1)\eta}T(t_m, f)\} \leq \tfrac{1}{4}M_2.$$

We assume a point z_0 on $L_{j+1m} \cap \overline{\Omega}(\theta_{k+1}+3\varepsilon, \theta_{k+2}-3\varepsilon; t_{m-1}, t_{m+1})$. Then, according to formulas (5.34), (5.41) and (5.42), we have

$$M_2 \leq |a_i - b_{j+1}| \leq |f(z_0) - a_i| + |f(z_0) - b_{j+1}|$$
$$\leq \tfrac{1}{4}M_2 + \tfrac{1}{4}M_2 = \tfrac{1}{2}M_2,$$

meaning that a contradiction is derived.

3) Suppose that L_{j+1m} does not intersect at

$$\overline{\Omega}(\theta_{k+1} - 3\varepsilon, \theta_{k+s} - 3\varepsilon; t_{m-1}, t_{m+1}) \quad (2 \leq s < q)$$

but at $\overline{\Omega}(\theta_{k+s} - 3\varepsilon, \theta_{k+s+1} - 3\varepsilon; t_{m-1}, t_{m+1})$. We prove that when m is sufficiently large, there exists a point z_0 in

$$\Omega(\theta_{k+1} - 2\varepsilon, \theta_{k+s-1} + 8\varepsilon; e^{-10\varepsilon}t_m, e^{10\varepsilon}t_m),$$

such that $|f(z_1) - a_i| \geq 1$. In fact, otherwise, when $z \in \Omega(\theta_{k+1} - 2\varepsilon, \theta_{k+s-1} + 8\varepsilon; e^{-10\varepsilon}t_m, e^{10\varepsilon}t_m)$, then

$$\log|f(z) - a_i| \leq 0. \tag{5.44}$$

According to formula (5.38) and the selection of η, applying Lemma 5.2, where we let $\theta = \frac{1}{2}(\theta_{k+1} - \theta_k) - \varepsilon$, $R_1 = e^{-30\varepsilon}t_m$, $R = t_m$, $R_2 = e^{30\varepsilon}t_m$, $E_\alpha = \{t_m e^{i\varphi} \mid \varphi \in E_i(t_m) \cap [\theta_k + 3\varepsilon, \theta_{k+1} - 3\varepsilon]\}$, $H = \frac{K}{2q} \geq \frac{\varepsilon}{2}$, $\alpha = a_i$, $N = \frac{\delta}{4}T(t_m, f)$ and $\nu = 0$, we conclude that there exists value m_{10}, $m_{10} \geq m_9$, such that when $m \geq m_{10}$ and $z \in \overline{\Omega}(\theta_k + 2\varepsilon, \theta_{k+1} - 2\varepsilon; e^{-30\varepsilon}t_m, e^{30\varepsilon}t_m)$,

$$\log|f(z) - a_i| \leq -AT(t_m, f),$$

where $A > 0$ is a constant independent of m. Particularly, when $z \in \Delta(\theta_{k+1} - 2\varepsilon; e^{-10\varepsilon}t_m, e^{10\varepsilon}t_m)$,

$$\log|f(z) - a_i| \leq -AT(t_m, f). \tag{5.45}$$

We construct the transformation

$$\zeta = \frac{1}{10\varepsilon}\{\log z - \log t_m - i(\theta_{k+1} - 2\varepsilon)\}$$

and let $g(\zeta) = f(z)$. Then $g(\zeta)$ is regular on the upper semicircle: $|\zeta| < 1$ with $I_m\zeta > 0$. Moreover, according to formula (5.44), $\log|g(\zeta) - a_i| \leq 0$, and from formula (5.45), when $|\zeta| < 1$ with $I_m\zeta = 0$, $\log|g(\zeta) - a_i| \leq -AT(t_m, f)$. Furthermore, applying Lemma 3.2, we conclude that when $|\zeta| \leq \frac{1}{2}$ with $\arg\zeta = \frac{\pi}{2}$,

$$\log|g(\zeta) - a_i| \leq \frac{1}{3}(-AT(t_m, f)) \leq -AT(t_m, f), \qquad A > 0.$$

Hence, when $m \geq m_{10}$ and $z \in \Gamma(\theta_{k+1} + 2\varepsilon, \theta_{k+1} + 3\varepsilon; t_m)$,

$$\log|f(z) - a_i| \leq -AT(t_m, f),$$

where $A > 0$ is a constant independent of m. Analogously, applying Lemma 5.2, we conclude that there exists value m_{11}, $m_{11} \geq m_{10}$, such that when $m \geq m_{11}$ and

$$z \in \overline{\Omega}(\theta_{k+1} + 2\varepsilon, \theta_{k+2} - 2\varepsilon; e^{-30\varepsilon}t_m, e^{30\varepsilon}t_m),$$

$$\log|f(z) - a_i| \leq -AT(t_m, f),$$

where $A > 0$ is a constant independent of m. Repeating the above discussion $s - 1$ times, we conclude that there exists value m_{12}, $m_{12} \geq m_{11}$, such that when $m \geq m_{12}$ and $z \in \overline{\Omega}(\theta_{k+s-1} + 2\varepsilon, \theta_{k+s} - 2\varepsilon; e^{-30\varepsilon}t_m, e^{30\varepsilon}t_m)$,

$$\log|f(z) - a_i| \leq -AT(t_m, f),$$

where $A > 0$ is a constant independent of m. Applying further Lemma 5.2, we conclude that when $m \geq m_{12}$ and

$$z \in \overline{\Omega}(\theta_{k+s-1} + 2\varepsilon, \theta_{k+s} - 2\varepsilon; t_{m-1}, t_{m+1}),$$

$$\log|f(z) - a_i| \le -At_m^{-((31\times8\pi)/\omega+1)\eta}T(t_m, f),$$

where $A > 0$ is a constant independent of m. In the following, analogous to the discussion of 1) and 2), we conclude that L_{j+1m} cannot intersect at $\overline{\Omega}(\theta_{k+s} - 3\varepsilon, \theta_{k+s+1} - 3\varepsilon; t_{m-1}, t_{m+1})$, and hence, we derive a contradiction. Hence, when $m \ge m_{12}$, there exists a point z_1 in

$$\Omega(\theta_{k+1} - 2\varepsilon, \theta_{k+s-1} + 8\varepsilon; e^{-10\varepsilon}t_m, e^{10\varepsilon}t_m)$$

such that

$$|f(z_1) - a_i| \ge 1. \tag{5.46}$$

We may assume that when $z \in \Omega(\theta_{k+2} - 2\varepsilon, \arg z_1; e^{-10\varepsilon}t_m, e^{10\varepsilon}t_m)$, $\log|f(z) - a_i| \le 0$. On the other hand, according to the above discussion, we may find that if point z_1 is in $\Omega(\theta_{k+s'} - 2\varepsilon, \theta_{k+s'+1} - 2\varepsilon; e^{-10\varepsilon}t_m, e^{10\varepsilon}t_m)$ $(1 \le s' \le s - 1)$, then point z_1 must be in

$$\overline{\Omega}(\theta_{k+s'} - 2\varepsilon, \theta_{k+s'} + 8\varepsilon; e^{-30\varepsilon}t_m, e^{30\varepsilon}t_m).$$

4) We prove that when m is sufficiently large, there exists a point z_{jms_j+1} in $\overline{\Omega}(\theta_{k+s'} - 2\varepsilon, \theta_{k+s'} + 18\varepsilon; e^{-30\varepsilon}t_m, e^{10\varepsilon}t_m)$ such that

$$\log|f(z_{jms_j+1})| \ge AT(t_m, f),$$

where $A > 0$ is a constant independent of m.

First, when $m \ge m_{12}$ and $z \in \Delta(\theta_{k+s'} - 2\varepsilon; e^{-30\varepsilon}t_m, e^{30\varepsilon}t_m)$,

$$\log|f(z) - a_i| \le -AT(t_m, f), A > 0. \tag{5.47}$$

Next, we assume a point $z_{jms_j+1} \in \overline{\Omega}(\theta_{k+s'} - 2\varepsilon, \theta_{k+s'} + 18\varepsilon; e^{-30\varepsilon}t_m, e^{30\varepsilon}t_m)$ such that when $z \in \overline{\Omega}(\theta_{k+s'} - 2\varepsilon, \theta_{k+s'} + 18\varepsilon; e^{-30\varepsilon}t_m, e^{30\varepsilon}t_m)$,

$$|f(z) - a_i| \le |f(z_{jms_j+1}) - a_i|. \tag{5.48}$$

We construct the transformation

$$\zeta = \frac{1}{20\varepsilon}\{\log z - \log|z_1| - i(\theta_{k+s'} - 2\varepsilon)\}$$

and let $g(\zeta) = f(z)$, $g(\zeta_1) = f(z_1)$, where ζ_1 is the image point of point z_1 on the ζ-plane. Then $g(\zeta)$ is regular on the upper semicircle: $|\zeta| < 1$ with $I_m\zeta > 0$. Moreover, according to formula (5.48), we have

$$\log|g(\zeta) - a_i| \le \log|f(z_{jms_j'+1}) - a_i|.$$

On the other hand, from formula (5.47), when $|\zeta| < 1$ with $I_m\zeta = 0$,

$$\log|g(\zeta) - a_i| \le -AT(t_m, f), \qquad A > 0.$$

Meanwhile, according to formula (5.46), $|\log(g(\zeta_1) - a_i| \ge 0$. Notice when $|\zeta_1| \le \frac{1}{2}$ with $\arg\zeta_1 = \frac{\pi}{2}$; applying Lemma 3.2, we conclude that

$$0 \le \log|g(\zeta_1) - a_i| \le \log|f(z_{jms_j+1}) - a_i| + \frac{1}{3}(-AT(t_m, f))$$

$$\le \log^+|f(z_{jms_j+1})| + \log 2M_1 - AT(t_m, f),$$

where $A > 0$ is a constant independent of m. Hence, there exists value m_{13}, $m_{13} \geq m_{12}$, such that when $m \geq m_{13}$, $\log|f(z_{jms_j+1})| \geq AT(t_m, f)$, where $A > 0$ is a constant independent of m. Up to now, we have proved that there exists a point z_{jms_j+1} in $D_{jms_j+1} \cap (e^{-30\varepsilon}t_m \leq |z| \leq e^{30\varepsilon}t_m)$ and a constant $A > 0$ which is independent of m, such that

$$\log|f(z_{jms_j+1})| \geq AT(t_{m_1}, f). \tag{5.49}$$

5) According to the above discussion, we find that the determination of value m_{13} and value A in formula (5.49), apart from relying on some invariant parameters M_0, M_1, M_2, d, δ, ω, η and ε, depends mainly on s' and s. Moreover, the larger the value of s' and s, the larger the value m_{13} and the smaller the value A. However, for each domain $D_{jm\nu}$ $(1 \leq \nu \leq s_j + 1, \ s_j \neq 0, \ 1 \leq j \leq l_1)$, the corresponding values s' and s satisfy $1 \leq s' \leq s - 1 < q < +\infty$. Hence, there exists value m_{14}, $m_{14} \geq m_{13}$ and value A_0, $0 < A_1 < 1$, such that when $m \geq m_{14}$,

$$\left.\begin{array}{l} \log|f(z_{jm\nu})| \geq A_0 T(t_m, f), \\[4pt] z_{jm\nu} \in D_{jm\nu} \cap (e^{-30^\varepsilon}t_m \leq |z| \leq e^{30^\varepsilon}t_m), \\[4pt] \nu = 1, 2, \ldots, s_j + 1, \ s_j \neq 0, \\[4pt] j = 1, 2, \ldots, l_1. \end{array}\right\} \tag{5.50}$$

(5) Suppose that $s_j \neq 0$ $(1 \leq j \leq l_1)$ and there exists value $A_0' T(t_m, f)$ in the interval $[\frac{1}{4}A_0 T(t_m, f), \frac{1}{2}A_0 T(t_m, f)]$ $(m \geq m_{14})$ such that the level curve $\log|f(z)| = A_0' T(t_m, f)$ is analytic. We consider the set

$$E = \{z \,|\, \log|f(z)| > A_0' T(t_m, f), \ |z| < t_{m+1}\}$$

and let $\Omega'_{jm\nu}$ be the connected branch containing point $z_{jm\nu}$ $(1 \leq \nu \leq s_j + 1)$. According to the maximum modulus principle, we conclude that $\overline{\Omega}'_{jm\nu} \cap (|z| = t_{m+1})$ is not an empty set and it contains at least one interior point.

1) We prove that when m is sufficiently large, $\Omega'_{jm\nu} \subset D_{jm\nu}$ $(j = 1, 2, \ldots, l_1; \ \nu = 1, 2, \ldots, s_j + 1, \ s_j \neq 0)$. Hence, $\{\Omega'_{jm\nu}\}$ do not intersect among one another.

First, according to $p_1 \geq 1$, $f(z)$ has at least two deficient values a_1 and ∞. Hence, according to Lemma 3.7, there exists a positive number $\tau = \tau(\delta(a_1, f), \ \delta(\infty, f))$ and for any arbitrary value σ, $1 \leq \sigma < +\infty$,

$$\varliminf_{t \to +\infty} \frac{T(\sigma t, f)}{T(t, f)} \geq \sigma^\tau.$$

If we let $\sigma = \sqrt[\tau]{24/A_0}$, $\sigma t = t_m$, then there exists value m_{15}, $m_{15} \geq m_{14}$, such that when $m \geq m_{15}$,

$$T(t_m, f) \geq \frac{12}{A_0} T(\sigma^{-1}t_m, f) > \frac{12}{A_0} T(2t_{m-1}, f).$$

And hence,

$$A_0'T(t_m, f) \geq \frac{A_0}{4}T(t_m, f) \geq 3T(2t_{m-1}, f) \qquad (5.51)$$
$$\geq \log M(t_{m-1}, f).$$

On the other hand, according to formulas (5.39)–(5.42), when $m \geq m_{15}$, and

$$z \in \bigcup_{j=1}^{l_1} L_{jm} \bigcup_{i=1}^{P_1} \Delta\left(\frac{\theta_{k_i} + \theta_{k_i+1}}{2}; t_{m-1}, t_{m+1}\right),$$

$\log|f(z)| \leq \log 2M_1$. Obviously, there exists value m_{16}, $m_{16} \geq m_{15}$, such that when $m \geq m_{16}$,

$$A_0'T(t_{m_1}, f) \geq \frac{A_0}{4}T(t_m, f) > \log 2M_1. \qquad (5.52)$$

Hence, according to formulas (5.51) and (5.52), as well as the definition of $\Omega_{jm\nu}'$, we conclude that when $m \geq m_{16}$, $\Omega_{jm\nu}' \subset D_{jm\nu}$.

2) We prove that when m is sufficiently large, $\Omega_{jm\nu}' \subset D_j(t_{m+1})$. First, there exists value m_{17}, $m_{17} \geq m_{16}$, such that when $m \geq m_{17}$,

$$A_0'T(t_m, f) \geq \frac{A_0}{4}T(t_m, f) > \log M_0. \qquad (5.53)$$

Hence, according to formula (5.32), we conclude that $\Omega_{jm\nu}'$ does not intersect L_j ($j = 1, 2, \ldots, l_1$). Next, according to $\Omega_{jm\nu}' \subset D_{jm\nu}$, we conclude that $\overline{\Omega}_{jm\nu}' \cap (|z| = t_{m+1}) \subset \overline{D}_{jm\nu} \cap (|z| = t_{m+1}) = \overline{D}_j(t_{m+1}) \cap (|z| = t_{m+1})$. Besides, since $\overline{\Omega}_{jm\nu}' \cap (|z| = t_{m+1})$ is not an empty set and it contains interior points, we conclude that there exists point $z_2 \in \Omega_{jm\nu}' \cap D_j(t_{m+1})$. Hence, when $m \geq m_{17}$, $\Omega_{jm\nu}' \subset D_j(t_{m+1})$.

3) Let $\Omega_{jm\nu} \subset \Omega_{jm\nu}'$ be the connected component of $\Omega_{jm\nu}'$ in the disk $|z| < \frac{1}{2}t_{m+1}$ and containing point $z_{jm\nu}$. Then $\{\Omega_{jm\nu}\}$ do not intersect among one another. We also let $r\theta_{jm\nu}(r)$ be the linear measure of circumference $|z| = r$ ($|z_{jm\nu}| \leq r \leq \frac{1}{4}t_{m+1}$; $m \geq m_{17}$) in the part of $\Omega_{jm\nu}$. Applying Theorem 3.1, we have

$$A_0T(t_m, f) \leq \log|f(z_{jm\nu})| \leq \frac{A_0}{2}T(t_m, f)$$
$$+ 9\sqrt{2}\exp\left\{-\pi\int_{2|z_{jm\nu}|}^{t_{m+1}/4}\frac{dr}{r\theta_{jm\nu}(r)}\right\}\log M\left(\frac{1}{2}t_{m+1}, f\right).$$

Furthermore, according to formula (5.50), we get

$$\pi\int_{2e^{30\varepsilon}t_m}^{\frac{1}{4}t_{m+1}}\frac{dr}{r\theta_{jm\nu}(r)} \leq \log T(t_{m+1}, f)$$
$$- \log T(t_m, f) + \log\frac{54\sqrt{2}}{A_0}. \qquad (5.54)$$

4) When $m \geq m_{17}$, according to formula (5.53), $\Omega_{jm\nu}$ may lie entirely in $\Omega_j(\frac{1}{2}t_{m+1})$, or it may have no intersection with $\Omega_j(\frac{1}{2}t_{m+1})$. Suppose that there exists domain $\Omega_{jm\nu}$ $(1 \leq \nu_j \leq s_j + 1)$, such that $\Omega_{jm\nu_j}$ lies entirely in $\Omega_j(\frac{1}{2}t_{m+1})$. We simply let $\Omega_{jm\nu_j} = \Omega_{jm}$ and $\theta_{jm\nu_j}(r) = \theta_{jm}(r)$; then applying Theorem 3.1, we have

$$A_0 T(t_m, f) \leq \log |f(z_{jm\nu_j})|$$

$$\leq \frac{A_0}{2} T(t_m, f) + 9\sqrt{2} \exp\left\{ -\pi \int_{2e^{30}t_m}^{\frac{1}{4}t_{m+1}} \frac{dt}{r\theta_{jm}(r)} \right\}$$

$$\times \log M\left\{ \overline{\Omega}_j\left(\frac{1}{2}t_{m+1}\right) \cap \left(|z| = \frac{1}{2}t_{m+1}\right), f \right\}.$$

Furthermore, we get

$$\pi \int_{2e^{30\varepsilon}t_m}^{\frac{1}{2}t_{m+1}} \frac{dr}{r\theta_{jm}(r)} \leq \log\log M$$

$$\times \left\{ \overline{\Omega}_j\left(\frac{1}{2}t_{m+1}\right) \cap \left(|z| + \frac{1}{2}t_{m+1}\right), f \right\} \tag{5.55}$$

$$- \log\log M\left\{ \overline{\Omega}_j\left(\frac{1}{2}t_m\right) \cap \left(|z| = \frac{1}{2}t_m\right), f \right\} + \log\frac{54\sqrt{2}}{A_0}.$$

5) Suppose that the $s_j + 1$ domains $\Omega_{jm\nu}$ $(\nu = 1, 2, \ldots, s_j + 1)$ have no intersection with $\Omega_j(\frac{1}{2}t_{m+1})$. First, we assume arbitrarily a value ν_j, $1 \leq \nu_j \leq s_j + 1$. Then, we determine a new domain Ω_{jm} to replace $\Omega_{jm\nu_j}$. Concretely speaking, the definition of Ω_{jm} is as follows: We first assume a point $z_{jm} \in \overline{\Omega}_j(t_m) \cap (|z| = t_m)$ such that

$$|f(z_{jm})| = M\{\overline{\Omega}_j(t_m) \cap (|z| = t_m)f\}.$$

Obviously, point z_{jm} is an interior point of $\Omega_j(\frac{1}{2}t_{m+1})$. Then, we assume value A_0'' in the interval

$$\left[\sqrt[4]{M\{\overline{\Omega}_j(t_m) \cap (|z| = t_m), f\}}, \quad \sqrt{M\{\overline{\Omega}_j(t_m) \cap (|z| = t_m), f\}} \right],$$

such that the level curve $|f(z)| = A_0''$ is analytic. We consider the set

$$E = \left\{ z \mid |f(z)| > A_0'', \ |z| < \frac{1}{2}t_{m+1} \right\}$$

and let Ω_{jm} be the connected component containing point z_{jm}; then according to the maximum modulus principle, the set $\overline{\Omega}_{jm} \cap (|z| = \frac{1}{2}t_{m+1})$ is not an empty set and it contains an interior point. On the other hand, based on formula (5.36), there exists value m_{18}, $m_{18} \geq m_{17}$, such that when $m \geq m_{18}$,

$$A_0'' \geq \sqrt[4]{M\{\overline{\Omega}_j(t_m) \cap (|z| = t_m), f\}} > M_0.$$

Hence, when $m \geq m_{18}$, $\Omega_{jm} \subset \Omega_j(\frac{1}{2}t_m)$. Moreover, Ω_{jm} has no intersection with $\Omega_{jm\nu}$ $(1 \leq \nu \leq s_j + 1)$. We let $r\theta_{jm}(r)$ be the linear measure of circumference $|z| = r$ $(t_m \leq r \leq \frac{1}{4}t_{m+1})$ in the part of Ω_{jm}. Applying Theorem 3.1, we have

$$
\log M\{\overline{\Omega}_j(t_m) \cap (|z| = t_m), f\}
$$
$$
= \log|f(z_{jm})|
$$
$$
\leq \tfrac{1}{2} \log M\{\overline{\Omega}_j(t_m) \cap (|z| = t_m), f\}
$$
$$
+ 9\sqrt{2} \exp\left\{-\pi \int_{2t_m}^{t_{m+1}/4} \frac{dr}{r\theta_{jm}(r)}\right\} \log M
$$
$$
\times \left\{\overline{\Omega}_j\left(\tfrac{1}{2}t_{m+1}\right) \cap \left(|z| = \tfrac{1}{2}t_{m+1}\right), f\right\}.
$$

Furthermore, we get

$$
\pi \int_{2e^{30\varepsilon}t_m}^{t_{m+1}/4} \frac{dr}{r\theta_{jm}(r)} \leq \log\log M
$$
$$
\times \left\{\overline{\Omega}_j\left(\tfrac{1}{2}t_{m+1}\right) \cap \left(|z| = \tfrac{1}{2}t_{m+1}\right), f\right\}
$$
$$
- \log\log M\{\overline{\Omega}_j(t_m) \cap (|z| = t_m), f\} + \log 18\sqrt{2}
$$
$$
\leq \log\log M \left\{\overline{\Omega}_j\left(\tfrac{1}{2}t_{m+1}\right) \cap \left(|z| = \tfrac{1}{2}t_{m+1}\right), f\right\}
$$
$$
- \log\log M\{\overline{\Omega}_j(t_m) \cap (|z| = t_m), f\}
$$
$$
+ \log \frac{54\sqrt{2}}{A_0}.
$$

Hence, we re-obtain formula (5.55) in form.

From now on, when $m \geq m_{18}$ and $s_j \neq 0$ $(1 \leq j \leq l_1)$, if we let $1 \leq \nu \leq \nu_j \leq s_j + 1$, then we assume formula (5.54); if we let $\nu = \nu_j$, then we assume formula (5.55).

6) Let $s_j = 0$ $(1 \leq j \leq l_1)$. Analogously, we may define region $\Omega_{jm} \subset \Omega_j(\frac{1}{2}t_{m+1}) \subset \Omega_j(t_{m+1}) \subset D_j(t_{m+1})$ and derive the following inequality:

$$
\pi \int_{2e^{30\varepsilon}t_m}^{t_{m+1}/4} \frac{dr}{r\theta_{jm}(r)} \leq \log\log M
$$
$$
\times \left\{\overline{\Omega}_j\left(\tfrac{1}{2}t_{m+1}\right) \cap \left(|z| = \tfrac{1}{2}t_{m+1}\right), f\right\} \quad (5.56)
$$
$$
- \log\log M\{\overline{\Omega}_j(t_m) \cap (|z| = t_m), f\}
$$
$$
+ \log \frac{54\sqrt{2}}{A_0}.
$$

7) According to the above discussion, we find that when $m \geq m_{18}$ and

$$2e^{30\varepsilon} t_m \leq r \leq \frac{1}{4} t_{m+1},$$

if we let

$$\sum_{1 \leq j \leq l_1} \theta_{jm}(r) = \sum_{\substack{1 \leq j \leq l_i \\ s_j = 0}} \theta_{jm}(r) + \sum_{\substack{1 \leq j \leq l_1 \\ s_j \neq 0 \\ \nu = \nu_j}} \theta_{jm\nu}(r),$$

then

$$\sum_{1 \leq j \leq l_1} \theta_{jm}(r) + \sum_{\substack{1 \leq j \leq l_1 \\ s_j \neq 0 \\ 1 \leq \nu \neq \nu_j \leq s_j + 1}} \theta_{jm\nu}(r) \leq 2\pi.$$

Hence, we obtain

$$(p_1 + l_1)^2 = \left\{ \sum_{\substack{1 \leq j \leq l_1 \\ s_j \neq 0 \\ 1 \leq \nu \neq \nu_j \leq s_j + 1}} \sqrt{\theta_{jm\nu}(r)} \cdot \frac{1}{\sqrt{\theta_{jm\nu}(r)}} \right.$$

$$\left. + \sum_{1 \leq j \leq l_1} \sqrt{\theta_{jm}(r)} \cdot \frac{1}{\sqrt{\theta_{jm}(r)}} \right\}^2$$

$$\leq 2\pi \left\{ \sum_{\substack{1 \leq j \leq l_1 \\ s_j \neq 0 \\ 1 \leq \nu \neq \nu_j \leq s_j + 1}} \frac{1}{\theta_{jm}(r)} + \sum_{1 \leq j \leq l_1} \frac{1}{\theta_{jm}(r)} \right\},$$

$$\frac{1}{2} (p_1 + l_1)^2 \int_{2e^{30\varepsilon} t_m}^{t_{m+1}/4} \frac{dr}{r}$$

$$\leq \sum_{\substack{1 \leq j \leq l_1 \\ s_j \neq 0 \\ 1 \leq \nu \neq \nu_j \leq s_j + 1}} \pi \int_{2e^{30\varepsilon} t_m}^{t_{m+1}/4} \frac{dr}{r \theta_{jm\nu}(r)}$$

$$+ \sum_{1 \leq j \leq l_1} \pi \int_{2e^{30\varepsilon} t_m}^{t_{m+1}/4} \frac{dr}{r \theta_{jm}(r)}.$$

Moreover, according to formulas (5.54)–(5.56), we derive

$$\frac{1}{2}(p_1 + l_1)^2 \left(\log \frac{t_{m+1}}{t_m} - \log 8e^{30\varepsilon} \right)$$

$$\leq p_1 (\log T(t_{m+1}, f) - \log T(t_m, f)) + p_1 \log \frac{54\sqrt{2}}{A_0}$$

$$+ \sum_{1 \leq j \leq l_1} \left\{ \log\log M \left\{ \overline{\Omega}_j \left(\frac{1}{2} t_{m+1} \right) \cap \left(|z| = \frac{1}{2} t_{m+1} \right), f \right\} \right. \qquad (5.57)$$

$$\left. - \log\log M \left\{ \overline{\Omega}_j \left(\frac{1}{2} t_m \right) \cap \left(|z| = \frac{1}{2} t_m \right), f \right\} \right\}$$

$$+ l_1 \log \frac{54\sqrt{2}}{A_0}.$$

Now, suppose that sequence R_n $(n = 1, 2, \dots)$ satisfies the condition

$$\lim_{n \to +\infty} \frac{\log T(R_n f)}{\log R_n} = \mu.$$

Also, for each sufficiently large value n, there exists value m_n such that

$$t_{m_n - 1} \leq R_n \leq t_{m_n}. \qquad (5.58)$$

Then, we have

$$r_0^{(1+\eta)^{2m_n - 2}} = r_{m_n - 1} \leq t_{m_n - 1} \leq R_n,$$

$$m_n \leq \frac{\log\log R_n - \log\log r_0}{2\log(1+\eta)} + 1. \qquad (5.59)$$

Meanwhile, we have

$$t_{m_n - 1} \geq r_0^{(1+\eta)^{2m_n - 2}} = \left(\frac{1}{2} \right)^{1/(1+\eta)^3} (2r_{m_n}^{1+\eta})^{1/(1+\eta)^3}$$

$$\geq \left(\frac{1}{2} \right)^{1/(1+\eta)^3} (t_m)^{1/(1+\eta)^3} \geq \left(\frac{1}{2} R_n \right)^{1/(1+\eta)^3}. \qquad (5.60)$$

Let $m = m_{18}, m_{18} + 1, \dots, m_n - 2$. Then according to formulas (5.57)–(5.60), we obtain

$$\frac{1}{2}(p_1 + l_1)^2 \left\{ \log \frac{t_{m_n - 1}}{t_{m_{18}}} - (m_n - m_{18} - 1) \log 8e^{30\varepsilon} \right\}$$

$$\leq p_1 \{ \log T(t_{m_n - 1}, f) - \log T(t_{m_{18}}, f) \}$$

$$+ \sum_{1 \leq j \leq l_1} \left\{ \log\log M \left\{ \overline{\Omega}_j \left(\frac{1}{2} t_{m_n - 1} \right) \cap \left(|z| = \frac{1}{2} t_{m_n - 1} \right), f \right\} \right.$$

$$\left. - \log\log M \left\{ \overline{\Omega}_j \left(\frac{1}{2} t_{m_{18}} \right) \cap \left(|z| = \frac{1}{2} t_{m_{18}} \right), f \right\} \right\}$$

$$+ (p_1 + l_1)(m_n - m_{18} - 1) \log \frac{54\sqrt{2}}{A_0},$$

$$\frac{1}{2}(p_1 + l_1)^2 \left\{ \frac{1}{(1+\eta)^3} \log R_n - \frac{1}{(1+\eta)^2} \log 2 \right.$$

$$\left. - \log t_{m_{18}} - \frac{\log \log R_n - \log \log r_0}{2 \log(1+\eta)} \log 8 e^{30\varepsilon} \right\}$$

$$\leq (p_1 + l_1) \log T(R_n, f) + l_1 \log 3 + (p_1 + l_1)$$

$$\times \left(\frac{\log \log R_n - \log \log r_0}{2 \log(1+\eta)} + 1 \right) \log \frac{54\sqrt{2}}{A_0}.$$

We use $\log R_n$ to divide the two sides of the last inequality, and then when $n \to +\infty$, we conclude that

$$\frac{(p_1 + l_1)^2}{2(1+\eta)^3} \leq (p_1 + l_1)\mu \quad \text{or} \quad (p_1 + l_1) \leq 2(1+\eta)^3 \mu.$$

Finally, we let $\eta \to 0$; then $p_1 + l_1 \leq 2\mu$. However, this contradicts formula (5.31). Hence, Theorem 5.6 is proved completely.

§5.4. Extremal length and Ahlfors Distortion Theorem[5]

5.4.1. Extremal length. Let Γ be a family of curves γ on the open plane $|z| < +\infty$, and each $\gamma \in \Gamma$ is made up of countable open arcs or closed arcs or closed curves. Meanwhile, each closed subarc is rectifiable.

Let $\rho(z)$ be defined as a real valued function on the whole plane $|z| \leq +\infty$. If $\rho(z)$ $(z = x + iy)$ satisfies the conditions

(1) $\rho(z) \geq 0$ is a measurable function,

(2) $A(\rho) = \iint_{|z| \leq +\infty} \rho^2(z) \, dx \, dy \neq 0, \infty$,

then $\rho(z)$ is called an admissible function.

Let $\rho(z)$ be an admissible function, if $\rho(z)$, as a function of the arc length of γ, is measurable, then we define

$$L_\gamma(\rho) = \int_\gamma \rho |dz|.$$

Otherwise, we define $L_\gamma(\rho) = +\infty$. Furthermore, we let

$$L(\Gamma, \rho) = \inf_{\gamma \in \Gamma} L_\gamma(\rho)$$

and define

$$\lambda(\Gamma) = \sup_\rho \frac{L^2(\Gamma, \rho)}{A(\rho)},$$

where ρ assumes the whole admissible functions. In the sequel, we call $\lambda(\Gamma)$ the extremal length of Γ.

---

[5] The content of this section is derived from books [1e, 1f].

THEOREM 5.7. *If each $\gamma_2 \in \Gamma_2$ contains one $\gamma_1 \in \Gamma_1$, then $\lambda(\Gamma_2) \geq \lambda(\Gamma_1)$.*

PROOF. Obviously, if $\gamma_2 \supset \gamma_1$, then

$$L_{\gamma_1}(\rho) \leq L_{\gamma_2}(\rho), \qquad L(\Gamma_1, \rho) \leq L(\Gamma_2, \rho).$$

Hence, $\lambda(\Gamma_1) \leq \lambda(\Gamma_2)$.

EXAMPLE 1. Let R be a closed rectangle with length $a > 0$ and width $b > 0$, and Γ be a set of all arcs linking the two opposite sides on R, with the length of each side being b. Then for any admissible function ρ, we have

$$L(\Gamma, \rho) \leq \int_0^a \rho(x + iy)\, dx,$$

$$bL(\Gamma, \rho) \leq \iint_R \rho(x + iy)\, dx\, dy,$$

$$b^2 L^2(\Gamma, \rho) \leq ab \iint_R \rho^2\, dx\, dy \leq ab A(\rho),$$

$$\frac{L^2(\Gamma, \rho)}{A(\rho)} \leq \frac{a}{b}.$$

Hence, $\lambda(\Gamma) \leq \frac{a}{b}$. On the other hand, we assume

$$\rho(z) = \begin{cases} 1, & z \in R, \\ 0, & z \notin R. \end{cases}$$

Then $L(\Gamma, \rho) = a$ and $A(\rho) = ab$. Accordingly, $\lambda(\Gamma) \geq \frac{a}{b}$. Hence, we have $\lambda(\Gamma) = \frac{a}{b}$.

EXAMPLE 2. Let Γ be the set of all arcs linking two circumferences $|z| = r_1$ and $|z| = r_2$ in the closed annulus $r_1 \leq |z| \leq r_2$. Hence, for any admissible function ρ, it follows that

$$L(\Gamma, \rho) \leq \int_{r_1}^{r_2} \rho\, dr,$$

$$2\pi L(\Gamma, \rho) \leq \iint \rho\, dr\, d\theta,$$

$$4\pi^2 L^2(\Gamma, \rho) \leq 2\pi \log \frac{r_2}{r_i} \iint \rho^2 r\, dr\, d\theta \leq 2\pi \log \frac{r_2}{r_1} A(\rho),$$

$$\frac{L^2(\Gamma, \rho)}{A(\rho)} \leq \frac{1}{2\pi} \log \frac{r_2}{r_1}.$$

And hence,

$$\lambda(\Gamma) \leq \frac{1}{2\pi} \log \frac{r_2}{r_1}.$$

On the other hand, we assume $\rho = \frac{1}{r}$; then $\lambda(\Gamma) \geq \frac{1}{2\pi} \log \frac{r_2}{r_1}$. Hence,

$$\lambda(\Gamma) = \frac{1}{2\pi} \log \frac{r_2}{r_1}.$$

EXAMPLE 3. Let G be a doubly-connected domain on the open plane $|z| < +\infty$, and C_1 and C_2 be the bounded and unbounded components of the two complementary sets of G, respectively. If for some closed curve γ in G corresponding to some point in C_1 has a nonzero winding number, then we call γ dividing C_1 and C_2. Let Γ be the set of all closed curves dividing C_1 and C_2 in G. Particularly, when G stands for annulus $r_1 < |z| < r_2$, for any admissible function ρ, we have

$$L(\Gamma, \rho) \leq \int_0^{2\pi} \rho(re^{i\theta}) r \, d\theta,$$

$$\frac{L(\Gamma, \rho)}{r} \leq \int_0^{2\pi} \rho(re^{i\theta}) \, d\theta,$$

$$L(\Gamma, \rho) \log \frac{r_2}{r_1} \leq \int_{r_1}^{r_2} \int_0^{2\pi} \rho(re^{i\theta}) \, d\theta \, dr,$$

$$L^2(\Gamma, \rho) \left(\log \frac{r_2}{r_1} \right)^2 \leq 2\pi \log \frac{r_2}{r_1} \iint \rho^2 \, dx \, dy$$

$$\leq 2\pi \log \frac{r_2}{r_1} A(\rho),$$

$$\frac{L^2(\Gamma, \rho)}{A(\rho)} \leq \frac{2\pi}{\log(r_2/r_1)}.$$

Hence,

$$\lambda(\Gamma) \leq \frac{2\pi}{\log(r_2/r_1)}.$$

On the other hand, we assume $\rho = \frac{1}{2\pi r}$; then $A(\rho) + \frac{1}{2\pi} \log(r_2/r_1)$, and according to the winding number of $\gamma \in \Gamma$ about $z = 0$,

$$n(\gamma, 0) = \frac{1}{2\pi} \int_\gamma \frac{dz}{z} \neq 0,$$

we have

$$1 \leq \left| \frac{1}{2\pi} \int_\gamma \frac{dz}{z} \right| \leq \frac{1}{2\pi} \int_\gamma \frac{|dz|}{r} = L_\gamma(\rho).$$

Particularly, we take that $\gamma' \in \Gamma$ to be circumference $|z| = r$, $r_1 < r < r_2$; then $L_{\gamma'}(\rho) = 1$. Hence, $L(\Gamma, \rho) = 1$, and therefore, we conclude that

$$\lambda(\Gamma) \geq \frac{2\pi}{\log(r_2/r_1)}.$$

Accordingly, we obtain

$$\lambda(\Gamma) = \frac{2\pi}{\log(r_2/r_1)}.$$

In the following, we prove the most important property regarding the extremal length $\lambda(\Gamma)$.

THEOREM 5.8. $\lambda(\Gamma)$ *is a conformal invariant.*

PROOF. Let Γ be in a certain region Ω, and $\zeta = \xi + i\eta = \varphi(z)$ maps Ω conformally onto region Ω'. Let Γ' be the image set of Γ. For any arbitrarily assumed admissible function ρ, we define

$$\rho_1(\zeta) = \begin{cases} \rho(\varphi^{-1}(\zeta))|d\varphi^{-1}(\zeta)/d\zeta|, & \zeta \in \Omega', \\ 0, & \zeta \notin \Omega'. \end{cases}$$

And hence,

$$\iint_{\Omega'} \rho_1^2 \, d\xi \, d\eta = \iint_{\Omega} \rho^2 \, dx \, dy \le A(\rho),$$

$$\int_{\gamma'} \rho_1 |d\zeta| = \int_{\gamma} \rho |dz|,$$

where l' is the image of $\gamma \in \Gamma$. Therefore, we conclude that $\lambda(\Gamma') \ge \lambda(\Gamma)$. On the other hand, for a given $\rho_1(\zeta)$, taking into account the definition

$$\rho(z) = \begin{cases} \rho_1(\varphi(z))|d\varphi(z)/dz|, & z \in \Omega, \\ 0, & z \notin \Omega, \end{cases}$$

we conclude that $\lambda(\Gamma) \ge \lambda(\Gamma')$. Hence, we have $\lambda(\Gamma) = \lambda(\Gamma')$, meaning that $\lambda(\Gamma)$ is a conformal invariant.

5.4.2. Rule of composition and the symmetry principle. Let Ω_1 and Ω_2 be two open sets not intersecting with each other, and Γ_1 and Γ_2 are two sets of arcs in Ω_1 and Ω_2. We also let Γ be the third set of arcs. We prove the following rule of composition:

THEOREM 5.9. *If each* $\gamma \in \Gamma$ *contains one* $\gamma_1 \in \Gamma_1$ *and one* $\gamma_2 \in \Gamma_2$, *then*

$$\lambda(\Gamma) \ge \lambda(\Gamma_1) + \lambda(\Gamma_2). \tag{5.61}$$

If each $\gamma_1 \in \Gamma_1$ *and each* $\gamma_2 \in \Gamma_2$ *contains one* $\gamma \in \Gamma$, *then*

$$\frac{1}{\lambda(\Gamma)} \ge \frac{1}{\lambda(\Gamma_1)} + \frac{1}{\lambda(\Gamma_2)}. \tag{5.62}$$

PROOF. We may assume that $\lambda(\Gamma_1)$ and $\lambda(\Gamma_2)$ are not zeros and ∞. Otherwise, formulas (5.61) and (5.62) are direct corollaries of Theorem 5.7.

First, we prove formula (5.61). Generally, for any arbitrarily assumed admissible function $\rho(z)$ and family of curves $\widetilde{\Gamma}$, if $L(\widetilde{\Gamma}, \rho) \ne 0, \infty$, we let

$$\rho'(z) = K\rho(z), \qquad K = \frac{L(\widetilde{\Gamma}, \rho)}{A(\rho)};$$

then $L(\widetilde{\Gamma}, \rho') = A(\rho')$. Hence, for the arbitrarily selected admissible functions $\rho_1(z)$ and $\rho_2(z)$ satisfying the conditions $L(\Gamma_1, \rho_1) \ne 0, \infty$ and

$L(\Gamma_2, \rho_2) \neq 0, \infty$, if we let

$$\rho_1'(z) = K_1\rho_1(z), \qquad K_1 = \frac{L(\Gamma_1, \rho_1)}{A(\rho_1)},$$

$$\rho_2'(z) = K_2\rho_1(z), \qquad K_2 = \frac{L(\Gamma_2, \rho_2)}{A(\rho_2)},$$

$$\rho'(z) = \max[\rho_1'(z), \rho_2'(z)],$$

then

$$L(\Gamma, \rho') \geq L(\Gamma_1, \rho_1') + L_2(\Gamma_2, \rho_2') = A(\rho_1') + A(\rho_2'),$$
$$A(\rho') \leq A(\rho_1') + A(\rho_2'),$$
$$\lambda(\Gamma) \geq A(\rho_1') + A(\rho_2') = K_1^2 A(\rho_1) + K_2^2 A(\rho_2)$$
$$= \frac{L^2(\Gamma_1, \rho_1)}{A(\rho_1)} + \frac{L^2(\Gamma_2, \rho_2)}{A(\rho_2)}.$$

When $L(\Gamma_1, \rho_1)$ or $L(\Gamma_2, \rho_2)$ equals zero or ∞, the last inequality above is obvious. Therefore, according to the arbitrary selection of $\rho_1(z)$ and $\rho_2(z)$, we conclude that $\lambda(\Gamma) \geq \lambda(\Gamma_1) + \lambda(\Gamma_2)$, meaning that formula (5.61) holds.

Next, we prove formula (5.62). We may assume that $\lambda(\Gamma) \neq 0$. For the arbitrarily selected admissible function, when $L(\Gamma, \rho) > 0$, we define

$$\rho_1(z) = \begin{cases} \rho(z), & z \in \Omega_1, \\ 0, & z \notin \Omega_1, \end{cases}$$

$$\rho_2(z) = \begin{cases} \rho(z), & z \in \Omega_2, \\ 0, & z \notin \Omega_2; \end{cases}$$

then

$$L(\Gamma, \rho) \leq L(\Gamma_1, \rho_1), \qquad L(\Gamma, \rho) \leq L(\Gamma_2, \rho_2),$$
$$A(\rho) \geq A(\rho_1) + A(\rho_2).$$

And hence,

$$\frac{A(\rho)}{L^2(\Gamma, \rho)} \geq \frac{A(\rho_1)}{L^2(\Gamma_1, \rho_1)} \frac{A(\rho_2)}{L^2(\Gamma_2, \rho_2)}.$$

When $L(\Gamma, \rho) = 0$, this inequality is obvious. Hence, according to the arbitrary selection of $\rho(z)$, we conclude that

$$\frac{1}{\lambda(\Gamma)} \geq \frac{1}{\lambda(\Gamma_2)} + \frac{1}{\lambda(\Gamma_2)},$$

meaning that formula (5.62) holds.

Finally, we prove a symmetry principle. Let $\bar{\gamma}$ be the mirror symmetry of γ with respect to the real axis, and γ^+ be the part of $\gamma \cup \bar{\gamma}$ on the upper half-plane: Im $z \geq 0$. Obviously, $\gamma \cup \bar{\gamma} = \gamma^+ \cup (\overline{\gamma^+})$. If Γ stands for the set $\{\gamma\}$, then $\bar{\Gamma}$ and Γ^+ represent the sets $\{\bar{\gamma}\}$ and $\{\gamma^+\}$, respectively.

THEOREM 5.10. *If* $\overline{\Gamma} = \Gamma$, *then* $\lambda(\Gamma) = \frac{1}{2}\lambda(\Gamma^+)$.

PROOF. Let $\rho(z)$ be an arbitrarily selected admissible function and

$$\hat{\rho}(z) = \max\{\rho(z), \rho(\bar{z})\}.$$

Then

$$L_\gamma(\hat{\rho}) = L_{\gamma^+}(\hat{\rho}) \geq L_{\gamma^+}(\rho) \geq L(\Gamma^+, \rho),$$

$$A(\hat{\rho}) \leq A(\rho) + A(\bar{\rho}) = 2A(\rho),$$

$$\frac{L^2(\Gamma^+, \rho)}{A(\rho)} \leq 2\frac{L^2(\Gamma, \hat{\rho})}{A(\hat{\rho})} \leq 2\lambda(\Gamma).$$

Hence, $\lambda(\Gamma^+) \leq 2\lambda(\Gamma)$. On the other hand, for an arbitrarily given admissible function $\rho(z)$, we let

$$\rho^+(z) = \begin{cases} \rho(z) + \rho(\bar{z}), & \text{Im } z \geq 0, \\ 0, & \text{Im } z < 0. \end{cases}$$

Then

$$L_{\gamma^+}(\rho^+) = L_{\gamma^+ \cup \overline{(\gamma^+)}}(\rho) = L_{\gamma \cup \bar{\gamma}}(\rho)$$

$$= L_\gamma(\rho) + L_{\bar{\gamma}}(\rho) \geq 2L(\Gamma, \rho),$$

$$A(\rho^+) \leq 2 \iint_{|z| \leq +\infty} (\rho^2 + \bar{\rho}^2) \, dx \, dy = 2A(\rho).$$

And hence,

$$\frac{L^2(\Gamma, \rho)}{A(\rho)} \leq \frac{1}{2} \cdot \frac{L_{\gamma^+}^2(\rho^+)}{A(\rho^+)} \leq \frac{1}{2}\lambda(\Gamma^+),$$

$$\lambda(\Gamma) \leq \frac{1}{2}\lambda(\Gamma^+).$$

Therefore, we have $\lambda(\Gamma) = \frac{1}{2}\lambda(\Gamma^+)$, meaning that Theorem 5.10 is proved.

5.4.3. Two problems on extremals. Let Ω be a domain on the open plane $|z| < +\infty$, and E_1 and E_2 be two sets in Ω's complementary set. Moreover, their intersections with the boundary of Ω are nonempty. We also let Γ be the set of connected arcs linking E_1 and E_2 in Ω, meaning that each $\gamma \in \Gamma$, except for the two endpoints, is in Ω. Meanwhile, the two endpoints of γ are on E_1 and E_2, respectively. In the sequel, we call $\lambda(\Gamma)$ the extremal distance of E_1 and E_2. Moreover, we denote it as $d_\Omega(E_1, E_2)$.

Let G be a doubly-connected domain on the open plane $|z| < +\infty$, and C_1 and C_2 are the bounded and unbounded components of the two complementary sets of G, respectively. We also let Γ be the set of all closed curves dividing C_1 and C_2 in G. In the sequel, we call $\lambda^{-1}(\Gamma)$ a module of G, denoting it as $M(G)$. If we map G conformally onto an annulus, then according to Examples 2 and 3, we conclude that $M(G) = d_\Omega(C_1, C_2)$. Below we find the largest value of $M(G)$, if one of the following conditions is satisfied:

(1) C_1 is the unit disk $|z| \leq 1$, and C_2 contains point $z = R > 1$.

(2) C_1 contains points $z = 0$ and $z = -1$. Meanwhile, C_2 contains point $z = P > 0$.

H. Grötzoch proved the following result:

THEOREM 5.11. *Let the complementary set of the doubly-connected domain* G_0 *be the unit disk* $|z| \leq 1$ *and straight line segment* $[R, +\infty]$. *Then when* G *satisfies condition* (1), *we have* $M(G) \leq M(G_0) = M(G_0, R)$.

PROOF. Let Γ be a set of closed curves dividing C_1 and C_2 in G, and $\lambda(\Gamma) = M(G)^{-1}$. We also let $\widetilde{\Gamma}$ be a set of closed curves in the complementary set of $C_1 \cup \{R\}$. Moreover, for each $\tilde{\gamma} \in \widetilde{\Gamma}$ its winding numbers about point $z = R$ is zero, and is nonzero about the origin $z = 0$. Obviously, $\Gamma \subset \widetilde{\Gamma}$; therefore, $\lambda(\Gamma) \geq \lambda(\widetilde{\Gamma})$. Notice that set $\widetilde{\Gamma}$ is symmetric, therefore, according to Theorem 5.10, we conclude that $\lambda(\widetilde{\Gamma}) = \frac{1}{2}(\widetilde{\Gamma}^+)$. Analogously, let Γ_0 be a set of closed curves dividing unit disk $|z| \leq 1$ and line $[R, +\infty]$ in G_0. According to the symmetry of G_0, we have $\lambda(\Gamma_0) = \frac{1}{2}(\Gamma_0^+)$.

In the following, we prove that $\widetilde{\Gamma}^+ = \Gamma_0^+$. In fact, each $\tilde{\gamma} \in \widetilde{\Gamma}$ intersects line segment $(-\infty, 1)$ and $(1, R)$ at points P_1 and P_2, respectively. Hence, points P_1 and P_2 divide $\tilde{\gamma}$ into two parts which are $\tilde{\gamma}_1$ and $\tilde{\gamma}_2$, meaning that $\tilde{\gamma} = \tilde{\gamma}_1 \cup \tilde{\gamma}_2$. Notice that $\tilde{\gamma}^+ = \tilde{\gamma}_1^+ \cup \tilde{\gamma}_2^+ = (\tilde{\gamma}_1 \cup \overline{\tilde{\gamma}_2^+})^+$, and $\tilde{\gamma}_1 \cup \overline{\gamma_2^+} \in \Gamma_0$; therefore, we conclude that $\tilde{\gamma}^+ \in \Gamma_0^+$, meaning that $\widetilde{\Gamma}^+ \subset \Gamma_0^+$. On the other hand, it is obvious that $\Gamma_0^+ \subset \widetilde{\Gamma}^+$. Therefore, $\widetilde{\Gamma}^+ = \Gamma_0^+$. Hence, we conclude that $\lambda(\Gamma) \geq \lambda(\widetilde{\Gamma}) = \lambda(\Gamma_0)$, meaning that $M(G) \leq M(G_0) = M(G_0R)$.

O. Teichmüller proved the following result:

THEOREM 5.12. *Let the complementary set of the doubly-connected domain* G_1 *be two straight line segments* $[-1, 0]$ *and* $[P, +\infty]$. *Then when* G *satisfies condition* (2), *we have* $M(G) \leq M(G_1) = M(G_1P)$.

PROOF. Let $z = f(\zeta)$ map the unit disk $|\zeta| < 1$ conformally onto $C_1 \cup G$. Moreover, $f(0) = 0$. We also let C_1' be the preimage of C_1. According to the Koebe $\frac{1}{4}$ covering theorem, we conclude that $|f'(0)| \leq 4P$. Furthermore, let a point α on C_1' satisfy $f(\alpha) = -1$. According to the distortion theorem, we conclude that

$$1 = |f(\alpha)| \leq \frac{|\alpha| \, |f'(0)|}{(1 - |\alpha|)^2} \leq \frac{4P|\alpha|}{(1 - |\alpha|)^2}.$$

On the other hand, Koebe function $z = f_1(\zeta) = \frac{\zeta}{(1+\zeta)^2}$ maps the unit disk $|\zeta| < 1$ conformally onto the complementary set of the straight line segment $[\frac{1}{4}, +\infty]$. Hence, $F(\zeta) = 4Pf_1(\zeta)$ maps the unit disk $|\zeta| < 1$ conformally onto $G_1 \cup [-1, 0]$. Obviously, the preimage of the line segment $[-1, 0]$ on the z-plane is the straight line segment $[\alpha_1, 0]$ on ζ-plane. Moreover,

$\alpha_1 < 0$ and $F(\alpha_1) = -1$. Hence,

$$1 = |F(\alpha_1)| = \frac{4P|\alpha_1|}{(1 - |\alpha|)^2}.$$

Since $\frac{t}{(1-t)^2}$ is an increasing function of t, therefore, $|\alpha_1| < |\alpha|$.

By means of the conformal mapping $z = \frac{1}{\zeta}$ and Theorem 5.11, we conclude that $M(G) \leq M(G_0 \frac{1}{|\alpha|})$, and $M(G_1, P) = M(G_0, \frac{1}{|\alpha_1|})$. Furthermore, according to $\frac{1}{|\alpha|} \leq \frac{1}{|\alpha_1|}$, we obtain $M(G_0, \frac{1}{|\alpha|}) \leq M(G_0, \frac{1}{|\alpha_1|})$. Hence, $M(G) \leq M(G, P)$, meaning that Theorem 5.12 holds.

We simply denote $M(G_1 P) = \Lambda(P) = \Lambda$. In the following, we make an estimation on $\Lambda(P)$. Let ω_1 and ω_2 be two finite complex numbers. Moreover, $\text{Im}\frac{\omega_2}{\omega_1} > 0$. Then the function

$$W = \wp(z) = \frac{1}{z^2} + \sum_{\omega \neq 0} \left\{ \frac{1}{(z - \omega)^2} - \frac{1}{\omega^2} \right\}, \quad \omega = m\omega_1 + n\omega_2, \tag{5.63}$$

$$m, n = 0, \pm 1, \pm 2, \ldots$$

is called the Weierstrass's $\wp$-function function. Obviously, $\wp(z)$ is a double periodic meromorphic function, $\wp(z) = \wp(-z)$, and the order of the poles $z = \omega$ of $\wp(z)$ are all equal to 2. Hence, $\wp(z)$ assumes each complex number twice in the periodic parallelogram. We have further,

$$\wp'(z) = -\frac{2}{z^3} - \sum_{\omega \neq 0} \frac{2}{(z - \omega)^3},$$

$$\wp'(z) = -\wp'(-z), \qquad \wp'(z + \omega) = \wp(z).$$

Particularly,

$$\wp'\left(\frac{\omega}{2}\right) = -\wp'\left(-\frac{\omega}{2}\right) = -\wp'\left(\frac{\omega}{2}\right).$$

Hence, we conclude that $\wp'(\frac{\omega}{2}) = 0$. Hence, if we let $e_1 = \rho(\frac{\omega_1}{2})$, $e_2 = \wp(\frac{\omega_2}{2})$ and $e_3 = \wp(\frac{\omega_1 + \omega_2}{2})$, then e_1, e_2 and e_3 are all multiple values of multiplicity 2 of $\wp(z)$. Moreover, they are distinct among one another.

Now, we assume, in particular, $\omega_1 = 1$ and $\omega_2 = 2i\Lambda$; then according to formula (5.63), when $x = 0$ or $\frac{1}{2}$, and when $y = 0$ or Λ, $w = \wp(z)$ assumes real values. Hence, e_1, e_2 and e_3 are all on the real axis of the w-plane. Moreover, $w = \wp(z)$ maps the boundary of the rectangle with $z = 0$, $\frac{\omega_1}{2}$, $\frac{\omega_1 + \omega_2}{2}$ and $\frac{\omega_2}{2}$ being its vertices onto the whole real axis. Furthermore, according to $\wp(z) = \wp(-z)$, we conclude that $w = \wp(z)$ is a univalent, conformal mapping in such a rectangle. If we examine the transformations of $w = \wp(z)$ around a neighborhood of the origin $z = 0$, then we may conclude that $w = \wp(z)$ maps the internal part of such a rectangle conformally onto the lower half-plane: $\text{Im } z < 0$. Moreover, according to the preserving of the directions of a conformal mapping, it must be that $e_2 < e_3 < e_1$. Hence, the line segments $[e_1 + \infty]$ and $[e_2, e_3]$ on the z-plane correspond to the line segments: $y = 0$, $0 \leq x \leq \frac{1}{2}$ and $y = \Lambda(P)$, $0 \leq x \leq \frac{1}{2}$ on the z-plane, respectively. If we continue to make a fractional linear transformation $\zeta = \frac{w - e_1}{e_3 - e_2}$, then the lower half-plane: $\operatorname{Im} w < 0$ is transformed into the lower half-plane: $\operatorname{Im} \zeta < 0$. Meanwhile, the line segments $[e_1, +\infty]$ and $[e_2, e_3]$ on the w-plane is transformed into the line segments $[P, +\infty]$ and $[-1, 0]$ on the ζ-plane, respectively, where

$$P = \frac{e_1 - e_3}{e_3 - e_2}.$$

Notice that the complementary set of the two line segments $[P, +\infty]$ and $[-1, 0]$ is just Teichmüller extremal region G_1. Moreover, the extremal distance between these two straight line segments corresponding to G_1 is just $\Lambda(P)$. We conclude that $M(G_1 P) = \Lambda(P)$.

In the following, we find the formula regarding the relation between $\Lambda(P)$ and P. Generally, we consider function

$$\frac{\wp(z) - \wp(u)}{\wp(z) - \wp(v)};$$

then $z = \pm u + m\omega_1 + n\omega_2$ and $z = \pm v + m\omega_1 + n\omega_2$ are the zeros and poles of this function. We then construct a function

$$F(z) = \prod_{n=-\infty}^{+\infty} \frac{(1 - e^{2\pi i(n\omega_2 + u - z)/\omega_1})(1 - e^{2\pi i(n\omega_2 - u - z)/\omega_1})}{(1 - e^{2\pi i(n\omega_2 + v - z)/\omega_1})(1 - e^{2\pi i(n\omega_2 - v - z)/\omega_1})},$$

and easily verify that this infinite product is convergent. We assume $\omega_1 = 1$, $\omega_2 = 2i\Lambda(P)$ and let $q = e^{-2\pi\Lambda} < 1$. Accordingly, we have

$$F(z) = \frac{(1 - e^{2\pi i(u-z)})(1 - e^{-2\pi i(u+z)})}{(1 - e^{2\pi i(v-z)})(1 - e^{-2\pi i(v+z)})}$$

$$\times \prod_{n=1}^{+\infty} \frac{(1 - q^{2n} e^{2\pi i(u+z)})(1 - q^{2n} e^{2\pi i(u-z)})}{(1 - q^{2n} e^{2\pi i(v+z)})(1 - q^{2n} e^{2\pi i(v-z)})}$$

$$\times \frac{(1 - q^{2n} e^{-2\pi i(u-z)})}{(1 - q^{2n} e^{-2\pi i(v-z)})} \cdot \frac{(1 - q^{2n} e^{-2\pi i(u+z)})}{(1 - q^{2n} e^{-2\pi i(v+z)})}.$$

Obviously, $F(z)$ and $\frac{\wp(z)-\wp(u)}{\wp(z)-\wp(v)}$ have the same zeros, poles, and periods. Hence,

$$\frac{\wp(z) - \wp(u)}{\wp(z) - \wp(v)} = \frac{F(z)}{F(0)}.$$

If we take $u = \frac{\omega_1}{2} = \frac{1}{2}$, $v = \frac{\omega_2}{2} = i\Lambda$ and $z = \frac{\omega_1 + \omega_2}{2} = \frac{1}{2} + i\Lambda$, then we conclude that

$$P = \frac{e_1 - e_3}{e_3 - e_2} = -\frac{F(z)}{F(0)}.$$

Notice that

$$e^{2\pi iz} = -q, \quad e^{2\pi iu} = -1, \quad e^{2\pi iv} = q.$$

It follows that

$$F(z) = \frac{1-q^{-1}}{1+1} \cdot \frac{1-q^{-1}}{1+q^{-2}} \prod_{n=1}^{+\infty} \frac{(1-q^{2n+1})^2(1'-q^{2n-1})^2}{(1+q^{2n+2})(1+q^{2n})^2(1+q^{2n-2})}$$

$$= \frac{1}{4} \prod_{n=1}^{+\infty} \left(\frac{1-q^{2n-1}}{1+q^{2n}} \right)^4 ,$$

$$F(0) = \frac{1+1}{1-q} \cdot \frac{1+1}{1-q^{-1}} \prod_{n=1}^{+\infty} \frac{(1+q^{2n})^4}{(1-q^{2n+1})^2(1-q^{2n-1})^2}$$

$$= -4q \prod_{n=1}^{+\infty} \left(\frac{1+q^{2n}}{1-q^{2n-1}} \right)^4 .$$

And therefore,

$$P = -\frac{F(z)}{F(0)} = \frac{1}{16q} \cdot \prod_{n=1}^{+\infty} \left(\frac{1-q^{2n-1}}{1+q^{2n}} \right)^8 . \tag{5.64}$$

Analogously, if we take $u = \frac{\omega_1}{2} = \frac{1}{2}$, $v = \frac{\omega_1+\omega_2}{2} = \frac{1}{2} + i\Lambda$, $z + \frac{\omega_2}{z} = i\Lambda$, then we conclude that

$$P + 1 = \frac{e_2 - e_1}{e_2^- - e_3} = \frac{1}{16q} \prod_{n=1}^{+\infty} \left(\frac{1+q^{2n-1}}{1+q^{2n}} \right)^8 . \tag{5.65}$$

Hence, from formulas (5.64) and (5.65), we have the estimate

$$16P \le e^{2\pi\Lambda(P)} \le 16(P+1). \tag{5.66}$$

5.4.4. Ahlfors' Distortion Theorem. Let Ω be a strip domain bounded by two curves on the open plane $|z| < +\infty$, and let it tend to ∞ along the two curves. We denote the upper boundary of Ω by B_1 and the lower boundary of Ω by B_2. We also let E_1 and E_2 be the two line segments linking B_1 and B_2 on lines $x = a$ and $x = b$ $(a < b)$, respectively. Hence, a region $\Omega_{12} \subset \Omega$ is formed by E_1 and E_2, as well as parts of B_1 and B_2 that between E_1 and E_2. Besides, we use $\theta(t)$ $(-\infty < t < +\infty)$ to represent the length of the part of the line $x = t$ in Ω. Under the circumstance that B_1 and B_2 satisfy considerably wide-ranging conditions, we may prove that $\theta(t)$ is a lower semicontinuous function, and hence, $\frac{1}{\theta(t)}$ is a measurable function.

We map Ω conformally onto a horizontal strip domain Ω' with its width equal to 1, and B_1 and B_2 correspond to horizontal lines $B_1': y = 0$ and $B_2': y = 1$, respectively. Moreover, Ω_{12} corresponds to Ω_{12}', and E_1 and E_2 correspond to E_1' and E_2', respectively.

We assume

$$\rho(z) = \rho(x, y) = \begin{cases} 1/\theta(x), & z \in \Omega \cap (a < \operatorname{Re} z < b), \\ 0, & z \notin \Omega \cap (a < \operatorname{Re} z < b). \end{cases}$$

Then for any arc γ linking E_1 and E_2 in Ω_{12},

$$\int_\gamma \rho(z)|dz| \geq \int_a^b \frac{dx}{\theta(x)}.$$

Meanwhile, we have

$$A(\rho) = \iint \rho^2 \, dx \, dy = \int_a^b \frac{1}{\theta(x)} \, dx.$$

And hence,

$$\int_a^b \frac{dx}{\theta(x)} \leq d_{\Omega_{12}}(E_1, E_2) = d_{\Omega_{12}'}(E_1', E_2').$$

Let

$$d = \max_{z \in E_1'}\{x\}, \qquad \beta = \min_{z \in E_2'}\{x\}.$$

For region Ω' we undergo a mirror symmetry with respect to the real axis; then Ω', the whole real axis and the mirror symmetry of Ω' constitute a region $\widehat{\Omega}$. Correspondingly, we denote them as $\widehat{\Omega}_{12}$, $\widehat{E}_1$ and $\widehat{E}_2$. Obviously, they are all symmetric with respect to the real axis. Hence, according to Theorem 5.10, we have

$$d_{\widehat{\Omega}_{12}}(\widehat{E}_1, \widehat{E}_2) = \frac{1}{2} d_{\Omega_{12}'}(E_1', E_2').$$

Next, by using $e^{\pi z}$, we map $\widehat{\Omega}$ into the whole plane, and $\widehat{E}_1$ and $\widehat{E}_2$ are mapped into $\widehat{E}_1'$ and $\widehat{E}_2'$, respectively. Obviously, $\widehat{E}_1'$ and $\widehat{E}_2'$ are closed curves, and hence, a doubly-connected region $\widehat{\Omega}_{12}'$ is formed. Therefore, $\widehat{\Omega}_{12}'$ corresponds to $\widehat{\Omega}_{12}$. Hence $d_{\widehat{\Omega}_{12}}(\widehat{E}_1, \widehat{E}_2) = d_{\widehat{\Omega}_{12}'}(\widehat{E}_1', \widehat{E}_2')$. Notice that the bounded part enclosed by $\widehat{E}_1'$ contains the origin and a point which is $e^{\alpha\pi}$ away from the origin and is passed by $\widehat{E}_1'$. Meanwhile, $\widehat{E}_2'$ divides $\widehat{E}_1'$ and ∞, and $\widehat{E}_2'$ passes through a point which is $e^{\beta\pi}$ away from the origin. According to Theorem 5.12, we conclude that

$$d_{\widehat{\Omega}_{12}'}(\widehat{E}_1', \widehat{E}_2') \leq M(G_1 e^{\pi(\beta-\alpha)}) = \Lambda(e^{\pi(\beta-\alpha)}).$$

And hence, we prove the following result:

THEOREM 5.13. *Let* $\theta(x)$ *be defined as above. Then*

$$\int_a^b \frac{dx}{\theta(x)} \leq 2\Lambda(e^{\pi(\beta-\alpha)}).$$

Now, by applying formula (5.66) we obtain a very useful formula:

COROLLARY. *If* $\int_a^b \frac{dx}{\theta(x)} \geq 2$, *then*

$$\beta - \alpha \geq \int_a^b \frac{dx}{\theta(x)} - \frac{1}{\pi} \log 32. \tag{5.67}$$

In fact, if we assume $P = 1$ in formula (5.66), then

$$e^{2\pi\Lambda(1)} \leq 32,$$

$$\Lambda(1) \leq \frac{1}{2\pi}\log 32 = \frac{1}{2\pi}\log 2 + \frac{2}{\pi}\log 2 \leq \frac{5}{2\pi} < 1.$$

On the other hand, if $\int_a^b 1/\theta(x)\,dx \geq 2$, then according to Theorem 5.13 we have $\Lambda(e^{\pi(\beta-\alpha)}) \geq 1$. Accordingly, we conclude that $e^{\pi(\beta-\alpha)} \geq 1$. Hence, when we assume $P = e^{\pi(\beta-\alpha)}$ in formula (5.66); it follows that

$$2\Lambda(e^{\pi(\beta-\alpha)}) \leq \frac{1}{\pi}\log\{32e^{\pi(\beta-\alpha)}\} = \beta - \alpha + \frac{1}{\pi}\log 32.$$

Furthermore, according to Theorem 5.13, we conclude that formula (5.67) holds.

§5.5. On entire functions with zeros distributed on a finite number of half lines [43j]

THEOREM 5.14. *Let $f(z)$ be an entire function (on the open plane $|z| < +\infty$) and $\Delta(\theta_k)$ $(k = 1, 2, \ldots, q;\ 0 \leq \theta_1 < \theta_2 < \cdots < \theta_q < 2\pi)$ q half lines on the z-plane. Moreover, for any arbitrarily selected number $\varepsilon > 0$, we have*

$$\varlimsup_{r\to+\infty} \frac{\log^+ n\left\{\bigcup_{k=1}^q \overline{\Omega}(\theta_k + \varepsilon,\, \theta_{k+1} - \varepsilon;\, r),\, f = 0\right\}}{\log r} \tag{5.68}$$
$$= 0,\quad \theta_{q+1} = \theta_1 + 2\pi.$$

Let l be the number of distinct finite asymptotic values of $f(z)$ and p be the number of finite nonzero deficient values, where l' deficient values are simultaneously asymptotic values. If $q < +\infty$, then $p - l' + l \leq 2\mu$, where μ is the lower order of $f(z)$.

PROOF. When $\mu = +\infty$, it is obvious that Theorem 5.14 holds. Hence, we need only to consider the case when $\mu < +\infty$. On the other hand, when $p - l' = 0$, according to Theorem 4.7, we conclude that Theorem 5.14 holds. Hence, we need only to consider the case when $p - l' \geq 1$. When $\mu < +\infty$ and $p - l' \geq 1$, according to Theorem 3.6, we conclude that the order λ of $f(z)$ is $< +\infty$. Meanwhile, from the corollary of Theorem 4.4, we have $\mu \geq \frac{1}{2}$.

In the following, we need only to prove that when $\frac{1}{2} \leq \mu \leq \lambda < +\infty$ and $p - l' \geq 1$, Theorem 5.14 holds. In fact, otherwise, then there exists integer p_1, $1 \leq p_1 \leq p - l'$, $p_1 < +\infty$ and integer l_1, $0 \leq l_1 \leq l$, $l_1 < +\infty$, such that

$$p_1 + l_1 \geq [2\mu] + 1. \tag{5.69}$$

(1) We select p_1 nonasymptotic, finite, and nonzero deficient values a_i $(i = 1, 2, \ldots, p_1)$ of $f(z)$, with their corresponding deficiencies $\delta(a_i, f) = \delta_i > 0$, and l_1 distinct finite asymptotic values b_j $(j = 1, 2, \ldots, l_1)$ of

$f(z)$, with their corresponding paths of fixed values L_j. Without any loss of generality, we may assume that L_j $(1 \leq j \leq l_1)$ is a simply continuous curve originating from the origin and tending to ∞, and it is also a line segment in the part of disk $|z| \leq 2$. Moreover, l_1 curves L_j $(j = 1, 2, \ldots, l_1)$ have no intersection points except at the origin. L_j and L_{j+1} $(1 \leq j \leq l_1$, $L_{l_1+1} \equiv L_1)$ are adjacent, bounding a simply connected domain D_j. Let

$$M_0 = \sup_{z \in \bigcup_{j=1}^{l_1} L_j} |f(z)| < +\infty, \tag{5.70}$$

$$M_1 = \max\{1, |a_1|, \ldots, |a_{p_1}|, |b_1|, \ldots, |b_{l_1}|\}, \tag{5.71}$$

$$M_2 = \min\{1, |a_i|, |a_i - a_{i'}|, |b_j - b_{j'}| \, |a_i - b_j| \tag{5.72}$$
$$1 \leq i \neq i' \leq p_1; \; 1 \leq j \neq j' \leq l_1, b_j \neq b_{j'}\}.$$

According to Theorem 4.3, we conclude that there exists a continuous curve Γ_j tending to ∞ in D_j $(1 \leq j \leq l_1)$ such that

$$\varlimsup_{\substack{|z| \to +\infty \\ z \in \Gamma_j}} \frac{\log \log |f(z)|}{\log |z|} \geq \frac{1}{2}. \tag{5.73}$$

Hence, we may select a point z_j on Γ_j $(1 \leq j \leq l_1)$ such that $|f(z'_j)| \geq M_0$. Let

$$r'_0 = \max\{|z'_1|, \ldots, |z'_{l_1}|\}.$$

When $r > r'_0$, we let $\Omega_j(r)$ be the connected component of D_j $(1 \leq j \leq l_1)$ in the disk $|z| < r$ that contains point z'_j. Moreover, notice that

$$M\{\overline{\Omega}_j(r) \cap (|z| = r), f\} = \max_{z \in \overline{\Omega}_j(r) \cap (|z|=r)} |f(z)|, \qquad j = 1, 2, \ldots, l_1.$$

Then $M\{\overline{\Omega}_j(r) \cap (|z| = r), f\}$ is a monotonic increasing function of r $(r > r'_0)$. Moreover, according to formula (5.73),

$$\varlimsup_{r \to +\infty} \frac{\log \log M\{\overline{\Omega}_j(r) \cap (|z| = r), f\}}{\log r} \geq \frac{1}{2}. \tag{5.74}$$

On the other hand, we assume a point z''_j on $D_j \cap (|z| = 1)$ $(1 \leq j \leq l_1)$. Then, taking connectivity into account, there exists a continuous curve L'_j linking points z'_j and z''_j in D_j $(1 \leq j \leq l_1)$. We then assume value r''_0, $r''_0 > r'_0$, such that l_1 curves L'_j $(j = 1, 2, \ldots, l_1)$ are in disk $|z| \leq r''_0$. Hence, when $r > r''_0$, the connected component of D_j $(1 \leq j \leq l_1)$ in disk $|z| < r$ that contains point z''_j is $\Omega_j(r)$. Let $L_j(r)$ be the part of L_j $(1 \leq j \leq l_1)$ between the origin and the first intersection point of L_j $(1 \leq j \leq l_1)$ and circumference $|z| = r$. Then a simply connected domain $D_j(r)$ is bounded by $L_j(r)$ and $L_{j+1}(r)$ in disk $|z| < r$. Obviously, when $r \geq r''_0$, $\Omega_j(r) \subset D_j(r)$ $(j = 1, 2, \ldots, l_1)$.

(2) Let $F(z) = f(z) - a_1$. Then according to formula (3.28), there exists a positive number $\alpha = \alpha\{\delta(a_1, f), \delta(\infty, f)\} > 0$ depending on $\delta(a_1, f)$ and $\delta(\infty, f)$, such that

$$\lim_{R \to +\infty} \frac{RT'(R, F)}{T(R, F)} \geq \alpha. \tag{5.75}$$

Moreover, we let

$$\omega = \min_{1 \leq k \leq q} \{\theta_{k+1} - \theta_k\}, \quad \theta_{q+1} = \theta_1 + 2\pi,$$

$$\delta = \min_{1 \leq i \leq p_1} \{\delta_i\}$$

and select arbitrarily a number η,

$$0 < \eta < \min\left\{\frac{\alpha\omega}{8 \times 32\pi}, \frac{\omega}{2 \times 31 \times 19\pi + 36\pi}\right\}$$

and number r_0, $r_0 > r_0''$. Then, we construct a sequence $r_n = r_0^{(1+\eta)^{2m}}$, $m = 1, 2, \ldots$. According to Theorem 2.4 and Lemma 3.9 (where we let $\sigma = \log 2$), there exist integers m_0, $m_0 \geq 2$, such that when $r \geq r_{m_0}$,

$$T(r, f) \leq r^{\lambda+\eta}, \quad \log M(r, f) \leq r^{\lambda+\eta}, \tag{5.76}$$

and

$$T(r, f) \geq r^{\mu-\eta}. \tag{5.77}$$

Moreover, when $m \geq m_0$, there exists value t_m in the interval $[r_n, 2r_n^{1+\eta}]$ such that the measure of the set

$$E_i(t_m) = \left\{\theta \,\Big|\, \log \frac{1}{|f(t_m e^{i\theta}) - a_i|} \geq \frac{\delta}{4} T(t_{m_1}, f), 0 \leq \theta < 2\pi\right\}$$

$$(i = 1, \ldots, p_1)$$

is

$$\operatorname{meas} E_i(t_m) \geq K = K(\delta, \lambda, p_1, \eta) > 0.$$

Finally, we further select a fixed number ε,

$$0 < \varepsilon < \min\left\{\frac{K}{12q}, \frac{\omega}{40}\right\}$$

and let $\overline{\Omega} = \bigcup_{k=1}^q \overline{\Omega}(\theta_k + \varepsilon, \theta_{k+1} - \varepsilon)$. Then according to formula (5.68),

$$\varlimsup_{r \to +\infty} \frac{\log^+ n\{\overline{\Omega} \cap (|z| < r), f = 0\}}{\log r} = 0.$$

(3) When $m \geq m_0$, based on the selection of ε, for each value i $(1 \leq i \leq p_1)$, there exists at least one set $E_i(t_m) \cap [\theta_{k_i} + 3\varepsilon, \theta_{k_i+1} - 3\varepsilon]$ $(1 \leq k_i \leq q)$ among the q sets $E_i(t_m) \cap [\theta_k + 3\varepsilon, \theta_{k+1} - 3\varepsilon]$ $(k = 1, 2, \ldots, q)$ such that

$$\operatorname{meas}\{E_i(t_m) \cap [\theta_{k_i} + 3\varepsilon, \theta_{k_i+1} - 3\varepsilon]\} \geq \frac{K}{2q}.$$

Furthermore, according to formula (5.77) and the selection of η, applying Lemma 3.14, where we let $\theta = \frac{1}{2}(\theta_{k_i+1} - \theta_{k_i}) - \varepsilon$, $\lambda' = 0$, $R_{n1} = t_{m-1}$, $R_n = t_m$, $R_{n2} = t_{m+1}$, $E_n = \{t_m e^{i\varphi} \mid \varphi \in E_i(t_m) \cap [\theta_{k_i} + 3\varepsilon, \theta_{k_i+1} - 3\varepsilon]\}$, $\alpha = \frac{K}{2q}$, $a = a_i$, $N_n = \frac{\delta}{4}T(t_m, f)$, $\eta_0 = \eta$, then there exists value m_1, $m_1 \geq m_0$, such that when $m \geq m_1$, $z \in \overline{\Omega}(\theta_{k_i} + 2\varepsilon, \theta_{k_i+1} - 2\varepsilon; t_{m-1}, t_{m+1})$,

$$\log|f(z) - a_i| \leq -At_m^{-(\frac{31 \times 24\pi}{\omega}+1)\eta} T(t_m, f), \tag{5.78}$$

where $A > 0$ is a constant independent of m. Let

$$\varepsilon_m' = e^{-At_m^{-(\frac{31 \times 24\pi}{\omega}+1)\eta} T(t_m, f)}. \tag{5.79}$$

Then there exists value m_2, $m_2 \geq m_1$, such that when $m \geq m_2$,

$$\varepsilon_m' < \frac{1}{4}M_2 < 1. \tag{5.80}$$

Hence, the p_1 sets $\overline{\Omega}(\theta_{k_i} + 2\varepsilon, \theta_{k_i+1} - 2\varepsilon; t_{m-1}, t_{m+1})$ $(i = 1, 2, \ldots, p_1)$ are distinct among one another. Furthermore, if we again apply Lemma 3.14, where we let, in particular, $R_{n1} = t_m^{1-\eta^2}$, $R_n = t_m$, $R_{n2} = t_m^{1+\eta^2}$, then when $m \geq m_1$ and $z \in \overline{\Omega}(\theta_{k_i} + 2\varepsilon, \theta_{k_i+1} - 2\varepsilon; t_m^{1-\eta^2}, t_m^{1+\eta^2})$,

$$\log|f(z) - a_i| \leq -At_m^{-(\frac{30\pi}{\omega}+1)\eta^2} T(t_m, f), \tag{5.81}$$

where $A > 0$ is a constant independent of m. Obviously,

$$\varepsilon_m' \geq e^{-At_m^{-(30\pi/\omega+1)\eta^2}} T(t_m, f).$$

Let L_{jm} be the part of L_j $(1 \leq j \leq l_1)$ between the last intersection point of L_j $(1 \leq j \leq l_1)$ and circumference $|z| = t_{m+1}$ and the first intersection point of L_j $(1 \leq j \leq l_1)$ and circumference $|z| = t_{m+1}$. The l_1 curves L_{jm} $(j = 1, 2, \ldots, l_1)$ divide annulus $C_m: t_{m-1} \leq |z| \leq t_{m+1}$ into l_1 regions D_{jm} $(j = 1, 2, \ldots, l_1)$. Moreover, L_{jm} and L_{j+1m} are parts of the boundary of D_{jm}. Let

$$\varepsilon_m'' = \max_{1 \leq j \leq l_1} \left\{ \sup_{z \in L_j \cap C_m} |f(z) - b_j| \right\}. \tag{5.82}$$

Then when $m \to +\infty$, $\varepsilon_m'' \to 0$. Hence, there exists value m_3, $m_3 \geq m_2$, such that when $m \geq m_3$,

$$\varepsilon_m' < \frac{1}{4}M_2 < 1. \tag{5.83}$$

Therefore, the line segment $\Delta((\theta_{k_i} + \theta_{k_i+1})/2; t_{m-1}, t_{m+1})$ may lie entirely in D_{jm} $(1 \leq j \leq l_1)$, or it may have no intersection with D_{jm}. Suppose that there are s_j $(0 \leq s_j \leq p_1)$ segments: $\Delta((\theta_{k_i} + \theta_{k_i+1})/2; t_{m-1}, t_{m+1})$ $(i = 1, 2, \ldots, s_j)$ in D_{jm}. Then D_{jm} is divided into $s_j + 1$ regions $D_{jm\nu}$

$(1 \leq \nu \leq s_j + 1)$. We assume that L_{jm} is part of the boundary of D_{jm1}, while L_{j+1m} is part of the boundary of D_{jms_j+1}. It is apparent that $\sum_{j=1}^{l_1} s_j = p_1$.

(4) Let $s_j \neq 0$ $(1 \leq j \leq l_1)$. When $1 < \nu < s_j + 1$, we let $D'_{jm\nu} \equiv D_{jm\nu} \cap (t_m^{1-\eta^2} < |z| < t_m^{1+\eta^2})$. When $\nu = s_j + 1$, we let D'_{jms_j+1} be a region bounded by L_{j+1m}, $\Delta((\theta_{ks_j} + \theta_{ks_j+1})/2; t_{m-1}, t_{m+1})$ and two circumferences $|z| = t_m^{1-\eta^2}$ and $|z| = t_m^{1+\eta^2}$. Moreover, when $\nu = 1$, we let D'_{jm1} be another region bounded by L_{jm}, $\Delta((\theta_{k_1} + \theta_{k_1+1})/2; t_{m-1}, t_{m+1})$ and two circumferences $|z| = t_m^{1-\eta^2}$ and $|z| = t_m^{1+\eta^2}$. In the following, we prove that when m is sufficiently large, there exists a point $z_{jm\nu}$ in $D'_{jm\nu}$ $(1 \leq \nu \leq s_j + 1)$ and a constant $A > 0$ independent of m, such that

$$\log |f(z_{jm\nu})| \geq A t_m^{-(30\pi/\omega+1)\eta^2} T(t_m, f).$$

In fact, we need only to consider a typical case, which is to prove the result for region D'_{jms_j+1}. First, we notice that L_{j+1m} is part of the boundary of D_{jms_j+1}. Next, we let $\Delta((\theta_k + \theta_{k+1})/2; t_{m-1}, t_{m+1})$ $(1 \leq k \leq q)$ be another part of the boundary of D_{jms_j+1}. When $m \geq m_3$ and

$$z \in \overline{\Omega}(\theta_k + 2\varepsilon, \theta_{k+1} - 2\varepsilon; t_{m-1}, t_{m+1}),$$

according to formula (5.78),

$$\log |f(z) - a_i| \leq \log \varepsilon'_m = -A t_m^{-((31\times24\pi)/\omega+1)\eta} T(t_m, f) \quad (1 \leq i \leq p_1),$$

where $A > 0$ is a constant independent of m. When $m \geq m_3$ and $z \in \overline{\Omega}(\theta_k + 3\varepsilon, \theta_{k+1} - 2\varepsilon; t_m^{1-\eta^2}, t_m^{1+\eta^2})$, according to formula (5.81),

$$\log |f(z) - a_i| \leq -A t_m^{-(30\pi/\omega+1)\eta^2} T(t_m, f) \quad (1 \leq i \leq p_1), \qquad (5.84)$$

where $A > 0$ is a constant independent of m.

1) We extend $\Delta((\theta_k + \theta_{k+1})/2, t_{m-1}, t_{m+1})$ and L_{j+1m} along the two directions, respectively, such that the two simply curves B_1 and B_2 obtained in this way all originate from the origin $z = 0$ and tend to ∞. Moreover, except at the origin $z = 0$, there are no more intersection points between B_1 and B_2 on the open plane $|z| < +\infty$. Hence, a simply connected region $D \supset D_{jms_j+1}$ is formed by B_1 and B_2. We assume value t, such that there is no intersection between circumference $|z| = t$ and $\overline{D}_{jms_j+1}$. Then, starting from the intersection point between circumference $|z| = t$ and $\Delta((\theta_k + \theta_{k+1})/2; t_{m-1}, t_{m+1})$ along circumference $|z| = t$, and following the anti-clockwise direction until we come across L_{j+1m}, we denote this part of the circular arc as $l_z(t)$. We construct a transformation $\zeta = \sigma + i\tau = \log z$; then D is mapped conformally onto a strip domain D_ζ on the ζ-plane. Moreover, B_1 is mapped onto the lower boundary B'_1 of D_ζ, and B_2 is

mapped onto the upper boundary B_2' of D_ζ. We let $l_\zeta(\sigma)$ $(\sigma = \log t)$ be the image of $l_z(t)$. Furthermore, we construct a transformation $w = u + iv = W(\zeta)$ and map D_ζ conformally onto a horizontal strip domain D_w on the w-plane. Moreover, B_1' is mapped into the line $B_1'': v = 0$ and B_2' into the line $B_2'': v = 1$. We denote $l_w(\sigma)$ as the image of $l_\zeta(\sigma)$. Let

$$u^*(\sigma) = \max_{\zeta \in l_\omega(\sigma)} u(\zeta), \qquad u_*(\sigma) = \min_{\zeta \in l_\omega(\sigma)} u(\zeta).$$

Then applying the corollary of Theorem 5.13, we conclude that

$$u_*(\log t_m + 2\pi + \log 34) - u^*(\log t_m - 2\pi - \log 34)$$
$$\geq \int_{\log t_m - 2\pi - \log 34}^{\log t_m + 2\pi + \log 34} \frac{d\sigma}{\theta(\sigma)} - \frac{1}{\pi} \log 32.$$

If we notice the continuity of $f(z)$, then without changing the asymptotic value, we may transform B_2' suitably in order to guarantee that $\theta(\sigma)$ is a lower semicontinuous function, that is, to make sure that $1/\theta(\sigma)$ is a measurable function. Furthermore, notice that $\theta(\sigma) \leq 2\pi$. We have

$$u_*(\log t_m + 2\pi + \log 34) - u^*(\log t_m - 2\pi - \log 34)$$
$$\geq \frac{1}{2\pi}\{4\pi + 2\log 34\} - \frac{1}{\pi} \log 32 > 2. \tag{5.85}$$

Analogously, we have

$$u_*(\log t_m^{1+\eta^2}) - u^*(\log t_m + 2\pi + \log 34)$$
$$\geq \frac{1}{2\pi}\{\eta^2 \log t_m - 2\pi - \log 34\} - \frac{1}{\pi} \log 32 \tag{5.86}$$
$$\geq \frac{\eta^2}{2\pi} \log t_m - \left(1 + \frac{3}{2\pi} \log 34\right),$$

and

$$u_*(\log t_m - 2\pi - \log 34) - u^*(\log t_m^{1-\eta^2})$$
$$\geq \frac{1}{2\pi}\{\eta^2 \log t_m - 2\pi - \log 34\} - \frac{1}{\pi} \log 32 \tag{5.87}$$
$$\geq \frac{\eta^2}{2\pi} \log t_m - \left(1 + \frac{3}{2\pi} \log 34\right).$$

2) We prove that if we let $g(w) = f(z)$, then when m is sufficiently large, there must exist a point w_{jms_j+1} in the region: $u^*(\log t_m^{1-\eta^2}) < u < u_*(\log t_m^{1+\eta^2})$, $0 < v < 1$, such that

$$\log|g(w_{jms_j+1})| \geq A t_m^{-(30\pi\omega+1)\eta^2} T(t_m, f),$$

where $A > 0$ is a constant independent of m.

First, we prove that if we let $\Delta = \eta^2/4(\lambda + 1)$, then when m is sufficiently large, there must exist a point w^* in the region: $u^*(\log t_m^{1-\eta^2}) + 1 \leq u \leq$

$u_*(\log t_m^{1+\eta^2}) - 1$, $0 < v < 1 - \Delta$, such that $|g(w^*)| > 2M_1$. In fact, otherwise, then when point w is on the closed region: $u^*(\log t_m^{1-\eta^2}) + 1 \leq u \leq u_*(\log t_m^{1+\eta^2}) - 1$, $0 \leq v \leq 1 - \Delta$, it is always true that

$$|g(w)| \leq 2M_1. \tag{5.88}$$

On the other hand, when $m \geq m_3$ and point w is on the straight line segment: $u^*(\log t_m^{1-\eta^2}) + 1 \leq u \leq u_*(\log t_m^{1+\eta^2}) - 1$, $v = 1$, according to formulas (5.71), (5.82) and (5.83), we have

$$|g(w)| \leq |g(w) - b_{j+1}| + |b_{j+1}| \leq 1 + M_1 \leq 2M_1.$$

Hence, by applying the transformation e^w and Lemma 2.10, when $m \geq m_3$, we conclude that when point w is in the region: $u^*(\log t_m - 2\pi - \log 34) \leq u \leq u_*(\log t_m + 2\pi + \log 34)$, $1 - \Delta < v < 1$,

$\log|g(w)|$

$$\leq 2\log(2M_1) + \frac{4\left\{\dfrac{\exp\{u^*(\log t_m^{1-\eta^2})+1\}}{e^u}\right\}^{\frac{\pi}{\Delta}}}{\pi\left\{1 - \left(\dfrac{\exp\{u^*(\log t_m^{1-\eta^2})+1\}}{e^u}\right)^{\frac{2\pi}{\Delta}}\right\}} t_m^{\lambda+1}$$

$$+ \frac{4\left\{\dfrac{e^u}{\exp_*\{u(\log t_m^{1+\eta^2})-1\}}\right\}^{\frac{\pi}{\Delta}}}{\pi\left\{1 - \left(\dfrac{e^u}{\exp_*\{u(\log t_m^{1+\eta^2})-1\}}\right)^{\frac{2\pi}{\Delta}}\right\}} t_m^{(1+\eta^2)(\lambda+1)}$$

$$\leq 2\log(2M_1)$$

$$+ \frac{4\exp\{\{u^*(\log t_m^{1-\eta^2}) + 1 - u_*(\log t_m - 2\pi - \log 34)\}\}^{\frac{\pi}{\Delta}}}{\pi\{1 - \exp\{\{u^*(\log t_m^{1-\eta^2}) + 1 - u_*(\log t_m - 2\pi - \log 34)\}\frac{2\pi}{\Delta}\}\}} t_m^{\lambda+1}$$

$$+ \frac{4\exp\{\{u^*(\log t_m + 2\pi + \log 34) - u_*(\log t_m^{1+\eta^2}) + 1\}\}^{\frac{\pi}{\Delta}}}{\pi\{1 - \exp\{\{u^*(\log t_m + 2\pi + \log 34) - u_*(\log t_m^{1+\eta^2}) + 1\}\frac{2\pi}{\Delta}\}\}}$$

$$\times t_m^{(1+\eta^2)(\lambda+1)}.$$

Furthermore, according to formulas (5.86) and (5.87), we obtain

$$\log|g(w)| \leq 2\log(2M_1)$$

$$+ \frac{4\exp\{(2 + \frac{3}{2\pi}\log 34)\frac{\pi}{\Delta}\} \cdot t^{-\eta^2/2\Delta+\lambda+1}}{\pi\{1 - \exp\{(2 + \frac{3}{2\pi}\log 34)\frac{2\pi}{\Delta}\} \cdot f_m^{-\eta^2/\Delta}\}}$$

$$+ \frac{4\exp\{(2 + \frac{3}{2\pi}\log 34)\frac{\pi}{\Delta}\}t_m^{-\eta^2/\Delta+(1+\eta^2)(\lambda+1)}}{\pi\{1 - \exp\{(2 + \frac{3}{2\pi}\log 34)\frac{2\pi}{\Delta}\}t_m^{-\eta^2/\Delta}\}}.$$

Since $\Delta = \eta^2/4(\lambda + 1)$, there exists value m_4, $m_4 \geq m_3$, such that when $m \geq m_4$,

$$\log|g(w)| \leq 2\log(2M_1) + \log(2M_1) = 3\log(2M_1),$$

meaning that when point w is in the region: $u^*(\log t_m - 2\pi - \log 34) < u < u_*(\log t_m + 2\pi + \log 34)$, $1 - \Delta < v < 1$,

$$|g(w)| \leq (2M_1)^3.$$

Combining this with formula (5.88), we conclude that when $m \geq m_4$ and point w is in the region: $u^*(\log t_m - 2\pi - \log 34) < u < u_*(\log t_m + 2\pi + \log 34)$, $0 < v < 1$,

$$|g(w)| \leq (2M_1)^3.$$

According to formulas (5.79) and (5.85), there exists value m_5, $m_5 \geq m_4$, such that when $m \geq m_5$,

$$\{M_1 + (2M_1)^3\}\{\varepsilon_{\varepsilon'}^{1/3} + \varepsilon_m''^{1/3}\} < M_2.$$

Then, notice the condition of formula (5.85), by applying a suitable transformation and Theorem 3.2, we may conclude that $a_i = b_{j+1}$. However, this is impossible. Hence, we prove that when $m \geq m_5$, there exists a point w^* in the region: $u^*(\log t_m^{1-\eta^2}) + 1 \leq u \leq u_*(\log t_m^{1+\eta^2}) - 1$, $0 < v < 1 - \Delta$, such that

$$|g(w^*)| > 2M_1. \tag{5.89}$$

Now, we construct an upper semicircle C: $|w - \mathrm{Re}\, w^*| < 1$, $\mathrm{Im}\, w \geq 0$ with point $w = \mathrm{Re}\, w^*$ being the center and 1 being the radius. Obviously, C is in the closed region: $u^*(\log t_m^{1-\eta^2}) \leq u \leq u_*(\log t_m^{1+\eta^2})$, $0 \leq v < 1$. Let Γ be the line segment: $\mathrm{Re}\, w^* - 1 \leq u \leq \mathrm{Re}\, w^* + 1$, $v = 0$. Then according to Lemma 3.2, the harmonic measure of Γ corresponding to point $w \in C$ is

$$u(w, \Gamma, C) = \frac{2\varphi}{\pi} - 1,$$

where φ is an extended angle observing Γ from point w. Besides, according to formula (5.84), when $w \in \Gamma$,

$$\log|g(w) - a_i| \leq -At_m^{-(30\pi/\omega+1)\eta^2} T(t_m, f). \tag{5.90}$$

On the other hand, we assume a point w_{jms_j+1} on C such that

$$|g(w_{jms_j+1})| = \max_{w \in c}|g(w)|.$$

Hence, when $w \in C$,

$$|g(w) - a_i| \leq |g(w)| + |a_i| \leq |g(w_{jms_j+1})| + M_1. \tag{5.91}$$

Therefore, it follows from formulas (5.89)–(5.91) that

$$\log M_1 \leq \log |g(w^*) - a_i| \leq \log\{|g(w_{jms_j+1})| + M_1\}$$

$$+ u(w^*, \Gamma, C)\{-At_m^{-(30\pi/\omega+1)\eta^2} T(t_m, f)\},$$

$$\log^+ |g(w_{jms_j+1})| \geq u(w^*, \Gamma, C) \tag{5.92}$$

$$\times At_m^{-(30\pi/\omega+1)\eta^2} T(t_m, f) - \log(2M_1),$$

where $A > 0$ is a constant independent of m. In the following, we estimate $u(w^*, \Gamma, C)$. In fact, according to the equality

$$\frac{\pi}{4} = \left\{\frac{\varphi}{2} - \frac{\pi}{4}\right\} + \left\{\frac{\pi}{2} - \frac{\varphi}{2}\right\},$$

we have

$$1 = \frac{\tan\left(\frac{\varphi}{2} - \frac{\pi}{4}\right) + \tan\left(\frac{\pi}{2} - \frac{\varphi}{2}\right)}{1 - \tan\left(\frac{\varphi}{2} - \frac{\pi}{4}\right)\tan\left(\frac{\pi}{2} - \frac{\varphi}{2}\right)} = \frac{\tan\left(\frac{\varphi}{2} - \frac{\pi}{4}\right) + \operatorname{Im} w^*}{1 - \tan\left(\frac{\varphi}{2} - \frac{\pi}{4}\right)\operatorname{Im} w^*},$$

$$\tan\left(\frac{\varphi}{2} - \frac{\pi}{4}\right) = \frac{1 - \operatorname{Im} w^*}{1 + \operatorname{Im} w^*}.$$

Since $\operatorname{Im} w^* \leq 1 - \Delta$,

$$\tan\left(\frac{\varphi}{2} - \frac{\pi}{4}\right) \geq \frac{\Delta}{2}.$$

Furthermore, notice that $\frac{\pi}{2} \leq \varphi \leq \pi$; it follows that

$$\sin\left(\frac{\varphi}{2} - \frac{\pi}{4}\right) \geq \frac{\Delta}{2}\cos\left(\frac{\varphi}{2} - \frac{\pi}{4}\right) \geq \frac{\Delta}{2\sqrt{2}}.$$

Hence, we conclude that

$$\frac{\varphi}{2} - \frac{\pi}{4} \geq \frac{\Delta}{2\sqrt{2}},$$

$$u(w^*, \Gamma, C) = \frac{2\varphi}{\pi} - 1 = \frac{4}{\pi}\left(\frac{\varphi}{2} - \frac{\pi}{4}\right) \geq \frac{\sqrt{2}\Delta}{\pi} > 0.$$

Substituting this into formula (5.92), we assume value m_6, $m_6 \geq m_5$, such that when $m \geq m_6$,

$$\log^+ |g(w_{jms_j+1})| \geq At_m^{-(30\pi/\omega+1)\eta^2} T(t_m, f) > 2M_1,$$

where $A > 0$ is a constant independent of m. Therefore, point w_{jms_j+1} is in the region: $u^*(\log t_m^{1-\eta^2}) < u < u_*(\log t_m^{1+\eta^2})$, $0 < v < 1$.

Let z_{jms_j+1} be the preimage of point w_{jms_j+1} on the z-plane. Then point z_{jms_j+1} is in the region $D'_{jms_j+1} \subset D_{jms_j+1}$. Hence, we prove that when $m \geq m_6$ there exist a point z_{jms_j+1} and a constant $A > 0$ independent of m in D'_{jms_j+1}, such that

$$\log |f(z_{jms_j+1})| \geq At_m^{-(30\pi/\omega+1)\eta^2} T(t_m, f).$$

3) Since the number of $D_{jm\nu}$ $(1 \leq \nu \leq s_j + 1, \, s_j \neq 0, \, 1 \leq j \leq l_1)$ is $\leq p_1 + l_1$, there exists value m_7, $m_7 \geq m_6$ and a constant A_0, $0 \leq A_0 < 1$, independent of m, such that when $m \geq m_7$,

$$\log|f(z_{jm\nu})| \geq A_0 t_m^{-(30\pi/\omega+1)\eta^2} T(t_m, f),$$

$$z_{jm\nu} \in D'_{jm\nu} \subset D_{jm\nu}, \quad \nu = 1, 2, \ldots, s_j + 1, \, s_j \neq 0, \quad j = 1, 2, \ldots, l_1.$$

(5) Suppose $s_j \neq 0$ $(1 \leq j \leq l_1)$ and suppose that there exists value $A_0 t_m^{-(30\pi/\omega+1)\eta^2} T(t_m, f)$ in the interval $[\frac{1}{4} A_0 t_m^{-(30\pi/\omega+1)\eta^2} T(t_m, f),$ $\frac{1}{2} A_0 t_m^{-(30\pi/\omega+1)\eta^2} T(t_m, f)]$ $(m \geq m_7)$, such that the level curve $\log|f(z)| = A'_0 t_m^{-(30\pi/\omega+1)\eta^2} T(t_m, f)$ is analytic. We consider the set

$$E = \{z \mid \log|f(z)| > A'_0 t_m^{-(30\pi/\omega+1)\eta^2} T(t_m, f), \, |z| < t_{m+1}\}$$

and let $\Omega'_{jm\nu}$ be the connected component containing point $z_{jm\nu}$ $(1 \leq \nu \leq s_j + 1)$. According to the maximum modulus principle, we conclude that $\overline{\Omega}_{jm\nu} \cap (|z| = t_{m+1})$ is not an empty set and it contains interior points.

1) We prove that when m is sufficiently large, $\Omega'_{jm\nu} \subset D_{jm\nu}$ $(j = 1, 2, \ldots, l_1; \, \nu = 1, 2, \ldots, s_j + 1, \, s_j \neq 0)$. Hence, $\{\Omega'_{jm\nu}\}$ do not intersect among one another.

First, according to formula (5.75),

$$\lim_{m \to +\infty} \frac{\log T(t_m, F) - \log T(2t_{m-1}, F)}{\log t_m - \log 2t_{m-1}} \geq \lim_{R \to +\infty} \frac{RT'(R, F)}{T(R, F)} \geq \alpha,$$

where $F(z) = f(z) - a_1$. Besides, according to the First Fundamental Theorem, we have

$$T(t_m, F) \leq T(t_m, f) + O(1),$$
$$T(2t_{m-1}, F) \geq T(2t_{m-1}, f) - O(1).$$

And hence,

$$\lim_{m \to +\infty} \frac{\log T(t_m, f) - \log T(2t_{m-1}, f)}{\log t_m - \log 2t_{m-1}} \geq \alpha.$$

Therefore, there exists value m_8, $m_8 \geq m_7$, such that when $m \geq m_8$,

$$\frac{\log T(t_m, f) - \log T(2t_{m-1}, f)}{\log t_m - \log 2t_{m-1}} \geq \frac{\alpha}{2},$$

$$T(2t_{m-1}, f) \leq \left(\frac{2t_{m-1}}{t_m}\right)^{\alpha/2} T(t_m, f).$$

Furthermore, according to the selection of η, $t_{m-1} \leq 2t_{m-1}^{1+\eta}$, $t_m \geq r_m$ and $r_m = r_{m-1}^{(1+\eta)^2}$, we may conclude that there exists value m_9, $m_9 \geq m_8$, such that when $m \geq m_9$,

$$
\begin{aligned}
\log M(t_{m-1}, f) &\leq 3T(2t_{m-1}, f) \\
&\leq 3 \left(\frac{2t_{m-1}}{t_m} \right)^{\alpha/2} T(t_m, f) \\
&\leq \frac{A_0}{4} t_m^{-(30\pi/\omega+1)\eta^2} T(t_m, f) \\
&\leq A_0' t_m^{-(30\pi/\omega+1)\eta^2} T(t_m, f).
\end{aligned}
\tag{5.93}
$$

On the other hand, from formulas (5.78), (5.79), (5.80), (5.82) and (5.83), when $m \geq m_9 \geq m_3$ and

$$
z \in \bigcup_{J=1}^{l_1} L_{jm} \bigcup_{i=1}^{p_1} \Delta \left(\frac{\theta_{k_i} + \theta_{k_i+1}}{2}; t_{m-1}, t_m \right),
$$

$\log |f(z)| \leq \log 2M_1$. Obviously, there exists value m_{10}, $m_{10} \geq m_9$, such that when $m \geq m_{10}$,

$$
\begin{aligned}
A_0' t_m^{-(30\pi/\omega+1)\eta^2} T(t_m, f) &\geq \frac{A_0}{4} t_m^{-(30\pi/\omega+1)} T(t_m, f) \\
&> \log 2M_1.
\end{aligned}
\tag{5.94}
$$

Hence, according to formulas (5.93) and (5.94), and also the definition of $\Omega'_{jm\nu}$, we conclude that when $m \geq m_{10}$, $\Omega'_{jm\nu} \subset D_{jm\nu}$.

2) We prove that when m is sufficiently large, $\Omega'_{jm\nu} \subset D_j(t_{m+1})$. First, there exists value m_{11}, $m_{11} \geq m_{10}$, such that when $m \geq m_{11}$,

$$
\begin{aligned}
A_0' t_m^{-(30\pi/\omega+1)\eta^2} T(t_m, f) &\geq \frac{A_0}{4} t_m^{-(30\pi/\omega+1)\eta^2} T(t_m, f) \\
&> \log 2M_0.
\end{aligned}
\tag{5.95}
$$

Hence, according to formula (5.70), we conclude that there is no intersection between $\Omega'_{jm\nu}$ and L_j $(j = 1, 2, \ldots, l_1)$. Next, according to $\Omega'_{jm\nu} \subset D_{jm\nu}$, we conclude that

$$
\begin{aligned}
\Omega'_{jm\nu} \cap (|z| = t_{m+1}) &\subset \overline{D}_{jm\nu} \cap (|z| = t_{m+1}) \\
&\subset \overline{D}_{jm} \cap (|z| = t_{m+1}) = \overline{D}_{jm}(t_{m+1}) \cap (|z| = t_{m+1}).
\end{aligned}
$$

Besides, since $\Omega'_{jm\nu} \cap (|z| = t_{m+1})$ is not an empty set and it contains interior points, we conclude that there exists point $z_1 \in \Omega'_{jm\nu} \cap D_j(t_{m+1})$. Hence, when $m \geq m_{11}$, $\Omega'_{jm\nu} \subset D_j(t_{m+1})$.

3) Let $\Omega_{jm\nu} \subset \Omega'_{jm\nu}$ be the connected component of $\Omega'_{jm\nu}$ in disk $|z| < \frac{1}{2} t_{m+1}$, which contains point $z_{jm\nu}$. Then $\{\Omega_{jm\nu}\}$ do not intersect among

one another. Let $\theta_{jm\nu} = \overline{\Omega}_{jm\nu} \cap (|z| = \frac{1}{2}t_{m+1})$ and denote $u_{jm\nu}(z)$ as the harmonic measure of $\theta_{jm\nu}$ regarding $\Omega_{jm\nu}$ corresponding to point z. We also let $r\theta_{jm\nu}(r)$ be the linear measure of circumference

$$|z| = r \ (2r_m^{1+\eta^2} \le r \le \tfrac{1}{4}t_{m+1}; \ m \ge m_{11})$$

in $\Omega_{jm\nu}$. Hence, we have

$$A_0 t_m^{-(30\pi/\omega+1)\eta^2} T(t_m, f) \le \log|f(z_{jm\nu})|$$
$$\le \frac{A_0}{2} t_m^{-(30\pi/\omega+1)\eta^2} T(t_m, f) + u_{jm\nu}(z)\log M\left(\frac{1}{2}t_{m+1}\right). \tag{5.96}$$

Notice that point $z_{jm\nu} \in D'_{jm\nu} \subset D_{jm\nu}$ and $z_{jm\nu} \in \Omega_{jm\nu} \subset D_{jm\nu}$. We use a curve l_z in $\Omega_{jm\nu}$ to connect a point on points $z_{jm\nu}$ and $\theta_{jm\nu}$, then l_z must intersect circumference $|z| = t_m^{1+\eta^2}$. We assume the last intersection point; then such a point may determine a connected arc on circumference $|z| = t_m^{1+\eta^2}$, such that this arc is in $\Omega_{jm\nu}$. Moreover, it divides $\Omega_{jm\nu}$ into two disconnected regions. We assume a point $z^*_{jm\nu}$, $|z^*_{jm\nu}| = t_m^{1+\eta^2}$ on such an arc, such that $u_{jm\nu}(z^*_{jm\nu})$ reaches the largest value. Obviously, $z^*_{jm\nu} \in \Omega_{jm\nu}$. Moreover, according to the maximum modulus principle, we conclude that $u_{jm\nu}(z) \le u_{jm\nu}(z^*_{jm\nu})$. Accordingly, it follows from formula (5.96) and Theorem 3.1 that

$$\frac{A_0}{2} t_m^{-(30\pi/\omega+1)\eta^2} T(t_m, f) \le u_{jm\nu}(z^*_{jm\nu})\log M\left(\frac{1}{2}t_{m+1}, f\right)$$
$$\le 9\sqrt{2}e^{-\pi\int_{2t_m}^{1/2t_{m+1}} 1+\eta^2 \frac{dr}{r\theta_{jm\nu}(r)}} 3T(t_{m+1}, f),$$
$$\pi\int_{2t_m^{1+\eta^2}}^{\frac{1}{2}t_{m+1}} \frac{dr}{r\theta_{jm\nu}(r)} \le \log T(t_{m+1}, f) - \log T(t_m, f) \tag{5.97}$$
$$+ \left(\frac{30\pi}{\omega}+1\right)\eta^2\log t_m + \log\frac{54\sqrt{2}}{A_0}.$$

4) When $m \ge m_{11}$, according to formula (5.95), $\Omega_{jm\nu}$ may lie entirely in $\Omega_j(\frac{1}{2}t_{m+1})$, or it does not intersect with $\Omega_j(\frac{1}{2}t_{m+1})$. Suppose that there exists some domain $\Omega_{jm\nu_j}$ $(1 \le \nu_j \le s_j+1)$, such that $\Omega_{jm\nu_j}$ lies entirely in $\Omega_j(\frac{1}{2}t_{m+1})$. We simply denote $\Omega_{jm\nu_j} = \Omega_{jm}$ and $\theta_{jm\nu_j}(r) = \theta_{jm}(r)$. Then according to Theorem 3.1,

$$A_0 t_m^{-(30\pi/\omega+1)\eta^2} T(t_m, f) \le \log|f(z_{jmv_j})|$$

$$\le \frac{A_0}{2} t_m^{-(30\pi/\omega+1)\eta^2} T(t_m, f)$$

$$+ u_{jmv_j}(z_{jmv_j}) \log M\left\{\overline{\Omega}_j\left(\frac{1}{2}t_{m+1}\right) \cap \left(|z| = \frac{1}{2}t_{m+1}\right), f\right\},$$

$$\frac{A_0}{2} t_m^{-(30\pi/\omega+1)\eta^2} T(t_m, f)$$

$$\le u_{jmv_j}(z^*_{jmv_j}) \log M\left\{\overline{\Omega}_j\left(\frac{1}{2}t_{m+1}\right) \cap \left(|z| = \frac{1}{2}t_{m+1}\right), f\right\}$$

$$\le 9\sqrt{2} e^{-\pi \int_{2t_m}^{\frac{1}{2}t_{m+1}} 1+\eta^2 \frac{dr}{r\theta_{jm}(r)}} \log M\left\{\overline{\Omega}_j\left(\frac{1}{2}t_{m+1}\right) \cap \left(|z| = \frac{1}{2}t_{m+1}\right), f\right\},$$

$$\pi \int_{2t_m^{1+\eta^2}}^{\frac{1}{2}t_{m+1}} \frac{dr}{r\theta_{jm}(r)} \le \log\log M\left\{\overline{\Omega}_j\left(\frac{1}{2}t_{m+1}\right) \cap \left(|z| = \frac{1}{2}t_{m+1}\right), f\right\}$$

$$- \log\log M\left\{\overline{\Omega}_j\left(\frac{1}{2}t_m\right) \cap \left(|z| = \frac{1}{2}t_m\right), f\right\}$$

$$+ \left(\frac{30\pi}{\omega} + 1\right)\eta^2 \log t_m + \log \frac{54\sqrt{2}}{A_0}.$$

$$(5.98)$$

5) Suppose that the $s_j + 1$ domains Ω_{jmv} ($v = 1, 2, \dots, s_{j+1}$) do not intersect $\Omega_j(\frac{1}{2}t_{m+1})$. We first assume arbitrarily a value v_j ($1 \le v_j \le s_j+1$); then we define a new domain Ω_{jm} to replace Ω_{jmv_j}. Concretely speaking, we define Ω_{jm} as follows: We first assume a point $z_{jm} \in \overline{\Omega}_j(t_m) \cap (|z| = t_m)$ such that

$$|f(z_{jm})| = M\{\overline{\Omega}_j(t_m) \cap (|z| = t_m), f\}.$$

Obviously, point z_{jm} is an interior point of $\Omega_j(\frac{1}{2}t_{m+1})$. Then, we assume a value A_0'' in the interval

$$\left[\sqrt[4]{M\{\overline{\Omega}_j(t_m) \cap (|z| = t_m), f\}}, \sqrt{M\{\overline{\Omega}_j(t_m) \cap (|z| = t_m), f\}}\right]$$

such that the level curve $|f(z)| = A_0''$ is analytic. We consider the set

$$E = \left\{z \mid |f(z)| > A_0'', \ |z| < \frac{1}{2}t_{m+1}\right\},$$

and let Ω_{jm} be the connected component containing point z_{jm}. Then according to the maximum modulus principle, the set $\overline{\Omega}_{jm} \cap (|z| = \frac{1}{2}t_{m+1})$ is not an empty set and it contains interior points. On the other hand, according to formula (5.74), there exists value m_{12}, $m_{12} \ge m_{11}$, such that when $m \ge m_{12}$,

$$A_0'' \ge \sqrt[4]{M\{\overline{\Omega}_j(t_m) \cap (|z| = t_m), f\}} > M_0.$$

Hence, when $m \geq m_{12}$, $\Omega_{jm} \subset \Omega_j(\frac{1}{2}t_{m+1})$. Moreover, there is not intersection between Ω_{jm} and $\Omega_{jm\nu}$ $(1 \leq \nu \leq s_j + 1)$. We let $r\theta_{jm}(r)$ be the linear measure of circumference $|z| = r$ $(t_m \leq r \leq \frac{1}{4}t_{m+1})$ in Ω_{jm}. Applying Theorem 3.1, we get

$$\log M\{\overline{\Omega}_j(t_m) \cap (|z| = t_m), f\}$$
$$= \log |f(z_{jm})|$$
$$\leq \frac{1}{2} \log M\{\overline{\Omega}_j(t_m) \cap (|z| = t_m), f\}$$
$$+ 9\sqrt{2} \exp\left\{-\pi \int_{2t_m^{1+\eta 2}}^{t_{m+1}/4} \frac{dr}{r\theta_{jm}(r)}\right\} \log M$$
$$\times \left\{\overline{\Omega}_j\left(\frac{1}{2}t_{m+1}\right) \cap \left(|z| = \frac{1}{2}t_{m+1}\right), f\right\}.$$

Furthermore, we obtain

$$\pi \int_{2t_m^{1+\eta^2}}^{4_{m+1}/4} \frac{dr}{r\theta_{jm}(r)}$$
$$\leq \log \log M\left\{\overline{\Omega}_j\left(\frac{1}{2}t_{m+1}\right) \cap \left(|z| = \frac{1}{2}t_{m+1}\right), f\right\}$$
$$- \log \log M\{\overline{\Omega}_j(t_m) \cap (|z| = t_m), f\}$$
$$+ \left(\frac{30\pi}{\omega} + 1\right)\eta^2 \log t_m + \log \frac{54\sqrt{2}}{A_0}.$$

Hence, we obtain formula (5.98) once more in form.

From now on, when $m \geq m_{12}$ and $s_j \neq 0$ $(1 \leq j \leq l_1)$, if we let $1 \leq \nu \neq \nu_j \leq s_j + 1$, then we assume formula (5.97); and if we let $\nu = \nu_j$, then we assume formula (5.98).

6) Let $s_j = 0$ $(1 \leq j \leq l_1)$. Analogously, we may define region $\Omega_{jm} \subset \Omega_j(\frac{1}{2}t_{m+1}) \subset \Omega_j(t_{m+1}) \subset D_j(t_{m+1})$ and derive the inequality

$$\pi \int_{2t_m^{1+\eta^2}}^{\frac{1}{4}t_{m+1}} \frac{dr}{r\theta_{jm}(r)} \leq \log \log M$$
$$\times \left\{\overline{\Omega}_j\left(\frac{1}{2}t_{m+1}\right) \cap \left(|z| = \frac{1}{2}t_{m+1}\right), f\right\}$$
$$- \log \log M\left\{\overline{\Omega}_j\left(\frac{1}{2}t_m\right) \cap \left(|z| = \frac{1}{2}t_m\right), f\right\}, \tag{5.99}$$
$$+ \left(\frac{30\pi}{\omega} + 1\right)\eta^2 \log t_m + \log \frac{54\sqrt{2}}{A_0}.$$

7) According to the above discussion, we find that when $m \geq m_{12}$ and $2t_m^{1+\eta^2} \leq r \leq \frac{1}{4}t_{m+1}$, if we let

$$\sum_{1 \leq j \leq l_1} \theta_{jm}(r) = \sum_{\substack{1 \leq j \leq l \\ s_j = 0}} \theta_{jm}(r) + \sum_{\substack{1 \leq j \leq l_1 \\ s_j \neq 0 \\ \nu = \nu_j}} \theta_{jm\nu}(r),$$

then

$$\sum_{1 \leq j \leq l_1} \theta_{jm}(r) + \sum_{\substack{1 \leq j \leq l_1 \\ s_j \neq 0 \\ 1 \leq \nu \neq \nu_j \leq s_j + 1}} \theta_{jm\nu}(r) \leq 2\pi.$$

And hence,

$$(p_1 + l_1)^2 = \left\{ \sum_{\substack{1 \leq j \leq l_1 \\ s_j \neq 0 \\ 1 \leq \nu \neq \nu_j \leq s_j + 1}} \sqrt{\theta_{jm\nu}(r)} \cdot \frac{1}{\sqrt{\theta_{jm\nu}(r)}} + \sum_{1 \leq j \leq l_1} \sqrt{\theta_{jm}(r)} \frac{1}{\sqrt{\theta_{jm}(r)}} \right\}^2$$

$$\leq 2\pi \left\{ \sum_{\substack{1 \leq j \leq l_1 \\ s_j \neq 0 \\ 1 \leq \nu \neq \nu_j \leq s_j + 1}} \frac{1}{\theta_{jm\nu}(r)} + \sum_{1 \leq j \leq l_1} \frac{1}{\theta_{jm}(r)} \right\},$$

$$\frac{1}{2}(p_1 + l_1)^2 \int_{2t_m^{1+\eta^2}}^{t_{m+1}/4} \frac{dr}{r} \leq \sum_{\substack{1 \leq j \leq l_1 \\ s_j \neq 0 \\ 1 \leq \nu \neq \nu_j \leq s_j + 1}} \pi \int_{2t_m^{1+\eta^2}}^{t_{m+1}/4} \frac{dr}{r\theta_{jm\nu}(r)}$$

$$+ \sum_{1 \leq j \leq l_1} \pi \int_{2t_m^{1+\eta^2}}^{t_{m+1}/4} \frac{dr}{r\theta_{jm}(r)}.$$

Furthermore, we derive from formulas (5.97)–(5.99) that

$$\frac{1}{2}(p_1 + l_1)^2 \left\{ \log \frac{t_{m+1}}{t_m} - \eta^2 \log t_m - \log 8 \right\}$$

$$\leq p_1 \{ \log T(t_{m+1}, f) - \log T(t_m, f) \}$$

$$+ \left(\frac{30\pi}{\omega} + 1 \right) \eta^2 p_1 \log t_m + p_1 \log \frac{54\sqrt{2}}{A_0}$$

$$+ \sum_{1 \leq j \leq l_1} \left\{ \log \log M \left\{ \overline{\Omega}_j \left(\frac{1}{2}t_{m+1} \right) \cap \left(|z| = \frac{1}{2}t_{m+1} \right), f \right\} \right.$$

$$\left. - \log \log M \left\{ \overline{\Omega}_j \left(\frac{1}{2}t_m \right) \cap \left(|z| = \frac{1}{2}t_m \right), f \right\} \right\}$$

$$+ \left(\frac{30\pi}{\omega} + 1 \right) \eta^2 l_1 \log t_m + l_1 \log \frac{54\sqrt{2}}{A_0}.$$

$$(5.100)$$

Now, suppose that sequence R_n $(n = 1, 2, \ldots)$ satisfies the condition

$$\lim_{n \to +\infty} \frac{\log T(R_n, f)}{\log R_n} = \mu.$$

Also for each sufficiently large value n, there exists m_n such that

$$t_{m_n - 1} < R_n \le t_{m_n}. \tag{5.101}$$

Hence, we have

$$r_0^{(1+\eta)^{2m_n - 2}} = r_{m_n - 1} \le t_{m_n - 1} < R_n,$$

$$m_n \le \frac{\log \log R_n - \log \log r_0}{2 \log(1 + \eta)} + 1. \tag{5.102}$$

Meanwhile,

$$
\begin{aligned}
t_{m_n - 1} &\ge r_0^{(1+\eta)^{2m_n - 2}} \\
&= \left(\frac{1}{2}\right)^{1/(1+\eta)^3} (2 r_{m_n}^{1+\eta})^{1/(1+\eta)^3} \\
&\ge \left(\frac{1}{2}\right)^{1/(1+\eta)^3} t_{m_n}^{1/(1+\eta)^3} \\
&\ge \left(\frac{1}{2} R_n\right)^{1/(1+\eta)^3}.
\end{aligned}
\tag{5.103}
$$

Moreover,

$$
\begin{aligned}
\sum_{m=m_{12}}^{m_n - 2} \log t_m &\le \sum_{m=m_{12}}^{m_n - 2} \log(2 r_m^{1+\eta}) \\
&= \sum_{m=m_{12}}^{m_n - 2} \log r_0^{(1+\eta)^{2m+1}} + (m_n - m_{12} - 1) \log 2 \\
&= \sum_{m=m_{12}}^{m_n - 2} (1 + \eta)^{2m+1} \log r_0 + (m_n - m_{12} - 1) \log 2 \\
&= \log r_0 \frac{(1+\eta)^{2m_n - 1} - (1+\eta)^{2m_{12}+1}}{2(\eta + 1)\eta} + (m_n - m_{12} - 1) \log 2 \\
&\le \frac{(1+\eta) \log R_n - (1+\eta)^{2m_{12}+1} \log r_0}{(2+\eta)\eta} + (m_n - m_{12} - 1) \log 2.
\end{aligned}
\tag{5.104}
$$

Let $m = m_{12}, m_{12}+1, \ldots, m_n - 2$. Then according to formulas (5.100)–(5.104), we obtain

$$\frac{1}{2}(p_1 + l_1)^2 \left\{ \log \frac{t_{m_n-1}}{t_{m_{12}}} - \eta^2 \sum_{m=m_{12}}^{m_n-2} \log t_m - (m_n - m_{12} - 1) \log 8 \right\}$$

$$\leq p_1 \{ \log T(t_{m_n-1}, f) - \log T(t_{m_{12}}, f) \}$$

$$+ \sum_{1 \leq j \leq l_1} \left\{ \log \log M \left\{ \overline{\Omega}_j \left(\frac{1}{2} t_{m_n-1} \right) \cap \left(|z| = \frac{1}{2} t_{m_n-1} \right), f \right\} \right.$$

$$\left. - \log \log M \left\{ \overline{\Omega}_j \left(\frac{1}{2} t_{m_{12}} \right) \cap \left(|z| = \frac{1}{2} t_{m_{12}} \right), f \right\} \right\}$$

$$+ (p_1 + l_1) \left(\frac{30\pi}{\omega} + 1 \right) \eta^2 \sum_{m=m_{12}}^{m_n-2} \log t_m$$

$$+ (p_1 + l_1)(m_n - m_{12} - 1) \log \frac{54\sqrt{2}}{A_0},$$

$$\frac{1}{2}(p_1 + l_1)^2 \left\{ \frac{1}{(1+\eta)^3} \log R_n - \frac{1}{(1+\eta)^3} \log 2 - \log t_{m_{12}} \right.$$

$$- \frac{(1+\eta) \log R_n - (1+\eta)^{2m_{12}+1} \eta \log r_0}{2 + \eta}$$

$$\left. - \frac{\log \log R_n - \log \log r_0}{2 \log(1+\eta)} \cdot \log 16 \right\}$$

$$\leq (p_1 + l_1) \log T(R_n, f) + l_1 \log 3 + (p_1 + l_1) \left(\frac{30\pi}{\omega} + 1 \right)$$

$$\times \frac{(1+\eta)\eta \log R_n - (1+\eta)^{2m_{12}+1} \eta \log r_0}{2 + \eta}$$

$$+ (p_1 + l_1) \frac{\log \log R_n - \log \log r_0}{2 \log(1+\eta)} \log \frac{2^{(30\pi/\omega+1)\eta^2} \cdot 54\sqrt{2}}{A_0}.$$

We use $\log R_n$ to divide the two sides of the last inequaity, and then let $n \to +\infty$. We conclude that

$$\frac{1}{2}(p_1 + l_1)^2 \left\{ \frac{1}{(1+\eta)^3} - \frac{(1+\eta)\eta}{2+\eta} \right\}$$

$$\leq (p_1 + l_1)\mu + (p_1 + l_1) \left(\frac{30\pi}{\omega} + 1 \right) \frac{(1+\eta)\eta}{2+\eta},$$

$$(p_1 + l_1) \left\{ \frac{1}{(1+\eta)^2} - \frac{(1+\eta)\eta}{2+\eta} \right\} \leq 2\mu + 2 \left(\frac{30\pi}{\omega} + 1 \right) \frac{(1+\eta)\eta}{2+\eta}.$$

Finally, we let $\eta \to 0$; then

$$p_1 + l_1 \leq 2\mu.$$

However, this contradicts formula (5.69). Hence, Theorem 5.14 is proved completely.

CHAPTER 6

The Relationship between Deficient Values
of a Meromorphic Function
and Direct Transcendental Singularities
of its Inverse Functions

We have discussed the relationship between deficient values and asymptotic values of an entire function in Chapter 5. Now, we shall discuss for a meromorphic function the relationship between its deficient values and the direct transcendental singularities of its inverse functions.

§6.1. On meromorphic functions having deficiency sum two [43g]

THEOREM 6.1. *Let* $w = f(z)$ *be a meromorphic function on the open plane* $|z| < +\infty$, $z = g(w)$ *be an inverse function of* $f(z)$, l *be the number of distinct direct transcendental singularities of* $g(w)$, *and* p *be the number of deficient values of* $f(z)$ *of which* l' *deficient values are simultaneously direct transcendental singularities of* $g(w)$. *If the sum of deficiencies of* $f(z)$

$$\Delta(f) = \sum_a \delta(a, f), \qquad \delta(a, f) > 0,$$

is equal to 2, *then* $p - l' + l \le 2\mu$, *where* μ *is the lower order of* $f(z)$.

PROOF. When $\mu = +\infty$, it is obvious that Theorem 6.1 holds. Hence, we need only to consider the case when $\mu < +\infty$.

When $\mu < +\infty$, suppose that Theorem 6.1 does not hold. Then we have

$$p_1 - l' + l \ge [2\mu] + 1.$$

Then there exist integer $p_1 < +\infty$, $0 \le p_1 \le p - l'$, and integer $l_1 < +\infty$, $0 < l_1 \le l$, such that

$$p_1 + l_1 \ge [2\mu] + 1. \tag{6.1}$$

According to the assumption that $\Delta(f) = 2$, then it must be that $\rho \ge 2$. Hence, we may further request that $p_1 + l_1 \ge 2$. Besides, according to Theorem 3.4, $\mu > 0$. Now, we select l_1 distinct direct transcendental singularities b_j $(j = 1, 2, \ldots, l_1)$ and p_1 deficient values a_i $(i = 1, 2, \ldots, p_1)$ that are not direct transcendental singularities, with their corresponding deficiencies $\delta(a_i, f) = \delta_i > 0$. Without any loss of generality, we may assume

349

that a_i $(i = 1, 2, \ldots, p_1)$ and b_j $(j = 1, 2, \ldots, l_1)$ are all finite values. Meanwhile, $\delta(\infty, f) = 0$. Otherwise, we need only to construct a suitable fractional linear transformation.

(1) According to the definition of μ, there exists a sequence $R_n: R_n < R_{n+1} \to +\infty$ $(n \to +\infty)$ such that

$$\lim_{n \to +\infty} \frac{\log T(R_n, f)}{\log R_n} = \mu, \qquad 0 < \mu < +\infty.$$

Furthermore, applying Lemma 2.1, where we let $\nu = \mu$, $h_1 = 0$, we conclude that for any arbitrarily assumed number h $(h > 0)$, H $(H > \mu)$, if we let

$$K = hH,$$
$$E = \{t \mid T(te^h, f) \le e^K T(t, f), t \ge 1\}, \qquad (6.2)$$
$$E[1, R_n] = E \cap [1, R_n],$$

then

$$\lim_{R_n \to +\infty} \frac{1}{\log R_n} \int_{E[1, R_n]} \frac{dt}{t} \ge 1 - \frac{h\mu}{K}. \qquad (6.3)$$

Moreover, according to Lemma 3.9, we conclude that there exists a value r_0', $r_0' \ge 1$, such that when $t \ge r_0'$ and $t \in E \cap [r_0', +\infty)$, there must exist value R and a set $E_i(R)$ $(1 \le i \le p_1)$ corresponding to value θ $(0 \le \theta < 2\pi)$ in the interval $[t, te^\sigma]$ $(0 < \sigma < \frac{1}{5}h)$, such that when $\theta \in E_i(R)$,

$$\log \frac{1}{|f(Re^{i\theta}) - a_i|} \ge \frac{\delta}{4} T(R, f), \qquad \delta = \min_{1 \le i \le p_1} \{\delta_i\}, \qquad (6.4)$$

$$\log \frac{1}{|f'(Re^{i\theta})|} \ge \frac{\delta}{8} T(R, f). \qquad (6.5)$$

Moreover,

$$\text{meas } E_i(R) \ge M = M(\delta, h, H, \sigma) > 0. \qquad (6.6)$$

(2) Suppose that around a neighborhood of the origin $z = 0$, $f'(z)$ has the expansion

$$f'(z) = c_s z^s + c_{s+1} z^{s+1} + \cdots, \qquad c_s \ne 0,$$

and γ_k represents a zero of $f'(z)$. When $t \in E \cap [r_0', +\infty)$, let

$$\pi(z) = \prod_{0 < |\gamma_k| < te^{h-2\sigma}} \frac{te^{h-2\sigma}(z - \gamma_k)}{(te^{h-2\sigma})^2 - \overline{\gamma}_k z},$$

$$\alpha = \frac{c_s}{\pi(0)}, \qquad (6.7)$$

$$G(z) = \frac{\alpha z^s \pi(z)}{f'(z)},$$

then $G(z)$ is regular in disk $|z| < te^{h-2\sigma}$ and $G(0) = 1$. Moreover,[1]

$$\log^+ |\alpha| \leq \log^+ |c_s| + N\left(te^{h-2^\sigma}, \frac{1}{f'}\right)$$
$$- n(0, f' = 0)\log(te^{h-2\sigma}),$$
(6.8)

and

$$\log^+ \frac{1}{|\alpha|} \leq \log^+ \frac{1}{|c_5|}.$$
(6.9)

We assume value r_1', $r_1' \geq r_0'$, such that when $r \geq r_1'$,

$$T\left(r, \frac{1}{f - b_j}\right) < 2T(r, f),$$

$$T\left(r, \frac{1}{f'}\right) < 2T(r, f'),$$
(6.10)

$$\text{meas}[re^{h-\sigma}, re^h] \geq 3.$$

Moreover, according to Lemma 3.10, when $r \geq r_1'$ and $r \notin E_0$,

$$\left\{\frac{1}{\sigma} \log \frac{16e^{1+h-2\sigma}}{M} \cdot \frac{N(r, 1/f')}{T(r, f')} \cdot \frac{T(r, f')}{T(r, f)}\right.$$
$$\left. + \frac{\log^+(1/|c_s|)}{T(r, f)} + \frac{n(0, f' = 0)\log r}{T(r, f)}\right\} e^K < \frac{\delta}{32},$$
(6.11)

$$\left\{\frac{N(r, 1/f')}{T(r, f')} \cdot \frac{T(r, f')}{T(r, f)} + \frac{\log^+ |c_s|}{T(r, f)}\right\} 2e^K < \frac{\delta}{8 \times 16},$$

$$\frac{T(r, f')}{T(r, f)} + \frac{\log^+(1/|c_s|)}{T(r, f)} \leq 4,$$
(6.12)

where E_0 is determined by Lemma 3.10, meas $E_0 \leq 2$.

In the following, analogous to the proof of formulas (3.98) and (3.99), we may conclude that when $t \in E \cap [r_1' + \infty)$, there exists value θ_{iR} in the set $E_i(R)$ $(1 \leq i \leq p_i)$, such that when $z_{iR} = Re^{i\theta_{iR}}$,

$$\log |G(z_{iR})| \geq \frac{\delta}{16} T(R, f).$$
(6.13)

Moreover, there exists value $R' \notin E_0$ in the interval $[te^{h-\sigma}, te^h]$, such that

$$\log M(te^{h-4\sigma}, G) \leq 8 \cdot \frac{e^\sigma + 1}{e^\sigma - 1} T(R', f).$$
(6.14)

(3) Let

$$d = \min_{\substack{1<i=i'<p_1 \\ 1<j=j'<l_1}} \{|a_i - a_{i'}|, |a_i - b_j|, |b_j - b_{j'}| |b_j \neq b_{j'}|\} > 0$$

---

[1] See formulas (3.92) and (3.93).

and select value r_2', $r_2' \geq r_1'$, such that when $t \geq r_2'$,

$$e^{(\delta/16)T(t, f)} \geq 16,$$

$$2te^{h-4\sigma}\left\{1 + \sqrt{2}\pi\sqrt{\frac{16}{\sigma}e^K T(t, f)}\right\}e^{-(\delta/(8\cdot 16))T(t, f)} + e^{-(\delta/2)T(t, f)}$$

$$\leq e^{-(\delta/256)T(t, f)} < \frac{d}{4}. \tag{6.15}$$

When $t \in E \cap [r_2', +\infty)$, applying Lemma 3.5 for $G(z)$, where we let $R = te^{h-3\sigma}$, $r = te^{h-4\sigma}$, $z_0 = z_{iR}$, $A = e^{(\delta/16)T(t, f)} \geq 16$, then there exists value A' in the interval $[\sqrt[4]{A}, \sqrt{A}]$, such that $G'(z)$ has no zeros on the level curve $|G(z)| = A'$, meaning that the level curve is analytic. Moreover, for any point z on the closure $\overline{\Omega}(z_{iR})$, there exists a piecewise analytic curve L, with its length satisfying

$$\text{meas } L \leq 2te^{h-4\sigma} + 2\sqrt{2}\pi te^{h-4\sigma}\sqrt{\frac{1}{\sigma}T(te^{h-3\sigma}, G)}$$

that links points z and z_{iR}, where the closure $\overline{\Omega}(z_{iR})$ is the connected component of the set

$$\Omega(A') = E\{z \mid |G(z)| > A', |z| < te^{h-3^\sigma}\} \tag{6.16}$$

contains point z_{iR} and lies in the disk $|z| < te^{h-4\sigma}$. Meanwhile, when $z \in L$,

$$|G(z)| \geq \exp\left\{\frac{1}{4} \cdot \frac{\delta}{16}T(t, f)\right\}, \qquad |z| \leq te^{h-4\sigma}.$$

Hence, according to formulas (6.2), (6.7), (6.8) and (6.12), we conclude that[2]

$$\text{meas } L \leq 2te^{h-4\sigma}\left\{1 + \sqrt{2}\pi\sqrt{\frac{16}{\sigma}e^K T(t, f)}\right\},$$

and when $z \in L$, according to formulas (6.7), (6.8), (6.11) and (6.12), we conclude that[3]

$$|f'(z)| \leq \exp\left\{-\frac{\delta}{8 \times 6}T(t, f)\right\}.$$

Therefore, when $z \in \Omega(z_{iR})$, it yields from formula (6.4) that

$$|f(z) - a_i| \leq |f(z) - f(z_{iR})| + |f(z_{iR}) - a_i|$$

$$\leq \int_L |f'(z)| |dz| + e^{-(\delta/4)T(t, f)}$$

$$\leq 2te^{h-4\sigma}\left\{1 + \sqrt{2}\pi\sqrt{\frac{16}{\sigma}e^K T(t, f)}\right\}$$

$$\times e^{-\delta/(18 \times 16)} + e^{-(\delta/4)T(t, f)}.$$

---

[2] See formula (3.103).
[3] See formula (3.104).

Furthermore, according to formula (4.14), we conclude that

$$|f(z) - a_i| < e^{-(\delta/256)T(t,f)} < \frac{d}{4}.$$

Hence, the p_1 domains $\Omega(z_{iR})$ $(i = 1, 2, \ldots, p_1)$ do not intersect among one another.

On the other hand, according to the definition of a direct transcendental singularity, there exists value $\alpha > \frac{4}{d}$, such that the region D_j corresponding to b_j $(1 \le j \le l_1)$ on the z-plane satisfies the following conditions:

1) When $z \in D_j$,

$$\frac{1}{|f(z) - b_j|} > \alpha, \qquad f(z) \ne b_j.$$

2) The finite boundary part Γ_j of D_j is an analytic curve. Moreover, when $z \in \Gamma_j$,

$$\frac{1}{|f(z) - b_j|} = \alpha.$$

3) If $j \ne j'$, then $D_j \cap D_{j'} = \varnothing$.

According to the assumption that $\Delta(f) = 2$, it must be that $p \ge 2$. Hence, for each value b_j $(1 \le j \le l_1)$, there exist deficient values that are different from b_j. Hence, according to Theorem 4.20, we conclude that

$$\varliminf_{r \to +\infty} \frac{\log\log M\{D_j \cap (|z| = r), \frac{1}{f - b_j}\}}{\log r} \ge \frac{1}{2}, \qquad j = 1, 2, \ldots, l_1.$$

Furthermore, we conclude that there exist value r_3', $r_3' \ge r_2'$, such that when $r \ge r_3'$,

$$M\left\{D_j \cap (|z| = r), \frac{1}{f - b_j}\right\} > \alpha^4. \tag{6.17}$$

Meanwhile, there are intersections between circumference $|z| = r$ and D_j $(j = 1, 2, \ldots, l_1)$. Obviously, $M\{D_j \cap (|z| = r), \frac{1}{f - b_j}\}$ $(1 \le j \le l_1)$ is a monotonic increasing function of r. Moreover, when $r \ge r_3'$, if there exists point z_{jr}', such that

$$\frac{1}{|f(z_{jr}') - b_j|} = M\left\{D_j \cap (|z| = r), \frac{1}{f - b_j}\right\},$$

then z_{jr}' is an interior point of D_j.

When $t \in E \cap [r_3', +\infty]$, there exists value A'' in the interval

$$\left[\sqrt[4]{M\left\{D_j \cap (|z| = R), \frac{1}{f - b_j}\right\}}, \sqrt{M\left\{D_j \cap (|z| = R), \frac{1}{f - b_j}\right\}}\right]$$

$(t \leq R \leq te^{\sigma}, \ 1 \leq j \leq l_1)$, such that derivative $\{1/(f(z) - b_j)\}'$ has no zeros and poles on the level curve $1/|f(z) - b_j| = A''$, meaning that the level curve is analytic. We consider the set

$$\Omega(A') = \left\{ z \bigg| \frac{1}{|f(z) - b_j|} > A'' , \ |z| < he^{h-4\sigma} \right\} \tag{6.18}$$

and let $\Omega(z'_{jR})$ be the connected branch of $\Omega(A')$ in the disk $|z| < te^{h-4\sigma}$ and containing point z'_{iR}. According to formula (6.17), we conclude that $\Omega(z'_{jR}) \subset D_j$. On the other hand, according to the maximum modulus principle, we conclude that the intersection set $\Omega(z'_{jR}) \cap (|z| = te^{h-4\sigma})$ is not an empty set and it contains interior points. Obviously, $\Omega(z'_{jR})$ $(j = 1, 2, \ldots l_1)$ do not intersect among one another. In the following, we shall illustrate that $\Omega(z_{iR})$ $(1 \leq i \leq p_1)$ and $\Omega(z'_{jR})$ $(1 \leq j \leq l_1)$ do not intersect among one another as well. Otherwise, there exists point $z_0 \in \Omega(z_{iR}) \cap \Omega(z'_{jR})$. Analogously, according to formulas (6.4), (6.15) and (6.17), we conclude that

$$d \leq |a_i - b_j| \leq |a_i - f(z_{iR})| + |f(z_{iR}) - f(z_0)| + |f(z_0) - b_j|$$
$$\leq e^{-(\delta/4)T(R,f)} + \int_{L_i} |f'(z)| \, |dz| + \frac{1}{\alpha} \leq \frac{d}{2},$$

and hence a contradiction is derived.

Notice that $p_1 + l_1 \leq 2$, therefore circumference $|z| = r$ $(r \geq R)$ cannot entirely lie in $\Omega(z_{iR})$ $(1 \leq i \leq p_1)$ as well as $\Omega(z'_{jR})$ $(1 \leq j \leq l_1)$. Let θ_{ir} be the part of circumference $|z| = r$ in $\Omega(z_{iR})$, θ'_{jr} be the part in $\Omega(z'_{jR})$, $r\theta_i(r)$ and $r\theta_j(r)$ represent the linear measures of θ_{ir} and θ'_{jr}, respectively. Hence, applying Theorem 3.1 and noting that $t \leq R \leq te^{\sigma}$, according to formulas (6.13) and (6.14), we have

$$\log |G(z_{iR})| \leq \frac{\delta}{2 \times 16} T(R, f)$$

$$+ 9\sqrt{2} \exp\left\{ -\pi \int_{2R}^{\frac{1}{2}te^{h-4\sigma}} \frac{dr}{r\theta_i(r)} \right\} \log M(te^{h-4\sigma}, G),$$

$$\pi \int_{2te^{\sigma}}^{te^{h-4\sigma}/2} \frac{dr}{r\theta_i(r)} \leq \log T(te^h, f) - \log T(t, f) \tag{6.19}$$

$$+ \log\left\{ \frac{6 \times 32 \times 9\sqrt{2}}{\delta} \cdot \frac{e^{\sigma} + 1}{e^{\sigma} - 1} \right\}, \qquad i = 1, 2, \ldots, p_1.$$

Moreover, according to formula (6.18),

$$\log \frac{1}{|f(z'_{jR}) - b_j|} \le \frac{1}{2} \log M \left\{ D_j \cap (|z| = R), \frac{1}{f - b_j} \right\}$$

$$+ 9\sqrt{2} \exp \left\{ -\pi \int_{2R}^{te^{h-4\sigma}/2} \frac{1}{r\theta'_j(r)} \right\} \log M$$

$$\times \left\{ D_j \cap (|z| = te^{h-4\sigma}), \frac{1}{f - b_j} \right\},$$

$$\pi \int_{2te^\sigma}^{te^{h-4\sigma}/2} \frac{dr}{r\theta'_j(r)} \le \log\log M \left\{ D_j \cap (|z| = te^{h-4\sigma}), \frac{1}{f - b_j} \right\}$$

$$- \log\log M \left\{ D_j \cap (|z| = t), \frac{1}{f - b_j} \right\} \tag{6.20}$$

$$+ \log(18\sqrt{2}), \qquad j = 1, 2, \ldots, l_1.$$

(4) According to the assumption, there exists a sequence $\{r_n\}$ tending monotonically to ∞, such that

$$\lim_{n \to +\infty} \frac{\log T(r_n, f)}{\log r_n} = \mu.$$

According to formula (6.3), when n is sufficiently large, the set of values t in the interval $[r'_3, r_n]$ satisfying the inequality

$$T(te^h, f) \le e^K T(t, f) \tag{6.21}$$

is not empty. We assume arbitrarily a sufficiently large value n. Let t_1 be the smallest value satisfying formula (6.21) in the interval $[r'_3, r_n]$, t_2 be the smallest value satisfying formula (6.21) in the interval $[t'_1, r_n]$ ($t'_1 = t_1 e^n$), and t_3 be the smallest value satisfying formula (6.21) in the interval $[t'_2, r_n]$ ($t'_2 = t_2 e^h$). After repeating the process m times we get

$$t'_m \le r_n < t'_{m+1}. \tag{6.22}$$

We assume $t = t_k$ ($k = 1, 2, \ldots, m$) in the formulas (6.19) and (6.20). Moreover, we let $\theta_i(r) = \theta_{ik}(r)$ and $\theta'_j(r) = \theta'_{jk}(r)$; then by adding k we obtain

$$\sum_{k=1}^m \pi \int_{2t_k e^\sigma}^{t_k e^{h-4\sigma}/2} \frac{dr}{r\theta_{ik}(r)} \le \log T(t_m e^h, f)$$

$$+ m \log \left\{ \frac{6 \times 32 \times 9\sqrt{2}}{\delta} \cdot \frac{e^\sigma + 1}{e^\sigma - 1} \right\}, \tag{6.23}$$

$$i = 1, 2, \ldots, p_1,$$

$$\sum_{k=1}^{m} \pi \int_{2t_k e^{\sigma}}^{t_k e^{h-4\sigma}/2} \frac{dt}{r\theta'_{ik}(r)} \le \log\log M$$

$$\times \left\{ D_j \cap (|z| = t_m e^{h-4\sigma}), \frac{1}{f - b_j} \right\} \tag{6.24}$$

$$+ m\log(18\sqrt{2}), \qquad j = 1, 2, \ldots, l_1.$$

Applying Lemma 3.8, where we let $f(z) = 1/(f(z) - b_j)$, $(1 \le j \le l_1)$, $R = t_m e^{h-\sigma}$, $R' = t_m e^{h-2\sigma}$, $r = t_m e^{h-3\sigma}$,

$$H = \frac{e^{\sigma} - 1}{8e} t_m e^{h-4\sigma},$$

then there exists value R'' in the interval $[t_m e^{h-4\sigma}, t_m e^{h-3\sigma}]$ such that circumference $|z| = R'' \notin (\gamma)$. Moreover,

$$\log M \left\{ D_j \cap (|z| = t_m e^{h-4\sigma}), \frac{1}{f - b_j} \right\}$$

$$\le \log M \left\{ D_j \cap (|z| = R''), \frac{1}{f - b_j} \right\}$$

$$\le \left\{ \frac{t_m e^{h-2\sigma} + t_m e^{h-3\sigma}}{t_m e^{h-2\sigma} - t_m e^{h-3\sigma}} + \frac{\log \frac{8e2t_m e^{h-2\sigma}}{(e^{\sigma}-1)t_m e^{h-4\sigma}}}{\log \frac{t_m e^{h-\sigma}}{t_m e^{h-2\sigma}}} \right\}$$

$$\times T \left(t_m e^{h-\sigma}, \frac{1}{f - b_j} \right).$$

Furthermore, from formula (6.10) we derive

$$\log M \left\{ D_j \cap (|z| = t_m e^{h-4\sigma}), \frac{1}{f - b_j} \right\}$$

$$\le 2 \left\{ \frac{e^{\sigma} + 1}{e^{\sigma} - 1} + \frac{1}{\sigma} \log \frac{16e^{2\sigma+1}}{e^{\sigma} - 1} \right\} T(t_m e^h, f).$$

Substituting the above into (6.24) we get

$$\sum_{k=1}^{m} \pi \int_{2t_k e^{\sigma}}^{t_k e^{h-4\sigma}/2} \frac{dr}{r\theta'_{jk}(r)} \le \log T(t_m e^h, f) + m\log(18\sqrt{2})$$

$$+ \log \left\{ 2\frac{e^{\sigma} + 1}{e^{\sigma} - 1} + \frac{2}{\sigma} \log \frac{16e^{2\sigma+1}}{e^{\sigma} - 1} \right\}, \tag{6.25}$$

$$j = 1, 2, \ldots, l_1.$$

According to formulas (6.22), (6.23) and (6.25), when

$$r \in \bigcup_{k=1}^{m} \left[2t_k e^{\sigma}, \frac{1}{2} t_k e^{h-4\sigma} \right],$$

we have

$$(p_1 + l_1)^2 = \left\{ \sum_{i=1}^{p_1} \sqrt{\theta_{ik}(r)} \cdot \frac{1}{\sqrt{\theta_{ik}(r)}} + \sum_{j=1}^{l_1} \sqrt{\theta'_{jk}(r)} \frac{1}{\sqrt{\theta'_{jk}(r)}} \right\}^2$$

$$\leq \left(\sum_{i=1}^{p_1} \theta_{ik}(r) + \sum_{j=1}^{l_1} \theta'_{jk}(r) \right) \left\{ \sum_{i=1}^{p_1} \frac{1}{\theta_{ik}(r)} + \sum_{j=1}^{l_1} \frac{1}{\theta_{jk}(r)} \right\}$$

$$\leq 2\pi \left\{ \sum_{i=1}^{p_1} \frac{1}{\theta_{ik}(r)} + \sum_{j=1}^{l_1} \frac{1}{\theta_{jk}(r)} \right\},$$

$$\sum_{k=1}^{m} \int_{2t_k e^{\sigma}}^{t_k e^{h-4\sigma}/2} \frac{(p_1 + l_1)^2}{r} \, dr$$

$$\leq 2 \left\{ \sum_{i=1}^{p_1} \sum_{k=1}^{m} \pi \int_{2t_k e^{\sigma}}^{t_k e^{h-4\sigma}/2} \frac{dr}{r\theta_{ik}(r)} + \sum_{j=1}^{l_1} \sum_{k=1}^{m} \pi \int_{2t_k e^{\sigma}}^{t_k e^{h-4\sigma}/2} \frac{dr}{r\theta'_{jk}(r)} \right\}$$

$$\leq 2 \left\{ p_1 \log T(r_n, f) + mp_1 \log \left[\frac{6 \times 32 \times 9\sqrt{2}}{\delta} \cdot \frac{e^{\sigma} + 1}{e^{\sigma} - 1} \right] \right.$$

$$+ l_1 \log T(r_n, f) + ml_1 \log(18\sqrt{2})$$

$$\left. + \log \left[2\frac{e^{\sigma} + 1}{e^{\sigma} - 1} + \frac{2}{\sigma} \log \frac{16e^{2\sigma+1}}{e^{\sigma} + 1} \right] \right\},$$

$$(p_1 + l_1)\{mh - m(\log 4 + 5\sigma)\}$$
$$\leq 2\{\log T(r_n, f) + m \cdot O(1)\}.$$

On the other hand, it yields from formulas (6.3) and (6.22) that

$$(m + 1)h \geq \int_{E[r_3', r_n]} \frac{dr}{r}, \qquad mh \leq \log \frac{r_n}{r_3'}.$$

Hence,

$$(p_1 + l_1) \left\{ \frac{1}{\log r_n} \int_{E[r_3', r_n]} \frac{dr}{r} - \frac{1}{\log r_n} - \frac{\log 4 + 5\sigma}{h} \cdot \frac{\log r_n - \log r_3'}{\log r_n} \right\}$$

$$\leq 2 \left\{ \frac{\log T(r_n, f)}{\log r_n} + \frac{O(1)}{h} \cdot \frac{\log r_n - \log r_3'}{\log r_n} \right\}.$$

When $n \to +\infty$, according to formula (6.3), we conclude that

$$(p_1 + l_1) \left(1 - \frac{h\mu}{K} - \frac{\log 4 + 5\sigma}{h} \right) \leq 2 \left(\mu + \frac{O(1)}{h} \right).$$

Furthermore, we let $K \to +\infty$ and $h \to +\infty$; then we have $p_1 + l_1 \leq 2\mu$. However, this contradicts formula (6.1) and, in turn, Theorem 6.1 is proved completely.

§6.2. On meromorphic functions of finite lower order [43c]

THEOREM 6.2. *Let $w = f(z)$ be a meromorphic function on the open plane $|z| < +\infty$ of lower order $\mu < +\infty$, $z = g(w)$ be the inverse function of $f(z)$, q be the number of Julia directions of $f(z)$, l be the number of distinct direct transcendental singularities of $g(w)$, and p be the number of deficiencies of $f(z)$, where l' of the deficient values are simultaneously direct transcendental singularities of $g(w)$. Then we have the following formula:*

$$p - l' + l \le q.$$

PROOF. (1) First we consider the following case: $q < +\infty$, $p < +\infty$, and $l < +\infty$. Moreover, there does not exist $p = 0$, $l = 0$; or $p = 0$, $l = 1$; or $p = 1$, $l = 0$ at the same time.

Under the above case, we may have $p \ge 2$ or $l \ge 2$, or $p \ge 1$ and $l \ge 1$. Hence, according to Theorem 3.4 and Corollary 1 of Theorem 4.20, we conclude that $\mu > 0$. However, the following case is exceptional: $p = 1$ and $l = 1$. Moreover, for the corresponding deficient value a and direct transcendental singularity b, we have $a = b$. However, it is obvious that Theorem 6.2 holds for such an exceptional case. On the other hand, according to Theorem 2.15, we conclude that $q \ge 1$.

In the following, we let a_i $(i = 1, 2, \ldots, p)$ be the p deficient values of $w = f(z)$, with their corresponding deficiencies $\delta(a_i, f) = \delta_i > 0$, and also b_j $(j = 1, 2, \ldots, l)$ be the l distinct direct transcendental singularities of the inverse function $z = g(w)$. Since it is impossible to have $p = 0$, $l = 0$; or $p = 0$, $l = 1$; or $p = 1$, $l = 0$, according to Lemma 4.11, then each b_j $(1 \le j \le l)$ corresponds to one region Ω_j. Finally, we let $\Delta(\theta_k)$ $(k = 1, 2, \ldots, q; 0 \le \theta_1 < \theta_2 < \cdots < \theta_q < 2\pi)$ be the q Julia directions of $f(z)$. Let

$$\omega = \min_{1 \le k \le q} \{\theta_{k+1} - \theta_k\}, \quad \theta_{q+1} = \theta_1 + 2\pi, \quad \delta = \min_{1 \le i \le p} \{\delta_i\}$$

and assume arbitrarily a number η satisfying

$$0 < \eta < \frac{\nu\omega}{80\pi}, \qquad \nu = \min\left\{\frac{1}{2}, \mu\right\}, \tag{6.26}$$

Then according to Lemma 2.2 (where we let $h_1 = 0$, $h = 3$) and Lemma 3.9 (where we let $\sigma = \log 2$, $H = \mu$), there must exist two sequences r_n $(n = 1, 2, \ldots)$ and t'_n $(n = 1, 2, \ldots,)$ such that[4]

$$\lim_{n \to +\infty} \frac{\log T(r_n, f)}{\log r_n} = \mu, \tag{6.27}$$

$$r_n \le t'_n \le 2r_n, \tag{6.28}$$

---

[4] When $p = 0$, we do not need to apply Lemmas 2.2 and 3.9. We can consider directly a sequence r_n $(n = 1, 2, \ldots)$ that satisfies formula (6.27).

as well as

$$\text{meas } E_i^n \geq K_1 = K_1(\delta, \mu, p) > 0, \qquad i = 1, 2, \ldots, p.$$

Here E_i^n represents the set of θ $(0 \leq \theta < 2\pi)$ that enables the following inequalities

$$\begin{cases} \log \dfrac{1}{|f(t_n' e^{i\theta}) - a_i|} \geq \dfrac{\delta}{4} T(t_n', f) & (a_i \neq \infty), \\[2mm] \log |f(t_n' e^{i\theta})| \geq \frac{\delta}{4} T(t_n', f) & (a = \infty) \end{cases}$$

to be held. According to the definition of lower order as well as formulas (6.26) and (6.28), when n is sufficiently large,

$$T(t_n', f) \geq T(r_n, f) \geq r_n^{\mu - \eta}. \tag{6.29}$$

On the other hand, for each direct transcendental singularity b_j $(j = 1, 2, \ldots, l)$ of $g(w)$, according to Lemma 4.11, when n is sufficiently large, there exists value t_{jn}'' in the interval $[r_n^{1-\eta}, r_n]$ such that

$$\theta(t_{jn}'') \geq K_2 = K_2(\eta, \mu) > 0, \qquad j = 1, 2, \ldots, l.$$

Here $\theta(t_{jn}'')$ contains the following implication: If we denote $\theta_{t_{jn}''}$ the part of circumference $|z| = t_{jn}''$ in Ω_j, then $t_{jn}'' \theta(t_{jn}'')$ represents the linear measure of $\theta_{t_{jn}''}$. Besides, when $z \in \theta_{t_{jn}''}$,

$$\begin{aligned} \log \dfrac{1}{|f(z) - b_j|} &\geq |z|^{1/2 - \eta} & (b_j \neq \infty), \\[2mm] \log |f(z)| &\geq |z|^{1/2 - \eta} & (b_j = \infty). \end{aligned} \tag{6.30}$$

We then select a number ε,

$$0 < \varepsilon < \min\left\{\frac{\omega}{4}, \frac{K}{8q}\right\}, \qquad K = \min\{K_1, K_2\}.$$

Let

$$\overline{\Omega} = \bigcup_{k=1}^{q} \overline{\Omega}(\theta_k + \varepsilon, \theta_{k+1} - \varepsilon).$$

Then according to Theorem 2.13 and the finite covering theorem, we conclude that there exist three distinct values α, β and γ, such that

$$\varlimsup_{r \to +\infty} \frac{\log^+ n(r, \overline{\Omega}, f = X)}{\log r} = 0, \qquad X = \alpha, \beta, \gamma.$$

Moreover, the spherical distances among α, β and γ are greater than d, $0 < d < \frac{1}{2}$.

(2) When n is sufficiently large, each deficient value a_i $(1 \leq i \leq p)$ corresponds to one set E_i^n, $\text{meas } E_i^n \geq K$. According to the selection of ε,

there exists at least one domain $\Omega(\theta_k + 2\varepsilon, \theta_{k_i+1} - 2\varepsilon; r_n^{1-\eta}, 2r_n)$ among the q domains $\Omega(\theta_k + 2\varepsilon, \theta_{k+1} - 2\varepsilon; r_n^{1-\eta}, 2r_n)$ $(k = 1, 2, \ldots, q)$ such that

$$\text{meas}\{E_i^n \cap [\theta_{k_i} + 2\varepsilon, \theta_{k_i+1} - 2\varepsilon]\} \geq \frac{K}{2q}.$$

Following this way, we may enable each deficient value a_i to correspond to one domain $\Omega(\theta_{k_i} + 2\varepsilon, \theta_{k_i+1} - 2\varepsilon; r_n^{1-\eta}, 2r_n)$ $(i = 1, 2, \ldots, p)$ and p deficient values a_i $(i = 1, 2, \ldots, p)$ to correspond to p domains $\Omega(\theta_{k_i} + 2\varepsilon, \theta_{k_i+1} - 2\varepsilon; r_n^{1-\eta}, 2r_n)$ $(i = 1, 2, \ldots, p)$. In the following, we prove any two of the domains in the set

$$\{\Omega(\theta_{k_i} + 2\varepsilon, \theta_{k_i+1} - 2\varepsilon; r_n^{1-\eta}, 2r_n) | i = 1, 2, \ldots, p\}$$

do not coincide with each other.

In fact, if there exist two domains $\Omega(\theta_{k_i} + 2\varepsilon; \theta_{k_i+1} - 2\varepsilon; r_n^{1-\eta}, 2r_n)$ and $\Omega(\theta_{k_{i'}} + 2\varepsilon, \theta_{k_{i'}+1} - 2\varepsilon; r_n^{1-\eta}, 2r_n)$ coincide with each other, then we have $k_i = k_{i'} = k$. However, $i \neq i'$. Applying Lemma 5.1, where we let $\theta = \frac{1}{2}(\theta_{k+1} - \theta_k) - \varepsilon$, $R = t_n'$, $R_1 = r_n^{1-\eta}$, $R_2 = 2r_n$,

$$E_\alpha = \{t_n' e^{i\varphi} | \varphi \in E_i^n \cap [\theta_k + 2\varepsilon, \theta_{k+1} - 2\varepsilon]\},$$

$H = \frac{K}{29} \geq \frac{\varepsilon}{2}$, $\alpha = a_i$ $(^5)$ $N = \frac{\delta}{4}T(t_n', f)$ and $\nu = 0$, then when n is sufficiently large, for the point z on $\Omega(\theta_k + 2\varepsilon, \theta_{k+1} - 2\varepsilon; r_n^{1-\eta}, 2r_n)$ and outside some circles (γ), we have

$$\log^+ \frac{1}{|f(z) - a_i|}$$

$$\geq A(\varepsilon, \theta) \left\{ (\log 2r_n)^{-1} \cdot 2^{-4\pi/(\theta_k+1 - \theta_k - 2\varepsilon)} \right.$$

$$\left. \times r_n^{-4\pi/(\theta_{k+1} - \theta_k - 2\varepsilon)} \frac{\delta}{4} T(t_n', f) \right\}$$

$$- B(\varepsilon, \theta, d) \cdot 2^{4\pi/(\theta_{k+1} - \theta_k - 2\varepsilon)} r_n^{4\pi/(\theta_{k+1} - \theta_k - 2\varepsilon)}$$

$$\times \{2^{2\eta} \cdot r_n^{2\eta^2} \cdot 2^\eta r_n^\eta \log(2r_n^\eta) + \log^+ |a_i|\}$$

$$\geq A(\varepsilon, \theta, \eta, \delta) \bar{r}_n^{-8\pi\eta/\omega - \eta}$$

$$\times T(t_n', f) - B(\varepsilon, \theta, d, \eta) r_n^{16\pi\eta/\omega + \eta}.$$

According to the selection of η, we get

$$-\frac{8\pi\eta}{\omega} - \eta + (\nu - \eta) > \frac{\nu}{2}, \qquad \frac{16\pi\eta}{\omega} + 4\eta + \frac{8\pi\eta}{\omega} + \eta - (\nu - \eta) < 0.$$

Furthermore, we conclude from formula (6.29) that

$$\log^+ \frac{1}{|f(z) - a_i|} \geq r_n^{\nu/2}$$

$(^5)$ When $a_i = \infty$, we need only to consider function $1/f(z)$.

or

$$|f(z) - a_i| \leq e^{-r_n^{\nu/2}}. \tag{6.31}$$

Besides, the sum of the radii of (γ) does not exceed $\frac{\varepsilon}{4} r_n^{1-\eta}$. Hence, according to the condition

$$\text{meas}\{E_{i'}^n \cap [\theta_{k_{i'}} + 2\varepsilon, \theta_{k_{i'}+1} - 2\varepsilon]\} \geq \frac{K}{2q} > \varepsilon,$$

we conclude that there exists point z' in $\Omega(\theta_k + 2\varepsilon, \theta_{k+1} - 2\varepsilon; r_n^{1-\eta}, 2r_n)$ and outside circles (γ), such that

$$\log \frac{1}{|f(z') - a_{i'}|} \geq \frac{\delta}{4} T(t_n', f).$$

Furthermore, according to formula (6.29) and the selection of η, when n is sufficiently large, we obtain

$$\log \frac{1}{|f(z') - a_{i'}|} \geq r_n^{\nu/2},$$

or

$$|f(z') - a_{i'}| \leq e^{-r_n^{\nu/2}}. \tag{6.32}$$

Since $i \neq i'$, therefore $a_i \neq a_{i'}$. However, according to formulas (6.31) and (6.32), we have

$$|a_i - a_{i'}| \leq |f(z') - a_i| + |f(z') - a_{i'}| \leq 2e^{-r_n^{\nu/2}}.$$

Hence, when n is sufficiently large, we shall derive a contradiction.

(3) When n is sufficiently large, each direct transcendental singularity b_j $(1 \leq j \leq l)$ of $g(w)$ corresponds to one set $\theta_{t_{jn}''}$. Moreover, $\theta(t_{jn}'') \geq K$ $(j = 1, 2, \ldots, l)$. According to the selection of ε, there exists at least one domain $\Omega(\theta_{k_j} + 2\varepsilon, \theta_{k_j+1} - 2\varepsilon; r_n^{1-\eta}, 2r_n)$ among the q domains $\Omega(\theta_k + 2\varepsilon, \theta_{k+1} - 2\varepsilon; r_n^{1-\eta}, 2r_n)$ $(k = 1, 2, \ldots, q)$ such that

$$\text{meas}\{\theta_{t_{jn}''} \cap \Gamma(\theta_{k_j} + 2\varepsilon, \theta_{k_j+1} - 2\varepsilon; t_{jn}'')\} \geq \frac{K}{2q} t_{jn}''.$$

Following this way, we may enable each direct transcendental singularity b_j to correspond to one domain $\Omega(\theta_{k_j} + 2\varepsilon, \theta_{k_j+1} - 2\varepsilon; r_n^{1-\eta}, 2r_n)$ $(j = 1, 2, \ldots, l)$. In the following, we prove that any two domains in the set $\{\Omega(\theta_{k_j} + 2\varepsilon, \theta_{k_j+1} - 2\varepsilon; r_n^{1-\eta}, 2r_n)|j = 1, 2, \ldots, l\}$ do not coincide with each other.

In fact, if there exist two domains $\Omega(\theta_{k_j} + 2\varepsilon, \theta_{k_j+1} - 2\varepsilon; r_n^{1-\eta}, 2r_n)$ and $\Omega(\theta_{k_j'} + 2\varepsilon, \theta_{k_j'+1} - 2\varepsilon; r_n^{1-\eta}, 2r_n)$ that coincide with each other, then we have $k_j = k_{j'} = k$. However, $j \neq j'$. Applying Lemma 5.1, where we let $\theta = \frac{1}{2}(\theta_{k+1} - \theta_k) - \varepsilon$, $R = t_{jn}''$, $R_1 = r_n^{1-\eta}$, $R_2 = 2r_n$, $E_\alpha = \theta_{t_{j'n}''}$,

$H = \frac{K}{2q} > \varepsilon$, $\alpha = b_j$, $N = t_{jn}''^{1/2-\eta} \geq t_{jn}''^{\nu-\eta}$, then when n is sufficiently large, for the point z in $\overline{\Omega}(\theta_k + 2\varepsilon, \theta_{k+1} - 2\varepsilon; r_n^{1-\eta}, 2r_n)$ and outside some circles (γ), we have

$$|f(z) - b_j| \leq e^{-r_n^{\nu/2}},$$

with the sum of the radii of (γ) not exceeding $\varepsilon/4r_n^{1-\eta}$. Hence, there exists a point z_n' on $\theta_{t_{jn}''}$ and outside circles (r), such that

$$|f(z_n') - b_j| \leq e^{-r_n^{\nu/2}},$$

as well as a point z_n'' on $\theta_{t_{jn}''}''$ and outside circles (γ), such that

$$|f(z_n'') - b_j| \leq e^{-r_n^{\nu/2}}. \tag{6.33}$$

Besides, according to formula (6.30), we have

$$|f(z_n'') - b_{j'}| \leq e^{-|z_n''|^{1/2} - \eta} \leq e^{-r_n^{(1-\eta)(\nu-\eta)}} \leq e^{-r_n^{\nu/2}}. \tag{6.34}$$

From formulas (6.33) and (6.34), we obtain

$$|b_j - b_{j'}| \leq |f(z_n'') - b_j| + |f(z_n'') - b_j| \leq 2e^{-r_n^{\nu/2}}.$$

Hence, when n is sufficiently large, $b_j = b_j = b$. We use a straight line to connect points z_n' and z_n''. If we come across circles (γ), then we replace them by circular arcs. Hence, we find a continuous curve L_n linking points z_n' and z_n''. Moreover, when $z \in L_n$,

$$|f(z) - b| \leq e^{-r_n^{\nu/2}}.$$

Notice that

$$\lim_{n \to +\infty} f(z_n') = b_j, \qquad \lim_{n \to +\infty} f(z_n'') = b_j,$$

$$\lim_{\substack{n \to +\infty \\ z \in L_n}} f(z) = b, \qquad b = b_j = b_{j'}.$$

We conclude that b_j and $b_{j'}$ are nondistinct direct transcendental singularities, and hence, a contradiction is derived.

(4) We prove that between the two sets

$$\{\Omega(\theta_{k_i} + 2\varepsilon, \theta_{k_i+1} - 2\varepsilon; r_n^{1-\eta}, 2r_n) | i = 1, 2, \ldots, p\}$$

and $\{\Omega(\theta_{k_j} + 2\varepsilon, \theta_{k_j+1} - 2\varepsilon; r_n^{1-\eta}, 2r_n) | j = 1, 2, \ldots l\}$ there are at most l' domains coinciding with one another.

In fact, if there are more than l' domains coinciding with one another between the two sets $\{\Omega(\theta_{k_i} + 2\varepsilon, \theta_{k_i+1} - 2\varepsilon; r_n^{1-\eta}, 2r_n) | i = 1, 2, \ldots p\}$ and $\{\Omega(\theta_{k_j} + 2\varepsilon, \theta_{k_j+1} - 2\varepsilon; r_n^{1-\eta}, 2r_n) | j = 1, 2, \ldots, l\}$, then among these domains, there must exist domain $\Omega(\theta_k + 2\varepsilon; \theta_{k+1} - 2\varepsilon; r_n^{1-\eta}, 2r_n)$ $(k = k_i =$

k_j), such that $a_i = b_j$. On the other hand, analogously, applying Lemma 5.1, we may conclude that $a_i = b_j$. Hence, we derive a contradiction.

Since there are at most l' domains between the two sets

$$\{\Omega(\theta_{k_i} + 2\varepsilon, \theta_{k_i+1} - 2\varepsilon; r_n^{1-\eta}, 2r_n) | i = 1, 2, \ldots p\}$$

and $\{\Omega(\theta_{k_j} + 2\varepsilon, \theta_{k_j+1} - 2\varepsilon; r_n^{1-\eta}, 2r_n) | j = 1, 2, \ldots l\}$, we conclude that

$$p - l' + l \le q.$$

(5) Finally, we consider other cases:

When $q = +\infty$ or p and l assume zeros simultaneously, it is obvious that Theorem 6.2 holds. When $p = 0, l = 1$, or $p = 1, l = 0$, then according to Theorem 2.15, we conclude $q \ge 1$, and hence the theorem holds.

When $q < +\infty$, it is impossible that $p = +\infty$ or $l = +\infty$. In fact, when $p = +\infty$, $l < +\infty$, we choose $p' < +\infty$ such that

$$p' - l' + l > q.$$

However, according to the proof of case (1) that we first considered, it must be that

$$p' - l' + l \le q,$$

and hence, a contradiction is derived. When $p < +\infty$, $l = +\infty$ or $p = +\infty$, $l = +\infty$, we may derive analogously a contradiction. Hence, Theorem 6.2 is proved completely.

Theorem 6.2 has the following corollaries:

COROLLARY 1. $l \le q$.

COROLLARY 2. $p \le q$.

COROLLARY 3. *If $p = q$, then $l = l'$.*

COROLLARY 4. *If $l = q$, then $p = l'$.*

Can we construct a meromorphic function of finite lower order μ which has q Julia directions, l distinct direct transcendental singularities that correspond to its inverse function, and p deficient values, where l' deficient values are simultaneously the direct transcendental singularities of its inverse function, such that the equality

$$p - l' + l = q$$

holds?

THEOREM 6.3. *Let $w = f(z)$ be a meromorphic function of $f(z)$ of lower order μ, $0 < \mu < +\infty$, on the open plane $|z| < +\infty$, and $z = g(w)$ be the inverse function of $f(z)$. We also let q be the number of Julia directions of $f(z)$, l be the number of distinct direct transcendental singularities of $g(w)$, and p be the number of deficient values of $f(z)$, where l' deficient values are* *simultaneously the direct transcendental singularities of $g(w)$. Furthermore, we assume that $q < +\infty$ and $p - l' + l = q$. Then each deficient value of $f(z)$ is at the same time an asymptotic value.*

PROOF. (1) First, according to the condition that $p - l' + l = q < +\infty$, we conclude that $p < +\infty$ and $l < +\infty$. Next, we may assume that $p \geq 1$. Otherwise, Theorem 6.3 will lose its meaning. When $p \geq 1$, according to Theorem 6.2, we have $q \geq 1$. Besides, from Theorem 3.7, we conclude that $\lambda < +\infty$.

In the following, we let a_i $(i = 1, 2, \ldots, p)$ be the p deficient values of $w = f(z)$, with their corresponding deficiencies $\delta(a_i, f) = \delta_i > 0$. Without any loss of generality, we let a_i $(i = 1, 2, \ldots, l')$ be the direct transcendental singularities of $z = g(w)$. We also let b_j $(j = 1, 2, \ldots, l)$ be the l distinct direct transcendental singularities of $g(w)$, applying Lemma 4.12, then each b_j $(1 \leq j \leq l)$ corresponds to one region Ω_j. Finally, we assume that $\Delta(\theta_k)$ $(k = 1, 2, \ldots, q; 0 \leq \theta_1 < \theta_2 < \cdots < \theta_q < 2\pi)$ are the q Julia directions of $f(z)$. Let

$$\omega = \min_{1 \leq k \leq q}(\theta_{k+1} - \theta_k), \qquad \theta_{q+1} = \theta_1 + 2\pi, \qquad \delta = \min_{1 \leq i \leq p}\{\delta_i\}$$

and assume arbitrarily a number η,

$$0 < \eta < \frac{\nu\omega}{80\pi}, \qquad \nu = \min\left\{\frac{1}{2}, \mu\right\}.$$

Then, we construct a sequence $r_n = 2^{(1+\eta)^n}$ $(n = 1, 2, \ldots)$. According to Lemma 2.5 and Lemma 3.9, provided that n_0 is sufficiently large, then when $n \geq n_0$, there must exist a value t'_n in the interval $[r_n^{1-\eta}, 2r_n]$ which satisfies the following property: E_i^n $(i = 1, 2, \ldots, p)$ denotes the θ-value set $(0 \leq \theta < 2\pi)$ that makes it possible for the inequality

$$\begin{cases} \log\dfrac{1}{|f(t'_n e^{i\theta}) - a_i|} \geq \dfrac{\delta}{4}T(t'_n, f) & (a_i \neq \infty), \\[2ex] \log|f(t'_n e^{i\theta})| \geq \dfrac{\delta}{4}T(t'_n, f) & (a_i = \infty) \end{cases}$$

to hold, then,

$$\operatorname{meas} E_i^n \geq K_1 = K(\delta, \lambda, p, \eta) > 0, \qquad i = 1, 2, \ldots, p.$$

Besides, we also have

$$T(t'_n, f) \geq r_n^{(1-\eta)(\mu-\eta)} \geq r_n^{(1-\eta)(\nu-\eta)}.$$

On the other hand, for each direct transcendental singularity b_j $(1 \leq j \leq l)$ of $g(w)$, according to Lemma 4.12, provided that n_0 is sufficiently large, then when $n \geq n_0$, there exists value t''_{jn} in the interval $[r_n^{1-\eta}, r_n]$ such that

$$\theta(t''_{jn}) \geq K_2 = K(\eta, \lambda) > 0, \qquad j = 1, 2, \ldots, l.$$

Here, $\theta(t''_{jn})$ contains the following implications: If we use $\theta_{t''_{jn}}$ to represent the part of circumference $|z| = t''_{jn}$ in Ω_j, then $t''_{jn}\theta(t''_{jn})$ represents the linear measure of $\theta_{t''_{jn}}$. Besides, when $z \in \theta_{t''_{jn}}$,

$$\log \frac{1}{|f(z) - b_j|} \geq |z|^{1/2-\eta} \qquad (b_j \neq \infty),$$

$$\log |f(z)| \geq |z|^{1/2-\eta} \qquad (b_j = \infty).$$

Finally, we assume a number ε,

$$0 < \varepsilon < \min\left\{\frac{\omega}{4}, \frac{K}{8q}\right\}, \qquad K = \min\{K_1, K_2\}.$$

Let

$$\overline{\Omega} = \bigcup_{k=1}^{q} \overline{\Omega}(\theta_k + \varepsilon, \theta_{k+1} - \varepsilon).$$

Then according to Theorem 2.13 and the finite covering theorem, we conclude that there exist three distinct values α, β and γ, such that

$$\varlimsup_{r \to +\infty} \frac{\log^+ n(r, \overline{\Omega}, f = X)}{\log r} = 0, \qquad X = \alpha, \beta, \gamma.$$

Moreover, the spherical distances among α, β and γ are greater than d, $0 < d < \frac{1}{2}$.

(2) Analogous to the proof of Theorem 6.2, provided that n_0 is sufficiently large, for each value n $(n = n_0, n_0+1, \ldots)$, we may let each deficient value a_i $(1 \leq i \leq p)$ correspond to one domain $\Omega(\theta_{k_{in}} + 2\varepsilon, \theta_{k_{in}+1} - 2\varepsilon; r_n^{1-\eta}, 2r_n)$. a_i corresponding to domain $\Omega(\theta_{k_{in}} + 2\varepsilon, \theta_{k_{in}+1} - 2\varepsilon; r_n^{1-\eta}, 2r_n)$ means that

$$\text{meas}\{E_i^n \cap [\theta_{k_{in}} + 2\varepsilon, \theta_{k_{in}+1} - 2\varepsilon]\} \geq \frac{K}{2q}.$$

Similarly, we may prove that any two domains in the set

$$\{\Omega(\theta_{k_{in}} + 2\varepsilon, \theta_{k_{in}+1} - 2\varepsilon; r_n^{1-\eta}, 2r_n) | i = 1, 2, \ldots p\}$$

do not coincide with each other.

Analogous to the proof of Theorem 6.2, provided that n_0 is sufficiently large, for each value n $(n = n_0, n_0 + 1, \ldots)$, we may let each direct transcendental singularity b_j $(1 \leq j \leq l)$ corresponds to one domain

$$\Omega(\theta_{k_{jn}} + 2\varepsilon, \theta_{k_{jn}+1} - 2\varepsilon; r_n^{1-\eta}, 2r_n).$$

b_j corresponding to domain $\Omega(\theta_{k_{jn}} + 2\varepsilon, \theta_{k_{jn}+1} - 2\varepsilon; r_n^{1-\eta}, 2r_n)$ means that

$$\text{meas}\{\theta_{t''_{jn}} \cap \Gamma(\theta_{k_{jn}} + 2\varepsilon, \theta_{k_{jn}+1} - 2\varepsilon; t''_{jn})\} \geq \frac{K}{2q} \cdot t''_{jn}.$$

Similarly, we may prove that any two domains in the set

$$\{\Omega(\theta_{k_{jn}} + 2\varepsilon, \theta_{k_{jn}+1} - 2\varepsilon; r_n^{1-\eta}, 2r_n) | j = 1, 2, \ldots, l\}$$

do not coincide with each other.

Finally, analogous to the proof of Theorem 6.2, we may conclude that there are at most l' domains coinciding with one another between the two sets $\{\Omega(\theta_{k_{in}} + 2\varepsilon, \theta_{k_{in}+1} - 2\varepsilon; r_n^{1-\eta}, 2r_n) | i = 1, 2, \ldots, p\}$ and

$$\{\Omega(\theta_{k_{jn}} + 2\varepsilon, \theta_{k_{jn}+1} - 2\varepsilon; r_n^{1-\eta}, 2r_n) | j = 1, 2, \ldots, l\}.$$

Hence, $p - l' + l \leq q$.

According to the assumption that $p - l' + l = q$, we conclude that each domain $\Omega(\theta_k + 2\varepsilon, \theta_{k+1} - 2\varepsilon; r_n^{1-\eta}, 2r_n)$ $(k = 1, 2, \ldots, q)$ must belong to the set $\{\Omega(\theta_{k_{in}} + 2\varepsilon, \theta_{k_{in}+1} - 2\varepsilon; r_n^{1-\eta}, 2r_n) | i = 1, 2, \ldots, p\}$ or the set $\{\Omega(\theta_{k_{jn}} + 2\varepsilon, \theta_{k_{jn}+1} - 2\varepsilon; r_n^{1-\eta}, 2r_n) j = 1, 2, \ldots, l\}$. Otherwise, it must be that $p - l' + l \leq q - 1$. Hence, a contradiction is derived.

Now, we prove that each deficient value a_i $(l' + 1 \leq i \leq p)$ can only correspond to one domain $\Omega(\theta_{k_{in}} + 2\varepsilon, \theta_{k_{in}+1} - 2\varepsilon; r_n^{1-\eta}, 2r_n)$. In fact, otherwise, then there exists a certain deficient value $a_{i'}$ $(l' + 1 \leq i' \leq p)$ such that

$$\text{meas}\{E_i^n \cap [\theta_{k_{i'n}} + 2\varepsilon, \theta_{k_{i'n}+1} - 2\varepsilon]\} \geq \frac{K}{2q},$$

and

$$\text{meas}\{E_i^n \cap [\theta_{k_{i'n}'} + 2\varepsilon, \theta_{k_{i'n}'+1} - 2\varepsilon]\} \geq \frac{K}{2q}.$$

Moreover, $k_{in} \neq k_{i'n}'$. Hence, similarly, we may prove that any two domains in the set

$$\Omega(\theta_{k_{i'n}'} + 2\varepsilon, \theta_{k_{i'n}'+1} - 2\varepsilon; r_n^{1-\eta}, 2r_n),$$

$$\Omega(\theta_{k_{in}} + 2\varepsilon, \theta_{k_{in}+1} - 2\varepsilon; r_n^{1-\eta}, 2r_n) \quad (i = 1, 2, \ldots, p)$$

do not coincide with each other. Moreover, there are at most l' domains coinciding with one another between this set and the set

$$\{\Omega(\theta_{k_{jn}} + 2\varepsilon, \theta_{k_{jn}+1} - 2\varepsilon; r_n^{1-\eta}, 2r_n) | j = 1, 2, \ldots, l\}.$$

Therefore,

$$(p + 1) - l' + l \leq q.$$

However, this contradicts to the assumption that $p - l' + l = q$.

(3) Now, we prove that when $n \geq n_0$, $k_{in} = k_{in+1}$ $(l' + 1 \leq i \leq p)$.

First, notice that each domain $\Omega(\theta_k + 2\varepsilon, \theta_{k+1} - 2\varepsilon; r_n^{1-\eta}, 2r_n)$ $(k = 1, 2, \ldots, q)$ must belong to

$$\{\Omega(\theta_{k_{in}} + 2\varepsilon, \theta_{k_{in}+1} - 2\varepsilon; r_n^{1-\eta}, 2r_n) | i = 1, 2, \ldots, p\}$$

or $\{\Omega(\theta_{k_{jn}} + 2\varepsilon, \theta_{k_{jn}+1} - 2\varepsilon; r_n^{1-\eta}, 2r_n)|j = 1, 2, \ldots, l\}$. Moreover, each domain $\Omega(\theta_k + 2\varepsilon, \theta_{k+1} - 2\varepsilon; r_{n+1}^{1-\eta}, 2r_{n+1})$ $(k = 1, 2, \ldots, q\}$ must belong to $\{\Omega(\theta_{k_{in+1}} + 2\varepsilon, \theta_{k_{in+1}+1} - 2\varepsilon; r_{n+1}^{1-\eta}, 2r_{n+1})|i = 1, 2, \ldots, p\}$ or

$$\{\Omega(\theta_{k_{jn+1}} + 2\varepsilon, \theta_{k_{jn+1}+1} - 2\varepsilon; r_{n+1}^{1-\eta}, 2r_{n+1})|j = 1, 2, \ldots, l\}.$$

Hence, if for a certain i $(l' + 1 \le i \le p)$, we have $k_{in} \ne k_{in+1}$, then it must be that $k_{in} = k_{i'n+1}$ $(i' \ne i)$ or $k_{in} = k_{jn+1}$.

If $k_{in} = k_{i'n+1}$. Analogous to the proof of Theorem 6.2, for the point z on $\overline{\Omega}(\theta_{k_{in}} + 2\varepsilon, \theta_{k_{in}+1} - 2\varepsilon; r_n^{1-\eta}, 2r_n)$ and outside some circles $(\gamma)_{i'}$, then

$$|f(z) - a_{i'}| \le e^{-r_{n+1}^{\nu/2}},$$

where the sum of the radii of circles $(\gamma)_{i'}$ does not exceed $\frac{\varepsilon}{4}r_{n+1}^{1-\eta}$. Moreover, for the point z on $\overline{\Omega}(\theta_{k_{i'n}+1} + 2\varepsilon, \theta_{k_{i'n+1}+1} - 2\varepsilon; r_{n+1}^{1-\eta}, 2r_n)$ and outside some circles $(\gamma)_{i'}$, we have

$$|f(z) - a_{i'}| \le e^{-r_{n+1}^{\nu/2}},$$

where the sum of the radii of $(\gamma)_{i'}$ does not exceed $\frac{\varepsilon}{4}r_{n+1}^{1-\eta}$. Notice that

$$k_{in} = k_{i'n+1}, \qquad r_{n+1}^{1-\eta} = r_n^{1-\eta^2} < r_n,$$

and also the sum of the radii of circles $(\gamma) = (\gamma)_{i'} \cup (\gamma)_i$ does not exceed $\frac{\varepsilon}{4}(r_n^{1-\eta} + r_{n+1}^{1-\eta}) \le \frac{\varepsilon}{2}r_n$. We conclude that there exists a point $z_0 \notin (\gamma)$ on $\Gamma(\theta_{k_{in}} + 2\varepsilon, \theta_{k_{in+1}} - 2\varepsilon; r_n)$. Hence, on the one hand we have

$$|f(z_0) - a_i| \le e^{-r_n^{\nu/2}},$$

and on the other hand

$$|f(z_0) - a_{i'}| \le e^{-r_n^{\nu/2}}.$$

However, according to $i \ne i'$, therefore, $a_i \ne a_{i'}$. Hence, provided that n_0 is sufficiently large, a contradiction will be derived.

If $k_i = k_{jn+1}$. Notice that $a_i \ne b_j$; then a contradiction is analogously derived.

Let $k_i = k_{in}$ $(n = n_0, n_0 + 1, \ldots)$. Then there must exist a continuous curve L_i tending to ∞ in $\Omega(\theta_{k_i} + 2\varepsilon, \theta_{k_i+1} - 2\varepsilon)$, such that

$$\lim_{\substack{|z| \to +\infty \\ z \in L_i}} f(z) = a_i, \qquad l' + 1 \le i \le p,$$

meaning that a_i is an asymptotic value. Besides, a_i $(i = 1, 2, \ldots, l')$ is a direct transcendental singularity of $g(w)$. Hence, a_i $(i = 1, 2, \ldots, l')$ is an asymptotic value of $f(z)$. Accordingly, Theorem 6.3 is proved completely.

If in the proof of Theorem 6.2 and Theorem 6.3, we replace Lemma 5.1 with Lemma 3.13, then analogously we may prove the following result:

THEOREM 6.4. *Let* $w = f(z)$ *be a meromorphic function of order* $\lambda <$ $+\infty$ *on the open plane* $|z| < +\infty$, $g(w)$ *be the inverse function of* $f(z)$. *We also assume that there exist* q ($1 \leq q < +\infty$) *half lines* $\Delta(\theta_k)$ ($k = 1, 2, \ldots, q$; $0 \leq \theta_1 < \theta_2 < \cdots < \theta_q < 2\pi$, $\theta_{q+1} = \theta_1 + 2\pi$), *such that for any arbitrary number* $\varepsilon > 0$ *we have*

$$\varlimsup_{r \to +\infty} \frac{\log^+ n\{\bigcup_{k=1}^q \Omega(\theta_k + \varepsilon, \theta_{k+1} - \varepsilon; r), f = X\}}{\log r} = 0, \qquad X = 0, \infty.$$

Let l *be the number of distinct, finite, nonzero direct transcendental singularities of* $g(w)$, *and* p *be the number of finite, nonzero deficient values of* $f(z)$, *where* l' *of the deficient values are simultaneously the direct transcendental singularities of* $g(w)$. *Then we have the formula* $p - l' + l \leq q$.

THEOREM 6.5. *Let* $w = f(z)$ *be a meromorphic function of lower order* μ, $0 < \mu < +\infty$,([6]) *on the open plane* $|z| < +\infty$, *and* $z = g(w)$ *be the inverse function of* $f(z)$. *We also assume that there exist* q ($1 \leq q < +\infty$) *half lines* $\Delta(\theta_k)$ ($k = 1, 2, \ldots, q$; $0 \leq \theta_1 < \theta_2 < \cdots < \theta_q < 2\pi$, $\theta_{q+1} = 2\pi + \theta_1$) *such that for any arbitrary number* $\varepsilon > 0$,

$$\varlimsup_{r \to +\infty} \frac{\log^+ n\{\bigcup_{k=1}^q \Omega(\theta_k + \varepsilon, \theta_{k+1} - \varepsilon; r), f = X\}}{\log r} = 0, \qquad X = 0, \infty.$$

Let l *be the number of distinct, finite, nonzero direct transcendental singularities of* $g(w)$, *and* p *be the number of finite nonzero deficient values of* $f(z)$, *where* l' *of the deficient values are simultaneously the direct transcendental singularities of* $g(w)$. *We further assume that* $p - l' + l = q$. *Then each deficient value of* $f(z)$ *is simultaneously an asymptotic value.*

Finally, we illustrate the following two points:

(1) When $q \geq 2$, according to the condition that $p - l' + l = q$, we may conclude that lower order $\mu > 0$. Hence, when $q \geq 2$, we may neglect the hypothetical condition $\mu > 0$ as in Theorem 6.3 and Theorem 6.5.

(2) Under the condition of Theorem 6.3 and Theorem 6.5, we may prove that $f(z)$ tends uniformly to a_i or b_j, except that it may neglect at most some point sets whose measure is small on each domain $\Omega(\theta_k + 2\varepsilon, \theta_{k+1} - 2\varepsilon)$ ($k = 1, 2, \ldots, q$).

([6]) When $p = 0$, Theorem 6.5 loses its meaning. When $p \geq 1$, according to Theorem 3.6, we conclude that the order λ of $f(z)$ is $< +\infty$.

Some Supplementary Results

Ever since the publication of the book, several interesting and significant results concerning the numerical relationships among the order, lower order, numbers of deficient values, asymptotic values, and singular directions of an entire or a meromorphic function and its derivatives have been obtained. The following list provides some of the more significant ones.

In 1988, L. Yang [*Deficient values and angular distribution of entire functions*, Trans. Amer. Math. Soc. **308** (1988), 583–601] gave the following definition: Let $f(z)$ be an entire-function of lower order μ, where $0 < \mu < +\infty$. A ray $\arg z = \theta_0$ $(0 \le \theta_0 < 2\pi)$ is called a Borel direction of order $\ge \mu$ of $f(z)$, if for any positive number ε, the inequality

$$\varlimsup_{r \to +\infty} \frac{\log n(r, \theta_0 - \varepsilon, \theta_0 + \varepsilon, f = a)}{\log r} \ge \mu$$

holds for any finite complex value a, with possibly one exceptional value, where $n(r, \theta_0 - \varepsilon, \theta_0 + \varepsilon, f = a)$ denotes the number of zeros of $f(z) - a$ in the region $(|z| \le r) \cap (\theta_0 - \varepsilon \le \arg z \le \theta_0 + \varepsilon)$ multiple zeros being counted according to their multiplicities.

Meanwhile, L. Yang also proved the following result.

THEOREM. *Let $f(z)$ be an entire function of lower order μ, where $0 < \mu < +\infty$. If $q < +\infty$ is the number of Borel directions of order $\ge \mu$ of $f(z)$ and $p_l(l = 0, -1, -2, \ldots)$ denotes the number of finite nonzero deficient values of $f^{(l)}(z)$ $(l = 0, -1, -2, \ldots, f^{(0)}(z) \equiv f(z))$, when $l < 0$, $f^{(l)}(z)$ is the primitive of order $|l|$ of $f(z)$, then we have*

$$\sum_{l=0}^{-\infty} p_l \le 2\mu,$$

More recently, as an extension over the above result, P. C. Wu obtained the following result in "Angular distribution of entire functions and its deficient values of each order derivative", preprint.

THEOREM. *Let $f(z)$ be an entire function of lower order μ, where $0 < \mu < +\infty$. If $q < +\infty$ is the number of Borel directions of order $\ge \mu$ of $f(z)$*

and $p_l (l = 0, -1, -2, \ldots)$ denotes the number of finite nonzero deficient values of $f^{(l)}(z)(l = 0, -1, -2, \ldots f^{(0)}(z) = f(z))$, then we have

$$\sum_{l=0}^{-\infty} p_l < 2\mu.$$

S. J. Wu also researched this topic in his thesis "Angular distribution and Borel theorem of entire and meromorphic functions". For instance, the following two results were obtained.

THEOREM. *Suppose that* $f(z)$ *is an entire function of finite lower order* μ *and satisfies* $p = \frac{q}{2}$ *where* $p(1 \leq p < +\infty)$ *denotes the number of finite deficient values and* q *denotes the number of Borel directions of order* $\geq \mu$ *of* $f(z)$. *Then for every deficient value* a_j $(j = 1, 2, \ldots, p)$, *there exists a corresponding angular domain* $\Omega(\theta_{k_j}, \theta_{k_j+1})$ *such that for every* $\varepsilon > 0$ *the inequality*

$$\log \frac{1}{|f(z) - a_j|} > A(\theta_{k_j}, \theta_{k_j+1}, \varepsilon, \delta(a_j, f)) T(|z|, f)$$

holds for $z \in \Omega(\theta_{k_j} + \varepsilon, \theta_{k_j+1} - \varepsilon, r_\varepsilon, +\infty)$, *where* $A(\theta_{k_j}, \theta_{k_j+1}, \varepsilon, \delta(a_j, f))$ *is a positive constant depending only on* $\theta_{k_j}, \theta_{k_j+1}, \varepsilon$, *and* $\delta(a_j, f)$. *In particular, every deficient value of* $f(t)$ *is also its deficient value.*

THEOREM. *Under the hypothesis of the above theorem, we have*

(i) *The order* λ *of* $f(z)$ *equals* μ,
(ii) *Every asymptotic value of* $f(z)$ *is also its deficient value,*
(iii) $\sum_{a \in \mathbb{C}} \delta(a, f) \leq 1 - k(\mu)$,

where

$$k(\mu) = \begin{cases} \dfrac{|\sin \mu\pi|}{q + |\sin \mu\pi|}, & q \leq \mu \leq q + \frac{1}{2}, \\[3mm] \dfrac{|\sin \mu\pi|}{q + 1}, & q + \frac{1}{2} < \mu \leq q + 1. \end{cases}$$

References

[1] L. V. Ahlfors,
[a]   *Beiträge der meromorphen Funktionen*, Congr. Math. Scand. Oslo, vol. 7 (1929), pp. 84–88.
[b]   *Untersuchungen zur Theorie der konformen Abbildung und der Theorie der ganzen Funktionen*, Acta Soc. Sci. Fenn **1** (1930), 1–40.
[c]   *Über die asymptotischen Werte der meromorphen Funktionen endlicher Ordnung*, Acta Acad. Aboensis. Math. et Phys. **6** (1932).
[d]   *Zur Theorie der Überlagerungsflächen*, Acta Math. **65** (1935), 157–194.
[e]   *Lectures on quasiconformal mapping*, Van Nostrand, Princeton, NJ, 1966.
[f]   *Conformal invariants: Topics in geometric function theory*, McGraw-Hill, NY, 1973.

[2] J. M. Anderson, K. F. Barth, and D. A. Brannan,
[a]   *Research problems in complex analysis*, Bull. London Math. Soc. **9** (1977), 129–162.

[3] N. V. Arakeljan,
[a]   *Entire functions of finite order with an infinite set of deficient values*, Dokl. Akad. Nauk SSSR **107** (1966), 999–1002.

[4] A. Baernstein,
[a]   *Proof of Edrei's spread conjecture*, Proc. London Math. Soc. **26** (1973), 418–434.
[b]   *Integral means, univalent functions and circular symmetrization*, Acta Math. **133** (1974), 139–169.

[5] K. F. Barth, D. A. Brannan, and W. K. Hayman,
[a]   *The growth of plane harmonic functions along an asymptotic path*, Proc. London Math. Soc. **37** (1978), 363–384.

[6] E. Borel,
[a]   *Sur les zéros des fonctions entières*, Acta Math. **20** (1897), 357–396.

[7] F. Bureau,
[a]   *Mémoire sur les fonctions uniformes à point singulier essentiel isolé*, Mém. Soc. Roy. Sci. Liége **17** (1932).

[8] D. M. Campbell, J. G. Clunie, and W. K. Hayman,
[a]   *Research problems in complex analysis*, Aspect of Contemporary Complex Analysis (D. A. Brannan and J. G. Clunie, eds.), Academic Press, NY, 1980.

[9] H. Cartan,
[a]   *Sur les systèmes de fonctions holomorphes à variétés linéaries lacunaires et leurs applications*, Ann. Sci. École Norm. Sup. **45** (1928), 255–346.
[b]   *Sur la fonction de croissance attachée à une fonction méromorphe de deux variables et ses applications aux fonctions méromorphes d'une variable*, C. R. Acad. Sci. Paris **189** (1929), 521–523.

[10] Chuang Chi-tai (or Zhang Chi-tai),
[a]   *Un théorème relatif aux directions de Borel des fonctions méromorphes d'ordre fini*, C. R. Acad. Sci. **204** (1937), 951–952.
[b]   *Une généralisation d'une inégalité de Nevanlinna*, Sci. Sinica **13** (1964), 887–895.
[c]   *Singular directions of meromorphic functions*, Science Press, 1982.

[11] A. Denjoy,
[a]   *Sur les fonctions entiéres de genre fini*, C. R. Acad. Sci. Paris **145** (1907), 106–109.

[12] D. Drasin,
[a]   *On asymptotic curves of functions extremal for Denjoy's conjecture*, Proc. London Math. Soc. **26** (1973), 142–166.
[b]   *The inverse problem of the Nevanlinna theory*, Acta Math. **138** (1977), 83–151.
[c]   *Proof of a conjecture of F. Nevanlinna concerning functions which have deficiency sum two*, Acta Math. **158** (1987), 1–94.

[13] D. Drasin and A. Weitsman,
[a]   *Meromorphic functions with large sums of deficiencies*, Adv. in Math. **15** (1974), 93–126.
[b]   *On the Julia directions and Borel directions of entire functions*, Proc. London Math. Soc. **32** (1976), 199–212.

[14] A. Edrei,
[a]   *Sums of deficiencies of meromorphic functions*, J. Analyse Math., (I) **14** (1965), 79–107; (II) **19** (1967), 53–74.
[b]   *Solution of the deficiency problem for functions of small lower order*, Proc. London Math. Soc. **26** (1973), 435–445.

[15] A. Edrei and W. H. J. Fuchs,
[a]   *Valeurs déficientes et valeurs asymptotiques des fonctions méromorphes*, Comment. Math. Helv. **33** (1959), 258–295.
[b]   *On the growth of meromorphic functions with several deficient values*, Trans. Amer. Math. Soc. **93** (1959), 292–328.
[c]   *The deficiencies of meromorphic functions of order less than one*, Duke Math. J. **27** (1960), 233–249.
[d]   *Bounds for the number of deficient values of certain classes of meromorphic functions*, Proc. London Math. Soc. **12** (1962), 315–344.
[e]   *On meromorphic functions with regions free of poles and zeros*, Acta Math. **108** (1962), 113–145.

[16] W. H. J. Fuchs,
[a]   *A Phragmén-Lindelöf theorem conjectured by D. J. Newman*, Trans. Amer. Math. Soc. **267** (1981), 285–293.
[b]   *The development of the theory of deficient values since Nevanlinna*, Ann. Acad. Sci. Fenn. Ser. A. I. Math. **7** (1982), 33–48.

[17] W. H. J. Fuchs and W. K. Hayman,
[a]   *An entire function with assigned deficiencies*, Studies in Mathematical Analysis and Related Topics Essays in Honor of George Pólya, Stanford Univ. Press, Stanford, CA, 1962, pp. 117–125.

[18] A. A. Gol'dberg,
[a]   *Defects of meromorphic functions*, Dokl. Akad. Nauk SSSR **98** (1954), 893–895.
[b]   *The possible magnitude of the lower order of an entire function with a finite deficient value*, Dokl. Akad. Nauk SSSR **159** (1964), 968–970.

[19] A. A. Gol'dberg and A. E. Eremenko,

[20] G. M. Golyzun,

[21] W. K. Hayman,
[a]   *Multivalent functions*, Cambridge Univ. Press, 1958.

[b]   *Picard values of meromorphic functions and their derivatives*, Ann. Math. **70** (1959), 9–42.

[c]   *Meromorphic functions*, Oxford Univ. Press, 1964.

[d]   *Research problems in function theory*, Athlone Press, London, 1967.

[e]   *Angular value distribution of power series with gaps*, Proc. London Math. Soc. **24** (1972), 590–624.

[f]   *Research problems in function theory*, Symposium on Complex Analysis, Canterbury, 1973 (J. G. Clunie and W. K. Hayman, eds.), Cambridge Univ. Press, NY, 1974.

[g]   *On Iversen's theorem for meromorphic functions with few poles*, Acta Math. **141** (1978), 115–145.

[22] Hiong King-lai,

[a]   *Sur les fonctions méromorphes et les fonctions algébroides*, Mém. Sci. Math. Fasc., vol. 139, Paris, 1957.

[23] A. Huber,

[a]   *On subharmonic function and differential geometry in the large*, Comment. Math. Helv. **32** (1957), 13–72.

[24] F. Iversen,

[a]   *Recherches sur les fonctions inverses des fonctions méromorphes*, Thèse de Helsingfors, 1914.

[25] G. Julia,

[a]   *Sur quelques propriétés nouvelles des fonctions entières ou méromorphes*, Ann. École Norm. Sup. **36** (1919), 93–125; **37** (1920), 165–218.

[26] P. B. Kennedy,

[a]   *On a conjecture of Heins*, Proc. London Math. Soc. **5** (1955), 22–47.

[b]   *A class of integral functions bounded on certain curves*, Proc. London Math. Soc. **6** (1956), 518–547.

[27] J. Lewis, J. Boss, and A. Weitsman,

[a]   *On the growth of subharmonic functions along paths*, Preprint.

[28] E. Lindelöf,

[a]   *Sur un principle générale de l'analysde et ses applications à la théorie de la representation conforme*, Acta Soc. Sci. Fenn. **46** (1915).

[29] Lü Yi-nian and Zhang Guang-hou,

[a]   *On Nevanlinna direction of a meromorphic function*, Sci. Simica **26** (1983), 607–617.

[30] H. Milloux,

[a]   *Le théorème de Picard, suites de fonctions holomorphes; fonctions holomorphes et fonctions entieres*, J. de Math. **3** (1924), 345–401.

[b]   *Sur une extension d'un théorème de P. Boutroux-H. Cartan*, Bull. Soc. Math. France **65** (1937), 65–75.

[c]   *Extension d'un théorème de M. R. Nevanlinna et applications*, Act. Sci. et Ind., No. 888, 1940.

[d]   *Sur les directions de Borel des fonctions entières, de leurs dérivées et de leurs integrales*, J. Analyse Math. **1** (1951), 244–330.

[31] F. Nevanlinna,

[a]   *Über eine Klasse meromorpher Funktionen*, C. R. Congr. Math. Scand., Oslo, 1929, pp. 81–83.

[32] R. Nevanlinna,

[a]   *Le théorème de Picard-Borel et la théorie des fonctions méromorphes*, Paris, 1929.

[b]   *Analytic functions*, Springer, Berlin, 1970.

[33] A. Ostrowski,

[a]   *Über Folgen analytischer Funktionen und einige Verschärfungen des Picardschen Satzes*, Math. Z. **24** (1926), 215–258.

[34] A. Pfluger,
[a] *Zur Defektrelation ganzer Funktionen endlicher Ordnung*, Comment. Math. Helv. **19** (1946), 91–104.

[35] A. Rauch,
[a] *Extension de théorèmes relatifs aux directions de Borel des fonctions méromorphes*, J. Math. Pures Appl. **12** (1933), 109–171.
[b] *Cas où une direction de Borel d'une fonction entière $f(z)$ d'ordre fini est aussi direction de Borel pour $f'(z)$*, C. R. Acad. Sci., Paris **199** (1934), 1014–1016.

[36] T. Shimizu,
[a] *On the theory of meromorphic functions*, Japan J. Math. **6** (1929).

[37] O. Teichmüller,
[a] *Vermutungen und sätze über die Wertverteilung gebrochener Funktionen endlicher Ordnung*, Deutsche Math. **4** (1939), 163–190.

[38] M. Tsuji,
[a] *Potential theory in modern function theory*, Maruzen, Tokyo, 1959.

[39] G. Valiron,
[a] *Lectures on the general theory of integral functions*, Edouard Privat, Toulouse, 1923.
[b] *Recherches sur le théorème de M. Morel dans la théorie des fonctions méromorphes*, Acta Math. **52** (1928), 67–92.
[c] *Sur les directions de Borel des fonctions entières*, Annali di Mat. **9** (1931), 273–285.
[d] *Directions de Borel des fonctions méromorphes*, Mém. Sci. Math., vol. 89, Paris, 1938.
[e] *Sur les valeurs déficientes des fonctions méromorphes d'ordre nul*, C. R. Acad. Sci. Paris **230** (1950), 40–42.

[40] A. Weitsman,
[a] *Meromorphic functions with maximal deficiency sum and a conjecture of F. Nevanlinna*, Acta. Math. **123** (1969), 115–139.
[b] *A growth property of the Nevanlinna characteristic*, Proc. Amer. Math. Soc. **26** (1970), 65–70.
[c] *A theorem on Nevanlinna deficiencies*, Acta Math. **128** (1972), 41–52.

[41] A. Wiman,
[a] *Sur une extension d'un théorème de M. Hadamard*, Ark. Mat. Ast. Phys. **2** (1905).

[42] Yang Lo and Zhang Guang-hou,
[a] *Sur la distribution des directions de Borel des fonctions méromorphes*, Sci. Sinica **16** (1973), 465–482.
[b] *On the total number of deficient values of an entire function*, Acta Math. Sinica **18** (1975), 35–53.
[c] *Recherches sur le nombre des valeurs déficientes et le nombre des directions de Borel des fonctions méromorphes*, Sci. Sinica **18** (1975), 23–37.
[d] *Sur la construction des fonctions méromorphes ayant des directions singulières données*, Sci. Sinica **19** (1976), 445–459.

[43] Zhang Guang-hou,
[a] *On the researches of common Borel directions of a meromorphic function and its derivatives or anti-derivatives*, Acta Math. Sinica **20** (1977; (I) 73–98; (II) 157–177; (III) 237–247).
[b] *A symptotic values of entire and meromorphic functions*, Sci. Sinica **20** (1977), 720–739.
[c] *Research of the relations among the deficient values asymptotic values, and Julia directions of entire and meromorphic functions*, Sci. Sinica add. I (1978), 1–80, Chinese.
[d] *The length of an asymptotic path of an entire function*, Sci. Sinica **22** (1979), 991–999.
[e] *On entire function extremal for Denjoy's conjecture*, Sci. Sinica, (I) **24** (1981), 885–898; (II) **11** (1982), 981–994.
[f] *Direct transcendental singularities for inverse functions of meromorphic functions and of their derivatives*, Sci. Sinica **25** (1982), 797–807.

[g]   *Meromorphic functions with maximal deficiency sum*, Sci. Sinica **4** (1983), 293–305.
[h]   *Entire functions with a finite number of Julia directions*, Sci. Sinica **9** (1983), 775–786.
[i]   *Some general theorems of meromorphic functions and their applications*, Acta. Math. Sinica **30** (1987), no. 3, 317–354.
[j]   *Entire functions with their zero and poles distributed on a finite number of rays*, Sci. Sinica (Ser. A) **10** (1986), 1009–1021.

[44] Zhang Guang-hou and Wu Peng-cheng,
[a]   *On order of meromorphic functions*, Sci. Sinica Ser. A **28** (1985), no. 8, 785–800.

Recent Titles in This Series

(Continued from the front of this publication)

(See the AMS catalog for earlier titles)